T0234931

Communications in Computer and Information Science   **800**

Commenced Publication in 2007
Founding and Former Series Editors:
Alfredo Cuzzocrea, Xiaoyong Du, Orhun Kara, Ting Liu, Dominik Ślęzak,
and Xiaokang Yang

More information about this series at http://www.springer.com/series/7899

Alexander Dudin · Anatoly Nazarov
Alexander Kirpichnikov (Eds.)

Information Technologies and Mathematical Modelling

Queueing Theory and Applications

16th International Conference, ITMM 2017
Named After A.F. Terpugov
Kazan, Russia, September 29 – October 3, 2017
Proceedings

 Springer

Editors
Alexander Dudin
Belarusian State University
Minsk
Belarus

Anatoly Nazarov
Tomsk State University
Tomsk
Russia

Alexander Kirpichnikov
Kazan National Research Technological
  University
Kazan
Russia

ISSN 1865-0929   ISSN 1865-0937 (electronic)
Communications in Computer and Information Science
ISBN 978-3-319-68068-2   ISBN 978-3-319-68069-9 (eBook)
DOI 10.1007/978-3-319-68069-9

Library of Congress Control Number: 2017954893

This Springer imprint is published by Springer Nature
The registered company is Springer International Publishing AG
The registered company address is: Gewerbestrasse 11, 6330 Cham, Switzerland

Preface

The series of scientific conferences Information Technologies and Mathematical Modelling (ITMM) was started in 2002. In 2012, the series acquired an international status, and selected revised papers have been published in *Communication in Computer and Information Science* since 2014. The conference series is named after Alexander Terpugov, one of the first organizers of the conference, an outstanding scientist of Tomsk State University, a leader of the famous Siberian school on applied probability, queueing theory, and applications.

Traditionally, the conferences have about 10 sections in various fields of mathematical modelling and information technologies. Throughout the years, the sections on probabilistic methods and models, queueing theory, and communication networks have been the most popular ones at the conference. These sections gather many scientists from different countries. During the last years, we accepted participants from Austria, Azerbaijan, Belarus, Bulgaria, China, Hungary, India, Italy, Kazakhstan, Korea, The Netherlands, Poland, United States. Many of these foreign participants come to this Siberia conference every year because we have a warm acceptance and serious scientific discussions here. This year the conference was held in Kazan, the capital of Tatarstan Republic, whose universities, research institutes and engineering firms are well-known in the world.

This volume presents selected papers from 16th ITMM conference. The papers are devoted to new results in the queueing theory and its applications. It is targeting to be used by specialists in probabilistic theory, random processes, mathematical modelling as well as engineers engaged into logical and technical design and operational management of data processing systems, communication and computer networks.

September 2017

Alexander Dudin
Anatoly Nazarov
Alexander Kirpichnikov

Organization

The conference was organized by the National Research Tomsk State University, Kazan National Research Technological University, Peoples' Friendship University of Russia (RUDN University), Trapeznikov Institute of Control Sciences of Russian Academy of Sciences.

International Program Committee

A. Dudin, Belarus, Chair
A. Nazarov, Russia, Co-chair
A. Kirpichnikov, Russia, Co-chair
I. Atencia, Spain
P. Cabral, Portugal
P.F. i Casas, Spain
S. Chakravarthy, USA
B.D. Choi, South Korea
T. Czachórski, Poland
R. Dinis, Portugal
D. Efrosinin, Austria
M. Farhadov, Russia
Yu. Gaydamaka, Russia
E. Gelenbe, UK
A. Gortsev, Russia
A. Khalid, UK
C.S. Kim, South Korea
U. Krieger, Germany

A. Krishnamurthy, India
Q.-L. Li, China
Yu. Malinkovsky, Belarus
A. Melikov, Azerbaijan
A. Moiseev, Russia
P. Montezuma-Carvalho, Portugal
R. Nobel, The Netherlands
M. Pagano, Italy
T.B. Preußer, Germany
V. Rykov, Russia
K. Samuylov, Russia
S. Suschenko, Russia
D. Stamate, UK
J. Sztrik, Hungary
H. Tijms, The Netherlands
O. Tikhonenko, Poland
V. Vishnevsky, Russia

Local Organizing Committee

S. Moiseeva, Chair
R. Gainullin, Co-chair
A. Zamyatin, Co-chair
V. Broner
V. Bukreev
V. Devyatkov
A. Elizarov
L. Emaletdinova
E. Fedorova
D. Flaks

V. Ivnitskii
Ya. Izmaylova
E. Lisovskaya
A. Litovtsev
M. Matalytski
N. Nuriev
S. Paul
V. Pesoshin
S. Rozhkova
D. Semenova

M. Shklennik
I. Shmyrin
M. Slamovich
E. Sopin

M. Talanov
G. Tsitsiashvili
V. Zadorozhny
A. Zorin
K. Voytikov

Contents

On a $BMAP/G/1$ Retrial System with Two Types of Search of Customers from the Orbit

Alexander Dudin[1](✉), T.G. Deepak[2], Varghese C. Joshua[3],
Achyutha Krishnamoorthy[3], and Vladimir Vishnevsky[4]

[1] Department of Applied Mathematics and Computer Science,
Belarusian State University, 220030 Minsk, Belarus
dudin@bsu.by

[2] Department of Mathematics, Indian Institute of Space Science and Technology,
Thiruvananthapuram, India
deepak@iist.ac.in

[3] Department of Mathematics, CMS College, Kottayam, India
{vcjoshua,krishnamoorthy}@cmscollege.ac.in

[4] Institute of Control Sciences, Russian Academy of Sciences,
Moscow, Russia
vishn@inbox.ru

Abstract. A single server retrial queueing model, in which customers arrive according to a batch Markovian arrival process (BMAP), is considered. An arriving batch, finding server busy, enters an orbit. Otherwise one customer from the arriving batch enters for service immediately while the rest join the orbit. The customers from the orbit try to reach the server subsequently and the inter-retrial times are exponentially distributed. Additionally, at each service completion epoch, two different search mechanisms are switched-on. Thus, when the server is idle, a competition takes place between primary customers, the customers coming by retrial and the two types of searches. It is assumed that if the type II search reaches the service facility ahead of the rest, all customers in the orbit are taken for service simultaneously, while in the other two cases, only a single customer is qualified to enter the service. We assume that the service times of the four types of customers namely, primary, repeated and those by the two types of searches are arbitrarily distributed with different distributions. Steady state analysis of the model is performed.

Keywords: Batch Markovian arrival process · Orbit · Retrials · Customers search · Group service

1   Intoduction

Retrial queues represent an important, challenging and complicated for mathematical analysis class of queueing systems. A retrial queueing system is characterised by the fact that a customer arriving when all servers accessible for him/her are busy, leaves the service area and joins a group of unsatisfied

© Springer International Publishing AG 2017
A. Dudin et al. (Eds.): ITMM 2017, CCIS 800, pp. 1–12, 2017.
DOI: 10.1007/978-3-319-68069-9_1

customers called orbit, but after a random amount of time he/she returns and repeats his/her demand for service. Retrial queueing systems arise frequently in the stochastic modelling of telecommunications, computer systems, contact centers, etc. Review of retrial queueing literature could be found in [1, 2, 24, 25, 27, 32]. In the retrial set up, each service is preceded and followed by the server(s) idle time because of the ignorance of the status of the server(s) and orbital customers by each other.

We are interested in designing retrial queueing models that reduce the server(s) idle time. One way to achieve this is by the introduction of search of orbital customers immediately after a service completion. Search for orbital customers was introduced in [3] and the paper [20] generalizes the result in [3] by introducing a search time, two types of services to customers (primary/orbital) and by assuming the arrival process to be the batch Markovian process. The queueing model with customers search in the buffer (not in the orbit) was considered in [31] where after each service completion the server starts searching of a customer in the buffer and the rate of the exponentially distributed search time is proportional to the number of customers presenting in the system.

This paper generalizes the model discussed in [20] by introducing two types of search and different types of services to primary/orbital customers (retrial/type I/type II searches) retaining the assumption that the arrival process is batch Markovian process $(BMAP)$. A particular case of the proposed model with batch Poisson arrival process has been considered in [11]. A retrial model with two types of search, in which the number of customers taken for service depends on the orbit size, and with the batch Poisson arrival process is considered in [12]. However, namely $BMAP$ suits well for modelling the correlated bursty traffic in the modern communication networks. Approximation of such flows in terms of the stationary Poisson process can cause huge errors in the evaluation of performance characteristics of the networks. Therefore, analysis of queueing models with the $BMAP$ is of a great importance. Chakravarthy S.R. in [7] provides a review of queueing models with the batch Markovian arrivals. Retrial models with $BMAP$ have been investigated, e.g., in the papers [4, 8, 9, 16–18, 26].

The present model is motivated, e.g. by the following practical situation: In Airport/Bus stations/Railway stations passengers individually get into transport vehicles to destinations. Also it is common that travel agencies arrange for bulk transport for all the customers. Broadcasting of information simultaneously to many customers is possible in various wireless communication networks. More motivations of group service can be found e.g. in [5, 6, 8, 23].

2   The Mathematical Model

We consider a single server queueing system in which the arrivals occur according to a $BMAP$. The $BMAP$, a special class of tractable Markov renewal process, is a rich class of point processes that includes many well known processes such as Poisson, PH-renewal processes and Markov-modulated Poisson process. One of the most significant features of the $BMAP$ is the underlying Markovian structure and it fits ideally in the context of matrix-analytic solutions to stochastic models. As is well known, Poisson processes are the simplest and most tractable ones used extensively in stochastic modelling. The idea of the $BMAP$ is to significantly generalize the Poisson processes and still keep the tractability for modelling purposes.

The $BMAP$ is described as follows. Let the underlying Markov chain $\{\nu_t, t \geq 0\}$ be irreducible and let $Q^* = (q_{ij})$ be the generator of this Markov chain with state space $\{1, 2, \cdots, m\}$. At the end of a sojourn time in state i, that is exponentially distributed with parameter $\lambda_i \geq -q_{ii}$, one of the following two events could occur: with probability $P_{ij}(l), 1 \leq i, j \leq m$, the transition corresponds to an arrival of group size $l \geq 1$, and the underlying Markov chain $\{\nu_t, t \geq 0\}$ is in state j; with probability $P_{ij}(0)$, the transition corresponds to no arrival and the state of the process $\{\nu_t, t \geq 0\}$ is j, $j \neq i$. Note that the Markov chain $\{\nu_t, t \geq 0\}$ can go from state i to state i only through an arrival. For $l \geq 0$, define matrices $D_l = (d_{ij}(l))$ such that $d_{ii}(0) = -\lambda_i$, $1 \leq i \leq m$; $d_{ij}(0) = \lambda_i P_{ij}(0)$, for $j \neq i$, $1 \leq i, j \leq m$, and $d_{ij}(l) = \lambda_i P_{ij}(l)$. Assuming D_0 to be a non-singular matrix, the interarrival times will be finite with probability one and the arrival process does not terminate. Hence, we see that D_0 is a stable matrix. The generator Q^* is then given by $Q^* = \sum_{l=0}^{\infty} D_l$. Let $D(z)$ be the matrix generating function of D_l. That is, $D(z) = \sum_{l=0}^{\infty} z^l D_l$.

Thus, the $BMAP$ is described by the matrices $\{D_l\}$ with D_0 governing the transitions corresponding to no arrival and D_l governing those corresponding to arrivals of group size $l, l \geq 1$. The point process described by the $BMAP$ is a special class of semi-Markov processes with transition probability matrix given by

$$\int_0^x e^{D_0 t} dt D_l = \left[I - e^{D_0 x} \right] (-D_0)^{-1} D_l, l \geq 1.$$

For use in the sequel, let $\mathbf{e}$, $\mathbf{0}$ and I denote, respectively, the (column) vector of dimension m consisting of 1's, the (row) vector of dimension m consisting of 0's, and the identity matrix of order m.

Let $\boldsymbol{\theta}$ be the stationary probability vector of the associated Markov process with generator Q^*. That is, $\boldsymbol{\theta}$ is the unique (positive) probability vector satisfying $\boldsymbol{\theta} Q^* = 0$, $\boldsymbol{\theta} \mathbf{e} = 1$. The constant

$$\lambda = \boldsymbol{\theta} \sum_{k=1}^{\infty} k D_k \mathbf{e},$$

referred to as the *fundamental rate* gives the expected number of arrivals per unit of time in the stationary version of the $BMAP$. For further details on $BMAP$, its properties, particular cases and usefulness in stochastic modelling, we refer to [7, 29].

The service mechanism of the present system is described in the following manner. The primary unit who meets the server idle is served with service times having the distribution function $B_0(t)$, while the rest join the orbit. Each unit in the arriving batch finding the server busy, enters the orbit and retry to access the server with the time between two successive retrials, exponentially distributed having intensity $\alpha_i, i \geq 0$, when the number of customers in the orbit is i. Additionally, at a service completion epoch, two different search mechanism are switched on. Thus, if the server is idle, a competition takes place among primary customer, retrial customers and those resulting in the two types of searches to access the server. If a retrial customer reaches the idle server first, the customer entering the service is served with service times having the distribution function $B_3(t)$ while if the type I search turned out to be successful, the selected customer is served according to the distribution function $B_1(t)$. If type II search succeeded, all units present in the orbit are taken for service simultaneously and the service time of the whole group follows distribution function $B_2(t)$. Denote by $\beta_i(s) = \int\limits_0^\infty e^{-st} dB_i(t),\ Re\ s > 0$, the Laplace-Stieltjes transform (LST) and $b_r^{(i)}$ (assumed to be finite), the rth moment associated with the distribution function $B_i(t), i = 0, 1, 2, 3 : b_r^{(i)} = \int\limits_0^\infty t^r dB_i(t)$. The duration of the type I (type II) search is characterized by the distribution function $H_1(t)(H_2(t))$ with LST $h_1(s)(h_2(s))$ and finite expectations h_1 and h_2. Distribution functions $H_l(t),\ l = 1, 2$, may be arbitrary, however, we assume that duration of the searches cannot be both constant. Otherwise, the search with larger value of $h_l, l = 1, 2$, will never succeed to be finished earlier that the another search and the search with larger duration has to be excluded from consideration.

The presence of the additional search mechanism allows to minimize the idle time of the server. If holding cost (charge paid due to the customers stay in the system) and costs associated with the different types of the search and service of customers are introduced, optimal tuning of the parameters of search mechanism will be possible based on the analysis results of which are presented below.

3   The Stationary Distribution of the Embedded Markov Chain

Denote by t_n the nth service completion epoch; i_n the number of customers in the orbit and ν_n the state of the $BMAP$ process ν_t at the moment $t_n + 0$. Then

$$\zeta_n = \{(i_n, \nu_n),\ n \geq 1\}$$

is a two-dimensional Markov chain with state space $\{(l, \nu)\ ;\ l \geq 0, \nu = 1, 2, \cdots m\}$. In the sequel, we need the following auxiliary matrices. Define

$$\Phi_i = \int\limits_0^\infty e^{(D_0 - \alpha_i I)t}(1 - H_1(t))(1 - H_2(t))dt,\ i > 0,\ \Phi_0 = (-D_0)^{-1},$$

$$F_i^{(1)} = \int_0^\infty e^{(D_0 - \alpha_i I)t}(1 - H_2(t))dH_1(t), \; F_i^{(2)} = \int_0^\infty e^{(D_0 - \alpha_i I)t}(1 - H_1(t))dH_2(t),$$

$$F_i^{(3)} = \alpha_i \Phi_i, \text{ for } i > 0 \text{ and } F_0^{(1)} = F_0^{(2)} = F_0^{(3)} = (-D_0)^{-1}.$$

Here $F_i^{(r)}, r = 1, 2, 3$, give the matrices of probabilities that the idle period of the server expires through type I search or type II search or retrial.

Let $\Omega_k^{(r)}$ be the matrix of probabilities that exactly k arrivals occur during a service time of the rth type, $r = 0, 1, 2, 3$. It is well-known, see, e.g., [29] that these matrices can be obtained as coefficient matrices in the following matrix generating function:

$$\Omega_r(z) = \beta_r(-D(z)) = \sum_{k=0}^\infty \Omega_k^{(r)} z^k = \int_0^\infty e^{D(z)t} dB_r(t), r = 0, 1, 2, 3.$$

Let $P(i, l)$, for $i \geq 0, l \geq 0$, denote the matrix of the one-step transition probabilities of the Markov chain $\zeta_n, n \geq 1$, with the $(\nu, \nu')^{th}$ entry defined as

$$P\{i_{n+1} = l, \nu_{n+1} = \nu' | i_n = i, \nu_n = \nu\}, \nu, \nu' = 1, 2, \cdots, m.$$

The following lemma, whose proof follows immediately from the described customers access mechanism and the formula of total probability, gives expression for the matrices $P(i, l)$.

Lemma 1. *The matrices $P(i, l)$ are calculated as follows:*

$$P(0, l) = \Phi_0 \sum_{k=1}^{l+1} D_k \Omega_{l-k+1}^{(0)} \quad l \geq 0,$$

$$P(i, l) = \Phi_i \sum_{k=1}^{l-i+1} D_k \Omega_{l-i-k+1}^{(0)} + F_i^{(1)} \Omega_{l-i+1}^{(1)} + F_i^{(2)} \Omega_l^{(2)} + F_i^{(3)} \Omega_{l-i+1}^{(3)}, \quad i \geq 1, l \geq i - 1,$$

$$P(i, l) = F_i^{(2)} \Omega_l^{(2)}, \quad i \geq 1, 0 \leq l < i - 1.$$

From now on, we make the assumption that the retrial rate α_i does not depend on i. That is $\alpha_i = \alpha$ for $i > 0$. In this case, the matrices Φ_i, $F_i^{(r)}$, $r = 1, 2, 3$, do not depend on i and are denoted as Φ, $F^{(r)}$, $r = 1, 2, 3$, correspondingly.

It can be shown that, due to the possibility of simultaneous service of all customers presenting in the system, the system is stable under any set of the system parameters, therefore the stationary probabilities of the chain always exist.

Let us denote these probabilities by

$$\pi(i, \nu) = \lim_{n \to \infty} P\{i_n = i, \nu_n = \nu\}, \; \nu = 1, \ldots, m,$$

and let us introduce the following row vectors:

$$\boldsymbol{\pi}_i = (\pi(i, 1), \cdots \cdots, \pi(i, m)), \; i \geq 0.$$

Using the obtained transition probabilities, we get the system of linear algebraic equations (equilibrium equations) for the steady state probabilities as given below:

$$\pi_l = \pi_0 \Phi_0 \sum_{k=1}^{l+1} D_k \Omega^{(0)}_{l-k+1} + \sum_{i=1}^{\infty} \pi_i F^{(2)} \Omega^{(2)}_l$$

$$+ \sum_{i=1}^{l+1} \pi_i \left[\Phi \sum_{k=1}^{l-i+1} D_k \Omega^{(0)}_{l-i-k+1} + F^{(1)} \Omega^{(1)}_{l-i+1} F^{(3)} \Omega^{(3)}_{l-i+1} \right], \quad l \geq 0. \quad (1)$$

To solve this infinite system of equations, we introduce the vector probability generating function $\pi(z) = \sum_{l=0}^{\infty} \pi_l z^l$, $|z| < 1$.

Multiplying each of the equations in (1) by the corresponding power of z, summing up and rearranging the terms, we get

$$z\pi(z) = \pi_0 \Phi_0 (D(z) - D_0)\Omega_0(z) + z(\pi(1) - \pi_0)F^{(2)}\Omega_2(z) + (\pi(z) - \pi_0)Y(z)$$

where

$$Y(z) = \Phi(D(z) - D_0)\Omega_0(z) + F^{(1)}\Omega_1(z) + F^{(3)}\Omega_3(z).$$

Thus, the vector generating function $\pi(z)$ satisfies the following vector functional equation

$$\pi(z)(zI - Y(z))$$
$$= \pi(0)(\Phi_0(D(z) - D_0)\Omega_0(z) - zF^{(2)}\Omega_2(z) - Y(z)) + z\pi(1)F^{(2)}\Omega_2(z). \quad (2)$$

This equation includes the unknown vector generating function $\pi(z)$ at three points: z, 0 and 1. Next we make an attempt to eliminate the unknown vector $\pi(1)$ from (2) in the trivial way, i.e., by substituting $z = 1$ in (2). Then we obtain the following relation between the vectors $\pi(1)$ and $\pi(0) = \pi_0$:

$$\pi(1)(I - Y(1) - F^{(2)}\Omega_2(1)) = \pi_0(\Phi_0 [D(1) - D_0]. \quad (3)$$

However, we cannot eliminate the vector $\pi(1)$ directly from Eq. (3) because it is possible to show that the matrix

$$A = Y(1) + F^{(2)}\Omega_2(1)$$

is irreducible stochastic and, consequently, the matrix $I - A$ in the right hand side of (3) is singular.

To overcome this difficulty, we apply the well-known trick by M. Neuts. Let ρ be the left probability eigenvector of the matrix A, i.e., it satisfies the equations

$$\rho A = \rho, \quad \rho e = 1.$$

Adding the vector $\pi(1)e\rho$ to both sides of (3), observing that the matrix $I - A + e\rho$ is nonsingular and that

$$\rho(I - A + e\rho)^{-1} = \rho,$$

we obtain from (3) that

$$\boldsymbol{\pi}(1) = \boldsymbol{\rho} + \boldsymbol{\pi}_0(\Phi_0\left[D(1) - D_0\right]\Omega_0(1) - A)C,$$

where

$$C = (I - A + \mathbf{e}\boldsymbol{\rho})^{-1}.$$

Then, the vector functional Eq. (2) transforms into equation

$$\boldsymbol{\pi}(z)(zI - Y(z)) = \boldsymbol{\pi}_0\left[\Phi_0(D(z) - D_0)\Omega_0(z) - zF^{(2)}\Omega_2(z) - Y(z))\right.$$

$$\left. + z(\Phi_0(D(1) - D_0)\Omega_0(1) - A)CF^{(2)}\Omega_2(z)\right] + z\boldsymbol{\rho}F^{(2)}\Omega_2(z)$$

$$(4)$$

which includes the unknown vector generating function $\boldsymbol{\pi}(z)$ only at two points: z and 0.

The methodologies for solving equations of type (4) in the case when the matrix $Y(1)$ is stochastic are well-known. One of them is based on the use of M. Neuts' approach (see [30]) that exploits the matrix G which is the solution of the nonlinear matrix equation $G = Y(G)$. Another one uses reasonings of analyticity of the vector generating function $\boldsymbol{\pi}(z)$ in the unit disk of the complex plane, see, e.g. [15].

However, in (4) the matrix $Y(1)$ is the sub-stochastic, but not stochastic. Solution of Eq. (4) in this case can be derived using the results obtained during the analysis of $BMAP/SM/1$ queue with so called disasters, see [13, 21] where the analyticity approach is properly adjusted or the papers [14, 22] where the M. Neuts' approach is generalized to the corresponding class of multi-dimensional Markov chains. Disasters have the same effect (removal of all customers from the system) as simultaneous service of all customers from the orbit after type II search succeeds to win in competition with type I search and primary or orbital customers.

4 Stationary Distributions of the Number of Customers in the Orbit and in the System at Arbitrary Time

Denote by $\mathbf{p}(i,r), i \geq 0, r = 0, \ldots, 4$, the steady state probability vector that at an arbitrary time there are i customers in the system, and the current service is in the rth mode. Note that $r = 4$ corresponds to the case when the server is idle. The following theorem gives expression for the steady state probability vectors.

Theorem 1. The stationary probability vector $\mathbf{p}(i,r)$ are calculated as follows:

$$\mathbf{p}(0,4) = \tau^{-1}\boldsymbol{\pi}_0(-D_0)^{-1},$$

$$\mathbf{p}(i,4) = \tau^{-1}\boldsymbol{\pi}_i\Phi, \quad i \geq 1,$$

$$\mathbf{p}(i,3) = \tau^{-1} \sum_{l=1}^{i} \boldsymbol{\pi}_l F^{(3)} \tilde{\Omega}_{i-l}^{(3)},$$

$$\mathbf{p}(i,2) = \tau^{-1} \sum_{l=1}^{i} \boldsymbol{\pi}_l F^{(2)} \tilde{\Omega}_{i-l}^{(2)},$$

$$\mathbf{p}(i,1) = \tau^{-1} \sum_{l=1}^{i} \boldsymbol{\pi}_l F^{(1)} \tilde{\Omega}_{i-l}^{(1)},$$

$$\mathbf{p}(i,0) = \tau^{-1} \boldsymbol{\pi}_0 \sum_{k=1}^{i} (-D_0)^{-1} D_k \tilde{\Omega}_{i-k}^{(0)} + \sum_{l=1}^{i-1} \boldsymbol{\pi}_l \sum_{k=1}^{i-l} \Phi D_k \tilde{\Omega}_{i-l-k}^{(0)}, \quad i > 0,$$

where the matrices $\tilde{\Omega}_m^{(r)}$ are the coefficients appearing in the matrix expansion

$$\tilde{\Omega}_r(z) = \sum_{k=0}^{\infty} \tilde{\Omega}_k^{(r)} z^k = \int_0^{\infty} e^{D(z)t} (1 - B_r(t)) dt, \quad r = 0, \dots, 3,$$

and the average inter-departure time, τ, is given by formula

$$\tau = \boldsymbol{\pi}_0 ((-D_0)^{-1} + b_1^{(0)} I) \mathbf{e} + \sum_{i=1}^{\infty} \boldsymbol{\pi}_i \left(\sum_{j=1}^{3} F^{(j)} b_1^{(j)} \mathbf{e} + \Phi (I - D_0 b_1^{(0)}) \mathbf{e} \right).$$

Proof follows from the theory of Markov renewal processes (see [10,31]).

In a similar manner, if we define $\mathbf{q}(i,r)$, $i \geq 0$, $r = 0, \dots, 4$, as the steady state probability vectors at an arbitrary time that there are i customers in the orbit and the current service is in the rth mode, we get the following result:

Theorem 2. Vectors $\mathbf{q}(i,r)$, $i \geq 0$, $r = 0, \dots, 4$, are computed as follows:

$$\mathbf{q}(0,4) = \tau^{-1} \boldsymbol{\pi}_0 (-D_0)^{-1},$$

$$\mathbf{q}(i,4) = \tau^{-1} \boldsymbol{\pi}_i \Phi, \quad i > 0,$$

$$\mathbf{q}(i,3) = \tau^{-1} \sum_{l=1}^{i+1} \boldsymbol{\pi}_l F^{(3)} \tilde{\Omega}_{i-l+1}^{(3)},$$

$$\mathbf{q}(i,2) = \tau^{-1} \sum_{l=1}^{\infty} \boldsymbol{\pi}_l F^{(2)} \tilde{\Omega}_i^{(2)},$$

$$\mathbf{q}(i,1) = \tau^{-1} \sum_{l=1}^{i+1} \boldsymbol{\pi}_l F^{(1)} \tilde{\Omega}_{i-l+1}^{(1)},$$

$$\mathbf{q}(i,0) = \tau^{-1} \left(\boldsymbol{\pi}_0 (-D_0)^{-1} \sum_{k=1}^{i+1} D_k \tilde{\Omega}_{i+1-k}^{(0)} + \sum_{l=1}^{i} \sum_{k=1}^{i+1-l} \Phi D_k \tilde{\Omega}_{i+1-l-k}^{(0)} \right), \quad i \geq 0.$$

5 Some Performance Measures

Let us introduce the following vector partial generating functions

$$\mathbf{P}(z,r) = \sum_{i=0}^{\infty} z^i \mathbf{p}(i,r), \ \mathbf{Q}(z,r) = \sum_{i=0}^{\infty} z^i \mathbf{q}(i,r), \ |z| < 1, \ r = 0,\dots,4.$$

Having computed the stationary distributions for both the system size and orbit size, we can calculate some important performance characteristics of the model as follows:

- Probability of the system being empty at an arbitrary moment is defined by $\mathbf{p}(0,4)\mathbf{e}$;
- Probability that the server is free at an arbitrary moment is defined by $\mathbf{P}(1,4)\mathbf{e}$;
- Probability that the server is working in the zero mode (primary customer service) at an arbitrary moment is defined by $\mathbf{P}(1,0)\mathbf{e}$;
- Probability that the server is working in mode 1 (orbital customer service after the type I search) at an arbitrary moment is defined by $\mathbf{P}(1,1)\mathbf{e}$;
- Probability that the server is working in mode 2 (orbital customer(s) service after the II search) at an arbitrary moment is defined by $\mathbf{P}(1,2)\mathbf{e}$;
- Probability that the server is working in mode 3 (retrial customer service) at an arbitrary moment is defined by $\mathbf{P}(1,3)\mathbf{e}$;
- Probability that the orbit is empty at an arbitrary moment is defined by $\sum_{j=0}^{4} \mathbf{q}(0,j)\mathbf{e}$;
- Probability of having i customers in the orbit at an arbitrary moment is defined by $\sum_{j=0}^{4} \mathbf{q}(i,j)\mathbf{e}$;
- Average number of customers in the system is defined by $\sum_{j=0}^{4} \mathbf{P}'(1,j)\mathbf{e}$;
- Average number of customers in the orbit is defined by $\sum_{j=0}^{4} \mathbf{Q}'(1,j)\mathbf{e}$.

Remark

Corresponding probabilities for an arbitrary batch arrival epochs are computed by the analogous formulas only the vector $\mathbf{e}$ has to be replaced with the vector $-D_0\mathbf{e}$. Probability of starting the service of an arbitrary customer immediately upon arrival (probability that the customer receives service in the system without visiting the orbit) is defined by

$$\lambda^{-1}\mathbf{P}(1,4)(-D_0)\mathbf{e}.$$

Some examples of computation of the key performance measures of the system for the considered model in case of the group Poisson arrival process are presented in [11]. Computations for the general case of the $BMAP$ are much more involved. But they can be successfully done based on the corresponding modules of software described in [19].

6 Conclusions

We considered retrial queueing model where the usual mechanism of customers access to the service via the competition of the primary and orbital customers is supplemented by the mechanisms of customers search in the orbit by the server. One option of the search leads to the individual service of a customer found in the orbit. Another one results in simultaneous service of all customers presenting in the orbit. Stationary distributions of the system states at the embedded service completion moments and arbitrary moments are computed along with some important performance measures of the system.

The results can be used for optimization of operation of the system if some cost criteria accounting the quality and cost of different kinds of customers service and access will be introduced. More types of customers search can be considered. The case when search times have a phase type distribution, in which the presented analytical results may be more easy implemented in the form of software, deserves more close consideration.

Extension of the analysis to the case when the total retrial intensity depends on the current number of customers in orbit is possible based on the results from [28] with modification that accounts possibility of emptying the system at the random moment, irrespectively to the current number of customers in the system.

Acknowledgments. This work has been financially supported by the Russian Science Foundation and the Department of Science and Technology (India) via grant No 16-49-02021 (INT/RUS/RSF/16) for the joint research project by the V.A. Trapeznikov Institute of Control Problems of the Russian Academy Sciences and the CMS College Kottayam.

References

1. Artalejo, J.R.: Accessible bibliography on retrial queues. Math. Comput. Modell. **30**, 1–6 (1999)
2. Artalejo, J.R.: A classified bibliography of research on retrial queues: progress in 1990–1999. TOP **7**, 187–211 (1999)
3. Artalejo, J.R., Joshua, V.C., Krishnamoorthy, A.: An $M/G/1$ retrial queue with orbital search by the server. In: Advances in Stochastic Modelling, pp. 41–54. Notable Publications Inc., New Jersey (2002)
4. Breuer, L., Dudin, A.N., Klimenok, V.I.: A retrial $BMAP/PH/N$ system. Queueing Syst. **40**, 431–455 (2002)

5. Brugno, A., D'Apice, C., Dudin, A.N., Manzo, R.: Analysis of a $MAP/PH/1$ queue with flexible group service. Appl. Math. Comput. Sci. **27**, 119–131 (2017)
6. Brugno, A., Dudin, A.N., Manzo, R.: Retrial queue with discipline of adaptive permanent pooling. Appl. Math. Model. **50**, 1–16 (2017)
7. Chakravarthy, S.R.: The batch Markovian arrival process: a review and future work. In: Krishnamoorthy, A., Raju, N., Ramaswami, V. (eds.) Advances in Probability Theory and Stochastic Processes, pp. 21–49. Notable Publications Inc., Neshanic Station, NJ (2001)
8. Chakravarthy, S.R., Dudin, A.N.: A multiserver retrial queue with $BMAP$ arrivals and group services. Queueing Syst. **42**, 5–31 (2002)
9. Choi, B.D., Chung, Y.H., Dudin, A.N.: The $BMAP/SM/1$ retrial queue with controllable operation modes. Eur. J. Oper. Res. **131**, 16–30 (2001)
10. Cinlar, E.: Introduction to Stochastic Processes. Prentice-Hall, Englewood Cliffs, NJ (1975)
11. Deepak, T.G., Dudin, A.N., Joshua, V.C., Krishnamoorthy, A.: On an $M^{(X)}/G/1$ retrial system with two types of search of customers from the orbit. Stochast. Anal. Appl. **31**, 92–107 (2013)
12. Deepak, T.G.: On a retrial queueing model with single/batch service and search of customers from the orbit. TOP **23**, 493–520 (2015)
13. Dudin, A.N., Karolik, A.V.: $BMAP/SM/1$ queue with Markovian input of disasters and non-instantaneous recovery. Perform. Eval. **45**, 19–32 (2001)
14. Dudin, A.N., Kim, C.S., Klimenok, V.I.: Markov chains with hybrid repeated rows - upper-Hessenberg quasi-Toeplitz structure of block transition probability matrix. J. Appl. Probab. **45**(1), 211–225 (2008)
15. Dudin, A.N., Klimenok, V.I.: Multi-dimensional quasi-Toeplitz Markov chains. J. Appl. Math. Stochast. Anal. **12**, 393–415 (1999)
16. Dudin, A., Klimenok, V.: Queueing system $BMAP/G/1$ with repeated calls. Math. Comput. Modell. **30**, 115–128 (1999)
17. Dudin, A., Klimenok, V.: A retrial $BMAP/SM/1$ system with linear repeated requests. Queueing Systems **34**, 47–66 (2000)
18. Dudin, A., Klimenok, V.: $BMAP/SM/1$ model with Markov modulated retrials. TOP **7**, 267–278 (1999)
19. Dudin, A.N., et al.: Software SIRIUS+ for evaluation and optimization of queues with the $BMAP$ input. In: Latouche, G., Taylor, P. (eds.) Advances in Algorithmic Methods in Stochastic Models, pp. 115–133. Notable Publication Inc., Neshanic Station, NJ (2000)
20. Dudin, A.N., Krishnamoorthy, A., Joshua, V.C., Tsarankov, G.V.: Analysis of the $BMAP/G/1$ retrial system with search of customers from the orbit. Eur. J. Oper. Res. **157**, 169–179 (2004)
21. Dudin, A.N., Nishimura, S.: A $BMAP/SM/1$ queueing system with Markovian arrival input of disasters. J. Appl. Probab. **36**, 868–881 (1999)
22. Dudin, A.N., Semenova, O.V.: Stable algorithm for stationary distribution calculation for a $BMAP/SM/1$ queueing system with Markovian input of disasters. J. Appl. Probab. **42**, 547–556 (2004)
23. Dudin, A.N., Piscopo, R., Manzo, R.: Queue with group admission of customers. Comput. Oper. Res. **61**, 89–99 (2015)
24. Falin, G.I.: A survey of retrial queues. Queueing Syst. **7**, 127–167 (1990)
25. Falin, G.I., Templeton, J.G.C.: Retrial Queues. Chapman and Hall, London (1997)
26. He, Q.M., Li, H., Zhao, Y.Q.: Ergodicity of the $BMAP/PH/S/S+K$ retrial queue with PH-retrial times. Queueing Syst. **35**, 323–347 (2000)

27. Kim, J., Kim, B.: A survey of retrial queueing systems. Ann. Oper. Res. **247**(1), 3–36 (2016)
28. Klimenok, V., Dudin, A.: Multi-dimensional asymptotically quasi-Toeplitz Markov chains and their application in queueing theory. Queueing Syst. **54**, 245–259 (2006)
29. Lucantoni, D.M.: New results on the single server queue with a batch Markovian arrival process. Commun. Stat. Stochast. Models **7**, 1–46 (1991)
30. Neuts, M.F.: Structured Stochastic Matrices of $M/G/1$ Type and Their Applications. Marcel Dekker, New York (1989)
31. Neuts, M.F., Ramalhoto, M.F.: A service model in which the server is required to search for customers. J. Appl. Probab. **21**, 157–166 (1984)
32. Yang, T., Templeton, J.G.: A survey on retrial queues. Queueing Syst. **2**, 201–233 (1987)

Reliability Analysis of a Two-Server Heterogeneous Unreliable Queueing System with a Threshold Control Policy

Dmitry Efrosinin[1,2]($\boxtimes$), Janos Sztrik[3], Mais Farkhadov[4],
and Natalia Stepanova[4]

[1] Peoples' Friendship University of Russia (RUDN University),
Miklukho-Maklaya Street 6, 117198 Moscow, Russia
[2] Johannes Kepler University Linz, Altenbergerstrasse 69, 4040 Linz, Austria
dmitry.efrosinin@jku.at
[3] University of Debrecen, Egyetem tér 1, Debrecen 4032, Hungary
sztrik.janos@inf.unideb.hu
[4] Institute of Control Sciences, RAS,
Profsoyuznaya Street 65, 117997 Moscow, Russia
http://www.rudn.ru, http://www.jku.at, http://www.unideb.hu,
http://www.ipu.ru

Abstract. Heterogeneous servers which can differ in service speed and reliability are getting more popular in modeling of modern communication systems. For a two-server queueing system with unreliable servers the allocation of customers between the servers is performed via a threshold control policy which prescribes to use the fastest server whenever it is free and the slower one only if the number of waiting customers exceeds some threshold level depending on the state of faster server. The main task of the paper consists in reliability analysis of the proposed system including evaluation of the stationary availability and reliability function. The effects of different parameters on introduced reliability characteristics are analyzed numerically.

Keywords: Reliability analysis · Quasi-birth-and-death process · Heterogeneous servers · Threshold policy · Matrix-geometric solution method

1 Introduction

To make modern communication systems superior in performance and reliability to the previous generation systems they can be supplied with heterogeneous communication links. Such links can differ in availability, link data throughputs,

D. Efrosinin—The publication was financially supported by the Ministry of Education and Science of the Russian Federation (the Agreement number 02.a03.21.0008), by the Russian Foundation for Basic Research, Project No. 16-37-60072 mol_a_dk and No. 15-08-08677 A, by the Austro-Hungarian Scientific Cooperation OMAA 96öu8.

© Springer International Publishing AG 2017
A. Dudin et al. (Eds.): ITMM 2017, CCIS 800, pp. 13–27, 2017.
DOI: 10.1007/978-3-319-68069-9_2

power consumption and reliability characteristics. To model the dynamic behaviour of the data transmission links subject to breakdowns a queueing system with non-reliable servers can be used. The analysis of multi-server queueing systems generally assume the servers to be homogeneous. Mitrany and Avi-Izhak [11] and Neuts and Lucantoni [13] have studied the $M/M/s$ queueing system with server breakdowns and repairs. In paper of Levi and Yechiali [9] the queue $M/M/s$ with servers' vacations was analyzed. A recent paper of Efrosinin et al. [3] deals with an stationary analysis performed on the busy period for the multi-server Markovian queueing system with simultaneous failures of servers. The queues with heterogeneous non-reliable servers occur quite rarely as a research subject. A queueing system with two heterogeneous servers and multiple vacations was studied by Kumar and Madheswari [6], who obtained the stationary queue length distribution by using matrix geometric method and provided analysis of busy period and waiting time. In Kumar et al. [7] the same authors have introduced the $M/M/2$ queueing system with heterogeneous servers subject to catastrophes and provided a transient solution for the system under study. A heterogeneous two-server queueing system with balking and server breakdowns has been studied by Yue et al. [16]. In their study, some stationary mean performance measures are obtained using the matrix-geometric solution method.

In heterogeneous queueing system with one common queue, especially in case of the service without preemption, when the customer can not change the server during a service time, the customer allocation mechanism between the servers must be specified. The majority of heterogeneous systems investigated use heuristic service policies (e.g. the Fastest Free Server (FFS) or Random Service Selection (RSS) policies). In fact these policies are not optimal, if e.g. the mean response time must be minimized. As it is already known, see. e.g. the results of Efrosinin [1], Koole [5], Legros and Jouini [8], Lin and Kumar [10], Rykov and Efrosinin [15], for the heterogeneous queueing systems the optimal allocation policy belongs to a class of threshold policies, where the less effective server must be used only if the number of customers in the queue has reached some pre-specified threshold level. The same result was confirmed for the queueing system with faster non-reliable server and absolutely reliable slower server in Efrosinin [2], Özkan and Kharoufeh [14] and for two non-reliable heterogeneous servers in system with a constant retrial discipline in Efrosinin and Sztrik [4]. In the latter paper it was shown that for the fixed threshold policy the corresponding Markov process is of the QBD (Quasi-birth-and-death) type with a tri-diagonal block infinitesimal matrix with a large number of bounding states.

While the first steps in performance analysis of controllable heterogeneous queueing systems have already been performed for completely reliable servers, a missing link to an applicability of heterogeneous models is a reliability analysis of such queues with servers subject to failures. In this paper we use a forward-elimination-backward-substitution method expressed in matrix form in terms of the Laplace-Stiltjes transforms (LST) combined with probability generating function (PGF) approach to evaluate reliability measures such as reliability function, which represents the complementary cumulative distribution function of the life time, and mean time to the first failure for each server separately and for the group of servers under the fixed threshold allocation control policy. The reliability functions are obtained in terms of the Laplace transform (LT) and numerical inversion algorithm is used to get the time dependent functions. Additionally a new discrete reliability metric in form of the distribution of the number of failures during a certain life time is introduced. We expect that the proposed results can be generalized to the case of an arbitrary controllable non-reliable queueing model with a QBD structure.

The rest of the paper is organized as follows. In Sect. 2, we describe the mathematical model and give a presentation of the stationary distribution of the system state using a matrix-geometric solution method. In Sect. 3, we develop computational analysis for the stationary reliability characteristics, for the reliability function and mean time to failure. The number of failures during a certain life time is investigated in Sect. 4. In Sect. 5, numerical illustrations are provided to highlight the effect of some parameters on the derived reliability characteristics.

Hereafter, the notations $\mathbf{e}(n)$, $\mathbf{e}_j(n)$, and I_n are used respectively for the column-vector consisting of 1's, the column vector with 1 in the j-th (beginning from 0-th) position and 0 elsewhere, and an identity matrix of the dimension n. When there is no need to emphasize the dimension of these vectors the suffix will be suppressed and dimension is determined by the context. The expressions $diag(a_1, \ldots, a_n)$, $diag^+(a_1, \ldots, a_n)$, and $diag^-(a_1, \ldots, a_n)$ denote respectively the diagonal matrix, the upper diagonal matrix, and the lower diagonal matrix with entries $a_1, \ldots, a_n$ that could be scalars or matrices.

2   Mathematical Model and Stationary Distribution

In the present paper we deal with a two-server heterogeneous non-reliable queueing model of the type $M/M/2$. The customers arrive according to a Poisson process with arrival rate λ. The service times are exponentially distributed with rates μ_1 and μ_2, where $\mu_1 \geq \mu_2$. We assume that the server fails respectively at an exponential rate α_1 and α_2. The servers can fail only if they are busy. The failed server is repaired immediately and the time required to repair it is exponentially distributed respectively with rate β_1 and β_2. The customer being served at the failure moment is left at this server during the repair time and can be served when the server becomes operational again. The allocation mechanism between two servers is based on a threshold policy: depending on the state of faster server the slower one is used whenever the number of customers in the queue exceeds a certain threshold level.

Let $Q(t)$ and $D(t) = \{D_1(t), D_2(t)\}$ denote, respectively, the number of customers in the queue and the vector state of servers at time t, where

$$D_j(t) = \begin{cases} 0, & \text{the server } j \text{ is idle,} \\ 1, & \text{the server } j \text{ is busy and operational,} \\ 2, & \text{the server } j \text{ is failed.} \end{cases}$$

The threshold policy $f = (q_1, q_2)$ is defined by two threshold levels $1 \leq q_2 \leq q_1 < \infty$. According to this policy server 1 must be activated whenever it is free and there are customers in the queue, whereas server 2 is used only if server 1 is in state 1 or 2 and the number of customers in the queue has reached the value q_1 or q_2. The process

$$\{X(t)\}_{t\geq 0} = \{Q(t), D(t)\}_{t\geq 0} \tag{1}$$

is a continuous-time Markov chain with a state space given by

$$E = \{x = (q, d_1, d_2); q \in \mathbb{N}_0, (d_1, d_2) \in E_D\}, \tag{2}$$

where E_D is a set of states of servers which is defined as

$$E_D = \left\{ (d_1, d_2); \begin{array}{l} d_j \in \{0,1,2\}, j \in \{1,2\}, q = 0 \\ d_1 \in \{1,2\}, d_2 \in \{0,1,2\}, 1 \leq q \leq q_2 - 1, \\ d_1 \in \{1,2\}, d_2 \in \{0,1,2\}, (d_1, d_2) \neq (2,0), q_2 \leq q \leq q_1 - 1, \\ d_j \in \{1,2\}, j \in \{1,2\}, q \geq q_1, \end{array} \right\}.$$

Next we partition E in blocks as follows,

$$(\mathbf{0}, \mathbf{0}) = \{(0,0,d_2); d_2 \in \{0,1,2\}\},$$

$$(\mathbf{q}, \mathbf{1}) = \begin{cases} \{(q,1,0),(q,2,0),(q,1,1),(q,2,1),(q,1,2),(q,2,2)\}, & 0 \leq q \leq q_2 - 1, \\ \{(q,1,0),(q,1,1),(q,2,1),(q,1,2),(q,2,2)\}, & q_2 \leq q \leq q_1 - 1, \\ \{(q,1,1),(q,2,1),(q,1,2),(q,2,2)\}, & q \geq q_1. \end{cases}$$

Due to above notation, the infinitesimal generator olude the rates of transition fromf the Markov chain $\{X(t)\}_{t\geq 0}$ has the block-tridiagonal structure,

$$\Lambda = [\lambda_{xy}]_{x,y \in E} = diag(Q_{1,0}, \underbrace{Q_{1,1}, \dots, Q_{1,1}}_{q_2 - 1}, Q_{1,2}, \underbrace{Q_{1,3}, \dots, Q_{1,3}}_{q_1 - q_2 - 1}, Q_{1,4}, Q_{1,5}, \dots)$$

$$+ \, diag^+(Q_{0,1}, \underbrace{Q_{0,2}, \dots, Q_{0,2}}_{q_2 - 1}, Q_{0,3}, \underbrace{Q_{0,4}, \dots, Q_{0,4}}_{q_1 - q_2 - 1}, Q_{0,5}, Q_{0,6}, \dots)$$

$$+ \, diag^-(Q_{2,1}, \underbrace{Q_{2,2}, \dots, Q_{2,2}}_{q_2 - 1}, Q_{2,3}, \underbrace{Q_{2,4}, \dots, Q_{2,4}}_{q_1 - q_2 - 1}, Q_{2,5}, Q_{2,6}, \dots).$$

The square matrices $Q_{1,n}, 0 \leq n \leq 5$, include the rates of the output from the current block of states,

$$Q_{1,0} = \begin{pmatrix} -\lambda & 0 & 0 \\ \mu_2 & -(\lambda + \alpha_2 + \mu_2) & \alpha_2 \\ 0 & \beta_2 & -(\lambda + \beta_2) \end{pmatrix},$$

$$Q_{1,1} = \begin{pmatrix} -(\lambda+\mu_1+\alpha_1) & \alpha_1 & 0 & 0 & 0 & 0 \\ \beta_1 & -(\lambda+\beta_1) & 0 & 0 & 0 & 0 \\ \mu_2 & 0 & -(\lambda+\mu+\alpha) & \alpha_1 & \alpha_2 & 0 \\ 0 & \mu_2 & \beta_1 & -(\lambda+\alpha_2+\beta_1+\mu_2) & 0 & \alpha_2 \\ 0 & 0 & \beta_2 & 0 & -(\lambda+\alpha_1+\beta_2+\mu_1) & \alpha_1 \\ 0 & 0 & 0 & \beta_2 & \beta_1 & -(\lambda+\beta) \end{pmatrix},$$

$$Q_{1,2} = Q_{1,1} + \lambda \mathbf{e}_1(6) \otimes \mathbf{e}_3'(6),$$

$$Q_{1,3} = \begin{pmatrix} -(\lambda+\mu_1+\alpha_1) & 0 & 0 & 0 & 0 \\ \mu_2 & -(\lambda+\mu+\alpha) & \alpha_1 & \alpha_2 & 0 \\ 0 & \beta_1 & -(\lambda+\alpha_2+\beta_1+\mu_2) & 0 & \alpha_2 \\ 0 & \beta_2 & 0 & -(\lambda+\alpha_1+\beta_2+\mu_1) & \alpha_1 \\ 0 & 0 & \beta_2 & \beta_1 & -(\lambda+\beta) \end{pmatrix},$$

$$Q_{1,4} = Q_{1,3} + \lambda \mathbf{e}_0(5) \otimes \mathbf{e}_1'(5),$$

$$Q_{1,5} = \begin{pmatrix} -(\lambda+\mu+\alpha) & \alpha_1 & \alpha_2 & 0 \\ \beta_1 & -(\lambda+\alpha_2+\beta_1+\mu_2) & 0 & \alpha_2 \\ \beta_2 & 0 & -(\lambda+\alpha_1+\beta_2+\mu_1) & \alpha_1 \\ 0 & \beta_2 & \beta_1 & -(\lambda+\beta) \end{pmatrix}.$$

The rectangular matrices $Q_{0,n}, 1 \le n \le 6$, include the rates of transitions from subsequent block to the current one,

$$Q_{0,1} = \lambda \begin{pmatrix} 1 & 0 & 0 & 0 & 0 & 0 \\ 0 & 0 & 1 & 0 & 0 & 0 \\ 0 & 0 & 0 & 0 & 1 & 0 \end{pmatrix}, \quad Q_{0,3} = \lambda \begin{pmatrix} 1 & 0 & 0 & 0 & 0 \\ 0 & 0 & 0 & 0 & 0 \\ 0 & 1 & 0 & 0 & 0 \\ 0 & 0 & 1 & 0 & 0 \\ 0 & 0 & 0 & 1 & 0 \\ 0 & 0 & 0 & 0 & 1 \end{pmatrix}, \quad Q_{0,5} = \lambda \begin{pmatrix} 0 & 0 & 0 & 0 \\ 1 & 0 & 0 & 0 \\ 0 & 1 & 0 & 0 \\ 0 & 0 & 1 & 0 \\ 0 & 0 & 0 & 1 \end{pmatrix},$$

$$Q_{0,2} = \lambda I_6, \quad Q_{0,4} = \lambda I_5, \quad Q_{0,6} = \lambda I_4, \quad \mu = \mu_1 + \mu_2, \quad \alpha = \alpha_1 + \alpha_2, \quad \beta = \beta_1 + \beta_2.$$

The rectangular matrices $Q_{2,n}, 1 \le n \le 6$, include the rates of transition from the previous block to the current one,

$$Q_{2,1} = \begin{pmatrix} \mu_1 & 0 & 0 \\ 0 & 0 & 0 \\ 0 & \mu_1 & 0 \\ 0 & 0 & 0 \\ 0 & 0 & \mu_1 \\ 0 & 0 & 0 \end{pmatrix}, \quad Q_{2,2} = \begin{pmatrix} \mu_1 & 0 & 0 & 0 & 0 & 0 \\ 0 & 0 & 0 & 0 & 0 & 0 \\ 0 & 0 & \mu_1 & 0 & 0 & 0 \\ 0 & 0 & 0 & 0 & 0 & 0 \\ 0 & 0 & 0 & 0 & \mu_1 & 0 \\ 0 & 0 & 0 & 0 & 0 & 0 \end{pmatrix}, \quad Q_{2,3} = \begin{pmatrix} \mu_1 & 0 & 0 & \alpha_1 & 0 & 0 \\ 0 & 0 & \mu_1 & 0 & 0 & 0 \\ 0 & 0 & 0 & \mu_2 & 0 & 0 \\ 0 & 0 & 0 & 0 & \mu_1 & 0 \\ 0 & 0 & 0 & 0 & 0 & 0 \end{pmatrix},$$

$$Q_{2,4} = \begin{pmatrix} \mu_1 & 0 & \alpha_1 & 0 & 0 \\ 0 & \mu_1 & 0 & 0 & 0 \\ 0 & 0 & \mu_2 & 0 & 0 \\ 0 & 0 & 0 & \mu_1 & 0 \\ 0 & 0 & 0 & 0 & 0 \end{pmatrix}, \quad Q_{2,5} = \begin{pmatrix} 0 & \mu & 0 & 0 & 0 \\ 0 & 0 & \mu_2 & 0 & 0 \\ 0 & 0 & 0 & \mu_1 & 0 \\ 0 & 0 & 0 & 0 & 0 \end{pmatrix}, \quad Q_{2,6} = \begin{pmatrix} \mu & 0 & 0 & 0 \\ 0 & \mu_2 & 0 & 0 \\ 0 & 0 & \mu_1 & 0 \\ 0 & 0 & 0 & 0 \end{pmatrix}.$$

Denote by $\boldsymbol{\pi} = (\boldsymbol{\pi}_{0,0}, \boldsymbol{\pi}_{0,1}, \boldsymbol{\pi}_{1,1}, \boldsymbol{\pi}_{2,1}, \dots)$ the stationary probability vector of Λ which satisfies

$$\boldsymbol{\pi}\Lambda = \mathbf{0}, \quad \boldsymbol{\pi}\mathbf{e} = 1. \tag{3}$$

The computation of the stationary distribution is reduced to solving a block-tridiagonal system. The process $\{X(t)\}_{t \ge 0}$ is in the format of a quasi-birth-and-death (QBD) process which allows to apply the matrix-analytic approach.

By [12, Theorem 3.1.1] it is well known that the stationary probability vector $\boldsymbol{\pi}$ of the QBD process exists if and only if

$$\mathbf{p}Q_{0,6}\mathbf{e}(4) < \mathbf{p}Q_{2,6}\mathbf{e}(4),$$

where $\mathbf{p} = (p_1, p_2, p_3, p_4)$ is the invariant probability of the matrix $Q_{0,6} + Q_{1,5} + Q_{2,6}$. This vector can be obtained by solving the system $\mathbf{p}(Q_{0,6} + Q_{1,5} + Q_{2,6}) = \mathbf{0}$ and $\mathbf{pe}(4) = 1$. After some routine manipulation we can obtain the condition

$$\rho = \frac{\lambda}{\sum_{j=1}^{2} \frac{\beta_j \mu_j}{\alpha_j + \beta_j}} < 1. \tag{4}$$

Theorem 1. *The vectors of stationary probabilities $\boldsymbol{\pi}_{q,i}$, $q \geq 0$, can be computed as follows,*

$$\boldsymbol{\pi}_{0,0} = \boldsymbol{\pi}_{q_1,1} \prod_{j=0}^{q_1} M_{q_1-j}, \tag{5}$$

$$\boldsymbol{\pi}_{q,1} = \boldsymbol{\pi}_{q_1,1} \prod_{j=0}^{q_1-q-1} M_{q_1-j}, \ 0 \leq q \leq q_1 - 1,$$

$$\boldsymbol{\pi}_{q,1} = \boldsymbol{\pi}_{q_1,1} R^{q-q_1}, \ q \geq q_1,$$

where the matrices $M_i, 0 \leq i \leq q_1$, are recursively defined

$$M_0 = -Q_{2,1}Q_{1,0}^{-1}, \ M_1 = -Q_{2,2}(M_0 Q_{0,1} + Q_{1,1})^{-1}, \tag{6}$$

$$M_q = -Q_{2,2}(M_{q-1}Q_{0,2} + Q_{1,1})^{-1}, \ 2 \leq q \leq q_2 - 1,$$

$$M_{q_2} = -Q_{2,3}(M_{q_2-1}Q_{0,2} + Q_{1,2})^{-1}, \ M_{q_2+1} = -Q_{2,4}(M_{q_2}Q_{0,3} + Q_{1,3})^{-1},$$

$$M_q = -Q_{2,4}(M_{q-1}Q_{0,4} + Q_{1,3})^{-1}, \ q_2 + 2 \leq q \leq q_1 - 1,$$

$$M_{q_1} = -Q_{2,5}(M_{q_1-1}Q_{0,4} + Q_{1,4})^{-1}.$$

The vector $\boldsymbol{\pi}_{q_1,1}$ is a unique solution of the system of equations

$$\boldsymbol{\pi}_{q_1,1}\left[\sum_{q=-1}^{q_1-1} \prod_{j=0}^{q_1-q-1} M_{q_1-j} + (I - R)^{-1}\right]\mathbf{e}(4) = 1, \tag{7}$$

$$\boldsymbol{\pi}_{q_1,1}(M_{q_1}Q_{0,5} + Q_{1,5} + RQ_{2,6}) = \mathbf{0}.$$

The matrix R is a minimal solution of the matrix quadratic equation,

$$R^2 Q_{2,6} + RQ_{1,5} + Q_{0,6} = 0. \tag{8}$$

Proof. The last row of (5) and equation $R^2 Q_{2,6} + RQ_{1,5} + Q_{0,6} = 0$ follow from the properties of the QBD process [12]. If the stability condition holds, then (3) yields the system,

$$\pi_{0,0}Q_{1,0} + \pi_{0,1}Q_{2,1} = \mathbf{0},$$

$$\pi_{q-1,1}Q_{0,1} + \pi_{q,1}Q_{1,1} + \pi_{q+1,1}Q_{2,2} = \mathbf{0},\ 2 \leq q \leq q_2 - 1,$$

$$\pi_{q_2-1,1}Q_{0,2} + \pi_{q_2,1}Q_{1,2} + \pi_{q_2+1,1}Q_{2,3} = \mathbf{0},$$

$$\pi_{q_2,1}Q_{0,3} + \pi_{q_2+1,1}Q_{1,3} + \pi_{q_2+2,1}Q_{2,4} = \mathbf{0},$$

$$\pi_{q-1,1}Q_{0,4} + \pi_{q,1}Q_{1,3} + \pi_{q+1,1}Q_{2,4} = \mathbf{0},\ q_2 + 2 \leq q \leq q_1 - 1,$$

$$\pi_{q_1-1,1}Q_{0,4} + \pi_{q_1}Q_{1,4} + \pi_{q_1+1}Q_{2,5} = \mathbf{0},$$

$$\pi_{q_1,1}R^{q-q_1-1}Q_{0,5} + \pi_{q_1,1}R^{q-q_1}Q_{1,5} + \pi_{q_1,1}R^{q-q_1+1}Q_{2,6} = \mathbf{0},\ q \geq q_1 + 1.$$

The routine of substitution applied to the previous system leads to recursive relations,

$$\pi_{0,0} = \pi_{0,1}M_0, \tag{9}$$

$$\pi_{q,1} = \pi_{q+1,1}M_{q+1},\ 1 \leq q \leq q_1 - 1,$$

where M_q is defined by (6). Hence it implies the first two rows of (5). Finally the vector $\pi_{q_1,1}$ is obviously a unique solution of the system of equations (7) which consists of the normalizing condition and the balance equation for the probability vector $\pi_{q_1,1}$ of the boundary states.

3   Reliability Characteristics of the System and Servers

In this section we consider some reliability quantities of the system and servers. Denote by

$$A_1(t) = \mathbb{P}[X(t) = (q, d_1, d_2); d_1 \neq 2 \vee d_2 \neq 2],$$
$$A_2(t) = \mathbb{P}[X(t) = (q, d_1, d_2); d_1 \neq 2 \wedge d_2 \neq 2],$$
$$A_3(t) = \mathbb{P}[X(t) = (q, d_1, d_2); d_1 \neq 2],$$
$$A_4(t) = \mathbb{P}[X(t) = (q, d_1, d_2); d_2 \neq 2],$$

the pointwise availability of the system and servers. The stationary availability in case $n, 1 \leq n \leq 4$, is defined as $A_n = \lim_{t \to \infty} A_n(t)$.

Corollary 1. *The stationary availability can be computed by*

$$A_n = \pi_{0,0}\mathbf{x}_{n,1} + \sum_{q=0}^{q_2-1} \pi_{q,1}\mathbf{x}_{n,2} + \sum_{q=q_2}^{q_1-1} \pi_{q,1}\mathbf{x}_{n,3} + \pi_{q_1,1}(I - R)^{-1}\mathbf{x}_{n,4},\ 1 \leq n \leq 4,$$

where $A_2 = A_3 + A_4 - A_1$ and

$$\mathbf{x}_{1,1} = \mathbf{e}(3), \ \mathbf{x}_{1,2} = \sum_{k=0}^{4} \mathbf{e}_k(6), \ \mathbf{x}_{1,3} = \sum_{k=0}^{3} \mathbf{e}_k(5), \ \mathbf{x}_{1,4} = \sum_{k=0}^{2} \mathbf{e}_k(4),$$

$$\mathbf{x}_{2,1} = \sum_{k=0}^{1} \mathbf{e}_k(3), \ \mathbf{x}_{2,2} = \sum_{k=0}^{1} \mathbf{e}_{2k}(6), \ \mathbf{x}_{2,3} = \sum_{k=0}^{1} \mathbf{e}_k(5), \ \mathbf{x}_{2,4} = \mathbf{e}_0(4),$$

$$\mathbf{x}_{3,1} = \mathbf{e}(3), \ \mathbf{x}_{3,2} = \sum_{k=0}^{2} \mathbf{e}_{2k}(6), \ \mathbf{x}_{3,3} = \mathbf{e}_0 + \sum_{k=0}^{1} \mathbf{e}_{2k+1}(5), \ \mathbf{x}_{3,4} = \sum_{k=0}^{1} \mathbf{e}_{2k}(4),$$

$$\mathbf{x}_{4,1} = \sum_{k=0}^{1} \mathbf{e}_k(3), \ \mathbf{x}_{4,2} = \sum_{k=0}^{3} \mathbf{e}_k(6), \ \mathbf{x}_{4,3} = \sum_{k=0}^{2} \mathbf{e}_k(5), \ \mathbf{x}_{4,4} = \sum_{k=0}^{1} \mathbf{e}_k(4).$$

Corollary 2. *The stationary failure frequency of the server $l \in \{1, 2\}$ can be computed by*

$$B_l = \alpha_l \pi_{0,0} \mathbf{y}_{l,1} + \sum_{q=0}^{q_2-1} \pi_{q,1} \mathbf{y}_{l,2} + \sum_{q=q_2}^{q_1-1} \pi_{q,1} \mathbf{y}_{l,3} + \pi_{q_1,1} (I - R)^{-1} \mathbf{y}_{l,4}, \ 1 \le l \le 2,$$

where

$$\mathbf{y}_{1,1} = \mathbf{0}, \ \mathbf{y}_{1,2} = \sum_{k=0}^{2} \mathbf{e}_{2k}(6), \ \mathbf{y}_{1,3} = \mathbf{e}_0(5) + \sum_{k=0}^{1} \mathbf{e}_{2k+1}(5), \ \mathbf{y}_{1,4} = \sum_{k=0}^{1} \mathbf{e}_{2k}(4),$$

$$\mathbf{y}_{2,1} = \mathbf{e}_1(3), \ \mathbf{y}_{2,2} = \sum_{k=2}^{3} \mathbf{e}_k(6), \ \mathbf{y}_{2,3} = \sum_{k=1}^{2} \mathbf{e}_k(5), \ \mathbf{y}_{2,4} = \sum_{k=0}^{1} \mathbf{e}_k(4).$$

Denote by T the random time to the first failure of one of server. The corresponding reliability function, which is the same as the complementary cumulative distribution function of the life time T, is then defined as

$$R(t) = \mathbb{P}[T > t].$$

In this section we intend to obtain this function in terms of the Laplace transform $\tilde{R}(s) = \int_0^\infty R(s) e^{-st} dt, Re[s] > 0$. In order to realize it we let the corresponding failure states be absorbing states. In this case we obtain new process which can be modelled by the auxiliary continuous-time absorbing Markov chains $\{\hat{X}(t)\}_{t \ge 0}$ with state space $\hat{E} = E \setminus \{x = (q, d_1, d_2); q \in \mathbb{N}_0, d_1 = 2 \vee d_2 = 2\}$. We describe two main approaches to get the function $\tilde{R}(s)$: By means of the transient solution of the absorbing Markov chain and using the remaining life time.

Theorem 2. *The Laplace transform of $R(t)$ is given by*

$$\tilde{R}(s) = \tilde{P}_{1,0}(s, 1) + \tilde{P}_{1,1}(s, 1) + \tilde{P}_{1,2}(s, 1), \tag{10}$$

where

$$\tilde{P}_{1,0}(s,1) = \frac{1 + \alpha_1\tilde{\pi}_{(0,0,0)}(s) - \lambda\tilde{\pi}_{(q_1-1,1,0)}(s) + \mu_2\tilde{P}_{1,1}(s,1)}{s+\alpha_1}, \tag{11}$$

$$\tilde{P}_{1,1}(s,1) = \frac{\alpha_1\tilde{\pi}_{(0,0,1)}(s) + \lambda(\tilde{\pi}_{(q_1-1,1,0)}(s) - \tilde{\pi}_{(q_1-1,1,1)}(s)) + \mu\tilde{\pi}_{(q_1,1,1)}(s)}{s+\alpha+\mu_2},$$

$$\tilde{P}_{1,2}(s,1) = \frac{\lambda\tilde{\pi}_{(q_1-1,1,1)}(s) - \mu\tilde{\pi}_{(q_1,1,1)}(s)}{s+\alpha},$$

the functions $\tilde{\pi}_x(s)$ are of the form,

$$\tilde{\pi}_{(q_1,1,1)}(s) = \frac{\lambda z(s)\tilde{L}_{q_1}(s)\mathbf{e}_1(2)}{\mu - \lambda z(s)\tilde{M}_{q_1}(s)\mathbf{e}_1(2)}, \tag{12}$$

$$(\tilde{\pi}_{(q_1-1,1,0)}(s), \tilde{\pi}_{(q_1-1,1,1)}(s)) = \tilde{\pi}_{(q_1,1,1)}(s)\tilde{M}_{q_1}(s) + \tilde{L}_{q_1}(s), \tag{13}$$

$$(\tilde{\pi}_{(0,0,0)}(s), \tilde{\pi}_{(0,0,1)}(s)) = \tilde{\pi}_{(q_1,1,1)}(s)\prod_{i=0}^{q_1}\tilde{M}_{q_1-i}(s) \tag{14}$$

$$+ \sum_{i=0}^{q_1}\tilde{L}_{q_1-i}(s)\prod_{j=i+1}^{q_1}\tilde{M}_{q_1-j}(s),$$

the matrices $\tilde{M}_i(s)$ and $\tilde{L}_i(s)$ are evaluated recursively,

$$\tilde{M}_0(s) = \mu_1\tilde{N}_0(s), \ \tilde{L}_0(s) = \mathbf{e}_0'(2)\tilde{N}_0(s), \ \tilde{N}_0(s) = -(\hat{Q}_{1,0} - sI_2)^{-1}, \tag{15}$$

$$\tilde{M}_q(s) = \mu_1\tilde{N}_q(s), \ \tilde{L}_q(s) = \lambda\tilde{L}_{q-1}(s)\tilde{N}_q(s), \ \tilde{N}_q(s) = -(\hat{Q}_{1,1} - sI_2 + \lambda\tilde{M}_{q-1}(s))^{-1} \ q = \overline{1,q_1-1},$$

$$\tilde{M}_{q_1}(s) = -\mu\mathbf{e}_1'(2)\tilde{N}_{q_1}(s), \ \tilde{L}_{q_1}(s) = -\lambda\tilde{L}_{q_1-1}\tilde{N}_{q_1}(s), \ \tilde{N}_{q_1}(s) = (\hat{Q}_{1,2} - sI_2 + \lambda\tilde{M}_{q_1-1}(s))^{-1},$$

the matrices $\hat{Q}_{1,0}, \hat{Q}_{1,1}$ and $\hat{Q}_{1,2}$ are of the form

$$\hat{Q}_{1,0} = \begin{pmatrix} -\lambda & 0 \\ \mu_2 & -(\lambda+\alpha_2+\mu_2) \end{pmatrix}, \ \hat{Q}_{1,1} = \begin{pmatrix} -(\lambda+\alpha_1+\mu_1) & 0 \\ \mu_2 & -(\lambda+\alpha+\mu) \end{pmatrix},$$

$$\hat{Q}_{1,2} = \begin{pmatrix} -(\lambda+\alpha_1+\mu_1) & \lambda \\ \mu_2 & -(\lambda+\alpha+\mu) \end{pmatrix},$$

the function $z(s)$ is defined as

$$z(s) = \frac{s+\alpha+\lambda+\mu}{2\lambda} - \sqrt{\left(\frac{s+\alpha+\lambda+\mu}{2\lambda}\right)^2 - \frac{\mu}{\lambda}}. \tag{16}$$

Proof. The absorbing states of the process $\{\hat{X}_2(t)\}$ are $x = (q,2,d_2), d_2 \in \{0,1,2\}$ and $x = (q,d_1,2), d_1 \in \{0,1,2\}$. Using the same notations as in previous section we can get the following set of Kolmogorov differential equations,

$$\pi'_{(0,0,0)}(t) = -\lambda\pi_{(0,0,0)}(t) + \mu_1\pi_{(0,1,0)}(t) + \mu_2\pi_{(0,0,1)}(t), \tag{17}$$

$$\pi'_{(q,1,0)}(t) = -(\alpha_1 + \lambda + \mu_1)\pi_{(q,1,0)} + \lambda\pi_{(q-1,1,0)}(t) + \mu_1\pi_{(q+1,1,0)}(t) + \mu_2\pi_{(q,1,1)}(t),$$

$$0 \leq q \leq q_1 - 2,$$

$$\pi'_{(q_1-1,1,0)}(t) = -(\alpha_1 + \lambda + \mu_1)\pi_{(q_1-1,1,0)} + \lambda\pi_{(q_1-2,1,0)}(t) + \mu_2\pi_{(q_1-1,1,1)},$$

$$\pi'_{(0,0,1)}(t) = -(\alpha_2 + \lambda + \mu_2)\pi_{(0,0,1)}(t) + \mu_1\pi_{(0,1,1)}(t),$$

$$\pi'_{(0,1,1)}(t) = -(\alpha_2 + \lambda + \mu_2)\pi_{(0,0,1)}(t) + \lambda\pi_{(0,0,1)}(t) + \mu_1\pi_{(0,1,1)}(t),$$

$$\pi'_{(q,1,1)}(t) = -(\alpha + \lambda + \mu)\pi_{(q,1,1)}(t) + \lambda\pi_{(q-1,1,1)}(t) + \mu_1\pi_{(q+1,1,1)}(t), \; 1 \leq q \leq q_1 - 2,$$

$$\pi'_{(q_1-1,1,1)}(t) = -(\alpha + \lambda + \mu)\pi_{(q_1-1,1,1)}(t) + \lambda\pi_{(q_1-1,1,0)}(t) + \lambda\pi_{(q_1-2,1,1)}(t) + \mu\pi_{(q_1,1,1)}(t)$$

with initial conditions $\pi_{(0,0,0)}(0) = 1$ and $\pi_x(0) = 0, x \in \hat{E}_2$. By taking Laplace transforms of these equations, where $\tilde{\pi}_x(s) = \int_0^\infty \pi_x(t)e^{-st}dt, Re[s] \geq 0$, and using then their partial generating functions,

$$\tilde{P}_{1,0}(s,z) = \tilde{\pi}_{(0,0,0)}(s) + \sum_{q=0}^{q_1-1} \tilde{\pi}_{(q,1,0)}(s)z^{i+1},$$

$$\tilde{P}_{1,1}(s,z) = \tilde{\pi}_{(0,0,1)}(s) + \sum_{q=0}^{q_1-1} \tilde{\pi}_{(q,1,1)}(s)z^{i+1},$$

$$\tilde{P}_{1,2}(s,z) = \sum_{q=q_1}^{\infty} \tilde{\pi}_{(q,1,1)}(s)z^{i+1}$$

for $|z| < 1$, after some manipulation the system (17) is transformed into the set of equations for the introduced double transforms,

$$\tilde{P}_{1,0}(s,z) = \frac{z + \tilde{\pi}_{(0,0,0)}(s)(\mu_1(z-1) + \alpha_1 z) - \lambda z^{q_1+2}\tilde{\pi}_{(q_1-1,1,0)}(s) + \mu_2 z\tilde{P}_{1,1}(s,z)}{-\lambda z^2 + (s + \alpha_1 + \lambda + \mu_1)z - \mu_1},$$

$$\tilde{P}_{1,1}(s,z) = \frac{\tilde{\pi}_{(0,0,0)}(s)(z(\alpha_1 + \mu_1) - \mu_1) + \lambda(\tilde{\pi}_{(q_1-1,1,0)}(s) - z^{q_1+1}\tilde{\pi}_{(q_1-1,1,1)}(s)) + \mu\tilde{\pi}_{(q_1,1,1)}(s)}{-\lambda z^2 + (s + \alpha + \lambda + \mu)z - \mu_1},$$

$$\tilde{P}_{1,2}(s,z) = \frac{z^{q_1+1}(\lambda z\tilde{\pi}_{(q_1-1,1,1)}(s) - \mu\tilde{\pi}_{(q_1,1,1)}(s))}{-\lambda z^2 + (s + \alpha + \lambda + \mu)z - \mu}.$$

Denote by $F(s,z) = -\lambda z^2 + (s + \alpha + \lambda + \mu)z - \mu$ the auxiliary function for the denominator of $\tilde{P}_{1,2}(s,z)$. It is easy to see that

$$F(s,0) = -\mu < 0, \quad F(s,1) = s + \alpha \geq 0.$$

Thus the square equation $F(s,z) = 0$ has for any $s > 0$ two roots and the minimal of them takes the value in the interval $[0,1]$. This root we denote by

$$z(s) = \frac{s + \alpha + \lambda + \mu}{2\lambda} - \sqrt{\left(\frac{s + \alpha + \lambda + \mu}{2\lambda}\right)^2 - \frac{\mu}{\lambda}}.$$

Since the function $\tilde{P}_{1,2}(s,z)$ is analytical, the numerator of this function must be zero at point $z = z(s)$ as well, i.e.

$$\lambda z(s)\tilde{\pi}_{(q_1-1,1,1)}(s) - \mu\tilde{\pi}_{(q_1,1,1)}(s) = 0. \tag{18}$$

To have a second equation for the boundary transforms $\tilde{\pi}_{(q_1-1,1,1)}(s)$ and $\tilde{\pi}_{(q_1,1,1)}(s)$ denote by

$$\tilde{\pi}_{0,0}(s) = (\tilde{\pi}_{(0,0,0)}(s), \tilde{\pi}_{(0,0,1)}(s)), \tilde{\pi}_{q,1}(s) = (\tilde{\pi}_{(q,1,0)}(s), \tilde{\pi}_{(q,1,1)}(s)), 1 \le q \le q_1 - 1.$$

For the system of the Laplace transforms $\tilde{\pi}_x(s)$ obtained from (17) we can get the following relations in matrix form,

$$\tilde{\pi}_{0,0}(s) = -\mu_1 \tilde{\pi}_{0,1}(s)(\hat{Q}_{1,0} - sI_2)^{-1} - e_0'(2)(\hat{Q}_{1,0} - sI_2)^{-1} = \tilde{\pi}_{0,1}(s)\tilde{M}_0(s) + \tilde{L}_0(s).$$

The substitution of the last expression into the matrix relation for $\tilde{\pi}_{0,1}(s)$ yields

$$\tilde{\pi}_{0,1}(s) = -\mu_1 \tilde{\pi}_{1,1}(s)(\hat{Q}_{1,1} - sI_2 + \lambda\tilde{M}_0(s))^{-1} - \lambda\tilde{L}_0(s)(\hat{Q}_{1,1} - sI_2 + \lambda\tilde{M}_0(s))^{-1}$$
$$= \tilde{\pi}_{1,1}(s)\tilde{M}_1(s) + \tilde{L}_1(s).$$

Sequential application of such forward-elimination-backward-substitution method leads to the following recursive relations

$$\tilde{\pi}_{q-1,1}(s) = \tilde{\pi}_{q,1}(s)\tilde{M}_q(s) + \tilde{L}_q(s), 1 \le q \le q_1 - 2,$$
$$\tilde{\pi}_{q_1-1,1}(s) = \tilde{\pi}_{(q_1,1,1)}(s)\tilde{M}_{q_1}(s) + \tilde{L}_{q_1}(s),$$

where $\tilde{M}_q(s)$ and $\tilde{L}_q(s)$ can be calculated by (15). By combining the relation

$$\tilde{\pi}_{(q_1-1,1,1)}(s) = (\pi_{q_1,1,1}(s)\tilde{M}_{q_1}(s) + \tilde{L}_{q_1}(s))e_1(2)$$

and (18), we may express $\tilde{\pi}_{(q_1,1,1)}(s)$ in form (12). The transforms for the rest of boundary states can be hence evaluated as a functions of $\tilde{\pi}_{(q_1-1,1,1)}(s)$. Finally the double transforms are calculated at point $z = 1$ and substituted into (10).

4   Numerical Results

In this section we present some numerical examples to study the effect of system parameters on proposed reliability measures. First we fix the systemparameters at values

$$\lambda = 1.7, \ \mu_1 = 2.4, \ \mu_2 = 0.4, \ \alpha_1 = 0.1, \ \alpha_2 = 0.2,$$
$$\beta_1 = 0.3, \ \beta_2 = 0.3, \ \rho = 0.83, \ q_1 = 9, \ q_2 = 6.$$

In all cases presented below the parametric values are chosen in such a way that the ergodicity condition holds.

In Figs. 1 and 2 the stationary availabilities $A_i, 1 \le i \le 4$, are plotted against the arrival rate λ versus failure rates α_1, α_2 and repair rates β_1, β_2, respectively. As we expect, A_i decreases with increasing λ. The upper curves correspond to the lower value of α_1 and α_2 and to the higher value of β_1 and β_2. The availabilities A_1, A_2 and A_3 take different values by changing of failure and repair rates of servers. We notice that descriptor A_3 changes by varying α_1 and β_1 but it is insensitive to the change of α_2 and β_2. It happens since the parameters α_1 and

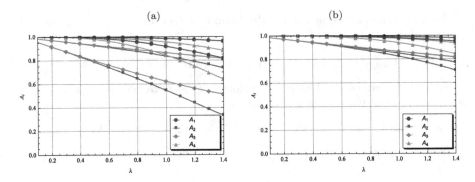

Fig. 1. The availability A_i, $1 \leq i \leq 4$, for $\alpha_1 = 0.1, 0.3$ (a) and $\alpha_2 = 0.1, 0.3$ (b) vs. λ

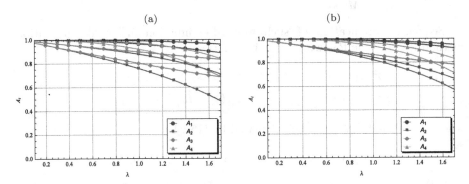

Fig. 2. The availability A_i, $1 \leq i \leq 4$, for $\beta_1 = 0.2, 0.4$ (a) and $\beta_2 = 0.2, 0.4$ (b) vs. λ

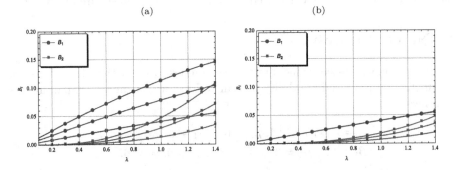

Fig. 3. The failure frequency B_i, $i = 1, 2$, for $\alpha_1 = 0.1, 0.2, 0.3$ (a) and $\alpha_2 = 0.1, 0.2, 0.3$ (b) vs. λ

β_1 influences the busy state of server 2 due to the threshold policy, which in turn makes a contribution to the availability A_3.

In Figs. 3 and 4 we plot the failure frequency B_l for

$$\alpha_l = \{0.1, 0.2, 0.3\} \text{ and } \beta_l = \{0.2, 0.3, 0.4\}, \ l = 1, 2,$$

(a)   (b)

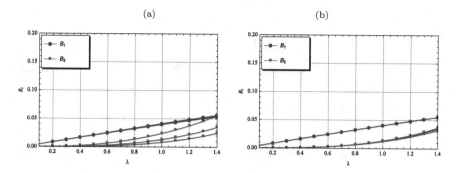

Fig. 4. The failure frequency B_i, $i = 1, 2$, for $\beta_1 = 0.2, 0.3, 0.4$ (a) and $\beta_2 = 0.2, 0.3, 0.4$ (b) vs. λ

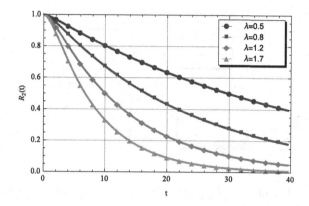

Fig. 5. The function $R(t)$ vs. λ

respectively. These characteristics monotonously increase by increasing of λ. Moreover we notice that $B_1 > B_2$, since the probability to be in state x with $d_1(x) = 1$ is higher than the probability for $d_2(x) = 1$, since server 2 is used according to the threshold control policy. We observe that the function B_1 is insensitive to changes of α_2, β_1 and β_2, and the function B_2 is almost insensitive to change of β_2.

In Fig. 5 we analyze the effect of the arrival rate λ to the reliability function $R(t)$. To evaluate this function we have used a numerical inversion algorithm for the corresponding Laplace transforms $\tilde{R}(s)$, which must be calculated in symbolic form. For the calculations we have used the program *Mathematica* of the Wolfram Research. This program has some limitation on the volume of symbolic representations. Due to this reason and in order to reduce the algorithm's evaluation time, we had to restrict the number of items of the sums in (10) by assuming that $q_1 = 2$ and $q_2 = 1$. We notice that the illustrated function for the higher values of λ exhibit heavier tails.

(a)   (b)

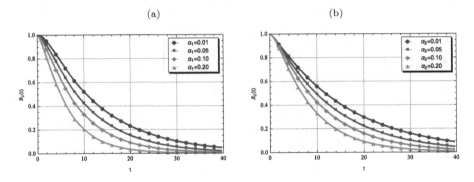

Fig. 6. The function $R(t)$ vs. α_1 (a) and α_2 (b)

(a)   (b)

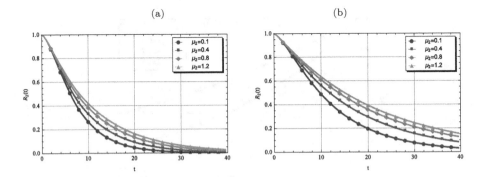

Fig. 7. The function $R(t)$ vs. μ_2 for $\mu_1 = 2.4$ (a) and $\mu_1 = 4.8$ (b)

In Figs. 6 and 7 we illustrate respectively the influence of α_1, α_2, μ_1 and μ_2 on the reliability function $R(t)$. Obviously, for

$$\alpha_1 = 0.01,\ \alpha_2 = 0.01,\ \mu_1 = 4.8,\ \mu_2 = 1.2$$

we observe that the corresponding distribution function exhibits a heavier tail. Finally, we calculate the moment of the life time $\mathbb{E}[T]$ by varying λ,

$$\lambda = \{0.5, 0.8, 1.2, 1.7\},\ \mathbb{E}[T] = \{42.81, 23.51, 13.81, 9.03\}.$$

As is to be expected, the mean life time is decreasing function of λ.

5   Conclusion

The paper provides reliability analysis of a two-server heterogeneous unreliable queueing system with a threshold control policy for the allocation of customers between the servers. The proposed results complement the classical performance analysis of the unreliable queueing models which can be described by the quasi-birth-and-death processes. The matrix-geometric solution method has been used to obtain the stationary state probabilities and some stationary reliability measures like availability and failure frequency. The combination of the forward-elimination-backward-substitution method for the boundary states with generating function approach for the states above the highest threshold level has led to a closed form solution in terms of Laplace transform for the reliability function and as a consequence for the mean time to the first failure. We finally performed numerical experiments to explore the effect of various system parameters on reliability of servers.

References

1. Efrosinin, D.: Controlled queueing systems with heterogeneous servers. In: Dynamic Optimization and Monotonicity Properties, Saarbrücken. VDM (2008)
2. Efrosinin, D.: Queueing model of a hybrid channel with faster link subject to partial and complete failures. Ann. Oper. Res. **202**, 75–102 (2013)
3. Efrosinin, D., Samouylov, K., Gudkova, I.: Busy period analysis of a queueing system with breakdowns and its application to wireless network under licensed shared access regime. In: Galinina, O., Balandin, S., Koucheryavy, Y. (eds.) NEW2AN/ruSMART -2016. LNCS, vol. 9870, pp. 426–439. Springer, Cham (2016). doi:10.1007/978-3-319-46301-8_36
4. Efrosinin, D., Sztrik, J.: Optimal control of a two-server heterogeneous queueing system with breakdowns and constant retrials. In: Dudin, A., Gortsev, A., Nazarov, A., Yakupov, R. (eds.) ITMM 2016. CCIS, vol. 638, pp. 57–72. Springer, Cham (2016). doi:10.1007/978-3-319-44615-8_5
5. Koole, G.: A simple proof of the optimality of a threshold policy in a two-server queueing system. Syst. Control Lett. **26**, 301–303 (1995)
6. Kumar, B.K., Madheswari, S.P.: An $M/M/2$ queueing system with heterogeneous servers and multiple vacations. Math. Comput. Model. **41**, 1415–1429 (2005)
7. Kumar, B.K., Madheswari, S.P., Venkatakrishnan, K.S.: Transient solution of an $M/M/2$ queue with heterogeneous servers subject to catastrophes. Inf. Manag. Sci. **18**(1), 63–80 (2007)
8. Legros, B., Jouini, O.: Routing in a queueing system with two heterogeneous servers in speed and in quality of resolution (2016). http://www.lgi.ecp.fr/~jouini
9. Levy, J., Yehiali, U.: An $M/M/s$ queue with servers' vacations. Inf. Syst. Oper. Res. **14**(2), 153–163 (1976)
10. Lin, W., Kumar, P.R.: Optimal control of a queueing system with two heterogeneous servers. IEEE Trans. Autom. Control **29**, 696–703 (1984)
11. Mitrany, I.L., Avi-Itzhak, B.: A many server queue with service interruptions. Oper. Res. **16**, 628–638 (1967)
12. Neuts, M.F.: Matrix-Geometric Solutions in Stochastic Models. The John Hopkins University Press, Baltimore (1981)
13. Neuts, M.F., Lucantoni, D.M.: A Markovian queue with N servers subject to breakdowns and repairs. Manag. Sci. **25**, 849–861 (1979)
14. Ozkan, E., Kharoufeh, J.: Optimal control of a two-server queueing system with failures. Probab. Eng. Inf. Sci. **28**, 489–527 (2014)
15. Rykov, V., Efrosinin, D.: On the slow server problem. Autom. Remote Control **70**(12), 2013–2013 (2009)
16. Yue, D., Yue, W., Yu, J., Tian, R.: A heterogeneous two-server queueing system with balking and server breakdowns. In: Proceedings of the Eighth International Symposium on Operations Research and Its Applications, China, pp. 230–244 (2009)

Heavy Outgoing Call Asymptotics for $MMPP/M/1/1$ Retrial Queue with Two-Way Communication

Anatoly Nazarov[1,2]([✉]), Tuan Phung-Duc[3], and Svetlana Paul[1]

[1] National Research Tomsk State University, Tomsk, Russia
nazarov.tsu@gmail.com, paulsv82@mail.ru
[2] Peoples Friendship University of Russia (RUDN University), Moscow, Russia
[3] Faculty of Engineering, Information and Systems,
University of Tsukuba, Tsukuba, Japan
tuan@sk.tsukuba.ac.jp

Abstract. In this paper, we consider an MMPP/M/1/1 retrial queue where incoming fresh calls arrive at the server according to a Markov modulated Poisson process. Upon arrival, an incoming call either occupies the server if it is idle or joins an orbit if the server is busy. From the orbit, an incoming call retries to occupy the server and behaves the same as a fresh incoming call. The server makes an outgoing call in its idle time. Our contribution is to derive the asymptotics of the number of calls in retrial queue under the conditions of high rate of making outgoing calls and low rate of service time of outgoing calls.

Keywords: Retrial queueing system · Incoming and outgoing calls · Asymptotic analysis method · Markov modulated Poisson process · Gaussian approximation · Gamma approximation

1 Introduction

In service systems idle time of an operator should be minimized to increase the productivity. An operator not only receives calls from outside but also makes outgoing calls in the idle time. The example of that could be the cellphone that is used for incoming and outgoing calls. In call centers operators could receive arriving calls but as soon as they have free time and are in standby mode they could make outgoing calls to sell packages and services of the center [1].

Retrial Queues with two-way communication have been extensively studied recently [2–7]. In these literatures the arrival process is Poisson process. However, it is well known that real traffic has a more complex structure. Markov modulated Poisson process (MMPP) can represent correlated traffic and thus it is more suitable for modelling real traffic.

A. Nazarov—The publication was financially supported by the Ministry of Education and Science of the Russian Federation (the Agreement number 02.a03.21.0008).

A. Dudin et al. (Eds.): ITMM 2017, CCIS 800, pp. 28–41, 2017.
DOI: 10.1007/978-3-319-68069-9_3

In this paper, we consider asymptotic analysis for the distribution of the number of customers in the system under two conditions: (i) high outgoing call rate and (ii) low service rate for outgoing calls. In case (i), the server makes an outgoing call as soon as it becomes idle while in case (ii), the duration of an outgoing call is extremely long.

In both cases, the number of incoming calls in the system explodes. However, using suitable scalings, we prove that the scaled version of the number of incoming calls in the system follow some simple distributions, i.e. Gaussian distribution [8] and Gamma distribution, respectively [9].

The remainder of the paper is presented as follows. In Sect. 2, we describe the model in detail and preliminaries for later asymptotic analysis. In Sects. 3 and 4, we present our main contribution for the model with Markov modulated Poisson process. In Sect. 5 we show the ranges of parameters under which our approximations are usable. Section 6 is devoted to concluding remark.

2 Model Description and Problem Definition

We consider a single server queueing model with two types of calls: incoming calls and outgoing calls. Incoming calls arrive at the system according to a Markov modulated Poisson process. The incoming call that finds the server idle receives a service for an exponentially distributed time with rate μ_1. Upon entering the system the call that finds the server being busy immediately joins the orbit, where it stays during a random time exponentially distributed with rate σ. If the server is idle (empty) it starts making outgoing calls to the outside with rate α. If the outgoing call finds the server free the call goes into service for an exponentially distributed time with rate μ_2. If upon entering the system the outgoing call finds the server being busy the call is lost and is not considered in the future. Let $i(t)$ denote the number of calls in the system at the time t, $k(t)$ denote the state of the server: 0 if the server is free, 1 if the server is busy serving an incoming call, 2 if the server is busy serving an outgoing call and $n(t)$ denote the state of the background process of the MMPP at time t. The infinitesimal generator of $n(t)$ is defined by matrix $\mathbf{Q}$. When $n(t) = n$, the arrival rate is given by λ_n $(n = 1, 2, ..., N)$. To determine the condition for the existence of a stationary regime, we define the matrix $\mathbf{\Lambda}$ in the form $\mathbf{\Lambda} = \frac{\rho \mu_1 \mathbf{\Lambda}_1}{\mathbf{r} \mathbf{\Lambda}_1 \mathbf{e}}$, where $\mathbf{\Lambda}_1$ is a diagonal matrix with nonnegative elements, and the condition for the existence of a stationary regime is the fulfillment of the inequalities $0 < \rho < 1$.

Under the current setting the three-dimensional process $\{k(t), n(t), i(t)\}$ is a Markov chain. Under the stability condition, the stationary probability distribution $P\{k(t) = k, n(t) = n, i(t) = i\} = P_k(n, i)$ is the unique solution of Kolmogorov system of equations:

$$-(\lambda_n + i\sigma + \alpha)P_0(n, i) + \mu_1 P_1(n, i+1) + \mu_2 P_2(n, i+1) + \sum_{v=1}^{N} P_0(v, i)q_{vn} = 0,$$

$$-(\lambda_n + \mu_1)P_1(n, i) + \lambda_n \left[P_1(n, i-1) + P_0(n, i-1)\right] + i\sigma P_0(n, i)$$
$$+ \sum_{v=1}^{N} P_1(v, i)q_{vn} = 0,$$

$$-(\lambda_n + \mu_2)P_2(n, i) + P_0(n, i-1)\alpha + P_2(n, i-1)\lambda_n + \sum_{v=1}^{N} P_2(v, i)q_{vn} = 0. \quad (1)$$

We introduce partial characteristic functions [10], denoting $j = \sqrt{-1}$:

$$H_0(n, u) = \sum_{i=0}^{\infty} e^{jui} P_0(n, i), \quad H_k(n, u) = \sum_{i=1}^{\infty} e^{jui} P_k(n, i), \quad k = 1, 2.$$

For $k = 1, 2$, there will be at least one call in the system. Rewriting system (1) in the following form:

$$-(\lambda_n + \alpha)H_0(n, u) + j\sigma \frac{\partial H_0(n,u)}{\partial u} + \mu_1 e^{-ju} H_1(n, u) + \mu_2 e^{-ju} H_2(n, u)$$
$$+ \sum_{v=1}^{N} H_0(v, u) q_{vn} = 0,$$

$$-(\lambda_n + \mu_1)H_1(n, u) + \lambda_n e^{ju} [H_1(n, u) + H_0(n, u)] - j\sigma \frac{\partial H_0(n,u)}{\partial u}$$
$$+ \sum_{v=1}^{N} H_1(v, u) q_{vn} = 0,$$

$$-(\lambda_n + \mu_2)H_2(n, u) + \alpha e^{ju} H_0(n, u) + \lambda_k e^{ju} H_2(n, u) + \sum_{v=1}^{N} H_2(v, u) q_{vn} = 0. \quad (2)$$

We define $\mathbf{I}$ - unit matrix, $\boldsymbol{\Lambda} = diag[\lambda_n]$,

$$\mathbf{H}(u) = \{H_k(1, u), H_k(2, u), \ldots, H_k(N, u),$$

$$\mathbf{H}'_k(u) = \left\{ \frac{\partial H_k(1, u)}{\partial u}, \frac{\partial H_k(2, u)}{\partial u}, \ldots, \frac{\partial H_k(N, u)}{\partial u} \right\}.$$

Let's write system (2) in a matrix form (3):

$$\mathbf{H}_0(u)(\mathbf{Q} - \boldsymbol{\Lambda} - \alpha \mathbf{I}) + j\sigma \mathbf{H}'_0(u) + \mu_1 e^{-ju} \mathbf{H}_1(u) + \mu_2 e^{-ju} \mathbf{H}_2(u) = 0,$$

$$\mathbf{H}_1(u) \left(\mathbf{Q} + \left(e^{ju} - 1 \right) \boldsymbol{\Lambda} - \mu_1 \mathbf{I} \right) + \mathbf{H}_0(u) e^{ju} \boldsymbol{\Lambda} - j\sigma \mathbf{H}'_0(u) = 0,$$

$$\mathbf{H}_2(u) \left(\mathbf{Q} + \left(e^{ju} - 1 \right) \boldsymbol{\Lambda} - \mu_2 \mathbf{I} \right) + \alpha e^{ju} \mathbf{H}_0(u) = 0. \quad (3)$$

Let's sum the equations of the system (3)

$$\mathbf{H}_0(u) \left[\mathbf{Q} + \left(e^{ju} - 1 \right) \left(\boldsymbol{\Lambda} + \alpha \mathbf{I} \right) \right] + \mathbf{H}_1(u) \left[\mathbf{Q} + \left(e^{ju} - 1 \right) \left(\boldsymbol{\Lambda} - \mu_1 e^{-ju} \mathbf{I} \right) \right]$$
$$+ \mathbf{H}_2(u) \left[\mathbf{Q} + \left(e^{ju} - 1 \right) \left(\boldsymbol{\Lambda} - \mu_2 e^{-ju} \mathbf{I} \right) \right] = 0.$$

Multiplying the last equation by a unit vector $\mathbf{e}$ and using $\mathbf{Qe} = \mathbf{0}$, we obtain

$$\mathbf{H}_0(u) \left(\boldsymbol{\Lambda} + \alpha \mathbf{I} \right) \mathbf{e} + H_1(u) \left(\boldsymbol{\Lambda} - \mu_1 e^{-ju} \mathbf{I} \right) \mathbf{e} + \mathbf{H}_2(u) \left(\boldsymbol{\Lambda} - \mu_2 e^{-ju} \mathbf{I} \right) \mathbf{e} = 0.$$

Multiplying the last equation by a e^{ju}:

$$\mathbf{H}_0(u) \left(e^{ju} \boldsymbol{\Lambda} + \alpha e^{ju} \mathbf{I} \right) \mathbf{e} + H_1(u) \left(e^{ju} \boldsymbol{\Lambda} - \mu_1 \mathbf{I} \right) \mathbf{e}$$
$$+ \mathbf{H}_2(u) \left(e^{ju} \boldsymbol{\Lambda} - \mu_2 \mathbf{I} \right) \mathbf{e} = 0. \quad (4)$$

We will consider the system (3) and the Eq. (4), i.e. a system of three matrix equations and one scalar equation:

$$\mathbf{H}_0(u)(\mathbf{Q} - \boldsymbol{\Lambda} - \alpha \mathbf{I}) + j\sigma \mathbf{H}'_0(u) + \mu_1 e^{-ju} \mathbf{H}_1(u) + \mu_2 e^{-ju} \mathbf{H}_2(u) = 0,$$

$$\mathbf{H}_1(u)\left(\mathbf{Q} + \left(e^{ju} - 1\right)\boldsymbol{\Lambda} - \mu_1\mathbf{I}\right) + \mathbf{H}_0(u)e^{ju}\boldsymbol{\Lambda} - j\sigma\mathbf{H}_0'(u) = 0,$$

$$\mathbf{H}_2(u)\left(\mathbf{Q} + \left(e^{ju} - 1\right)\boldsymbol{\Lambda} - \mu_2\mathbf{I}\right) + \alpha e^{ju}\mathbf{H}_0(u) = 0,$$

$$\begin{aligned}\mathbf{H}_0(u)\left(e^{ju}\boldsymbol{\Lambda} + \alpha e^{ju}\mathbf{I}\right)\mathbf{e} + \mathbf{H}_1(u)\left(e^{ju}\boldsymbol{\Lambda} - \mu_1\mathbf{I}\right)\mathbf{e} \\ + \mathbf{H}_2(u)\left(e^{ju}\boldsymbol{\Lambda} - \mu_2\mathbf{I}\right)\mathbf{e} = 0.\end{aligned} \qquad (5)$$

The characteristic function $H(u)$ of the number of incoming calls in the retrial queue is expressed through partial characteristic functions $\mathbf{H}_k(u)$ by the following equation

$$H(u) = Ee^{jui(t)} = (\mathbf{H}_0(u) + \mathbf{H}_1(u) + \mathbf{H}_2(u))\mathbf{e}.$$

We will find the characteristics of our retrial queue with two-way communication with Markov modulated Poisson input. The main content of this paper is the solution of system (5) by using an asymptotic analysis method in two limit conditions: of the high rate of making outgoing calls and the low rate of service time of outgoing calls.

3   Asymptotic Analysis of MMPP/M/1/1 Retrial Queue with Two-Way Communication Under the High Rate of Making Outgoing Calls ($\alpha \to \infty$)

We will investigate system (5) by asymptotic analysis method under the high rate of making outgoing calls.

3.1   First Order Asymptotic

Theorem 1. *Suppose $i(t)$ is the number of calls in the system of the stationary MMPP/M/1/1 retrial queue with outgoing calls, then the (6) holds*

$$\lim_{\alpha \to \infty} Ee^{jw\frac{i(t)}{\alpha}} = e^{jw\kappa_1}, \qquad (6)$$

where κ_1 is the positive root of the equation

$$r\left\{\kappa_1\sigma\left(\mu_1\boldsymbol{I} - \boldsymbol{Q}\right)^{-1} + \left(\mu_2\boldsymbol{I} - \boldsymbol{Q}\right)^{-1}\right\}^{-1}$$

$$\times \left\{\boldsymbol{I} + \kappa_1\sigma\left(\mu_1\boldsymbol{I} - \boldsymbol{Q}\right)^{-1}\left(\boldsymbol{\Lambda} - \mu_1\boldsymbol{I}\right) + \left(\mu_2\boldsymbol{I} - \boldsymbol{Q}\right)^{-1}\left(\boldsymbol{\Lambda} - \mu_2\boldsymbol{I}\right)\right\}\boldsymbol{e} = 0. \qquad (7)$$

The row vector r is the stationary probability distribution of the underlying process $n(t)$ which is given as the unique solution of the system $r\boldsymbol{Q} = 0$, $r\boldsymbol{e} = 1$.

Proof. We denote $\alpha = 1/\varepsilon$ in the system (5), and introduce the following notations

$$u = \varepsilon w, \qquad \mathbf{H}_0(u) = \varepsilon\mathbf{F}_0(w,\varepsilon), \qquad \mathbf{H}_k(u) = \mathbf{F}_k(w,\varepsilon), \qquad k = 1, 2,$$

in order to get the following system

$$\mathbf{F}_0(w,\varepsilon)(\varepsilon\mathbf{Q} - \varepsilon\mathbf{\Lambda} - \mathbf{I}) + j\sigma\frac{\partial\mathbf{F}_0(w,\varepsilon)}{\partial w} + \mu_1 e^{-j\varepsilon w}\mathbf{F}_1(w,\varepsilon) + \mu_2 e^{-j\varepsilon w}\mathbf{F}_2(w,\varepsilon) = 0,$$

$$\mathbf{F}_1(w,\varepsilon)\left(\mathbf{Q} + \left(e^{j\varepsilon w} - 1\right)\mathbf{\Lambda} - \mu_1\mathbf{I}\right) + \varepsilon e^{j\varepsilon w}\mathbf{F}_0(w,\varepsilon)\mathbf{\Lambda} - j\sigma\frac{\partial\mathbf{F}_0(w,\varepsilon)}{\partial w} = 0,$$

$$\mathbf{F}_2(w,\varepsilon)\left(\mathbf{Q} + \left(e^{j\varepsilon w} - 1\right)\mathbf{\Lambda} - \mu_2\mathbf{I}\right) + e^{j\varepsilon w}\mathbf{F}_0(w,\varepsilon) = 0,$$

$$\begin{aligned}\mathbf{F}_0(w,\varepsilon)\left(\varepsilon e^{j\varepsilon w}\mathbf{\Lambda} + e^{j\varepsilon w}\mathbf{I}\right)\mathbf{e} + \mathbf{F}_1(w,\varepsilon)\left(e^{j\varepsilon w}\mathbf{\Lambda} - \mu_1\mathbf{I}\right)\mathbf{e} \\ + \mathbf{F}_2(w,\varepsilon)\left(e^{j\varepsilon w}\mathbf{\Lambda} - \mu_2\mathbf{I}\right)\mathbf{e} = 0.\end{aligned} \tag{8}$$

Considering the limit as $\varepsilon \to 0$ in the system (8), then we will get

$$-\mathbf{F}_0(w) + j\sigma\mathbf{F}_0'(w) + \mu_1\mathbf{F}_1(w) + \mu_2\mathbf{F}_2(w) = 0,$$

$$\mathbf{F}_1(w)\left(\mathbf{Q} - \mu_1\mathbf{I}\right) - j\sigma\mathbf{F}_0'(w) = 0,$$

$$\mathbf{F}_2(w)\left(\mathbf{Q} - \mu_2\mathbf{I}\right) + \mathbf{F}_0(w) = 0,$$

$$\mathbf{F}_0(w)\mathbf{e} + \mathbf{F}_1(w)\left(\mathbf{\Lambda} - \mu_1\mathbf{I}\right)\mathbf{e} + \mathbf{F}_2(w)\left(\mathbf{\Lambda} - \mu_2\mathbf{I}\right)\mathbf{e} = 0. \tag{9}$$

Our idea is to find the solution of (9) in the form of

$$\mathbf{F}_k(w) = \Phi(w)\mathbf{r}_k. \tag{10}$$

Here $\mathbf{r}_k$, $k = 1, 2$ are vectors with components r_{kn}, where r_{kn} is the probability that the server is in state k, and the MMPP is in state n; $\mathbf{r}_0$ is a vector with components $\mathbf{r}_{0n}$, and has no sense of probability, since the probability that the server will be in the zero state (will be free) as $\alpha \to \infty$ is zero.

$$-\mathbf{r}_0 + j\sigma\frac{\Phi'(w)}{\Phi(w)}\mathbf{r}_0 + \mu_1\mathbf{r}_1 + \mu_2\mathbf{r}_2 = 0,$$

$$\mathbf{r}_1\left(\mathbf{Q} - \mu_1\mathbf{I}\right) - j\sigma\frac{\Phi'(w)}{\Phi(w)}\mathbf{r}_0 = 0,$$

$$\mathbf{r}_2\left(\mathbf{Q} - \mu_2\mathbf{I}\right) + \mathbf{r}_0 = 0,$$

$$\mathbf{r}_0\mathbf{e} + \mathbf{r}_1\left(\mathbf{\Lambda} - \mu_1\mathbf{I}\right)\mathbf{e} + \mathbf{r}_2\left(\mathbf{\Lambda} - \mu_2\mathbf{I}\right)\mathbf{e} = 0. \tag{11}$$

As the relation $j\frac{\Phi'(w)}{\Phi(w)}$ does not depend on w, the function is obtained in the following form

$$\Phi(w) = \exp\{jw\kappa_1\},$$

which coincides with (6). The value of the parameter κ_1 will be defined below. We rewrite the system (11) in the form

$$-\mathbf{r}_0 - \kappa_1\sigma\mathbf{r}_0 + \mu_1\mathbf{r}_1 + \mu_2\mathbf{r}_2 = 0,$$

$$\mathbf{r}_1\left(\mathbf{Q} - \mu_1\mathbf{I}\right) + \kappa_1\sigma\mathbf{r}_0 = 0,$$

$$\mathbf{r}_2 \left(\mathbf{Q} - \mu_2 \mathbf{I} \right) + \mathbf{r}_0 = 0,$$

$$\mathbf{r}_0 \mathbf{e} + \mathbf{r}_1 \left(\mathbf{\Lambda} - \mu_1 \mathbf{I} \right) \mathbf{e} + \mathbf{r}_2 \left(\mathbf{\Lambda} - \mu_2 \mathbf{I} \right) \mathbf{e} = 0. \tag{12}$$

Let's review the normalization condition for stationary server state probability distribution

$$\mathbf{r}_1 + \mathbf{r}_2 = \mathbf{r}.$$

The row vector $\mathbf{r}$ is the stationary probability distribution of the underlying process $n(t)$. Vector $\mathbf{r}$ is defined as the unique solution of the system $\mathbf{r}\mathbf{Q} = 0$, $\mathbf{r}\mathbf{e} = 1$. We have

$$\mathbf{r}_1 = \kappa_1 \sigma \mathbf{r}_0 \left(\mu_1 \mathbf{I} - \mathbf{Q} \right)^{-1},$$

$$\mathbf{r}_2 = \mathbf{r}_0 \left(\mu_2 \mathbf{I} - \mathbf{Q} \right)^{-1},$$

$$\mathbf{r}_1 + \mathbf{r}_2 = \mathbf{r}. \tag{13}$$

We substitute the values of the vectors $\mathbf{r}_k$, $k = 1, 2$ into the last equation of the system (13). We obtain an equation that determines the vector $\mathbf{r}_0$:

$$\mathbf{r}_0 = \mathbf{r} \left\{ \kappa_1 \sigma \left(\mu_1 \mathbf{I} - \mathbf{Q} \right)^{-1} + \left(\mu_2 \mathbf{I} - \mathbf{Q} \right)^{-1} \right\}^{-1}. \tag{14}$$

Now we substitute the first two equalities of the system (13) into the scalar equation of system (12). We obtain the equation that determines the value of $\mathbf{r}_0$:

$$\mathbf{r}_0 \left\{ \mathbf{I} + \kappa_1 \sigma \left(\mu_1 \mathbf{I} - \mathbf{Q} \right)^{-1} \left(\mathbf{\Lambda} - \mu_1 \mathbf{I} \right) + \left(\mu_2 \mathbf{I} - \mathbf{Q} \right)^{-1} \left(\mathbf{\Lambda} - \mu_2 \mathbf{I} \right) \right\} \mathbf{e} = 0.$$

Substituting this equality into Eq. (14), we obtain an equation for κ_1, which coincides with (7):

$$\mathbf{r} \left\{ \kappa_1 \sigma \left(\mu_1 \mathbf{I} - \mathbf{Q} \right)^{-1} + \left(\mu_2 \mathbf{I} - \mathbf{Q} \right)^{-1} \right\}^{-1}$$
$$\times \left\{ \mathbf{I} + \kappa_1 \sigma \left(\mu_1 \mathbf{I} - \mathbf{Q} \right)^{-1} \left(\mathbf{\Lambda} - \mu_1 \mathbf{I} \right) + \left(\mu_2 \mathbf{I} - \mathbf{Q} \right)^{-1} \left(\mathbf{\Lambda} - \mu_2 \mathbf{I} \right) \right\} \mathbf{e} = 0. \tag{15}$$

The first order asymptotic i.e. Theorem 1, only defines the mean asymptotic value $\kappa_1 \alpha$ of a number of calls in an system in prelimit situation of $\alpha \to 0$. For more detailed research of the number $i(t)$ of calls in an system let's consider the second order asymptotic.

3.2   Second Order Asymptotic

Theorem 2. *In the context of Theorem 1 the following equation is true*

$$\lim_{\alpha \to \infty} E \exp \left\{ jw \frac{\frac{1}{\alpha} i(t) - \kappa_1}{\sqrt{\alpha}} \right\} = e^{\frac{(jw)^2}{2} \kappa_2}, \tag{16}$$

where parameter κ_2 is given by

$$\kappa_2 = \frac{1}{\sigma} \cdot \frac{\mathbf{r}_0 \mathbf{e} + \mathbf{r}_1 \mathbf{\Lambda} \mathbf{e} + \mathbf{r}_2 \mathbf{\Lambda} \mathbf{e} + \left[\mathbf{y}_0 + \mathbf{y}_1 \left(\mathbf{\Lambda} - \mu_1 \mathbf{I} \right) + \mathbf{y}_2 \left(\mathbf{\Lambda} - \mu_2 \mathbf{I} \right) \right] \mathbf{e}}{\left[-\mathbf{g}_0 + \mathbf{g}_1 \left(\mu_1 \mathbf{I} - \mathbf{\Lambda} \right) + \mathbf{g}_2 \left(\mu_2 \mathbf{I} - \mathbf{\Lambda} \right) \right] \mathbf{e}}. \tag{17}$$

Here the vector of r_0 and the vectors of probabilities r_1, r_2 are defined above. The vectors g_0, g_1, g_2, y_0, y_1, y_2 are defined by the two systems:

$$g_0 \left[-I - \sigma\kappa_1 I + \mu_2 \left(\mu_2 I - Q \right)^{-1} + \mu_1 \sigma\kappa_1 \left(\mu_1 I - Q \right)^{-1} \right]$$
$$= r_0 - r_0 \mu_1 \left(\mu_1 I - Q \right)^{-1},$$

$$g_1 = \left(g_0 \sigma\kappa_1 + r_0 \right) \left(\mu_1 I - Q \right)^{-1},$$

$$g_2 = g_0 \left(\mu_2 I - Q \right)^{-1},$$

$$\left(g_0 + g_1 + g_2 \right) e = 0. \tag{18}$$

$$y_0 \left[\left(-I - \sigma\kappa_1 I \right) + \mu_1 \sigma\kappa_1 \left(\mu_1 I - Q \right)^{-1} + \mu_2 \left(\mu_2 I - Q \right)^{-1} \right]$$
$$= \mu_1 r_1 \left[I - \Lambda \left(\mu_1 I - Q \right)^{-1} \right] + \mu_2 \left[r_2 - \left(r_0 + r_2 \Lambda \right) \left(\mu_2 I - Q \right)^{-1} \right]$$

$$y_1 = \left(y_0 \sigma\kappa_1 + r_1 \Lambda \right) \left(\mu_1 I - Q \right)^{-1},$$

$$y_2 = \left(y_0 + r_0 + r_2 \Lambda \right) \left(\mu_2 I - Q \right)^{-1},$$

$$\left(y_0 + y_1 + y_2 \right) e = 0. \tag{19}$$

Proof. We introduce the following notations in the system (5)

$$\mathbf{H}_k(u) = \exp\left(j\alpha u\kappa_1 \right) \mathbf{H}_k^{(2)}(u), \tag{20}$$

and we get

$$\mathbf{H}_0^{(2)}(u)(\mathbf{Q} - \Lambda - \alpha I - \sigma\alpha\kappa_1) + \mu_1 e^{-ju} \mathbf{H}_1^{(2)}(u) + \mu_2 e^{-ju} \mathbf{H}_2^{(2)}(u)$$
$$+ j\sigma \frac{d\mathbf{H}_0^{(2)}(u)}{du} = 0,$$

$$\mathbf{H}_1^{(2)}(u) \left(\mathbf{Q} + \left(e^{ju} - 1 \right) \Lambda - \mu_1 I \right) + \mathbf{H}_0^{(2)}(u) \left(e^{ju} \Lambda + \sigma\alpha\kappa_1 \, I \right) - j\sigma \frac{d\mathbf{H}_0^{(2)}(u)}{du} = 0,$$

$$\mathbf{H}_2^{(2)}(u) \left(\mathbf{Q} + \left(e^{ju} - 1 \right) \Lambda - \mu_2 I \right) + \alpha e^{ju} \mathbf{H}_0^{(2)}(u) = 0,$$

$$\mathbf{H}_0^{(2)}(u) \left(e^{ju} \Lambda + \alpha e^{ju} I \right) e + \mathbf{H}_1^{(2)}(u) \left(e^{ju} \Lambda - \mu_1 I \right) e$$
$$+ \mathbf{H}_2^{(2)}(u) \left(e^{ju} \Lambda - \mu_2 I \right) e = 0. \tag{21}$$

Denoting $\alpha = 1/\varepsilon^2$, and introducing the following notations

$$u = \varepsilon w, \quad \mathbf{H}_0^2(u) = \varepsilon^2 \mathbf{F}_0^2(w, \varepsilon), \quad \mathbf{H}_k^2(u) = \mathbf{F}_k^2(w, \varepsilon), \quad k = 1, 2, \tag{22}$$

we obtain

$$\mathbf{F}_0^{(2)}(w, \varepsilon)(\varepsilon^2 \mathbf{Q} - \varepsilon^2 \Lambda - I - \sigma\kappa_1 I) + \mu_1 e^{-j\varepsilon w} \mathbf{F}_1^{(2)}(w, \varepsilon) + \mu_2 e^{-j\varepsilon w} \mathbf{F}_2^{(2)}(w, \varepsilon)$$
$$+ j\sigma\varepsilon \frac{\partial \mathbf{F}_0^{(2)}(w, \varepsilon)}{\partial w} = 0,$$

$$\mathbf{F}_1^{(2)}(w,\varepsilon)\left(\mathbf{Q}+\left(e^{j\varepsilon w}-1\right)\mathbf{\Lambda}-\mu_1\mathbf{I}\right)+\mathbf{F}_0^{(2)}(w,\varepsilon)\left(\varepsilon^2 e^{j\varepsilon w}\mathbf{\Lambda}+\sigma\kappa_1\mathbf{I}\right)$$
$$-j\sigma\varepsilon\frac{\partial\mathbf{F}_0^{(2)}(w,\varepsilon)}{\partial w}=0,$$

$$\mathbf{F}_2^{(2)}(w,\varepsilon)\left(\mathbf{Q}+\left(e^{j\varepsilon w}-1\right)\mathbf{\Lambda}-\mu_2\mathbf{I}\right)+e^{j\varepsilon w}\mathbf{F}_0^{(2)}(w,\varepsilon)=0,$$

$$\mathbf{F}_0^{(2)}(w,\varepsilon)e^{j\varepsilon w}\left(\varepsilon^2\mathbf{\Lambda}+\mathbf{I}\right)\mathbf{e}+\mathbf{F}_1^{(2)}(w,\varepsilon)\left(e^{j\varepsilon w}\mathbf{\Lambda}-\mu_1\mathbf{I}\right)\mathbf{e} \tag{23}$$
$$+\mathbf{F}_2^{(2)}(w,\varepsilon)\left(e^{j\varepsilon w}\mathbf{\Lambda}-\mu_2\ \mathbf{I}\right)\mathbf{e}=0.$$

Our idea is to seek a solution of the system (5) in the form

$$\mathbf{F}_k^{(2)}(w,\varepsilon)=\Phi_2(w)\left\{\mathbf{r}_k+j\varepsilon w\mathbf{f}_k\right\}+o\left(\varepsilon^2\right). \tag{24}$$

Substituting (24) to (23), we obtain

$$\mathbf{r}_0\left(-\mathbf{I}-\sigma\kappa_1\mathbf{I}\right)+\mu_1\mathbf{r}_1+\mu_2\mathbf{r}_2+j\varepsilon w\left[\mathbf{f}_0\left(-\mathbf{I}-\sigma\kappa_1\mathbf{I}\right)+\mu_1\left(\mathbf{f}_1-\mathbf{r}_1\right)+\mu_2\left(\mathbf{f}_2-\mathbf{r}_2\right)\right]$$
$$+j\sigma\varepsilon\frac{d\Phi_2(w)/dw}{\Phi_2(w)}\mathbf{r}_0=o\left(\varepsilon^2\right),$$
$$\mathbf{r}_1\left(\mathbf{Q}-\mu_1\mathbf{I}\right)+\mathbf{r}_0\sigma\kappa_1+j\varepsilon w\left[\mathbf{f}_1\left(\mathbf{Q}-\mu_1\mathbf{I}\right)+\mathbf{f}_0\sigma\kappa_1+\mathbf{r}_1\mathbf{\Lambda}\right]$$
$$-j\sigma\varepsilon\frac{d\Phi_2(w)/dw}{\Phi_2(w)}\mathbf{r}_0=o\left(\varepsilon^2\right),$$
$$\mathbf{r}_2\left(\mathbf{Q}-\mu_2\mathbf{I}\right)+\mathbf{r}_0+j\varepsilon w\left[\mathbf{f}_2\left(\mathbf{Q}-\mu_2\mathbf{I}\right)+\mathbf{r}_0+\mathbf{f}_0+\mathbf{r}_2\mathbf{\Lambda}\right]=o\left(\varepsilon^2\right),$$
$$\mathbf{r}_0\mathbf{e}+\mathbf{r}_1\left(\mathbf{\Lambda}-\mu_1\mathbf{I}\right)\mathbf{e}+\mathbf{r}_2\left(\mathbf{\Lambda}-\mu_2\mathbf{I}\right)\mathbf{e}$$
$$+j\varepsilon w\left[\mathbf{f}_0+\mathbf{f}_1\left(\mathbf{\Lambda}-\mu_1\mathbf{I}\right)+\mathbf{f}_2\left(\mathbf{\Lambda}-\mu_2\mathbf{I}\right)+\mathbf{r}_0+\mathbf{r}_1\mathbf{\Lambda}+\mathbf{r}_2\mathbf{\Lambda}\right]\mathbf{e}=o\left(\varepsilon^2\right).$$

Previously, the system of equations (12) was obtained. Taking this into account, we have

$$j\varepsilon\left[\mathbf{f}_0\left(-\mathbf{I}-\sigma\kappa_1\mathbf{I}\right)+\mu_1\left(\mathbf{f}_1-\mathbf{r}_1\right)+\mu_2\left(\mathbf{f}_2-\mathbf{r}_2\right)\right]+j\sigma\varepsilon\frac{d\Phi_2(w)/dw}{w\Phi_2(w)}\ \mathbf{r}_0=o\left(\varepsilon^2\right),$$

$$j\varepsilon\left[\mathbf{f}_1\left(\mathbf{Q}-\mu_1\mathbf{I}\right)+\mathbf{f}_0\sigma\kappa_1+\mathbf{r}_1\mathbf{\Lambda}\right]-j\sigma\varepsilon\frac{d\Phi_2(w)/dw}{w\Phi_2(w)}\mathbf{r}_0=o\left(\varepsilon^2\right),$$

$$j\varepsilon w\left[\mathbf{f}_2\left(\mathbf{Q}-\mu_2\mathbf{I}\right)+\mathbf{r}_0+\mathbf{f}_0+\mathbf{r}_2\mathbf{\Lambda}\right]=o\left(\varepsilon^2\right),$$

$$j\varepsilon w\left[\mathbf{f}_0+\mathbf{f}_1\left(\mathbf{\Lambda}-\mu_1\mathbf{I}\right)+\mathbf{f}_2\left(\mathbf{\Lambda}-\mu_2\mathbf{I}\right)+\mathbf{r}_0+\mathbf{r}_1\mathbf{\Lambda}+\mathbf{r}_2\mathbf{\Lambda}\right]\mathbf{e}=o\left(\varepsilon^2\right).$$

Dividing these equations by ε and taking the limit as $\varepsilon\to0$ yields

$$\mathbf{f}_0\left(-\mathbf{I}-\sigma\kappa_1\mathbf{I}\right)+\mu_1\left(\mathbf{f}_1-\mathbf{r}_1\right)+\mu_2\left(\mathbf{f}_2-\mathbf{r}_2\right)+\sigma\frac{d\Phi_2(w)/dw}{w\Phi_2(w)}\mathbf{r}_0=0,$$

$$\mathbf{f}_1\left(\mathbf{Q}-\mu_1\mathbf{I}\right)+\mathbf{f}_0\sigma\kappa_1+\mathbf{r}_1\mathbf{\Lambda}-\sigma\frac{d\Phi_2(w)/dw}{w\Phi_2(w)}\mathbf{r}_0=0,$$

$$\mathbf{f}_2\left(\mathbf{Q}-\mu_2\mathbf{I}\right)+\mathbf{r}_0+\mathbf{f}_0+\mathbf{r}_2\mathbf{\Lambda}=0,$$

$$\left[\mathbf{f}_0+\mathbf{f}_1\left(\mathbf{\Lambda}-\mu_1\mathbf{I}\right)+\mathbf{f}_2\left(\mathbf{\Lambda}-\mu_2\mathbf{I}\right)+\mathbf{r}_0+\mathbf{r}_1\mathbf{\Lambda}+\mathbf{r}_2\mathbf{\Lambda}\right]\mathbf{e}=0.$$

These equations imply that $\frac{\Phi_2'(w)}{w\Phi_2(w)}$ doesn't depend on w and thus the scalar function $\Phi_2(w)$ is given in the following form

$$\Phi_2(w) = \exp\frac{(jw)^2}{2}\kappa_2,$$

which coincides with (16). We have $\frac{\Phi_2'(w)}{w\Phi_2(w)} = -\kappa_2$ and then we obtain the system

$$\mathbf{f}_0\left(-\mathbf{I} - \sigma\kappa_1\mathbf{I}\right) + \mu_1\mathbf{f}_1 + \mu_2\mathbf{f}_2 = \sigma\kappa_2\mathbf{r}_0 + \mu_1\mathbf{r}_1 + \mu_2\mathbf{r}_2,$$

$$\mathbf{f}_1\left(\mathbf{Q} - \mu_1\mathbf{I}\right) + \mathbf{f}_0\sigma\kappa_1 = -\mathbf{r}_1\mathbf{\Lambda} - \sigma\kappa_2\mathbf{r}_0,$$

$$\mathbf{f}_2\left(\mathbf{Q} - \mu_2\mathbf{I}\right) + \mathbf{f}_0 = -\mathbf{r}_0 - \mathbf{r}_2\mathbf{\Lambda},$$

$$[\mathbf{f}_0 + \mathbf{f}_1\left(\mathbf{\Lambda} - \mu_1\mathbf{I}\right) + \mathbf{f}_2\left(\mathbf{\Lambda} - \mu_2\mathbf{I}\right)]\,\mathbf{e} = -\left(\mathbf{r}_0 + \mathbf{r}_1\mathbf{\Lambda} + r_2\mathbf{\Lambda}\right)\mathbf{e}. \quad (25)$$

System (25) is an inhomogeneous system of linear equations, with respect to the vectors $\mathbf{f}_0$, $\mathbf{f}_1$, $\mathbf{f}_2$. The determinant of the matrix of the system is zero (the sums of rows are all zero). The rank of the extended matrix and the rank of the matrix of coefficients coincide . Consider systems (12) and (25). System (12) is homogeneous, system (25) is inhomogeneous. Consequently, we can write the solution of the inhomogeneous system (25) in the form $\mathbf{f}_k = C\mathbf{r}_k + \kappa_2\sigma\mathbf{g}_k + \mathbf{y}_k$, where C is a constant, vectors $\mathbf{r}_n$ are defined above, vectors $\mathbf{g}_k$ and $\mathbf{y}_k$ are particular solutions of the system (25) and then

$$C\left[\mathbf{r}_0\left(-\mathbf{I} - \sigma\kappa_1\mathbf{I}\right) + \mu_1\mathbf{r}_1 + \mu_2\mathbf{r}_2\right] + \kappa_2\sigma\left[\mathbf{g}_0\left(-\mathbf{I} - \sigma\kappa_1\mathbf{I}\right) + \mu_1\mathbf{g}_1 + \mu_2\mathbf{g}_2\right]$$
$$+ \mu_1\mathbf{y}_1 + \mu_2\mathbf{y}_2 + \mathbf{y}_0\left(-\mathbf{I} - \sigma\kappa_1\mathbf{I}\right) = \sigma\kappa_2\mathbf{r}_0 + \mu_1\mathbf{r}_1 + \mu_2\mathbf{r}_2,$$
$$C\left[\mathbf{r}_1\left(\mathbf{Q} - \mu_1\mathbf{I}\right) + \mathbf{r}_0\sigma\kappa_1\right] + \kappa_2\sigma\left[\mathbf{g}_1\left(\mathbf{Q} - \mu_1\mathbf{I}\right) + \mathbf{g}_0\sigma\kappa_1\right] + \mathbf{y}_1\left(\mathbf{Q} - \mu_1\mathbf{I}\right) + \mathbf{y}_0\sigma\kappa_1$$
$$= -\mathbf{r}_1\mathbf{\Lambda} - \sigma\kappa_2\mathbf{r}_0,$$
$$C\left[\mathbf{r}_2\left(\mathbf{Q} - \mu_2\mathbf{I}\right) + \mathbf{r}_0\right] + \kappa_2\sigma\left[\mathbf{g}_2\left(\mathbf{Q} - \mu_2\mathbf{I}\right) + \mathbf{g}_0\right] + \mathbf{y}_2\left(\mathbf{Q} - \mu_2\mathbf{I}\right) + \mathbf{y}_0 = -\mathbf{r}_0 - \mathbf{r}_2\mathbf{\Lambda},$$
$$C\left[\mathbf{r}_0 + \mathbf{r}_1\left(\mathbf{\Lambda} - \mu_1 I\right) + \mathbf{r}_2\left(\mathbf{\Lambda} - \mu_2 I\right)\right]\mathbf{e}$$
$$+ \kappa_2\sigma\left[\mathbf{g}_0 + \mathbf{g}_1\left(\mathbf{\Lambda} - \mu_1\mathbf{I}\right) + \mathbf{g}_2\left(\mathbf{\Lambda} - \mu_2\mathbf{I}\right)\right]\mathbf{e}$$
$$+ \left[\mathbf{y}_0 + \mathbf{y}_1\left(\mathbf{\Lambda} - \mu_1\mathbf{I}\right) + \mathbf{y}_2\left(\mathbf{\Lambda} - \mu_2\mathbf{I}\right)\right]\mathbf{e} = -\left(\mathbf{r}_0 + \mathbf{r}_1\mathbf{\Lambda} + r_2\mathbf{\Lambda}\right)\mathbf{e}.$$

Previously, the system of Eq. (12) was obtained. Taking this into account, the coefficients of C are zeros and we can rewrite the last system in the form

$$\kappa_2\sigma\left[\mathbf{g}_0\left(-\mathbf{I} - \sigma\kappa_1\mathbf{I}\right) + \mu_1\mathbf{g}_1 + \mu_2\mathbf{g}_2\right] + \mu_1\mathbf{y}_1 + \mu_2\mathbf{y}_2 + \mathbf{y}_0\left(-\mathbf{I} - \sigma\kappa_1\mathbf{I}\right)$$
$$= \sigma\kappa_2\mathbf{r}_0 + \mu_1\mathbf{r}_1 + \mu_2\mathbf{r}_2,$$
$$\kappa_2\sigma\left[\mathbf{g}_1\left(\mathbf{Q} - \mu_1 I\right) + \mathbf{g}_0\sigma\kappa_1\right] + \mathbf{y}_1\left(\mathbf{Q} - \mu_1\mathbf{I}\right) + \mathbf{y}_0\sigma\kappa_1$$
$$= -\mathbf{r}_1\mathbf{\Lambda} - \sigma\kappa_2\mathbf{r}_0, \kappa_2\sigma\left[\mathbf{g}_2\left(\mathbf{Q} - \mu_2\mathbf{I}\right) + \mathbf{g}_0\right] + \mathbf{y}_2\left(\mathbf{Q} - \mu_2\mathbf{I}\right) + \mathbf{y}_0 = -\mathbf{r}_0 - \mathbf{r}_2\mathbf{\Lambda},$$

$$\kappa_2\sigma\left[\mathbf{g}_0 + \mathbf{g}_1\left(\mathbf{\Lambda} - \mu_1\mathbf{I}\right) + \mathbf{g}_2\left(\mathbf{\Lambda} - \mu_2\mathbf{I}\right)\right]\mathbf{e}$$
$$+ \left[\mathbf{y}_0 + \mathbf{y}_1\left(\mathbf{\Lambda} - \mu_1\mathbf{I}\right) + \mathbf{y}_2\left(\mathbf{\Lambda} - \mu_2\mathbf{I}\right)\right]\mathbf{e} = -\left(\mathbf{r}_0 + \mathbf{r}_1\mathbf{\Lambda} + r_2\mathbf{\Lambda}\right)\mathbf{e}. \quad (26)$$

We consider the first three equations of the system (26). We equate the corresponding coefficients for κ_2 to obtain

$$\mathbf{g}_0\left(-\mathbf{I} - \sigma\kappa_1\mathbf{I}\right) + \mu_1\mathbf{g}_1 + \mu_2\mathbf{g}_2 = \mathbf{r}_0,$$

$$\mathbf{g}_1\left(\mathbf{Q} - \mu_1\mathbf{I}\right) + \mathbf{g}_0\sigma\kappa_1 = -\mathbf{r}_0,$$

$$\mathbf{g}_2 \left(\mathbf{Q} - \mu_2\mathbf{I}\right) + \mathbf{g}_0 = 0, \tag{27}$$

and

$$\mu_1\mathbf{y}_1 + \mu_2\mathbf{y}_2 + \mathbf{y}_0 \left(-\mathbf{I} - \sigma\kappa_1\mathbf{I}\right) = \mu_1\mathbf{r}_1 + \mu_2\mathbf{r}_2,$$

$$\mathbf{y}_1 \left(\mathbf{Q} - \mu_1\mathbf{I}\right) + \mathbf{y}_0\sigma\kappa_1 = -\mathbf{r}_1\mathbf{\Lambda},$$

$$\mathbf{y}_2 \left(\mathbf{Q} - \mu_2\mathbf{I}\right) + \mathbf{y}_0 = -\mathbf{r}_0 - \mathbf{r}_2\mathbf{\Lambda}. \tag{28}$$

From systems (27) and (28) we obtain systems:

$$\mathbf{g}_0 \left[-\mathbf{I} - \sigma\kappa_1\mathbf{I} + \mu_2 \left(\mu_2\mathbf{I} - \mathbf{Q}\right)^{-1} + \mu_1\sigma\kappa_1 \left(\mu_1\mathbf{I} - \mathbf{Q}\right)^{-1}\right] = \mathbf{r}_0 - \mathbf{r}_0\mu_1 \left(\mu_1\mathbf{I} - \mathbf{Q}\right)^{-1},$$

$$\mathbf{g}_1 = \left(\mathbf{g}_0\sigma\kappa_1 + \mathbf{r}_0\right) \left(\mu_1\mathbf{I} - \mathbf{Q}\right)^{-1},$$

$$\mathbf{g}_2 = \mathbf{g}_0 \left(\mu_2\mathbf{I} - \mathbf{Q}\right)^{-1}. \tag{29}$$

$$\mathbf{y}_0 \left[\left(-\mathbf{I} - \sigma\kappa_1\mathbf{I}\right) + \mu_1\sigma\kappa_1 \left(\mu_1\mathbf{I} - \mathbf{Q}\right)^{-1} + \mu_2 \left(\mu_2\mathbf{I} - \mathbf{Q}\right)^{-1}\right]$$
$$= \mu_1\mathbf{r}_1 \left[\mathbf{I} - \mathbf{\Lambda} \left(\mu_1\mathbf{I} - \mathbf{Q}\right)^{-1}\right] + \mu_2 \left[\mathbf{r}_2 - \left(\mathbf{r}_0 + \mathbf{r}_2\mathbf{\Lambda}\right) \left(\mu_2\mathbf{I} - \mathbf{Q}\right)^{-1}\right],$$

$$\mathbf{y}_1 = \left(\mathbf{y}_0\sigma\kappa_1 + \mathbf{r}_1\mathbf{\Lambda}\right) \left(\mu_1\mathbf{I} - \mathbf{Q}\right)^{-1},$$

$$\mathbf{y}_2 = \left(\mathbf{y}_0 + \mathbf{r}_0 + \mathbf{r}_2\mathbf{\Lambda}\right) \left(\mu_2\mathbf{I} - \mathbf{Q}\right)^{-1}. \tag{30}$$

The determinants of the coefficient matrices systems (29) and (30) are zero. Then we define an additional condition for this systems of equations

$$\left(\mathbf{g}_0 + \mathbf{g}_1 + \mathbf{g}_2\right)\mathbf{e} = 0,$$

$$\left(\mathbf{y}_0 + \mathbf{y}_1 + \mathbf{y}_2\right)\mathbf{e} = 0.$$

Thus, the solutions of inhomogeneous systems for $\mathbf{g}_0$, $\mathbf{g}_1$, $\mathbf{g}_2$, $\mathbf{y}_0$, $\mathbf{y}_1$, $\mathbf{y}_2$ are uniquely determined. We obtain systems that coincide with the systems (18) and (19). Substituting values $\mathbf{g}_0$, $\mathbf{g}_1$, $\mathbf{g}_2$, $\mathbf{y}_0$, $\mathbf{y}_1$, $\mathbf{y}_2$ into the scalar equation of the system (26), we obtain

$$\kappa_2 = \frac{1}{\sigma} \cdot \frac{\mathbf{r}_0\mathbf{e} + \mathbf{r}_1\mathbf{\Lambda}\mathbf{e} + \mathbf{r}_2\mathbf{\Lambda}\mathbf{e} + \left[\mathbf{y}_0 + \mathbf{y}_1 \left(\mathbf{\Lambda} - \mu_1\mathbf{I}\right) + \mathbf{y}_2 \left(\mathbf{\Lambda} - \mu_2\mathbf{I}\right)\right]\mathbf{e}}{\left[-\mathbf{g}_0 + \mathbf{g}_1 \left(\mu_1\mathbf{I} - \mathbf{\Lambda}\right) + \mathbf{g}_2 \left(\mu_2\mathbf{I} - \mathbf{\Lambda}\right)\right]\mathbf{e}}.$$

This equality coincides with (17).

Second order asymptotic i.e. Theorem 2, shows that the asymptotic probability distribution of the number $i(t)$ of calls in a system is Gaussian with mean asymptotic $\kappa_1\alpha$ and dispersion $\kappa_2\alpha$.

4   Asymptotic Analysis of MMPP/M/1/1 Retrial Queue with Two-Way Communication Under the Low Rate of Service Time of Outgoing Calls ($\mu_2 \to 0$)

We will research system (5) by asymptotic analysis method under the low rate of service time of outgoing calls.

Theorem 3. *Suppose $i(t)$ is a number of calls in an system of stationary MMPP/M/1/1 retrial queue with two-way communication, then the following equation is true*

$$H(u) = \lim_{\mu_2 \to 0} E e^{jw\mu_2 i(t)} = \left(1 - j\frac{\rho\mu_1}{\mu_2}u\right)^{-\left(\frac{\alpha}{\sigma(1-\rho)}+1\right)}, \tag{31}$$

where $\rho\mu_1 = \mathbf{r}\boldsymbol{\Lambda}\mathbf{e}$ and ρ is the trafic intensity.

Proof. We denote $\mu_2 = \varepsilon$, let's substitute the following in the system (5)

$$u = \varepsilon w, \quad \mathbf{H}_0(u) = \varepsilon\mathbf{F}_0(w, \varepsilon), \quad \mathbf{H}_k(u) = \mathbf{F}_k(w, \varepsilon), \quad k = 1, 2.$$

We will get the system

$$\varepsilon\mathbf{F}_0(w,\varepsilon)(\mathbf{Q} - \boldsymbol{\Lambda} - \alpha\mathbf{I}) + j\sigma\frac{\partial\mathbf{F}_0(w,\varepsilon)}{\partial w} + \mu_1 e^{-j\varepsilon w}\mathbf{F}_1(w,\varepsilon)$$
$$+ \varepsilon e^{-j\varepsilon w}\mathbf{F}_2(w,\varepsilon) = 0,$$

$$\mathbf{F}_1(w,\varepsilon)\left(\mathbf{Q} + \left(e^{j\varepsilon w} - 1\right)\boldsymbol{\Lambda} - \mu_1\mathbf{I}\right) + \mathbf{F}_0(w,\varepsilon)\varepsilon e^{j\varepsilon w}\boldsymbol{\Lambda} - j\sigma\frac{\partial\mathbf{F}_0(w,\varepsilon)}{\partial w} = 0,$$

$$\mathbf{F}_2(w,\varepsilon)\left(\mathbf{Q} + \left(e^{j\varepsilon w} - 1\right)\boldsymbol{\Lambda} - \varepsilon\mathbf{I}\right) + \alpha\varepsilon e^{j\varepsilon w}\mathbf{F}_0(w,\varepsilon) = 0,$$

$$\mathbf{F}_0(w,\varepsilon)\varepsilon\left(e^{j\varepsilon w}\boldsymbol{\Lambda} + \alpha e^{j\varepsilon w}\mathbf{I}\right)\mathbf{e} + \mathbf{F}_1(w,\varepsilon)\left(e^{j\varepsilon w}\boldsymbol{\Lambda} - \mu_1\mathbf{I}\right)\mathbf{e}$$
$$+ \mathbf{F}_2(w,\varepsilon)\left(e^{j\varepsilon w}\boldsymbol{\Lambda} - \varepsilon\mathbf{I}\right)\mathbf{e} = 0. \tag{32}$$

Considering the limit as $\varepsilon \to 0$ in the system (32) then we will get

$$j\sigma\mathbf{F}_0'(w) + \mu_1\mathbf{F}_1(w) = 0,$$

$$\mathbf{F}_1(w)\left(\mathbf{Q} - \mu_1\mathbf{I}\right) - j\sigma\mathbf{F}_0'(w) = 0,$$

$$\mathbf{F}_2(w)\mathbf{Q} = 0,$$

$$\mathbf{F}_1(w)\left(\boldsymbol{\Lambda} - \mu_1 I\right)\mathbf{e} + \mathbf{F}_2(w)\boldsymbol{\Lambda}\mathbf{e} = 0. \tag{33}$$

From the first and second equations we obtain $\mathbf{F}_1(w)\mathbf{Q} = 0, \quad \mathbf{F}_2(w)\mathbf{Q} = 0$. We seek the solution of the system (33) in the form $\mathbf{F}_k(w) = \Phi_k(w)\mathbf{r}, \quad k = 1, 2$. Then, given the fact that $\mathbf{r}\boldsymbol{\Lambda}\mathbf{e} = \rho\mu_1$ and

$$j\sigma\mathbf{F}_0'(w) + \mu_1\Phi_1(w)\mathbf{r} = 0,$$

$$\Phi_1(w)\mathbf{r}\left(\mathbf{Q} - \mu_1\mathbf{I}\right) - j\sigma\mathbf{F}_0'(w) = 0,$$

$$\Phi_2(w)\mathbf{r}\mathbf{Q} = 0,$$

$$\Phi_1(w)\mathbf{r}\left(\mathbf{\Lambda} - \mu_1\mathbf{I}\right)\mathbf{e} + \Phi_2(w)\mathbf{r}\mathbf{\Lambda}\mathbf{e} = 0,$$

we have

$$j\sigma\mathbf{F}_0'(w) + \mu_1\Phi_1(w)\mathbf{r} = 0,$$

$$\Phi_1(w)\left(\rho - 1\right)\mu_1 + \Phi_2(w)\rho\mu_1 = 0.$$

We denote $\Phi_1(w) + \Phi_2(w) = \Phi(w)$, then $\Phi_1(w) = \rho\Phi(w)$, $\Phi_2(w) = (1 - \rho)\Phi(w)$. Furthermore,

$$F_1(w) = \rho\Phi(w)\mathbf{r}, \quad F_2(w) = (1 - \rho)\Phi(w)\mathbf{r}. \qquad (34)$$

Multiplying the third equation of system (32) by the unit vector $\mathbf{e}$, considering the limit as $\varepsilon \to 0$, we have

$$(1 - \rho)\Phi(w)\mathbf{r}\left(jw\mathbf{\Lambda} - \mathbf{I}\right)\mathbf{e} + \alpha\mathbf{F}_0(w)\mathbf{e} = 0.$$

We denote

$$\mathbf{F}_0(w)\mathbf{e} = \varphi(w). \qquad (35)$$

Then

$$\frac{\alpha}{(1 - \rho)\left(1 - jw\rho\mu_1\right)}\varphi(w) = \Phi(w). \qquad (36)$$

We consider the first equation of system (33), multiplying it by a unit vector $\mathbf{e}$ and taking into account (34), (35) and (36), we obtain

$$j\sigma\varphi'(w) + \frac{\alpha\mu_1\rho}{(1 - \rho)\left(1 - jw\rho\mu_1\right)}\varphi(w) = 0.$$

The solution of the differential equation has the form

$$\varphi(w) = C\left(1 - jw\rho\mu_1\right)^{-\frac{\alpha}{\sigma(1-\rho)}}.$$

Then

$$\Phi(w) = \left(1 - jw\rho\mu_1\right)^{-\left(\frac{\alpha}{\sigma(1-\rho)} + 1\right)}.$$

Making reverse substitutions, we obtain the characteristic function (31).

Theorem 3 shows that the asymptotic probability distribution of $i(t)$ of calls in the system under the low rate of service time of outgoing calls is Gamma.

5   Approximation Accuracy $P^{(2)}(i)$

The accuracy of the approximation $P^{(2)}(i)$ is defined by using Kolmogorov range $\Delta_2 = \max_{0 \le i \le N}\left|\sum_{v=0}^{i}\left(P(v) - P^{(2)}(v)\right)\right|$, which represents the difference between distributions $P(i)$ and $P^{(2)}(i)$, where $P(i)$ is obtained by using numerical algorithm for the MMPP/M/1/1 retrial queue and the approximation $P^{(2)}(i)$ is given by Gaussian and Gamma approximations.

Table 1. Kolmogorov range $\mu_1 = 1, \mu_2 = 2, \sigma = 1$

	$\alpha = 350$	$\alpha = 500$	$\alpha = 600$	$\alpha = 800$	$\alpha = 1000$
$\rho = 0.2$	0.054	0.045	0.041	0.036	0.032
$\rho = 0.4$	0.041	0.034	0.029	–	–

Table 2. Kolmogorov range, $\mu_1 = 1, \alpha = 1, \sigma = 1$

	$\mu_2 = 0.07$	$\mu_2 = 0.05$	$\mu_2 = 0.04$	$\mu_2 = 0.035$
$\rho = 0.5$	0.05	0.036	0.029	0.026
$\rho = 0.6$	0.058	0.042	0.034	0.030

Tables 1 contains the values of Δ_2 for various values of rate ρ and α for MMPP/M/1/1 retrial queue with two-way communication. We fix $\mu_1 = 1$, $\mu_2 = 2$ and $\sigma = 1$ in Table 1. Table 2 contains the values of Δ_2 for various values of rate ρ and μ_2 for MMPP/M/1/1 retrial queue with two-way communication. We fix $\mu_1 = 1$, $\alpha = 1$ and $\sigma = 1$ in Table 2. We observe in Table 1 that the approximation accuracy increases with the increase in α and in Table 2 that the approximation accuracy increases with the decrease in μ_2.

6   Conclusions

In this paper, we have considered retrial queue with two-way communication with MMPP input. We have found the first and the second order asymptotics of the number of calls in the system under the condition of the high rate of making outgoing calls. Based on the obtained asymptotics we have built the Gaussian approximation of the probability distribution of the number of calls in the system. Our numerical results have revealed that the accuracy of Gaussian approximation increases while increasing α. We have found the Gamma approximation of the number of calls in the system under the condition of the low service rate of outgoing calls. Our numerical results have revealed that the accuracy of Gamma approximation increases while decreasing μ_2. In future we plan to consider this retrial queueing system in other asymptotic conditions.

References

1. Bhulai, S., Koole, G.: A queueing model for call blending in call centers. IEEE Trans. Autom. Control **48**, 1434–1438 (2003)
2. Artalejo, J.R., Phung-Duc, T.: Markovian retrial queues with two way communication. J. Ind. Manag. Optim. **8**(4), 781–806 (2012)
3. Artalejo, J.R., Phung-Duc, T.: Single server retrial queues with two way communication. Appl. Math. Model. **37**(4), 1811–1822 (2013)

4. Phung-Duc, T., Kawanishi, K.: An efficient method for performance analysis of blended call centers with redial. Asia Pac. J. Oper. Res. **31**, 1440008 (2014). https://doi.org/10.1142/S0217595914400089
5. Sakurai, H., Phung-Duc, T.: Two-way communication retrial queues with multiple types of outgoing calls. First Top **23**, 466–492 (2014). doi:10.1007/s11750-014-0349-5
6. Sakurai, H., Phung-Duc, T.: Scaling limits for single server retrial queues with two-way communication. Ann. Oper. Res. **247**(1), 229–256 (2016)
7. Nazarov, A., Paul, S., Gudkova, I.: Asymptotic analysis of Markovian retrial queue with two-way communication under low rate of retrials condition. In: Proceedings of the 31st European Conference on Modelling and Simulation, ECMS, Budapest, pp. 687–693 (2017)
8. Nazarov, A.A., Terpugov, A.F.: Queueing theory - Tutorial. NTL, Tomsk (2010). (in Russian)
9. Nazarov, A.A., Moiseeva, S.P.: Method of asymptotic analysis in queueing theory - Tutorial. NTL, Tomsk (2006). (in Russian)
10. Nazarov, A., Paul, S.: A cyclic queueing system with priority customers and t-strategy of service. In: Vishnevskiy, V.M., Samouylov, K.E., Kozyrev, D.V. (eds.) DCCN 2016. CCIS, vol. 678, pp. 182–193. Springer, Cham (2016). doi:10.1007/978-3-319-51917-3_17

Two-Server Queueing System with Unreliable Servers and Markovian Arrival Process

Valentina Klimenok[✉]

Department of Applied Mathematics and Computer Science,
Belarusian State University, Minsk 220030, Belarus
klimenok@bsu.by

Abstract. In this paper, we investigate a queueing system consisting of an infinite buffer and two unreliable heterogeneous servers which fail alternately. If both servers are able to provide the service, they serve a customer in parallel, independently of each other. The service of a customer is completed when his/her service by any of two servers ends. The service times at the servers have PH-type (Phase-type) distributions. The input flow and the flow of breakdowns are described by the MAP (Markovian Arrival Process). An arriving breakdown is directed to the first server with some probability and to the second server with complementary probability. After a breakdown occurrence a server fails and the repair period starts immediately. A customer, whose service is interrupted by the breakdown, goes to another server if it is idle, or enters the queue otherwise We derive a condition for the stable operation of the system, calculate its stationary distribution and base performance measures. Illustrative numerical examples are presented.

Keywords: Unreliable queueing system · Markovian Arrival Process · Phase-type service time distribution · Stationary distribution · Performance measures

1 Introduction

At the present time, the requirements for the speed and reliability of information transmission in wireless communication systems have increased significantly. In recent years, the FSO - Free Space Optics technologies have become widespread due to their undoubted advantages. The main advantages of atmospheric optical (laser) communication link are high capacity and quality of communication. However, optical communication systems have also disadvantages, the main of which is the dependence of the communication channel on the weather conditions. The unfavorable weather conditions which reduce visibility significantly reduce the effectiveness of atmospheric optical communication link.

As it is mentioned in [1], one of the main directions of creating the ultra-high speed and reliable wireless means of communication is the development of hybrid communication systems based on laser and radio-wave technologies. Hybrid radio-optical equipment is based on the use of the FSO channel and a

© Springer International Publishing AG 2017
A. Dudin et al. (Eds.): ITMM 2017, CCIS 800, pp. 42–55, 2017.
DOI: 10.1007/978-3-319-68069-9_4

backup radio channel. Because the increased interest in hybrid communication systems, a considerable amount of studies of this class of systems have appeared in the last decade. It should be noted that most of the studies are devoted to simulation modeling, see, e.g. [2–4]. Among the works devoted to the mathematical modeling of hybrid communication systems, we note [5–8]. The papers [5,6] deals with hybrid communication channel with so called "hot" redundancy, where the backup IEEE 802.11n radio channel continuously transmits data along with the FSO channel, but, unlike the latter, at low speed. In the papers [7,8], the hybrid communication system with "cold" redundancy is considered, where the radiowave link is assumed to be absolutely reliable and backs up the atmospheric optical communication link only in cases when the latter interrupts its functioning because of the unfavorable weather conditions. It is assumed in [5,7] that an input flow is a stationary Poisson one and the service and repair times have exponential distributions. More realistic assumptions have been made in [6,8] where the $BMAP$ (Batch Markovian Arrival Process) and the PH service and repair times distributions are under consideration.

The paper [1] is devoted to the study of a hybrid communication system which consists of FSO channel and a millimeter-wave radio channel which is used as a backup one. The peculiarity of such reservation is that the unfavorable weather conditions for one of the channel do not affect the other one. The FSO channel is unable to transfer data in fog or mist and mm-wave radio channel is unable to transfer data in case of precipitation (rain, snow, etc.). Thus, the hybrid communication system is able to transfer data under almost any weather conditions. To model this hybrid channel, the authors consider two-channel queueing system with unreliable heterogeneous servers which fail alternately. An arriving breakdown is directed to the first server with some probability and to the second server with complementary probability. After a breakdown occurrence a server fails and the repair period (period of unfavorable weather conditions for this server) starts immediately. It is assumed that fault-free periods for both channels alternate with periods of repair period for one of the channels. At every moment, a customer is served by one of the fault-free channels. If this channel breaks down, the customer occupies the other server, if it is idle, or enters the queue otherwise. Customers and breakdowns arrive to the system according to the stationary Poisson flow, the service and repair times are exponentially distributed.

In the present paper, we consider more complicated queueing system which differs from the system considered in [1] in the following: (i) a customer is processed by two servers simultaneously if both servers are fault-free according to "hot" reservation technology. Otherwise, it is processed by a fault-free channel; (ii) the input flow and the flow of breakdowns are described by $MAPs$, the service and repair times have PH distributions. We describe operation of the system by a multi-dimensional continuous time Markov chain, derive the ergodicity condition for this Markov chain and give the brief description of the algorithm for computation of its stationary distribution. We derive formulas for computation of some performance measures of the system and present illustrative numerical examples.

2   Model Description

We consider a queueing system consisting of two unreliable servers (server 1 and server 2) and an infinite buffer. Customers arrive to the system according to the MAP. The MAP is defined by the underlying process ν_t, $t \geq 0$, which is an irreducible continuous-time Markov chain with the finite state space $\{0, \ldots, W\}$, and the $(W + 1) \times (W + 1)$ matrices D_k, $k = 0, 1$. The matrix D_1 (non-diagonal entries of the matrix D_0) define the rates of the process ν_t, $t \geq 0$, transitions which are accompanied by generating a customer. The matrix $D = D_0 + D_1$ is an infinitesimal generator of the process ν_t, $t \geq 0$. The fundamental rate of the MAP is defined as $\lambda = \boldsymbol{\theta} D_1 \mathbf{e}$ where the vector $\boldsymbol{\theta}$ is the unique solution of the system $\boldsymbol{\theta} D = \mathbf{0}$, $\boldsymbol{\theta} \mathbf{e} = 1$. Here and in the sequel $\mathbf{e}$ ($\mathbf{0}$) is a column (row) vector of appropriate size consisting of 1's (0's). The coefficient of variation of inter-arrival intervals is given by $c_{var}^2 = 2\lambda\boldsymbol{\theta}(-D_0)^{-1}\mathbf{e} - 1$ while the coefficient of correlation of intervals between successive arrivals is calculated as $c_{cor} = (\lambda\boldsymbol{\theta}(-D_0)^{-1}(D - D_0(-D_0)^{-1}\mathbf{e} - 1)/c_{var}^2$. For more information about MAP see, e.g. [9].

If an arriving customer or the first customer from the queue sees two servers idle and ready for service, he/she starts the service at both servers. If the servers are busy at an arrival epoch or one of the servers is busy while the other server is under repair, the customer is placed at the end of the queue in the buffer and is picked-up for a service later on, according the FIFO discipline. If one of the servers is under repair and the other server is idle, the idle server begins the service of the customer. If the service of a customer at one of the servers is not finished until the end of repair period on the other server, the latter server immediately connects to the service of the customer. The service of the customer is considered be completed when his/her service by any of two servers is finished.

Breakdowns arrive to the servers according to a MAP which is defined by the $(V + 1) \times (V + 1)$ matrices H_0 and H_1. An arriving breakdown is directed to the server 1 with probability p and to the server 2 with complementary probability $1 - p$. The breakdowns fundamental rate is calculated as $h = \boldsymbol{\vartheta} H_1 \mathbf{e}$ where the row vector $\boldsymbol{\vartheta}$ is the unique solution of the system $\boldsymbol{\vartheta}(H_0 + H_1) = \mathbf{0}$, $\boldsymbol{\vartheta}\mathbf{e} = 1$.

The service time of a customer by the kth server, $k = 1, 2$, has PH type distribution with an irreducible representation $(\boldsymbol{\beta}^{(k)}, \mathbf{S}^{(k)})$. The service process on the kth server is directed by the Markov chain $m_t^{(k)}$, $t \geq 0$, with state space $\{1, \ldots, M^{(k)}, M^{(k)} + 1\}$ where $M^{(k)} + 1$ is an absorbing state. The intensities of transitions into the absorbing state are defined by the vector $\mathbf{S}_0^{(k)} = -S^{(k)}\mathbf{e}$. The service rates are calculated as $\mu_k = -[\boldsymbol{\beta}^{(k)}(S^{(k)})^{-1}\mathbf{e}]^{-1}$. For more information about the PH type distribution, see, e.g., [10].

The repair period at the kth server, $k = 1, 2$, has PH type distribution with an irreducible representation $(\boldsymbol{\tau}^{(k)}, T^{(k)})$. The repair process at the kth server is directed by the Markov chain $r_t^{(k)}, t \geq 0$, with state space $\{1, \ldots, R_k, R_k + 1\}$ where $R_k + 1$ is an absorbing state. The intensities of transitions into the absorbing state are defined by the vector $\mathbf{T}_0^{(k)} = -T^{(k)}\mathbf{e}$. The repair rate is $\mathbf{æ}_k = -(\boldsymbol{\tau}^{(k)}(T^{(k)})^{-1}\mathbf{e})^{-1}$.

3   Process of the System States

Let at the moment t

- i_t be the number of customers in the system, $i_t \geq 0$,
- $n_t = 0$, if both server are fault-free, $n_t = k$, if the server k is under repair, $k = 1, 2$;
- $m_t^{(k)}$ be the state of the directing process of the service at the kth busy server, $m_t^{(k)} = \overline{1, M^{(k)}}$, $k = 1, 2$;
- $r_t^{(k)}$ be the state of the directing process of the repair time at the kth server, $r_t^{(k)} = \overline{1, R_k}$, $k = 1, 2$;
- ν_t and η_t be the states of the underlying process of the MAP of customers and the MAP of breakdowns respectively, $\nu_t = \overline{0, W}$, $\eta_t = \overline{0, V}$.

The process of the system states is described by the regular irreducible continuous time Markov chain, $\xi_t, t \geq 0$, with state space

$$\Omega = \{(0, n, \nu, \eta), \, n = 0, 1, 2, \, \nu = \overline{0, W}, \, \eta = \overline{0, V}\} \bigcup$$

$$\{(i, 0, \nu, \eta, m^{(1)}, m^{(2)}), \, i > 0, \nu = \overline{0, W}, \eta = \overline{0, V}, \, m^{(k)} = \overline{1, M_k}, k = 1, 2\} \bigcup$$

$$\{(i, 1, \nu, \eta, m^{(2)}, r^{(1)}), i > 0, \nu = \overline{0, W}, \eta = \overline{0, V}, m^{(2)} = \overline{1, M_2}, r^{(1)} = \overline{1, R_1}\} \bigcup$$

$$\{(i, 2, \nu, \eta, m^{(1)}, r^{(2)}), \, i > 0, \nu = \overline{0, W}, \eta = \overline{0, V}, m^{(1)} = \overline{1, M_1}, r^{(1)} = \overline{1, R_2}\}.$$

In the following, we assume that the states of the chain $\xi_t, t \geq 0$, are ordered as follows. Within the indicated above subsets of the set Ω the states of the chain are enumerated in the lexicographic order. Denote the obtained ranked sets as $\Omega(0, 0), \Omega(0, 1), \Omega(i, n, r), i \geq 1, n = 0, 1, 2$, and arrange these sets in the lexicographic order. Let $Q_{ij}, i, j \geq 0$, be the matrices formed by rates of the chain transition from the state corresponding to the value i of the component i_n to the state corresponding to the value j of this component. Denote as $Q = (Q_{ij})_{i,j \geq 0}$ the generator of the chain.

Lemma 1. *Infinitesimal generator Q of the Markov chain $\xi_t, t \geq 0$, has the following block structure*

$$Q = \begin{pmatrix} Q_{0,0} & Q_{0,1} & O & O & \cdots \\ Q_{1,0} & Q_1 & Q_2 & O & \cdots \\ O & Q_0 & Q_1 & Q_2 & \cdots \\ O & O & Q_0 & Q_1 & \cdots \\ \vdots & \vdots & \vdots & \vdots & \ddots \end{pmatrix}$$

where non-zero blocks have the following form:

$$Q_{0,0} = \begin{pmatrix} D_0 \oplus H_0 & I_{\bar{W}} \otimes pH_1 \otimes \boldsymbol{\tau}_1 & I_{\bar{W}} \otimes \bar{p}H_1 \otimes \boldsymbol{\tau}_2 \\ I_a \otimes \boldsymbol{T}_0^{(1)} & D_0 \oplus H \oplus T^{(1)} & O \\ I_a \otimes \boldsymbol{T}_0^{(2)} & O & D_0 \oplus H \oplus T^{(2)} \end{pmatrix},$$

$$Q_{0,1} = \begin{pmatrix} D_1 \otimes I_{\bar{V}} \otimes \boldsymbol{\beta}_1 \otimes \boldsymbol{\beta}_2 & O & O \\ O & D_1 \otimes I_{\bar{V}} \otimes \boldsymbol{\beta}_2 \otimes I_{R_1} & O \\ O & O & D_1 \otimes I_{\bar{V}} \otimes \boldsymbol{\beta}_1 \otimes I_{R_2} \end{pmatrix},$$

$$Q_{1,0} = \begin{pmatrix} I_a \otimes \tilde{\boldsymbol{S}}_0 & O & O \\ O & I_a \otimes \boldsymbol{S}_0^{(2)} \otimes I_{R_1} & O \\ O & O & I_a \otimes \boldsymbol{S}_0^{(1)} \otimes I_{R_2} \end{pmatrix},$$

$$Q_0 = \begin{pmatrix} I_a \otimes \tilde{\boldsymbol{S}}_0(\boldsymbol{\beta}_1 \otimes \boldsymbol{\beta}_2) & O & O \\ O & I_a \otimes \boldsymbol{S}_0^{(2)}\boldsymbol{\beta}_2 \otimes I_{R_1} & O \\ O & O & I_a \otimes \boldsymbol{S}_0^{(1)}\boldsymbol{\beta}_1 \otimes I_{R_2} \end{pmatrix},$$

$$Q_1 = \begin{pmatrix} D_0 \oplus H_0 \oplus S_1 \oplus S_2 & I_{\bar{W}} \otimes pH_1 \otimes \mathbf{e}_{M_1} \otimes I_{M_2} \otimes \boldsymbol{\tau}_1 & I_{\bar{W}} \otimes \bar{p}H_1 \otimes I_{M_1} \otimes \mathbf{e}_{M_2} \otimes \boldsymbol{\tau}_2 \\ I_a \otimes \boldsymbol{\beta}_1 \otimes I_{M_2} \otimes \boldsymbol{T}_0^{(1)} & D_0 \oplus H \oplus S_2 \oplus T^{(1)} & O \\ I_a \otimes I_{M_1} \otimes \boldsymbol{\beta}_2 \otimes \boldsymbol{T}_0^{(2)} & O & D_0 \oplus H \oplus S_1 \oplus T^{(2)} \end{pmatrix},$$

$$Q_2 = \begin{pmatrix} D_1 \otimes I_{\bar{V}M_1M_2} & O & O \\ O & D_1 \otimes I_{\bar{V}M_2R_1} & O \\ O & O & D_1 \otimes I_{\bar{V}M_1R_2} \end{pmatrix}$$

where $H = H_0 + H_1$, $\tilde{\boldsymbol{S}}_0 = -(S_1 \oplus S_2)\mathbf{e}$, $\otimes$, $\oplus$ are the symbols of Kronecker's product and sum of matrices, $\bar{W} = W + 1$, $\bar{V} = V + 1$, $a = \bar{W}\bar{V}, \bar{p} = 1 - p$.

The proof of the lemma is implemented by means of calculation of probabilities of transitions of the components of the Markov chain ξ_t during a time interval having infinitesimal length.

Corollary 1. *The Markov chain $\xi_t, t \geq 0$, belongs to the class of quasi-birth-and-death (QBD) processes, see [10].*

The proof of the corollary follows from the definition of QBD given in [10] and the structure of the generator Q.

4   Stationary Distribution

Theorem 1. *The necessary and sufficient condition for existence of the stationary distribution of the Markov chain ξ_t, $t \geq 0$, is the fulfillment of the inequality*

$$\lambda < \boldsymbol{\delta}_0 \tilde{\boldsymbol{S}}_0 + \hat{\boldsymbol{\delta}}_1 \boldsymbol{S}_0^{(2)} + \hat{\boldsymbol{\delta}}_2 \boldsymbol{S}_0^{(1)}, \tag{1}$$

where $\hat{\boldsymbol{\delta}}_0 = \boldsymbol{\delta}_0 (\mathbf{e}_{\bar{V}} \otimes I_{M_1 M_2})$, $\hat{\boldsymbol{\delta}}_1 = \boldsymbol{\delta}_1 (\mathbf{e}_{\bar{V}} \otimes I_{M_2} \otimes \mathbf{e}_{R_1})$, $\hat{\boldsymbol{\delta}}_2 = \boldsymbol{\delta}_2 (\mathbf{e}_{\bar{V}} \otimes I_{M_1} \otimes \mathbf{e}_{R_2})$, and the vector $\boldsymbol{\delta} = (\boldsymbol{\delta}_0, \boldsymbol{\delta}_1, \boldsymbol{\delta}_2)$ is the unique solution of the system

$$\boldsymbol{\delta}\Phi = \mathbf{0}, \;\; \boldsymbol{\delta}\mathbf{e} = 1 \tag{2}$$

where

$$\Phi = \begin{pmatrix} I_{\bar{V}} \otimes \tilde{S}_0(\boldsymbol{\beta}_1 \otimes \boldsymbol{\beta}_2) + H_0 \otimes I_{M_1 M_2} & pH_1 \otimes \mathbf{e}_{M_1} \otimes I_{M_2} \otimes \boldsymbol{\tau}_1 & \bar{p}H_1 \otimes I_{M_1} \otimes \mathbf{e}_{M_2} \otimes \boldsymbol{\tau}_2 \\ I_{\bar{V}} \otimes \boldsymbol{\beta}_1 \otimes I_{M_2} \otimes \boldsymbol{T}_0^{(1)} & I_{\bar{V}} \otimes \boldsymbol{S}_0^{(2)} \boldsymbol{\beta}_2 \otimes I_{R_1} & O \\ I_{\bar{V}} \otimes I_{M_1} \otimes \boldsymbol{\beta}_2 \otimes \boldsymbol{T}_0^{(2)} & O & I_{\bar{V}} \otimes \boldsymbol{S}_0^{(1)} \boldsymbol{\beta}_1 \otimes I_{R_2} \end{pmatrix}.$$

$$+ \; diag\{I_{\bar{V}} \otimes S_1 \oplus S_2, \; I_{\bar{V}} \otimes S_2 \oplus T^{(1)}, \; I_{\bar{V}} \otimes S_1 \oplus T^{(2)}\}.$$

Proof. It follows from [10], that a necessary and sufficient condition for the existence of the stationary distribution of the chain ξ_t, $t \geq 0$, is the fulfillment of the following inequality:

$$\mathbf{x} Q_2 \mathbf{e} < \mathbf{x} Q_0 \mathbf{e} \tag{3}$$

where the vector $\mathbf{x}$ is the unique solution of the system of linear algebraic equations

$$\mathbf{x}(Q_0 + Q_1 + Q_2) = \mathbf{0}, \tag{4}$$

$$\mathbf{x}\mathbf{e} = 1. \tag{5}$$

Let $\mathbf{x}$ be a stochastic vector separated into parts as

$$\mathbf{x} = (\mathbf{x}_0, \mathbf{x}_1, \mathbf{x}_2) \tag{6}$$

where the vectors $\mathbf{x}_0, \mathbf{x}_1, \mathbf{x}_2$ have sizes $aM_1 M_2$, $aM_2 R_1$ $aM_1 R_2$ respectively.
Then the system (4) can be written as

$$\mathbf{x}_0[I_a \otimes \tilde{S}_0(\boldsymbol{\beta}_1 \otimes \boldsymbol{\beta}_2) + D_0 \oplus H_0 \oplus S_1 \oplus S_2 + D_1 \otimes I_{\bar{V} M_1 M_2}] \tag{7}$$
$$+ \mathbf{x}_1[I_a \otimes \boldsymbol{\beta}_1 \otimes I_{M_2} \otimes \boldsymbol{T}_0^{(1)}] + \mathbf{x}_2[I_a \otimes I_{M_1} \otimes \boldsymbol{\beta}_2 \otimes \boldsymbol{T}_0^{(2)}] = \mathbf{0},$$

$$\mathbf{x}_0[I_{\bar{W}} \otimes pH_1 \otimes \mathbf{e}_{M_1} \otimes I_{M_2} \otimes \boldsymbol{\tau}_1] \tag{8}$$
$$+ \mathbf{x}_1[I_a \otimes \boldsymbol{S}_0^{(2)} \boldsymbol{\beta}_2 \otimes I_{R_1} + D_0 \oplus H \oplus S_2 \oplus T^{(1)} + D_1 \otimes I_{\bar{V} M_2 R_1}] = \mathbf{0},$$

$$\mathbf{x}_0[I_{\bar{W}} \otimes (1-p)H_1 \otimes I_{M_1} \otimes \mathbf{e}_{M_2} \otimes \boldsymbol{\tau}_2] \tag{9}$$
$$+ \mathbf{x}_2[I_a \otimes \boldsymbol{S}_0^{(1)} \boldsymbol{\beta}_1 \otimes I_{R_2} + D_0 \oplus H \oplus S_1 \oplus T^{(2)} + D_1 \otimes I_{\bar{V} M_1 R_2}] = \mathbf{0}.$$

Represent the vectors $\mathbf{x}_0$, $\mathbf{x}_1$, $\mathbf{x}_2$ in the form

$$\mathbf{x}_0 = \boldsymbol{\theta} \otimes \boldsymbol{\delta}_0, \ \mathbf{x}_1 = \boldsymbol{\theta} \otimes \boldsymbol{\delta}_1, \ \mathbf{x}_2 = \boldsymbol{\theta} \otimes \boldsymbol{\delta}_2. \tag{10}$$

Substituting the vector $\mathbf{x}$ of form (10) into Eqs. (7)–(9) and taking into account that $\boldsymbol{\theta}(D_0 + D_1) = 0$, and $\mathbf{x}$ is a stochastic vector, the system (7)–(9) reduces to the first equation in (2).

Further, substituting the vector $\mathbf{x}$ of form (10) into (3) and using the relation $\lambda = \boldsymbol{\theta}D_1 \mathbf{e}$, we reduce (3) to the following inequality:

$$\lambda < \boldsymbol{\delta}_0(\mathbf{e}_{\bar{V}} \otimes \tilde{S}_0) + \boldsymbol{\delta}_1(\mathbf{e}_{\bar{V}} \otimes S_0^{(2)} \otimes \mathbf{e}_{R_1}) + \boldsymbol{\delta}_2(\mathbf{e}_{\bar{V}} \otimes S_0^{(1)} \otimes \mathbf{e}_{R_2}).$$

Using notation for $\hat{\boldsymbol{\delta}}_n, n = 0,1,2$, introduced in the statement of the theorem, we obtain ergodicity condition (1). □

Remark 1. We can give the intuitive explanation of stability condition (3). The vectors $\hat{\boldsymbol{\delta}}_n$, $n = 0,1,2$, have the following sense: the vector $\hat{\boldsymbol{\delta}}_0$ describes the probabilities that both servers are fault-free and serve a customer, the vector $\hat{\boldsymbol{\delta}}_n$ describes the probabilities that only the server $n, n = 1,2$, serves a customer under overload condition. Than the right hand side of inequality (3) is the rate of customers leaving the system after service under overload condition while the left hand side of this inequality is the rate λ of customers arriving into the system. It is obvious that in steady state the latter rate must be less that the former one.

Remark 2. The stability condition (1) can be formulated in terms of the system load ρ as follows:

$$\rho = \frac{\lambda}{\hat{\delta}_0 \tilde{S}_0 + \hat{\delta}_1 S_0^{(2)} + \hat{\delta}_2 S_0^{(1)}} < 1.$$

Corollary 2. *In the case of stationary Poisson flow of breakdowns and exponential distribution of service and repair times, the stable condition (1)–(2) is reduced to the following inequality:*

$$\lambda < \frac{\text{æ}_1 \text{æ}_2}{\text{æ}_1 \text{æ}_1 + ph\text{æ}_2 + (1-p)h\text{æ}_1}\left(\mu_1 + \mu_2 + \frac{ph}{\text{æ}_1}\mu_2 + \frac{(1-p)h}{\text{æ}_2}\mu_1\right). \tag{11}$$

In what follows, we assume inequality (3) be fulfilled. Introduce the steady state probabilities of the chain under consideration

$$p_0^{(n)}(\nu, \eta) = \lim_{t \to \infty} P\{i_t = 0, n_t = n, \nu = \overline{0,W}, \eta = \overline{0,V}\}, \ n = 0,1,2,$$

$$p_i^{(0)} = \lim_{t \to \infty} P\{i_t = i, n_t = 0, \nu_t = \nu, \eta_t = \eta, m_t^{(1)} = m^{(1)}, m_t^{(2)} = m^{(2)}),$$

$$p_i^{(1)} = \lim_{t \to \infty} P\{i_t = i, n_t = 1, \nu_t = \nu, \eta_t = \eta, m_t^{(2)} = m^{(2)}, r_t^{(1)} = r^{(1)}),$$

$$p_i^{(2)} = \lim_{t \to \infty} P\{i_t = i, n_t = 2, \nu_t = \nu, \eta_t = \eta, m_t^{(1)} = m^{(1)}, r_t^{(2)} = r^{(2)}),$$

$$i > 0, \nu = \overline{0, W}, \eta = \overline{0, V}, m^{(k)} = \overline{1, M_k}, r^{(k)} = \overline{1, R_k}, k = 1, 2.\}.$$

Let us enumerate the steady state probabilities in accordance with the intro-duced above order of the states of the Markov chain ξ_t and form the row vectors $\mathbf{p}_i$ of steady state probabilities corresponding the value i of the first component of the chain, $i \geq 0$.

To calculate the vectors $\mathbf{p}_i$, $i \geq 0$, we use the algorithm for calculating the stationary distribution of QBD process, see [10].

Algorithm

1. Calculate the matrix R as the minimal nonnegative solution of the non-linear matrix equation

$$R^2 Q_0 + R Q_1 + Q_2 = O.$$

2. Calculate the vector $\mathbf{p}_1$ as the unique solution of the system

$$\mathbf{p}_1[Q_1 + Q_{1,0}(-Q_{0,0})^{-1}Q_{0,1} + RQ_0] = \mathbf{0},$$

$$\mathbf{p}_1[\mathbf{e} + Q_{1,0}(-Q_{0,0})^{-1}\mathbf{e} + R(I - R)^{-1}\mathbf{e}] = 1.$$

3. Calculate the vectors $\mathbf{p}_0$, $\mathbf{p}_i$, $i \geq 2$, as follows:

$$\mathbf{p}_0 = \mathbf{p}_1 Q_{1,0}(-Q_{0,0})^{-1}, \quad \mathbf{p}_i = \mathbf{p}_1 R^{i-1}, i \geq 2.$$

5   Stationary performance measures

Having the stationary distribution $\mathbf{p}_i$, $i \geq 0$, been calculated we can find a number of stationary performance measures of the system. Some of them are listed below.

- Throughput of the system

$$\varrho = \delta_0 \tilde{S}_0 + \hat{\delta}_1 S_0^{(2)} + \hat{\delta}_2 S_0^{(1)}.$$

In case of exponential distributions of service and repair times the throughput of the system is calculated as

$$\varrho = \frac{\text{æ}_1 \text{æ}_2}{\text{æ}_1 \text{æ}_1 + ph\text{æ}_2 + (1-p)h\text{æ}_1}(\mu_1 + \mu_2 + \frac{ph}{\text{æ}_1}\mu_2 + \frac{(1-p)h}{\text{æ}_1}\mu_1).$$

- Mean number of customers in the system $L = \sum_{i=1}^{\infty} i\mathbf{p}_i\mathbf{e}.$

- Variance of the number of customers in the system $V = \sum\limits_{i=1}^{\infty} i^2 \mathbf{p}_i \mathbf{e} - L^2$.
- Probability that the system is idle and both servers are fault-free

$$P_0^{(0)} = \mathbf{p}_0 \begin{pmatrix} \mathbf{e}_a \\ \mathbf{0}_{a(R_1+R_2)} \end{pmatrix},$$

- Probability that the system is idle and the server 1 (server 2) is under repair

$$P_0^{(1)} = \mathbf{p}_0 \begin{pmatrix} \mathbf{0}_a \\ \mathbf{e}_{aR_1} \\ \mathbf{0}_{aR_2} \end{pmatrix}, P_0^{(2)} = \mathbf{p}_0 \begin{pmatrix} \mathbf{0}_{a(1+R_1)} \\ \mathbf{e}_{aR_2} \end{pmatrix},$$

- Probabilities that there are $i, i > 0$, customers in the system and both server are fault-free (serve a customer) at an arbitrary time and at an arrival epoch

$$P_i^{(0)} = \mathbf{p}_i \begin{pmatrix} \mathbf{e}_{aM_1M_2} \\ \mathbf{0}_{a(M_2R_1+M_1R_2)} \end{pmatrix}, P_{arrival}^{(0,i)} = \lambda^{-1} \mathbf{p}_i \begin{pmatrix} I_{\bar{W}} \otimes \mathbf{e}_{\bar{V}M_1M_2} \\ \mathbf{0}_{a(M_2R_1+M_1R_2) \times \bar{W}} \end{pmatrix} D_1 \mathbf{e}.$$

- Probabilities that there are $i, i > 0$, customers in the system, the server 1 is under repair and the server 2 serves a customer at an arbitrary time and at an arrival epoch

$$P_i^{(1)} = \mathbf{p}_i \begin{pmatrix} \mathbf{0}_{aM_1M_2} \\ \mathbf{e}_{aM_2R_1} \\ \mathbf{0}_{aM_1R_2} \end{pmatrix}, P_{arrival}^{(1,i)} = \lambda^{-1} \mathbf{p}_i \begin{pmatrix} \mathbf{0}_{aM_1M_2 \times \bar{W}} \\ I_{\bar{W}} \otimes \mathbf{e}_{\bar{V}M_2R_1} \\ \mathbf{0}_{aM_1R_2 \times \bar{W}} \end{pmatrix} D_1 \mathbf{e}.$$

- Probability that there are $i, i > 0$, customers in the system, the server 2 is under repair and the server 1 serves a customer at an arbitrary time and at an arrival epoch

$$P_i^{(2)} = \mathbf{p}_i \begin{pmatrix} \mathbf{0}_{aM_1M_2} \\ \mathbf{0}_{aM_2R_1} \\ \mathbf{e}_{aM_1R_2} \end{pmatrix}, P_{arrival}^{(2,i)} = \lambda^{-1} \mathbf{p}_i \begin{pmatrix} \mathbf{0}_{a(M_1M_2+M_2R_1) \times \bar{W}} \\ I_{\bar{W}} \otimes \mathbf{e}_{\bar{V}M_1R_2} \end{pmatrix} D_1 \mathbf{e}.$$

- Probability that the servers are in the state n at an arbitrary time and at an arrival epoch

$$P_n = \sum\limits_{i=0}^{\infty} P_i^{(n)}, \hat{P}_n = \sum\limits_{i=0}^{\infty} P^{(n,i)}, n = 0,1,2.$$

- Probability that an arriving breakdown sees both servers fault-free and will be directed to the kth server

$$P_{break}^{(k)} = h^{-1}[\delta_{1,k}p + \delta_{2,k}(1-p)]\mathbf{p}_i \begin{pmatrix} \mathbf{e}_{\bar{W}} \otimes I_{\bar{V}} \otimes \mathbf{e}_{M_1M_2} \\ \mathbf{0}_{a(M_2R_1+M_1R_2) \times \bar{W}} \end{pmatrix} H_1 \mathbf{e}, k = 1,2,$$

where $\delta_{i,j}$ is Kronecker's symbol.

6   Numerical Examples

In this section, we demonstrate feasibility of the developed algorithms and investigate numerically the behavior of the mean number of customers in the system depending on the parameters of the system. To this end, we present the results of two numerical experiments .

Experiment 1. In this experiment, we investigate the behavior of the mean number of customers in the system, L, as a function of input rate λ under the different rates of breakdowns, h.

Define the parameters of the system.

The MAP of customers is defined by the following matrices:

$$D_0 = \begin{pmatrix} -1.349076 & 1.09082 \times 10^{-6} \\ 1.09082 \times 10^{-6} & -0.043891 \end{pmatrix}, \quad D_1 = \begin{pmatrix} 1.340137 & 0.008939 \\ 0.0244854 & 0.0194046 \end{pmatrix}.$$

This MAP has the coefficient of variation $c_{var} = 3.1$ and the coefficient of correlation $c_{cor} = 0.4$.

The MAP of breakdowns is defined by the following matrices:

$$H_0 = \begin{pmatrix} -8.110725 & 0 \\ 0 & -0.26325 \end{pmatrix}, \quad H_1 = \begin{pmatrix} 8.0568 & 0.053925 \\ 0.146625 & 0.116625 \end{pmatrix}.$$

For this MAP $c_{var} = 3.5$, $c_{cor} = 0.2$.

The service time distribution at the server 1 is assumed to be Erlangian of order 2 with parameter 20. This distribution is defined by the vector $\beta = (1,0)$ and the matrix $S = \begin{pmatrix} -20 & 20 \\ 0 & -20 \end{pmatrix}$. The service rate $\mu = 10$ and the coefficient of variation $c_{var} = 0.7$.

The service time distribution at the server 2 is assumed to be Erlangian of order 2 with parameter 150. This distribution is defined by the vector $\beta = (1,0)$ and the matrix $S = \begin{pmatrix} -15 & 15 \\ 0 & -15 \end{pmatrix}$. The service rate $\mu = 7.5$ and the coefficient of variation $c_{var} = 0.7$.

The repair time of the server 1 and the server 2 has hyper-exponential distribution of order 2. It is defined by the vector $(0.05, 0.95)$ and the matrix $T^{(1)} = \begin{pmatrix} -0.003 & 0 \\ 0 & -0.245 \end{pmatrix}$. The coefficient of repair time variation $c_{var} = 5$.

Let us vary the MAP fundamental rate λ by multiplying the matrices D_0, D_1 by a certain positive number. In this way, any desired value of λ can be obtained while the coefficient of correlation is not changed. Similarly, we obtain three different values of $h(h = 0.0001, h = 0.001, h = 0.001)$ by multiplying the matrices H_0, H_1 by a positive numbers.

Figure 1 shows the dependence of the mean number of customers in the system, L, on the fundamental rate λ under the different values of breakdown rate h. It is seen from Fig. 1, that the mean L expectable increases with λ and h increasing and the rate of increasing grows with increasing λ. It is worth to note

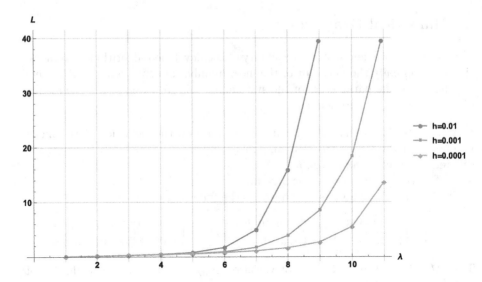

Fig. 1. Mean number of customers in the system, L, as a function of input rate λ for different values of breakdowns rate h

that the curves in Fig. 1 are terminated when λ approaches to the point where the load coefficient ρ becomes sufficiently large (greater than 0.78). To make clear this fact, we present in Table 1 the values of λ, L and the load coefficient ρ corresponding to the curves in Fig. 1.

Experiment 2. In this experiment, we are interested in how the correlation in the input flow impacts on the mean number of customers in the system. To this end, we consider three $MAPs$: MAP_1, MAP_2 and MAP_3 having different coefficients of correlation.

Table 1. Experiment 1: the values of λ, h, L, ρ

λ	1.0	2.0	3.0	4.0	5.0	6.0	7.0	8.0	9.0	10.0	11.0
$h = 0.0001$											
ρ	0.071	0.143	0.214	0.287	0.355	0.426	0.498	0.569	0.639	0.711	0.780
L	0.077	0.167	0.277	0.422	0.575	0.809	1.138	1.651	2.711	5.515	13.661
$h = 0.001$											
ρ	0.072	0.144	0.215	0. 298	0.358	0.430	0.502	0.574	0.646	0.718	0.782
L	0.078	0.169	0.280	0.484	0.618	0.958	1.741	3.883	8.519	18.347	41.433
$h = 0.01$											
ρ	0.075	0.151	0.226	0.301	0.377	0.452	0.528	0.603	0.679	0.754	0.830
L	0.084	0.187	0.318	0.501	0.817	1.730	4.955	15.889	40.159	90.742	210.924

MAP_1 is defined as the stationary Poisson process. It has the coefficient of variation of inter-arrival times $c_{var} = 1$ and the coefficient of correlation $c_{cor} = 0$. We define this process by the scalars $D_0 = -1$ and $D_1 = 1$.

MAP_2 has the coefficient of variation $c_{var} = 3.5$ and the coefficient of correlation $c_{cor} = 0.2$ and is defined by the matrices

$$D_0 = \begin{pmatrix} -1.3526 & 0 \\ 0 & -0.04391 \end{pmatrix}, \quad D = \begin{pmatrix} 1.3436 & 0.009 \\ 0.02446 & 0.01945 \end{pmatrix}.$$

MAP_3 is the same as MAP in Experiment 1. It has the coefficient of variation $c_{var} = 3.5$ and the coefficient of correlation $c_{cor} = 0.4$. The MAP of breakdowns and the PH distributions of service and repair times are assumed to be the same as in Experiment 1.

Figure 2 shows the dependence of the mean L on the fundamental rate λ under the different coefficients of correlation in the MAP of customer.

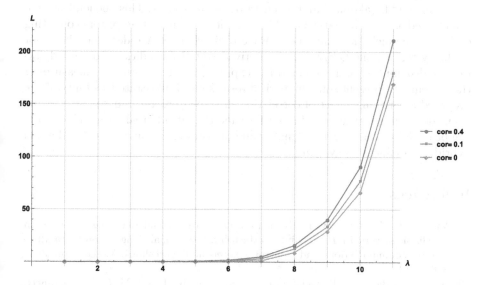

Fig. 2. Mean number of customers in the system, L, as a function of λ for MAPs with different coefficients of correlation

Besides, in Table 2 we present the values of λ, L and the load coefficient ρ corresponding to the curves in Fig. 2.

It is seen from Fig. 2, that the mean number of customers in the system, L, expectable increases when the input rate λ increases and, under the same value of λ, the mean L increases with the coefficient of correlation increasing although the load coefficient ρ does not depend on the correlation in the input flow (the latter follows from Remarks 1 and 2).

Table 2. Experiment 2: the values of λ, c_{cor}, L, ρ

λ	1.0	2.0	3.0	4.0	5.0	6.0	7.0	8.0	9.0	10.0	11.0
ρ	0.075	0.151	0.226	0.301	0.376	0.451	0.526	0.602	0.678	0.753	0.828
$BMAP_1 : c_{cor} = 0$											
L	0.082	0.177	0.288	0.423	0.597	0.846	1.393	9.254	29.481	66.723	169.774
$BMAP_2 : c_{cor} = 0.2$											
L	0.085	0.182	0.301	0.456	0.679	1.122	3.265	12.691	33.583	77.115	179.661
$BMAP_3 : c_{cor} = 0.4$											
L	0.094	0.187	0.318	0.501	0.818	1.730	4.955	15.889	40.159	90.742	210.924

7   Conclusion

In this paper, we investigate unreliable queueing system with Markovian flows of customers and breakdowns and two heterogeneous servers. This queueing system can be used as a mathematical model of hybrid communication system consisting of a laser channel and a millimeter-wave radio channel. We described behavior of the system by the QBD process, derived stability condition, computed stationary distribution and presented the expressions for performance measures of the system. We present some numerical results which illustrate feasibility of the proposed algorithms, effect of correlation in arrival process and behavior of the queue depending on the rate of input flow and the rate of breakdown. The results can be exploited for capacity planning and performance evaluations of real world hybrid communication systems.

References

1. Vishnevsky, V., Kozyrev, D., Semenova, O.V.: Redundant queueing system with unreliable servers. In: Proceedings of the 6th International Congress on Ultra Modern Telecommunications and Control Systems and Workshops, Moscow, pp. 383–386 (2014)
2. Arnon, S., Barry, J., Karagiannidis, G., Schober, R., Uysal, M. (eds.): Advanced Optical Wireless Communication Systems. Cambridge University Press, New York (2012)
3. Nadeem, F., Leitgeb, E., Kvicera, V., Grabner, M., Awan, M.S., Kandus, G.: Simulation and analysis of FSO/RF switch over for different armospheric effects. In: Proceedings of the 10th International Conference on Telecommunications, Zagreb, Croatia, pp. 39–43 (2009)
4. Letzepis, N., Nguyen, K.D., Guillen, I., Fabregas, A., Cowley, W.G.: Outage analysis of the hybrid free-space optical and radio-frequency channel. IEEE J. Sel. Areas Commun. **27**, 1709–1719 (2009)
5. Vishnevsky, V.M., Semenova, O.V., Sharov, S.Y.: Modeling and analysis of a hybrid communication channel based on free-space optical and radio-frequency technologies. Autom. Remote Control **72**, 345–352 (2013)

6. Dudin, A., Klimenok, V., Vishnevsky, V.: Analysis of unreliable single server queueing system with hot back-up server. Commun. Comput. Inf. Sci. **499**, 149–161 (2015)
7. Sharov, S.Y., Semenova, O.V.: Simulation model of wireless channel based on FSO and RF technologies. In: Distributed Computer and Communication Networks. Theory and Applications, pp. 368–374. Moscow (2010)
8. Klimenok, V., Vishnevsky, V.M.: Unreliable queueing system with cold redundancy. In: Gaj, P., Kwiecień, A., Stera, P. (eds.) CN 2015. Communications in Computer and Information, vol. 522, pp. 336–346. Springer, Cham (2015). doi:10.1007/978-3-319-19419-6_32
9. Lucantoni, D.M.: New results on the single server queue with a batch Markovian arrival process. Commun. Stat. Stoch. Models **7**, 1–46 (1991)
10. Neuts, M.: Matrix-geometric Solutions in Stochastic Models - An Algorithmic Approach. Johns Hopkins University Press, Baltimore (1981)

On the Total Customers' Capacity
in Multi-server Queues

Ekaterina Lisovskaya[1]([✉]), Svetlana Moiseeva[1], and Michele Pagano[2]

[1] Tomsk State University, Tomsk, Russia
{ekaterina_lisovs,smoiseeva}@mail.ru
[2] University of Pisa, Pisa, Italy
m.pagano@iet.unipi.it

Abstract. In this paper we consider a generalization of $M/GI/N/\infty$ queues, in which customer capacity is an additional parameter of the system and it is independent of the service time. In more detail we focus on the distributions of the total capacity of customers in the different elements of the queue (waiting line, service and entire system) and provide approximate expressions for the corresponding characteristic functions. To verify the goodness of the proposed approximation, several sets of simulations have been carried out, considering discrete and continuous distributions of the customer capacity and using the Kolmogorov distance as a measure of similarity.

Keywords: N-server queuing system · Customer with random capacity · Approximation of the probability distribution

1 Introduction

Queuing theory is one of the most relevant branches of probability theory and applied mathematics [3,6,12,13]. Indeed, queuing systems represent a powerful mathematical tool for performance analysis of a wide variety of real-life systems, including, for instance, telecommunication networks, financial markets, supply chain management and airplane traffic control.

In many application customers are simply characterized in terms of arrival and service processes [1,8,9]. For instance, in computer networks it is typically assumed that the customer volume (i.e., the packet length) is proportional to the service time (namely, the time needed to transmit the packet itself). In this work, we consider a more general model and assume that customer volume and service time are described by independent random variables with arbitrary distributions. As highlighted in the next section, customer capacity plays a relevant role in modeling new network architectures.

In more detail, we consider a queuing system with Poisson arrivals, N servers and unlimited capacity (such assumption is widely used in modeling for sake of analytical tractability). Extending the approach developed by some of the authors in [4] (in which an approximate expression for the distribution of the

© Springer International Publishing AG 2017
A. Dudin et al. (Eds.): ITMM 2017, CCIS 800, pp. 56–67, 2017.
DOI: 10.1007/978-3-319-68069-9_5

number of customers was derived), we will be able to find an explicit approximation for the distribution of the customers' total capacity in the queuing system as well as in its elements (waiting line and service).

The rest of the paper is organized as follows. In Sect. 2, we review the most relevant works on queuing systems with random capacity of customers and highlight the novelty of our contribution. Section 3 properly defines the analyzed queuing model and recalls an approximation for the probability distribution of the number of customers in the system, while Sect. 4 presents the original contribution of the paper, i.e. provides a general expression for the characteristic function of the total customers' capacity. Then, in Sect. 5 the goodness of the approximation (in terms of Kolmogorov distance) is verified through discrete-event simulations for different values of the system parameters. Finally, Sect. 6 concludes the paper with some final remarks.

2 Related Work

In recent years queuing systems with random customer capacity have attracted the interest of researchers for their applicability in different fields, mainly in the framework of computer networks. In this section some of the most relevant contributions are discussed.

In the paper [2] an efficient analytical model that evaluates the behavior of the downlink LTE (Long-Term Evolution) channel with CLA (Cross-Layer Adaptation) is presented. Since video traffic is resource–intensive, it is a challenging issue to stream video over low bandwidth networks, whereas video communication over LTE becomes an open research topic nowadays due to LTEs high throughput capabilities.

The paper [11] deals with a model of a multi-server queuing system with losses caused by lack of resources necessary to service claims. A claim accepted for servicing occupies a random amount of resources of several types with given distribution functions. Random vectors that define the requirements of claims for resources are independent of the processes of customer arrivals and servicing, mutually independent, and identically distributed. Under the assumptions of a Poisson arrival process and exponential service times, the authors analytically find the joint distribution of the number of customers in the system and the vector of amounts of resources occupied by them. Moreover, sample computations are presented to illustrate an application of the model to analyzing the characteristics of a videoconferencing service in an LTE wireless network.

In [10] the authors consider queuing systems, in which customers occupy some resources that are released after customer departure. Arriving customers are lost if there are not enough free resources required for their servicing. In such systems for each customer it is necessary to record the vector of occupied resources until its departure.

Multi-server queuing systems with AQM-type (Active Queue Management) mechanisms are considered in [16,17]. In more detail, in the first work M/M/n-type ($n \geq 1$) queuing systems with a bounded total volume and finite queue size are considered. It is assumed that the volumes of the arriving packets are generally distributed random variables. Moreover, an AQM-type (Active Queue Management) mechanism is used to control the actual buffer state: each of the arriving packets is dropped with a probability depending on its volume and the occupied volume of the system at the pre-arrival epoch. The stationary queue-size distribution and the loss probability are derived, and numerical examples illustrating theoretical results are also provided. Then, in [17] the analysis is extended to the case of arbitrary distribution of the service time.

The main aim of the paper [14] is to develop a simulation model for queuing systems with non-priority cyclic service RR (round robin) discipline and to compare, in terms of queuing performance, such service discipline with traditional FCFS (first come-first served).

Finally, the paper [15] investigates single server queuing systems with batch Poisson arrivals and without demands losses under assumption that each demand has some random capacity (generally, each demand is characterized by an l-dimensional indication vector). Service time of the demand arbitrary depends on its capacity (indications). The Laplace-Stieltjes transform of total capacities (random vector of sum of indications) of demands that were served during a busy period of the system is determined.

The main novelty of our approach is that it deals with systems without losses and, in this way, permits to dimension the system resources (in terms of buffer space) in order to have loss probabilities below any given threshold (as well-known in the literature, the complementary probability provides an upper bound for the loss probability in the corresponding finite-buffer system). Moreover, our approach is quite general and may be applied to any distribution (discrete or continuous) of the customer capacity, provided that its characteristic function is well-defined. Finally, we also provide the distribution of the overall capacity for the customers in the different components of the queue (waiting line and buffer); such information may be useful to dimension the different elements of the real system under analysis.

3 Approximation of Probability Distribution of the Customers' Number in the System

We consider the $M/GI/N/\infty$ queue. The arrival process is distributed by Poisson law with rate λ. The system has N servers and service times on each server are i.i.d. with distribution function $A(x)$. The arriving customer occupies any free server or goes to the queue in case of all servers are busy. Let each customer have some random capacity $v > 0$ with distribution function $G(y)$. Customers' capacities and service times are mutually independent and do not dependent on the epochs of customers' arrivals.

Denote by $i(t)$ and $V(t) = \sum_{i=1}^{i(t)} v_i$ the number of customers in the system at time t and their total capacity, respectively.

Let $P(i) = P\{i(t) = i\}$ be the stationary probability distribution of the number of customers in the system. We denote by π_i an approximation of $P(i)$, which is defined as a composite distribution [4]:

$$\pi_i = \begin{cases} C_1 P_1(i), 0 \leq i \leq N, \\ C_2 P_2(i - N + 1), i \geq N. \end{cases} \tag{1}$$

Note that the equality of the two expression for $i = N$ provides an additional condition to determine the constants C_1 and C_2.

The probabilities $P_1(i)$, where $0 \leq i \leq N$, are the probabilities of the number of occupied servers in an N-server M/GI/N/0 queue with customer losses when all servers are busy. Hence, they can be determined by the Erlang B formula:

$$P_1(i) = \frac{(\lambda a)^i}{i!} \left(\sum_{k=0}^{N} \frac{(\lambda a)^k}{k!} \right)^{-1}$$

where a is the mean service time.

The probabilities $P_2(i)$ refers to states in which all servers are busy. In this case, the block of occupied servers is considered as a single one, characterized by an equivalent service time distribution $B(x)$ and an equivalent mean service time b. Therefore, the probabilities $P_2(i)$, where $i \geq 1$, are defined as the probabilities of having i customers in a single-server queuing system with waiting (i.e., the classical M/GI/1 queue). Hence, they can be determined by the Pollaczek-Khinchin formula [4] and we can write

$$P_2(i) = (1 - \lambda b) \sum_{k=0}^{i} \alpha_k b_{i-k},$$

where the coefficients of the expansion are given by

$$\alpha_0 = \frac{1}{\beta_0}, \quad \alpha_n = \frac{1}{\beta_0} \left[\alpha_{n-1} - \sum_{k=0}^{n-1} \alpha_k \beta_{n-k} \right],$$

$$b_0 = \beta_0, \quad b_n = \beta_n - \beta_{n-1},$$

$$\beta_n = \int_0^{\infty} e^{-\lambda z} \frac{(\lambda z)^n}{n!} dB(z),$$

and the distribution function $B(x)$ has the form

$$B(x) = 1 - (1 - A(x)) \left(1 - \frac{1}{a} \int_0^x (1 - A(z)) dz \right)^{N-1}.$$

The constants C_1 and C_2 in (1) can be found from the normalization condition and the conditions of "stitching" [4]. So the expression (1) becomes:

$$\pi_i = \begin{cases} \dfrac{P_2(1)}{P_2(1) + P_1(N)(1 - (P_2(0) + P_2(N)))} P_1(i), 0 \leq i \leq N, \\ \dfrac{P_1(N)}{P_2(1) + P_1(N)(1 - (P_2(0) + P_2(N)))} P_2(i - N + 1), i \geq N. \end{cases} \tag{2}$$

4   Characteristic Function for the Total Capacity

Starting from the definition of conditional expectation, we can write the characteristic function of the total capacity in the form

$$
h(u) = M\left\{e^{juV(t)}\right\} = M\left\{M\left\{e^{ju\sum\limits_{k=1}^{i}v_k}\,\middle|\,i(t)=i\right\}\right\}
$$

$$
= \sum_{i=0}^{\infty} M\left\{e^{ju\sum\limits_{k=1}^{i}v_k}\right\}P\{i(t)=i\} = \sum_{i=0}^{\infty}\left(M\left\{e^{juv}\right\}\right)^{i}P\{i(t)=i\},
$$

where we took into account that for $i = 0$ the queue is empty and the sum at the exponent is 0.

Then, using approximation (2), the characteristic function can be rewritten as

$$
h(u) = \sum_{i=0}^{\infty}\left(M\left\{e^{juv}\right\}\right)^{i}\pi_i,
$$

and, taking the inverse Fourier transform, we obtain an approximation of the density function of the customers' total capacity in the M/GI/N/∞ queue:

$$
f_V(x) = \int_{-\infty}^{\infty} e^{-jux}h(u)du. \tag{3}
$$

Similarly, we can obtain the characteristic functions of the total capacity of the customers in the service and in the waiting line. These results have practical relevance since the customers in each element of the queue typically require specific resources (for instance, in routers there is a physical separation between input buffers and output ports).

In more detail, for the customers in the service we obtain:

$$
h_{serv}(u) = \sum_{i=0}^{N}\left(M\left\{e^{juv}\right\}\right)^{i}\frac{P_2(1)P_1(i)}{P_2(1)+P_1(N)\left(1-(P_2(0)+P_2(N))\right)},
$$

and for the customers in the waiting queue:

$$
h_{wait}(u) = \sum_{i=0}^{\infty}\left(M\left\{e^{juv}\right\}\right)^{i+N}\frac{P_1(N)P_2(i+1)}{P_2(1)+P_1(N)\left(1-(P_2(0)+P_2(N))\right)}.
$$

5   Simulation and Numerical Examples

Several simulation experiments, performed in the same way as [5], have been carried out to estimate the distribution function of the customers number and the total customers capacity and verify the goodness of the proposed approximation. To this aim, it was also necessary to calculate numerically the approximations (2) and (3) since a close-form solution is, in general, not available.

As a measure of the similarity between simulation and approximation results, we consider the Kolmogorov distance

$$\Delta = \sup_x |F(x) - D(x)|.$$

Here $F(x)$ represents the approximation based on (2) or (3), respectively for $i(t)$ and $V(t)$, and $D(x)$ is the cumulative distribution function built on the basis of the simulation results (in order to reduce the variance of the estimates 10^{10} arrivals have been generated). As typically done in the literature [7], we suppose that an approximation is applicable if its Kolmogorov distance is less than 0.03.

In the following we present the result for three different scenarios, in order to highlight the applicability of our approximation in different settings. Note that the parameters for the arrival and service processes were selected in such a way that the condition for the stationary regime existence is always met $(N > \lambda a)$.

Example 1. Let us consider the following parameters for the queue:

- arrival rate $\lambda = 25$
- number of servers $N = 10$
- exponential distribution (with parameter μ) of the service time, i.e.

$$A(x) = \begin{cases} 1 - e^{-\mu x}, & x \geq 0, \\ 0, & x < 0, \end{cases}$$

- uniform distribution (in the interval $[a, b]$) of customers' capacity, i.e.

$$G(y) = \begin{cases} 0, & y < a, \\ \dfrac{y - a}{b - a}, & a \leq y \leq b, \\ 1, & y > b. \end{cases}$$

Furthermore, we used the following numerical values: $\mu = 5$, $a = 0$ and $b = 1$. It is easy to verify that the distributions are very similar both for the customer numbers and the total capacity, as highlighted by Figs. 1 and 2; indeed, we obtain that $\Delta_i = 0.007$ and $\Delta_V = 0.012$, respectively for $i(t)$ and $V(t)$.

Example 2. In the second set of simulation we changed the distribution of the service time. In more detail, the parameters of the queuing system are as follows:

- arrival rate $\lambda = 25$
- number of servers $N = 10$

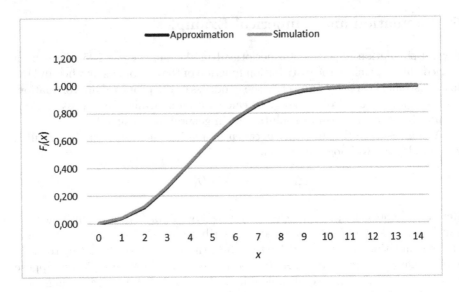

Fig. 1. Example 1 – Distributions of the customers number

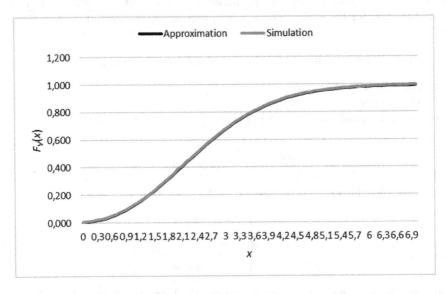

Fig. 2. Example 1 – Distributions of the total capacity

– gamma distribution (with parameters α and β) of the service time, i.e.

$$A(x) = \begin{cases} \dfrac{\gamma(\alpha, \beta x)}{\Gamma(\alpha)}, & x \geq 0, \\ 0, & x < 0, \end{cases}$$

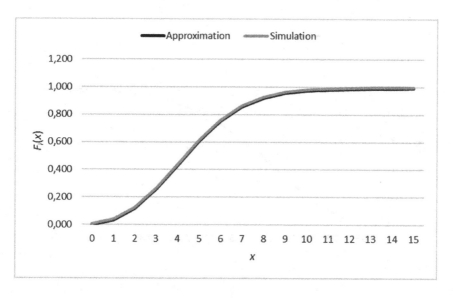

Fig. 3. Example 2 – Distributions of the customers number

- uniform distribution (in the interval $[a, b]$) of customers' capacity, i.e.

$$G(y) = \begin{cases} 0, & y < a, \\ \dfrac{y - a}{b - a}, & a \le y \le b, \\ 1, & y > b. \end{cases}$$

In this case (with $\alpha = 0.5$, $\beta = 2.5$ and, as before, $a = 0$, $b = 1$), the approximation is even closer since $\Delta_i = 0.009$ and $\Delta_V = 0.007$ (see Figs. 3 and 4).

Example 3. In the third set of simulations we verified the goodness of the approximation in case of discrete distribution of the customer capacity. In more detail, we considered the following set of parameters:

- arrival rate $\lambda = 45$
- number of servers $N = 6, 7, 8$
- gamma distribution (with parameters α and β) of the service time, i.e.

$$A(x) = \begin{cases} \dfrac{\gamma(\alpha, \beta x)}{\Gamma(\alpha)}, & x \ge 0, \\ 0, & x < 0, \end{cases}$$

- geometric distribution (in the form representing the number of failures before the first success, with parameter p) of customers' capacity:

$$G(y) = P\{v = y\} = p(1 - p)^y.$$

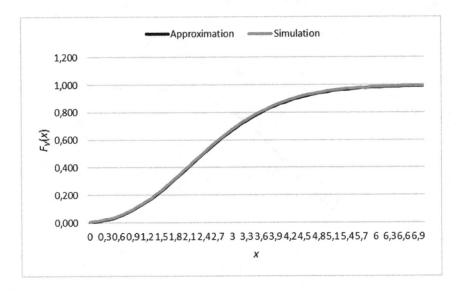

Fig. 4. Example 2 – Distributions of the total capacity

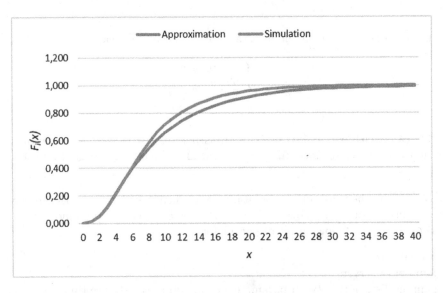

Fig. 5. Example 3 ($N = 6$) – Distributions of the customers number

In all the scenarios we assumed $\alpha = 3.5$, $\beta = 29.7$, $p = 0.4$ and checked how the value of N influences the goodness of the approximation. Figures 5 and 6 points out that the approximation is rather poor for $N = 6$ (indeed, in this case the values of the Kolmogorov distance are $\Delta_i = 0.064$ and $\Delta_V = 0.048$), while it improves when the number of servers is increased, as highlighted by the corresponding values of the Kolmogorov distance (namely $\Delta_i = 0.029$ and

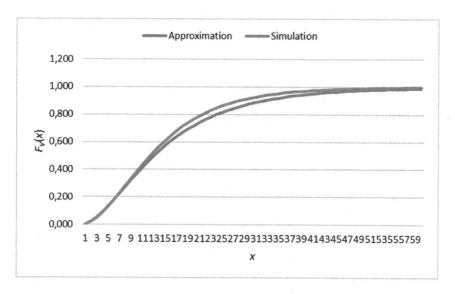

Fig. 6. Example 3 ($N = 6$) – Distributions of the total capacity

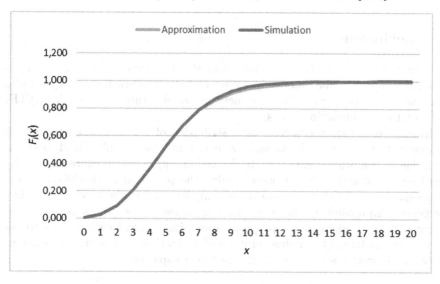

Fig. 7. Example 3 ($N = 8$) – Distributions of the customers number

$\Delta_V = 0.016$ for $N = 7$, $\Delta_i = 0.017$ and $\Delta_V = 0.005$ for $N = 8$) that are clearly below the admissibility threshold. Finally, a visual evidence of the goodness of the proposed approximation is provided by Figs. 7 and 8, referring to $N = 8$ (for sake of brevity, the graphs for $N = 7$ are omitted).

We can conclude that the accuracy of the total capacity approximation is suitable over a wide range of system parameters and improves with the increase of the number of servers in the system.

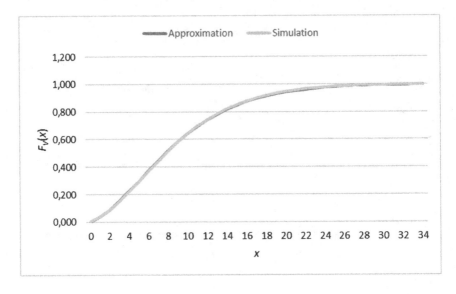

Fig. 8. Example 3 ($N = 8$) – Distributions of the total capacity

6   Conclusions

In this paper we analyzed a generalization of $M/GI/N/\infty$ queues with customers of random capacity. Such models present not only theoretical interest, but also practical relevance in modeling new network architectures (eg., CLA in LTE) and AQM mechanisms in queues.

In more detail we considered the distribution of the total capacity of customers in the system and, starting from our previous results in [4] and the definition of conditional expectation, derived an approximate expression for its characteristic function. Then, we extended the proposed methodology to the total capacity of the customers in the waiting line and in the service, providing the general expressions of the corresponding characteristic functions.

Finally, the goodness of the proposed approximation was verified (in terms of Kolmogorov distance) through several sets of simulations, considering continuous as well as discrete distributions of the customer capacity.

References

1. Apachidi, X.N., Katsman, Y.: Development of a queuing system with dynamic priorities. Key Eng. Mater. **685**, 934–938 (2016)
2. Efimushkina, T., Gabbouj, M., Samuylov, K.: Analytical model in discrete time for cross-layer video communication over LTE. Autom. Control Comput. Sci. **48**(6), 345–357 (2014)
3. Fedorova, E.: The second order asymptotic analysis under heavy load condition for retrial queueing system MMPP/M/1. In: Dudin, A., Nazarov, A., Yakupov, R. (eds.) ITMM 2015. CCIS, vol. 564, pp. 344–357. Springer, Cham (2015). doi:10. 1007/978-3-319-25861-4_29

4. Lisovskaya, E., Moiseeva, S.: Study of the Queuing Systems $M/GI/N/\infty$. Commun. Comput. Inf. Sci. **564**, 175–184 (2015)
5. Lisovskaya, E., Pagano, M.: Imitacionnoe modelirovanie sistemy massovogo obsluzhivaniya trebovanij sluchajnogo ob"ema. Problemy optimizacii slozhnyh sistem: Trudy 12-j Mezhdunarodnoj Aziatskoj shkoly-seminara, 352–357 (in Russian)(2016)
6. Moiseev, A., Nazarov, A.: Queueing network $MAP/(GI/\infty)^K$ with high-rate arrivals. Eur. J. Oper. Res. **254**(2), 161–168 (2016)
7. Moiseev, A., Sinyakov, M.: Razrabotka ob'ektno-orientirovannoj modeli sistemy imitacionnogo modelirovaniya processov massovogo obsluzhivaniya. Vestnik Tomskogo gosudarstvennogo universiteta. Upravlenie, vychislitel'naya tekhnika i informatika 1, 89–93 (In Russian)(2010)
8. Moiseev, A.: Asymptotic Analysis of the Queueing Network $SM/(GI/\infty)^K$. Commun. Comput. Inf. Sci. **564**, 73–84 (2015)
9. Moiseeva, S., Zadiranova, L.: Feedback in infinite-server queuing systems. In: Vishnevsky, V., Kozyrev, D. (eds.) DCCN 2015. CCIS, vol. 601, pp. 370–377. Springer, Cham (2016). doi:10.1007/978-3-319-30843-2_38
10. Naumov, V.A., Samuilov, K.E.: On Modeling Queueing Systems with Multiple Resources. Vestn. Ross. Univ. Druzhby Narodov, Ser. Mat. Informatika. Fiz. 3, 60–64 (2014)
11. Naumov, V.A., Samuilov, K.E., Samuilov, A.K.: On the total amount of resources occupied by serviced customers. Autom. Remote Control **77**(8), 1419–1427 (2016)
12. Nazarov, A., Broner, V.: Inventory management system with Erlang distribution of batch sizes. In: Dudin, A., Gortsev, A., Nazarov, A., Yakupov, R. (eds.) ITMM 2016. CCIS, vol. 638, pp. 273–280. Springer, Cham (2016). doi:10.1007/978-3-319-44615-8_24
13. Pankratova, E., Moiseeva, S.: Queueing system $GI/GI/\infty$ with n types of customers. Commun. Comput. Inf. Sci. **564**, 216–225 (2015)
14. Raspopov, A., Katsman, Y.Y.: Resource allocation algorithm modeling in queuing system based on quantization. Key Eng. Mater. **685**, 886–891 (2016)
15. Tikhonenko, O., Kawecka, M.: Busy period characteristics for single server queue with random capacity demands. In: Kwiecień, A., Gaj, P., Stera, P. (eds.) CN 2012. CCIS, vol. 291, pp. 393–400. Springer, Heidelberg (2012). doi:10.1007/978-3-642-31217-5_41
16. Tikhonenko, O., Kempa, W.M.: On the queue-size distribution in the multi-server system with bounded capacity and packet dropping. Kybernetika **49**(6), 855–867 (2013)
17. Tikhonenko, O., Kempa, W.M.: Performance evaluation of an $M/G/n$-type queue with bounded capacity and packet dropping. Int. J. Appl. Math. Comput. Sci. **26**(4), 841–854 (2016)

On the Problems of Queues in Mixed Type Queuing Systems with Random Quantity of Sources and Size-Limited Queues

Alexander Kirpichnikov and Anton Titovtsev[✉]

Kazan National Research Technological University,
K. Marksa str. 68, 420015 Kazan, Russia
kirpichnikov@kstu.ru, notna6683@mail.ru
http://www.kstu.ru

Abstract. The article proposes the technique to investigate the behavior of the moments of numerical characteristics of mixed-type queuing system with a random number of sources upon the change of demands input stream intensity and size-limited queues based on the calculation of boundary values of the number of servicing devices at which the mean squared deviation (MSD) of the investigated quantity does not exceed its mathematical expectation. For the first time the linear nature of behavior of boundary values of the number of service facilities with the change of the given intensity of demands input stream is determined numerically. The article also considers various types of queues arising in queuing systems. The concept of an N-th order queue is introduced, and generalized Little's formulas for N-th order queues in queuing systems of various types are presented.

Keywords: Queue · Physical queue · Real queue · Quality of service (QoS) · Queuing system · M/M/m/K · Service facility

1 Introduction

Issues of studying combined models of queuing originate from Cohen's works (Cohen J.W.) [1], where the combination of Erlang models and classical queuing system was considered for the first time. A number of formulae for probabilities of queuing system (QS) steady states, call loss probability, and first moments of demands number in a queue and waiting time in a queue are given in the paper.

Another specific case of a combined model is a mixed system with losses and expectation having some servers and finite memory, presented in the work of H. Takagi [2]. In this case there are two sources of demands in the system, thus demands from the first source will be lost if all servers are busy at the time of their arrival in the system. Demands from the second source are accepted in a queue only if the number of demands in it does not exceed some defined value K. Streams of demands arriving in the system also have a Poisson character. Formulae for probabilistic characteristics of the system and for the moments of

© Springer International Publishing AG 2017
A. Dudin et al. (Eds.): ITMM 2017, CCIS 800, pp. 68–82, 2017.
DOI: 10.1007/978-3-319-68069-9_6

n order of waiting time and common delay time in the system are given in the paper. In the specific case K $\to \infty$, this model is reduced to J. Cohen's model.

A more general model of a queuing system which is a combination of a multi-channel Erlang model, M/M/m/E model, and also multi-channel classical model (M/M/m models) is considered in the work of authors [3]. A complete formula derivation for probabilistic characteristics, and also for the first and second moments of numerical and temporary characteristics of this type of a queuing system is presented in work [4]; a general algorithm of queuing models mathematical formalization taken from monographs [5,6] is used.

A mathematical model of an open multi-channel system of queuing having m service facilities of identical efficiency with exponentially distributed service time is presented in this paper. A demand input stream in this case is a superposition of components'random number h, each of which represents a Poisson stream of demands served in the order of arrival. For each type of demands entering the system from the j-th source there is a specific size-limited queue ε_j where $\varepsilon_0 < \varepsilon_1 < \varepsilon_2 < \cdots < \varepsilon_h$.

A zero (Erlang) component contains demands which are served only if there is at least one free service facility, and they never stand in a queue. In the case, if at the time when the next similar demand arrives in the system there is no free service facility this demand is refused and leaves the system unserved. The model of a queuing system, containing one such component in an input stream, is the Erlang model; therefore we will call this component an Erlang component.

The first component includes demands which are served if there is a free service facility, or they stand in a queue if the number of demands in the queue is fewer than a particular number ε_1. In case when there is already available ε_1 or more demands in a queue, a newly arrived demand from the first source is refused and leaves the system unserved.

The second component contains demands which are served if there is a free service facility, or they stand in a queue if the number of demands in a queue is fewer than a particular number $\varepsilon_2 > \varepsilon_1$. In the case when ε_2 or more demands are already available in the queue, an arrived demand from this source is refused and leaves the system unserved, and so on.

In general, the h-th component includes demands which are served if there is a free service facility, or they stand in a queue if the number of demands in the queue are fewer than a particular number $\varepsilon_h > \varepsilon_{h-1} > \cdots > \varepsilon_1$. In case when there are already ε_h demands in the queue, a newly arrived demand from the h-th source is refused and leaves the system unserved.

Let us accept the following designations:

$$\varepsilon_0 = E_0 = 0; \; \varepsilon_1 = E_1; \; \varepsilon_2 = E_1 + E_2; \; \cdots \; \varepsilon_j = \sum_{i=0}^{j} E_i = \sum_{i=1}^{j} E_i; \text{ a size-limited}$$

queue (memory volume) for demands of the j-th component;

$$\Lambda_0 = \sum_{j=0}^{h} \lambda_j; \; \Lambda_1 = \sum_{j=1}^{h} \lambda_j; \; \Lambda_2 = \sum_{j=2}^{h} \lambda_j; \; \cdots \; \Lambda_h = \lambda_h; \text{ where } \lambda_j \text{ demand}$$

stream intensity of the j-th component;

$$R_0 = \sum_{j=0}^{h} \rho_j; \ R_1 = \sum_{j=1}^{h} \rho_j; \ R_2 = \sum_{j=2}^{h} \rho_j; \ \cdots \ R_h = \rho_h; \ R_i = \frac{\Lambda_i}{\mu}, \text{ where } \rho_j \text{ is}$$

the given demand stream intensity of the j-th component.

Demand streams arriving from each source are Poisson and have intensity λ_j; in this case total streams with intensities Λ_j also have, as we know, a Poisson character. Let us designate the mean intensity of demand service by one service facility as μ. In this case the intensity of an output stream of served demands before the m-th states is multiple μ and depends on the number of busy channels. After the m-th state the intensity of served demand stream is equal to $m\mu$. The served demand stream is also Poisson.

With accepted designations and assumptions taken into account, we will obtain a continuous-time Markov chain.

2   Probabilistic Characteristics of a Queuing System in a Steady-State Mode

We make up a set of Kolmogorov-Chapman equations for probabilities of QS states in a steady-state mode of its functioning. Adding the normalization condition $\sum_{i=0}^{m+\varepsilon_h} P_i = 1$, to this set of equations, we obtain a system that has a unique solution

$$P_0 = \left[e_m(R_0) + \frac{R_0^m}{m!} \sum_{g=1}^{h} \prod_{j=0}^{g-1} \left(\frac{R_j}{m} \right)^{E_j} \right.$$

$$\left. \times \begin{cases} \frac{R_g}{m-R_g} \left(1 - \left(\frac{R_g}{m} \right)^{E_g} \right), & R_g \neq m \\ E_g, & R_g = m \end{cases} \right]^{-1}; \qquad (1)$$

$$P_i = \begin{cases} \frac{R_0^i}{i!} P_0, & 0 < i \leq m, \\ \left(\frac{R_{j+1}}{m} \right)^{i-m-\varepsilon_j} \prod_{g=0}^{j} \left(\frac{R_g}{m} \right)^{E_g} \frac{R_0^m}{m!} P_0, & m + \varepsilon_j \leq i \leq m + \varepsilon_{j+1}, \\ & 0 \leq j \leq h-1, \end{cases} \qquad (2)$$

where the designation $e_m(R_0) = \sum_{i=0}^{m} \frac{R_0^i}{i!}$ is accepted - a non-complete exponential function. The solution (1) and (2) defines expressions for probabilities of all possible QS states of this type in a steady-state mode of its functioning.

For further calculations it is convenient to introduce the following basic probabilistic characteristics of QS of this type through which all other quantities are expressed:

- basic probability 1

$$P_{B1} = \sum_{i=m}^{m+\varepsilon_1-1} P_i = \frac{1 - \left(\frac{R_1}{m} \right)^{E_1}}{1 - \frac{R_1}{m}} \frac{R_0^m}{m!} P_0;$$

- basic probability 2

$$P_{B2} = \sum_{i=m+\varepsilon_1}^{m+\varepsilon_2-1} P_i = \frac{1 - \left(\frac{R_2}{m}\right)^{E_2}}{1 - \frac{R_2}{m}} \left(\frac{R_1}{m}\right)^{E_1} \frac{R_0^m}{m!} P_0;$$

$$\vdots$$

- basic probability h

$$P_{Bh} = \sum_{i=m+\varepsilon_{h-1}}^{m+\varepsilon_h-1} P_i = \frac{1 - \left(\frac{R_h}{m}\right)^{E_h}}{1 - \frac{R_h}{m}} \prod_{g=1}^{h-1} \left(\frac{R_g}{m}\right)^{E_g} \frac{R_0^m}{m!} P_0;$$

- congestion probability of the system

$$P_{m+\varepsilon_h} = \prod_{g=1}^{h} \left(\frac{R_g}{m}\right)^{E_g} \frac{R_0^m}{m!} P_0. \tag{3}$$

As a result, a general formula for basic probability is written in the form

$$P_{Bi} = \prod_{g=0}^{i-1} \left(\frac{R_g}{m}\right)^{E_g} \frac{R_0^m}{m!} P_0 \begin{cases} \frac{m}{m-R_i}\left(1 - \left(\frac{R_i}{m}\right)^{E_i}\right), & R_i \neq m \\ E_i, & R_i = m \end{cases}. \tag{4}$$

By means of the expression (4) it is possible to present traditional probabilistic characteristics of a queuing system in the most compact form:

- probability of a newly arrived demand service expectation in the queue

$$P_W = \frac{\Lambda_1}{\Lambda_0} \sum_{i=m}^{m+\varepsilon_1-1} P_i + \frac{\Lambda_2}{\Lambda_0} \sum_{i=m+\varepsilon_1}^{m+\varepsilon_2-1} P_i + \frac{\Lambda_3}{\Lambda_0} \sum_{i=m+\varepsilon_2}^{m+\varepsilon_3-1} P_i + \cdots$$

$$+ \frac{\Lambda_h}{\Lambda_0} \sum_{i=m+\varepsilon_{h-1}}^{m+\varepsilon_h-1} P_i = \frac{1}{R_0} \sum_{i=1}^{h} R_i P_{Bi};$$

- probability of a newly arrived demand service refusal (probability of demand loss)

$$P_L = \frac{\Lambda_0 - \Lambda_1}{\Lambda_0} \sum_{i=m}^{m+\varepsilon_1-1} P_i + \frac{\Lambda_0 - \Lambda_2}{\Lambda_0} \sum_{i=m+\varepsilon_1}^{m+\varepsilon_2-1} P_i + \frac{\Lambda_0 - \Lambda_3}{\Lambda_0} \sum_{i=m+\varepsilon_2}^{m+\varepsilon_3-1} P_i + \cdots$$

$$+ \frac{\Lambda_0 - \Lambda_h}{\Lambda_0} \sum_{i=m+\varepsilon_{h-1}}^{m+\varepsilon_h-1} P_i + P_{m+\varepsilon_h} = \frac{1}{R_0} \sum_{i=1}^{h} (R_0 - R_i) P_{Bi} + P_{m+\varepsilon_h}$$

$$= \sum_{i=1}^{h} P_{Bi} - P_W + P_{m+\varepsilon_h} = 1 - P_{IS} - P_W.$$

The probability of an immediate service of a newly arrived demand has, apparently, a form

$$P_{IS} = \sum_{i=0}^{m-1} P_i = e_{m-1}(R_0) P_0. \tag{5}$$

3   Numerical Characteristics of a Queuing System

By means of probabilistic characteristics of the system found above, it is possible to express all main features characterizing a steady-state mode of a queuing system functioning. So, through put capacity of a queuing system is a number of demands passing through the system per unit of time $A = \Lambda_0 q = \Lambda_0 (1 - P_L) = \Lambda_0 (P_{IS} + P_W)$. This number includes all demands from a general input stream except refused demands and those that did not get into the system. Relative through put capacity of the system, thus, is a share of demands passing through a queuing system from a general input stream of demands $q = 1 - P_L$. The average number of demands under service at the same time (or, that is the same, an average number of busy channels) with formulae (2)–(5) taken into account has a form

$$\bar{n} = \sum_{i=1}^{m-1} i P_i + m \sum_{i=m}^{m+\varepsilon_h} P_i = R_0 P_0 e_{m-2}(R_0) + m (P_W + P_L)$$

$$= R_0 P_0 e_{m-2}(R_0) + m \left(\sum_{i=1}^{h} P_{Bi} + P_{m+\varepsilon_h} \right).$$

The second initial moment of demands number under service is

$$\overline{n^2} = \sum_{i=1}^{m-1} i^2 P_i + m^2 \sum_{i=m}^{m+\varepsilon_h} P_i$$

$$= R_0 P_0 e_{m-2}(R_0) + R_0^2 P_0 e_{m-3}(R_0) + m^2 \left(\sum_{i=1}^{h} P_{Bi} + P_{m+\varepsilon_h} \right).$$

An average demands number in a queue (average queue length) are

$$\bar{l} = \sum_{i=m+1}^{m+\varepsilon_h} (i - m) P_i$$

$$= \sum_{i=1}^{h} \left\{ \begin{array}{ll} \frac{R_i}{m-R_i} [P_{Bi} - E_i P_{m+\varepsilon_i}], & R_i \neq m \\ \frac{E_i(E_i+1)}{2} P_{m+\varepsilon_i-1}, & R_i = m \end{array} \right\} + \sum_{i=2}^{h} \varepsilon_{i-1} \frac{R_i}{m} P_{Bi}.$$

The second initial moment of demands number in a queue is

$$\overline{l^2} = \sum_{i=m+1}^{m+\varepsilon_h} (i - m)^2 P_i$$

$$= \sum_{i=1}^{h} \left[\varepsilon_{i-1}^2 P_{Bi} + \left\{ \begin{array}{ll} \frac{R_i}{m-R_i} \left(\frac{m+R_i}{m-R_i} + 2\varepsilon_{i-1} \right) P_{Bi} - \\ -\frac{mE_i}{m-R_i} \left(E_i + \frac{2R_i}{m-R_i} + 2\varepsilon_{i-1} \right) P_{m+\varepsilon_i}, & R_i \neq m \\ (E_i - 1) E_i \left(\frac{2E_i-1}{6} + \varepsilon_{i-1} \right) P_{m+\varepsilon_i}, & R_i = m \end{array} \right\} \right]$$

$$+ \varepsilon_h^2 P_{m+\varepsilon_h}.$$

Further, in the considered queuing system, the queue is possible only when all service facilities are busy. Thus, the total stream of served demands of the whole system consists of service streams of each channel and has $m\mu$ intensity. In this case, the probability that the system serves i demands during t time in the event of queue, will be recorded in the form $B_i(t) = \frac{(m\mu t)^i}{i!} e^{-m\mu t}$.

The function of service waiting time distribution for one demand we will find according to a known dependence $F_W(t) = 1 - P(t_W \geq t)$, where $P(t_W \geq t)$ - the probability that waiting time in a queue for one demand is more than an advanced set time t. As it is easy to see, it is possible, firstly, in case when the queue is absent, but a newly arrived demand finds all service facilities in the system busy, and during t time none of facilities is released. Secondly, in case when one demand is already in a queue and during t time the system serves no more than one demand, or there are two demands in a queue, and during t time no more than two demands are served, and so on. In this case, according to the formula of full probability, we have

$$
q\left[1 - F_W(t)\right]
$$

$$
= \frac{\Lambda_1}{\Lambda}\left[B_0(t)\sum_{i=m}^{m+\varepsilon_1-1} P_i + B_1(t)\sum_{i=m+1}^{m+\varepsilon_1-1} P_i + \cdots \right.
$$

$$
\left. + B_{\varepsilon_1-1}(t) P_{m+\varepsilon_1-1}\right]
$$

$$
+ \frac{\Lambda_2}{\Lambda}\left[\sum_{i=0}^{\varepsilon_1} B_i(t)\sum_{i=m+\varepsilon_1}^{m+\varepsilon_2-1} P_i + B_{\varepsilon_1+1}(t)\sum_{i=m+\varepsilon_1+1}^{m+\varepsilon_2-1} P_i + \cdots \right.
$$

$$
\left. + B_{\varepsilon_2-1}(t) P_{m+\varepsilon_2-1}\right]
$$

$$
+ \frac{\Lambda_3}{\Lambda}\left[\sum_{i=0}^{\varepsilon_2} B_i(t)\sum_{i=m+\varepsilon_2}^{m+\varepsilon_3-1} P_i + B_{\varepsilon_2+1}(t)\sum_{i=m+\varepsilon_2+1}^{m+\varepsilon_3-1} P_i + \cdots \right.
$$

$$
\left. + B_{\varepsilon_3-1}(t) P_{m+\varepsilon_3-1}\right] + \cdots
$$

$$
+ \frac{\Lambda_h}{\Lambda}\left[\sum_{i=0}^{\varepsilon_{h-1}} B_i(t)\sum_{i=m+\varepsilon_{h-1}}^{m+\varepsilon_h-1} P_i + B_{\varepsilon_{h-1}+1}(t)\sum_{i=m+\varepsilon_{h-1}+1}^{m+\varepsilon_h-1} P_i + \cdots \right.
$$

$$
\left. + B_{\varepsilon_h-1}(t) P_{m+\varepsilon_h-1}\right]. \tag{6}
$$

After a number of intermediate calculations, it is possible to obtain the following expressions for finite-sums sequence in square brackets in the right-hand side of this ratio. As a result, substituting obtained ratios into the right member of a formula (6), we will finally find

$$F_W(t) = 1 - e^{-m\mu t} \frac{P_{m-1}}{q}$$

$$+ \left\{ \frac{R_1}{m - R_1} \left[e_{\varepsilon_1 - 1}(R_1 \mu t) - \left(\frac{R_1}{m}\right)^{E_1} e_{\varepsilon_1 - 1}(m\mu t) \right] \right.$$

$$+ \sum_{i=2}^{h} \frac{R_i}{m - R_i} \left[\prod_{g=1}^{i-1} \left(\frac{R_g}{m}\right)^{E_g} e_{\varepsilon_{i-1} - 1}(m\mu t) \right.$$

$$+ \prod_{g=1}^{i-1} \left(\frac{R_g}{R_i}\right)^{E_g} \left[e_{\varepsilon_i - 1}(R_i \mu t) - e_{\varepsilon_{i-1} - 1}(R_i \mu t) \right]$$

$$\left. \left. - \prod_{g=1}^{i} \left(\frac{R_g}{m}\right)^{E_g} e_{\varepsilon_i - 1}(m\mu t) \right] \right\} ;$$

Hence, the density of a demand waiting time distribution for service in a queue is

$$f_W(t) = \frac{dF_W(t)}{dt} = e^{-m\mu t} \frac{P_{m-1}}{q}$$

$$\times \left\{ \Lambda_1 e_{\varepsilon_1 - 1}(\Lambda_1 t) + \sum_{i=2}^{h} \Lambda_i \prod_{g=1}^{i-1} \left(\frac{R_g}{R_i}\right)^{E_g} \left[e_{\varepsilon_i - 1}(\Lambda_i t) - e_{\varepsilon_{i-1} - 1}(\Lambda_i t) \right] \right\} \quad (7)$$

and then, mean waiting time of demand service in a queue is

$$\bar{t}_W = \int_0^{\infty} t f_W(t)\, dt$$

$$= \frac{1}{\Lambda_0 q} \sum_{i=1}^{h} \left\{ \frac{R_i}{m - R_i} \left[P_{Bi} - E_i P_{m+\varepsilon_i} \right] + \frac{R_i}{m} \varepsilon_{i-1} P_{Bi} \right\} = \frac{\bar{l}}{A}$$

in compliance with J. Littl's formulae. In the same way the second initial moment of a demand waiting time in a queue is

$$\overline{t_W^2} = \int_0^{\infty} t^2 f_W(t)\, dt$$

$$= \frac{1}{\Lambda_0 q} \sum_{i=1}^{h} R_i \left\{ \begin{array}{l} \frac{2(P_{Bi} - E_i P_{m+\varepsilon_i})}{\mu(m-R_i)^2} \left[1 + \frac{\varepsilon_{i-1}}{m}(m - R_i) \right] \\ \frac{P_m}{3m^2 \mu} \prod_{g=0}^{i-1} \left(\frac{R_g}{R_i}\right)^{E_g} \end{array} \right.$$

$$\left. \begin{array}{l} + \frac{\varepsilon_{i-1}(\varepsilon_{i-1}+1)P_{Bi}}{m^2 \mu} - \frac{E_i(E_i+1)P_{m+\varepsilon_i}}{m\mu(m-R_i)}, \quad R_i \neq m \\ \times \left[\varepsilon_i(\varepsilon_i+1)(\varepsilon_i+2) - \varepsilon_{i-1}(\varepsilon_{i-1}+1)(\varepsilon_{i-1}+2) \right], \quad R_i = m \end{array} \right\} .$$

Let us note that the ratio (7) gives a possibility to calculate moments of any order as a demand waiting time in a queue for service.

4   Numerical Investigation of Queue Parameters Behavior in QS

In actual conditions of objects operating according to the principle of queuing systems, the problem of queues and delays in service is always topical. It naturally causes desire to organize the process of their exploitation in such a way that the operation of these objects and systems would proceed in more stable modes. It should be borne in mind that a single parameter which could be changed more or less quickly in actual practice for multi-channel devices in practice is the number of homogeneous service facilities m working in parallel. Therefore, we will set the task to study the work of QS in the following way.

Let us investigate the nature of behavior of the moments of queue length and waiting time of the demand in queue with the change of the number of service facilities m. For this purpose, let us formally replace factorial dependences m in formulas for probabilistic characteristics [7] through which the moments of the number of demands in the queue and waiting time are expressed with corresponding gamma-functions $G(m+1)$; m is conditionally regarded as a continuous quantity. Dependencies of mathematical expectation and variance of demands number waiting for service in the queue on the number of service facilities show that there is some boundary value m corresponding to a cross point of the moments of demands number in the queue which divides the axis m into two parts. The first part is the area in which the mean squared deviation (MSD) of the queue length is within the limits of mathematical expectation; the second part is the area in which the dispersion of demands number in the queue exceeds the mean value. The system functioning mode at which MSD of the queue length does not exceed its mean value is pretty stable and predictable from the point of view of operation.

In this case it is interesting to trace the dynamics of m change that is boundary when the given intensity components of demands input stream change and the queue length for corresponding components of input stream is limited.

A special program was developed to conduct a series of computational experiments to calculate m boundary according to the mathematical model with known as initial data of given intensity components of the demands input stream and corresponding size-limited queues. Varying the given components intensity of demands input stream ρ_i within 1 to 12, we found values $m1$ boundary for the moments of queue length and $m2$ boundary for the moments of demand servicing-waiting moments in a queue at various values of step size between queue length limitations for various components of demands input stream $E_i = 1$; 2; 5; 10.

As an example let us consider the queuing model with a two-component demand input stream and two queue length limits for each component. For this purpose let us set $\lambda_0 = 0$; $\mu = 1$; $E_0 = 0$; $h = 2$ in the program. As $\lambda_0 = 0$, the zero (Erlang) component in this model is absent. Here $\varepsilon_1 = E_1$ is queue length limit for demands of the first component of the input stream with the given intensity ρ_1, and $\varepsilon_2 = E_1 + E_2$ is queue length limit for demands of the second component of the input stream with the given intensity ρ_2.

The behavior of $m1$ and $m2$ boundary with the change of given intensity ρ_1 and ρ_2 is linearly increasing. We will call obtained straight lines limits of stability. Each point lying on the stability boundary corresponds to equal values of mathematical expectation and MSD of the queue length (for $m1$), and waiting time to service the demand in the queue (for $m2$) at a definite value of the given intensity of demand input stream. The coefficients of variation of the queue length and waiting time in the queue are equal to the unity. In fact, it is the border above which MSD exceeds mathematical expectation. The area below the straight line corresponds to the stable mode of system operation at which the mean squared deviation is within mathematical expectation.

When $\rho_1 > 1$ obtained straight lines divide the coordinate plane into 3 areas: the upper one corresponds to an unstable mode of system operation both according to the queue length and waiting time; the middle one corresponds to the stable mode as for the queue and unstable as for waiting time; the lower – to the stable mode on the queue and waiting time as well. It turns out that the set of values of the number of service facilities corresponding to the stable mode of system operation is limited from above by the stability boundary for waiting time. Both straight lines form a multiplicative strip of instability in regard to waiting time; its width enhances upon increasing of the given stream intensity ρ_1.

When the step between queue length limits for demands of different components is $E_1 = E_2 = 2$, stability boundaries on the queue length and waiting time when the given intensity of the first component of the stream is changed ρ_1, form a multiplicative instability strip of the system according to waiting time. In case the given intensity of the second component of the stream changes, ρ_2 form the additive instability strip of the system as for waiting time; its width does not practically change with the increase of ρ_2.

When the step between queue length limits for demands of different components is $E_1 = E_2 = 5$, the further narrowing of instability strips with regard to waiting time both for multiplicative at increase of the given intensity of the first component of stream ρ_1 and additive is observed when the given intensity of the second component ρ_2 changes.

Finally, when the step between queue length limits for demands of different components is $E_1 = E_2 = 10$, instability strips on waiting time practically disappear turning into a single boundary of the stability area both in queue and waiting time as well.

In case of a two-component service model with two queue length limits for each component of the demands input stream with intervals between limits $E_1 \geq 10$ and $E_2 \geq 10$, boundary values of the number of service facilities (inside of which MSD queue lengths and waiting time meet corresponding mathematical expectations) practically coincide. They are approximately equal to the sum of given intensity of all components of demands input stream. Also boundaries of stability on queue length and waiting time are straight lines and at $E_1 \geq 10$ and $E_2 \geq 10$ their slope angle makes $45°$.

For a two-component queuing model with queue limits there is an opportunity to investigate behavior of $m1$ and $m2$ boundary at simultaneous change of the given intensity of both components of demands input stream ρ_1 and ρ_2.

Having conducted a cycle of corresponding computational experiments at the step between queue length limits for demands of different components $E_1 = E_2 = 5$, we obtain hypersurfaces of stability on queue length and waiting time of the demand in a queue, very close to planes.

Obtained hypersurfaces break a coordinate space into 3 parts: upper is the space of system instability on queue and waiting time; low is the space of system behavior stability both on the queue length and demands waiting time in the queue; middle – the layer corresponding to an unstable operation mode of the system only on waiting time.

5   Higher Orders Queues

An N-th order queue will be called the queue calculated in case when there are N claims in the system as minimum, and some of them are in the memory. If $N = 0$ we have a usual mathematical queue, when $N = m$ where m - the number of channels in the service facility, we have a physical queue which is explicitly studied in work [8]. At $N = m + 1$ we have the so-called real queue [5], [6]; at all values $N > m + 1$ we have consequently higher orders queues [9].

Apparently, T. Saaty was the first to state the issue of real queues in his classical monograph [10]; it specified the value for the M/M/m system representing itself as an average number of demands which stay in the queue for some time to be served.

The physical sense of the real queue defined in the above-stated sense is that in this case a newly arrived into the system claim finds busy all service channels (all devices) and, at least, one more claim in the queue waiting for the service. Thus, the minimum mean length of a real queue (in case the intensity of an input stream of claims tends to zero) is unity but not zero as a general and well-studied mathematical queue has. As we see, the real queue is understood as the situation when there is at least one claim in the queue for the service on a par.

However, this numerical characteristic is not the only one to characterize real queues in queuing systems.

Along with real queues in the sense explained above, it is possible to consider another numerical characteristic of QS which, for example, in the standard report of the GPSS simulation system has the name "a queue without zero inputs". Here, zero input is understood as such arrival of the claim in the system at which there is, at least, one free service channel in the multi-channel device, and in this case the claim is served immediately. Let's emphasize that unlike the situation considered above, in this case we imply the situation when at the time of a new claim arrival in the system all service channels of the service facility are occupied, but the queue, as such, can be absent. In the latter case, the claim expecting service has no other service waiting claims before it; it is just before the service facility in which all channels are busy at that time. Thus defined the "queue without zero inputs" is calculated considering only those claims which really expected service, and without taking into account claims which did not have to wait as at the time of their arrival in the system one serving channel was free at least. Queue mean length without zero inputs is, apparently, longer than mean length known for all and more habitual mathematical queue but, in its turn, it is less than mean length of the real queue considered above. It is clear, that the minimum mean length of such queue is zero, as well as the usual mathematical queue is, i.e. on average such a queue, as well as a mathematical queue, can have any number of claims.

Thus, if the usual mathematical queue is calculated as the average for all claims which visited the system, then the queue without zero inputs is calculated as the average value minus those claims which were served immediately as they got into the system when, at least, one of service channels was free. The so-called real queue in this case is calculated as the average minus both those claims which were served without a queue, and those ones which found all service channels occupied but were the first in the service waiting list as there were no other claims in the system at this moment. In work [8] it was proposed to call the queues calculated without zero inputs as physical queues.

It is clear, that this result can be generalized if the concept of higher orders queues of systems with queues is introduced in the following way.

Let the queuing system have m serving channels with identical service intensity μ. In this case we will call the queue of a 0-th order the average queue calculated on condition that when a new claim enters the system, there can be any number of claims including the case when there are no claims at all, i.e. the system can be the completely free from claims. In this case we will call the queue of the 1-st order the average queue calculated on condition that when a new claim enters the system, it already contains at least one claim, and so on. It is clear, that upon this the physical queue means an average queue in all those cases that when the claim enters the system, there are at least m claims in it; thus according to this nomenclature, the physical queue is a queue of the m-th order, Then the real queue is a queue of the $m + 1$-th order, etc.

Thus, the N-th order queue is the average queue calculated on condition that when a new claim enters the system there are already Nclaims in it, and some of them can be in the memory. At the same time the case $N = 0$ corresponds to a usual mathematical queue; for $N = m$ we have a physical queue; let us remind that in the system of GPSS World simulation modeling this characteristic has the name "a queue without zero inputs". In case $N = m + 1$ we have a real queue; for those cases when $N > m + 1$ we have higher orders queues. In case all serving channels are busy, a newly arrived claim will have to expect service, the minimum quantity of claims in the physical queue is equal to zero in the memory; for the real queue it is equal to unity, and so on. It should be noted that physical and real queues have the greatest deviations from the known mathematical queue at small values of the intensity of claims stream entering the system.

As it is known, mean processing time of one claim in the system $\bar{t}_S$, mean staying time of claims in the queue $\bar{t}_W$ and the common mean staying time of the claim in the system in general $\bar{t}_T = \bar{t}_S + \bar{t}_W$ for Markov queuing systems are bound to corresponding discrete characteristics of QS by the following three formulas [5,6]:

$$\bar{t}_S = \bar{n}/A; \quad \bar{t}_W = \bar{l}/A; \quad \bar{t}_T = \bar{k}/A; \tag{8}$$

where A is throughout capacity of the system, i.e. an average number of claims served by the system in unity of time. Discrete characteristics of the system are understood respectively as an average number of busy channels $\bar{n}$, mean length of the queue $\bar{l}$, and an average number of claims in the system in general $\bar{k} = \bar{n} + \bar{l}$. Sometimes, these formulas are written in the form

$$\bar{t}_S = \bar{n}/\lambda; \quad \bar{t}_W = \bar{l}/\lambda; \quad \bar{t}_T = \bar{k}/\lambda,$$

when the total intensity of claims stream λ coming into the system is in the denominator.

In fact, however, the denominator of these formulas should not be made of the total intensity of claims stream but of that part only which corresponds to those claims that are really transferred through the system (more precisely, through the service facility), i.e. absolute throughout capacity of the system A.

Formulas (8) are commonly called Little's formulas. At first, the result which engineers used for a long time existed as several empirical formulas, i.e. in the form of some kind of "folkloric theorem", as it is said. Apparently, J. D. C. Little was the first person who gave it a strict formulation in 1961. The intuitive proof of Little's formulas comes to the fact that in a steady state mode the next demand entering the system finds in it the same average number of demands which remains in the system when this demand leaves it. This quantity is just equal to the product of claims stream intensity transferred through the system (or its any subsystem) multiplied by the mean time of their staying in this system (subsystem):

$$\bar{n} = A \,\bar{t}_S; \quad \bar{l} = A \,\bar{t}_W; \quad \bar{k} = A \,\bar{t}_T. \tag{9}$$

Direct mechanical analog of formulas (9) is a well-known relation for the way passed at a steady movement s based on moving velocity v and travel time t.

$$s = v \, t.$$

The case is somewhat different with QS numerical characteristics concerning a real queue and higher orders queues in these systems. Let us remind that the N-th order queue we have called the average queue calculated on condition that when a new claim enters the system there are already $2N$ claims in it, and some of them can be in the memory.

At the same time $N = 0$ corresponds to a usual mathematical queue; for $N = m$ we have a physical queue which in the system of GPSS World simulation modeling has the name "a queue without zero inputs".

In case $N = m + 1$ we have a real queue; for those cases when $N > m + 1$ we respectively have higher orders queues.

For a physical queue, as it is shown in work [8], the corresponding ratio has the form quite similar to (8):

$$\bar{t}_{Wphys} = \bar{l}_{phys}/A \tag{10}$$

It is possible to ascertain that the relation (10) is applicable for all types of queues from a mathematical to a physical queue, including the latter one, however for a real queue and higher orders queues this formula becomes unfair.

Somewhat different is the situation with numerical characteristics of QS concerning real queues and higher orders queues in regard to a real queue in these systems. In works [5,6] it was found out that the following relation is performed for the systems of M/M/m and M/M/m/E classes (however, all numerical characteristics of the first ones can be obtained by ultimate passing from numerical characteristics of the second ones)

$$\bar{t}_{Wreal} = \bar{l}_{real}/m\mu \tag{11}$$

as the real queue moves with velocity $m\mu$ to serve demands by the multi-channel device. It is possible to show that the same dependence will remain fair for all types of higher orders queues for which $N > m + 1$:

$$\bar{t}_{WN} = \bar{l}_N/m\mu \tag{12}$$

Relations (8)–(12) connect parameters of usual mathematical, physical and real queues in open queuing systems and parameters of higher orders queues in these systems as well. It is clear that these relations will be absolutely similar for close-loop queuing systems. At the same time the obtained system of formulas (8)–(12) may be called as *generalized Little's formulas*.

As we see, all higher orders queues in queuing systems of various types from the point of view of claims traveling velocity in these queues can be divided into two unequal classes. In this case, the first class will include all types of queues from mathematical to physical inclusive, which move with a transferring velocity of claims through system A. Thus $m + 1$ types of queues of various orders from zero to m-th are in the first class. The second, a more extensive class, includes a real queue and all higher orders queues in regard to a real queue for which, according to the definition, we have $N > m + 1$. All these queues move with the service velocity $m\mu$. The number of queues of various orders in this class is not limited.

Further, the work [8] provides formulas obtained for the mean length of a physical queue for queuing systems of various types. In particular, the expression for a mean length of a real queue of the system with an unlimited memory volume (within M. Kendall's symbolism – M/M/m model) is the following

$$\bar{l}_{phys} = \frac{\rho}{m - \rho}.$$

But for the M/M/m model $A = \lambda$, and then according to formulas (10) and (11) we have

$$\bar{t}_{Wphys} = \frac{\bar{l}_{phys}}{\lambda} = \frac{1}{\mu(m-\rho)};$$

$$\bar{t}_{Wreal} = \frac{\bar{l}_{real}}{m\mu} = \frac{1}{\mu(m-\rho)}.$$

i.e. for the model with an unlimited queue the mean staying time of one claim in a physical queue coincides with the mean staying time of the claim in a real queue: $\bar{t}_{Wphys} = \bar{t}_{Wreal}$. The obtained result can be called the theorem on physical and real queues in queuing systems with an unlimited memory volume.

6   Conclusion

Generalizing data of all computational experiments submitted in the work it is possible to draw the following conclusion.

In queuing systems of multicomponent streams stable operation modes of the system on the queue length and waiting time of demands are possible. Boundaries of these modes correspond to single coefficients of queue length variation and demands servicing-waiting in system. Regardless the number of components in demands input stream and values of the step between queue length limits for various components of the stream, boundary values of the number of service facilities depending on the given intensity of various stream components form straight lines described by the equation $m(\rho_i) = a + b\rho_i$ where ρ_i- given intensity of the i-th component of demands input stream. When the step between queue length limits for various components of demands input stream is $E_i \geq 10$, coefficients a and b accept values $a = \sum_{j \neq i} \rho_j$, $b = 1$. Thus, at $E_i \geq 10$ the boundary value of the number of service facilities is numerically equal to the sum of the given intensity of all input stream components. If above this limit, the operation mode of the system will be unstable both on the queue length and demand waiting time.

The proposed results of the work can be used to project and operate quite a wide class of objects and systems to assess their efficiency, and also to develop projects of modernization or construction of various technical objects working according to the principle of queuing systems.

References

1. Cohen, J.W.: Certain delay problems for a full availability trunk group loaded by two sources. Commun. News **16**(3), 105–113 (1956)
2. Takagi, H.: Explicit delay distribution in first-come first-served M/M/m/K and M/M/m/K/n queues and mixed loss-delay system. Int. J. Pure Appl. Math. **40**(2), 185–200 (2007)
3. Kirpichnikov, A.P., Titovtsev, A.S.: Open systems of multicomponent flows differentiated service. Ciência e Técnica Vitivinícola **29**(7), 108–122 (2014)

4. Titovtsev, A.: Sistemy differentsirovannogo obsluzhivaniya polikomponentnykh potokov. Modeli i kharakteristiki [Systems of differentiated services multicomponent flows. Models and specifications]. LAP LAMBERT Academic Publishing GmbH & Co. KG Publ., Saarbrücken (2012). (in Russian)
5. Kirpichnikov, A.P.: Prikladnaya teoriya massovogo obsluzhivaniya [Applied queuing theory]. Publishing office of KSU Publ., Kazan (2008). (in Russian)
6. Kirpichnikov, A.P.: Metody prikladnoy teorii massovogo obsluzhivaniya. Publishing office of KSU Publ., Kazan (2011). (in Russian)
7. Kirpichnikov, A., Titovtsev, A.: Mathematical model of a queuing system with arbitrary quantity of sources and size-limited queue. Int. J. Pure Appl. Math. **106**(2), 649–661 (2016)
8. Kirpichnikov, A., Titovtsev, A.: Physical and mathematical queues in the applied queuing theory. Int. J. Pure Appl. Math. **108**(2), 409–418 (2016)
9. Titovtsev, A.: The concept of higher orders queues in the queuing theory. Int. J. Pure Appl. Math. **109**(2), 451–457 (2016)
10. Saaty, T.L.: Elements of Queueing Theory with Applications. McGRAW-HILL book company Inc., New York, Toronto, London (1961)

Analysis of Perishable Queueing-Inventory System with Positive Service Time and $(S - 1, S)$ Replenishment Policy

Agassi Melikov and Mammad Shahmaliyev[✉]

Deparment of Computer Sciences,
National Aviation Academy of Azerbaijan, Mardakan pr. 30, Baku, Azerbaijan
agassi.melikov@gmail.com, mammad.shahmaliyev@gmail.com

Abstract. In this paper, model of inventory system with positive service time and perishable inventory is studied. It is assumed that some demands do not acquire the item after service completion and order replenishment lead time is a positive random variable. $(S - 1, S)$ order replenishment policy is applied. The exact and approximate methods are developed for calculation of joint distributions of the inventory level and number of customers in the system. The formulas for the system performance measures calculation are given as well. The high accuracy of formulas are confirmed by numerical experiments. The problem of choosing the optimal server rate to minimize the total cost is solved.

Keywords: Perishable inventory systems · Positive service time · $(S - 1, S)$ order replenishment policy · Calculation methods

1 Introduction

Service time of demands in classical models of Inventory Systems (IS) is usually assumed to be equal to zero (or inconsiderable). However, in real systems this assumption does not always hold. Therefore, IS models where demand service time is a positive quantity were introduced. These models with positive demand service time are called Queueing-Inventory Systems (QIS) and were first studied in [1,2]. Detailed review of QIS models is given in [3].

In QIS model, usually, it is assumed that after service completion the inventory level decreases. However, in works [4,5] are given the real systems where this condition does not hold and the models of such QIS are studied.

In this paper, studied QIS models are different from the models in [4,5] in following moments. Firstly, unlike in [4,5], the QIS models with perishable inventory are studied (Perishable QIS, PQIS). Secondly, we assume that the arrived demands enter the queue even when the inventory level is zero, while they become impatient in the queue. Thirdly, the mean service time for demands that acquire the item is different than for the demands do not acquiring the item. Finally, it is assumed that the replenishment orders are placed according to $(S - 1, S)$ policy.

© Springer International Publishing AG 2017
A. Dudin et al. (Eds.): ITMM 2017, CCIS 800, pp. 83–96, 2017.
DOI: 10.1007/978-3-319-68069-9_7

PQIS models have been extensively studied and developed in a peer-reviewed scientific literature. Numerous literature references on this subject are given in the review works [6–9], as well as, in monography [10]. The results of analysis of PQIS models with positive service time performed in [11–17] could be found in [18]. It should be noted that, the order replenishment policies (ORP) used in the most works belong to a (s, S) policy class.

At the same time, studying PQIS models with positive service time using different ORP in order to find the most optimal policy is a popular research subject. In this paper, $(S - 1, S)$ policy is used. According to this policy, when inventory level decreases (after demand service completion or inventory perishing) an order of unit size is placed.

Some serious results of PQIS analysis with $(S - 1, S)$ policy could be found in [19–21]. In these works, service time is assumed to be equal to zero, moreover in [19] the inventory level is right continuous. Analysis of available literature shows that the PQIS models with positive service time and $(S - 1, S)$ policy are not studied. Therefore, methods of exact and asymptotic analysis of PQIS model with finite queue length are given in this paper.

The paper is organized as follows. In Sect. 2, the description of the investigated PQIS model is presented and main performance measures are introduced. Exact and approximate methods to calculate the steady-state probabilities as well as performance measures are developed in Sect. 3. High accuracy of the developed approximate formulas by using numerical experiments are demonstrated in Sect. 4. The results of solution of the problem for choosing optimal server rate to minimize the total cost are shown as well. Conclusion remarks are given in Sect. 5.

2   Model Description and Problem Statement

The studied system has a finite storage warehouse of size S and continuous inventory level monitoring. Each inventory item independently perishes after a random time with exponential distribution function with parameter γ, $\gamma > 0$. At the same time, the item reserved for the demand service is not perishable. In other words, inventory level decreases not only after the demand service, but also because of the item perishing.

Demands are arriving into the system according to Poisson arrival process with the intensity λ for acquiring the inventory items. For the simplicity, we assume that the demands acquiring the item requires unit resource, that is, after service completion of such demands inventory level decreases for a single unit.

If the inventory level is positive upon arrival moment the demand is taken for the service with probability 1 if the server is idle by that time; otherwise, demand joins the queue. Demands are assumed to join the queue even if the inventory level is zero. If upon arrival moment of the demand the inventory level is zero, then according to Bernoulli trial with the parameter ϕ_1 it joins the queue and waits for inventory replenishment for a certain time, while with the probability ϕ_2 demand leaves the system being unserved, $\phi_1 + \phi_2 = 1$. In that cases, demands in queue are impatient, that is, if inventory level is zero every demand independently waits in the queue for an exponentially distributed random time with mean τ^{-1}.

Queues with finite length is studied in this paper. In the model with finite queue, it is assumed that if at the moment of demand arrival there are N demands in the system (including the one that is being served) then it is lost with probability 1.

After service completion according to Bernoulli trial with parameter σ_1 demand refuses to acquire the item, while with probability σ_2 acquires, where, $\sigma_1 + \sigma_1 = 1$. If the demand refuses to acquire the item its service time has exponential distribution with mean μ_1^{-1}; otherwise its service time is exponentially distributed with mean μ_2^{-1}, $\mu_2 < \mu_1$.

Inventory replenishment is performed according to $(S - 1, S)$ policy with delay, that is, the order lead time is a positive random quantity that has an exponential distribution with mean ν^{-1}. So, if the number of pending orders at the moment is n, then the replenishment rate is $n\nu$.

Problem is to find the joint distributions of inventory level and number of demands in the system. Solution of this problem will allow to calculate the performance measures of PQIS model, as well as, to perform its cost analysis. The main performance measures are the average values of the following quantities: inventory level S_{av}, inventory perishing rate Γ_{av}, average reorder rate RR, loss rate of the customers due to balking RL_b, loss rate of the demands due to reneging RL_r, average queue length L_{av}.

3 Methods for Calculation of the System Performance Measures

System is modeled by 2-D MC with the states (m, n), where m - is inventory level, n - is number of demands in the system. State Space (SS) of the system is defined as follows:

$$E = \{(m, n) : m = 0, 1, \ldots, S, n = 0, 1, \ldots, N\} \qquad (1)$$

Transition rate from the state $(m_1, n_1) \in E$ to $(m_2, n_2) \in E$ is denoted by $q((m_1, n_1), (m_2, n_2))$. All of these rates form generator matrix (Q-Matrix) of the given 2-D MC. Let's consider the problem of their calculation.

Transition between the states of the system are related to the following events: (i) demand arrival, (ii) service completion, (iii) product perishing, (iv) leaving the queue due to impatience and (v) inventory replenishment.

Taking into account assumed replenishment policy, following cases are considered while determining the initial state $(m_1, n_1) \in E$ of the system: (1) $m_1 > 0$; (2) $m_1 = 0$.

When $m_1 > 0$ transition from the state (m_1, n_1) because of the events (iv) is impossible, as in that case, demands in the queue are patient. Other transitions are defined as follows.

If the number of demands in system is less than N at the moment of demand arrival (event (i)) then the number of demands increases by one unit, that is, transition to the state $(m_1, n_1+1) \in E$ occurs; intensity of that transition is equal to λ. If after service completion the demand refuses to acquire the item (event of type (ii)), number of demands in the system is decreased by one, while inventory level remains unchanged, i.e. transition to the state $(m_1, n_1 - 1) \in E$ occurs and intensity of such transition is $\mu_1 \sigma_1$. If after service completion demand acquires the item (event of type (ii)), then both number of demands and inventory level decreases by one, that is, transition to the state $(m_1 - 1, n_1 - 1) \in E$ occurs; intensity of such transition is $\mu_2(1 - \sigma_1)$. After inventory item perishes (event of type (iii)) transition to the state $(m_1 - 1, n_1) \in E$ occurs, the intensity of such transition is equal to $m_1 \gamma$ for case $n_1 = 0$ and to $(m_1 - 1)\gamma$ for case $n_1 > 0$. At the moment of order replenishment (event of type (v)) transition to the state $(m_1 + 1, n_1) \in E$ occurs; intensity of such transition is equal to $(S - m_1)\nu$.

Consequently, for the case $m_1 > 0$, non-negative elements of Q-matrix are defined as follows:

$$q((m_1, n_1), (m_2, n_2)) = \begin{cases} \lambda, & \text{if } m_2 = m_1, \ n_2 = n_1 + 1 \\ \mu_1 \sigma_1, & \text{if } m_2 = m_1, \ n_2 = n_1 - 1 \\ \mu_2 \sigma_2, & \text{if } m_2 = m_1 - 1, \ n_2 = n_1 - 1 \\ m_1 \gamma, & \text{if } m_2 = m_1 - 1, \ n_2 = n_1 = 0 \\ (m_1 - 1)\gamma, & \text{if } m_2 = m_1 - 1, \ n_1 > 0, \ n_2 = n_1 \\ (S - m_1)\nu, & \text{if } m_2 = m_1 + 1, \ n_2 = n_1 \\ 0, & \text{otherwise} \end{cases} \qquad (2)$$

Now, let at the initial state $(m_1, n_1) \in E$ holds the condition $m_1 = 0$. In this case transition from the current state because of the events (ii) and (iii) is impossible, as in these states demand service could not be performed because of zero inventory level. In these states transitions for the events (i) and (v) are defined analogously as in (2) and the arrived demand joins the queue with the probability ϕ_1. Transition intensities because of demand impatience (event of type (iv)) are defined as follows: after the demand leaves the system because of impatience demand count decreases by one unit, while, inventory level remains unchanged, that is, transition to the state $(0, n_1 - 1) \in E$ occurs; intensity of such transition is equal to $n_1 \tau$. At the moment of the order replenishment transition to the state $(1, n_1)$ occurs; intensity of such transition is equal to $S\nu$. So, for the case $m_1 = 0$ non-negative elements of Q-matrix are defined as follows:

$$q((0, n_1), (m_2, n_2)) = \begin{cases} \lambda\phi_1, & \text{if } m_2 = 0, \ n_2 = n_1 + 1 \\ n_1 \tau, & \text{if } m_2 = 0, \ n_2 = n_1 - 1 \\ S\nu, & \text{if } m_2 = 1, \ n_2 = n_1 \\ 0, & \text{otherwise} \end{cases} \qquad (3)$$

It is clear from the formulas (2)–(3) that 2-D MC is irreducible and there exists stationary mode. Consequently, the steady-state probabilities $p(m, n)$,

$(m, n) \in E$ are the only solution of the system of balance equations (SBE), that are constructed based on the formulas (2) and (3). This SBE represents the set of linear equations of dimension $(S+1) \times (N+1)$. Due to its large dimension and obviousness the explicit form of SBE is not given in this work.

Required performance measures of the given PQIS are calculated through the steady state probabilities. So, the mean inventory level and the average number of demands in the system are calculated as the mathematical expectation of the corresponding random variables:

$$S_{av} = \sum_{m=1}^{S} m \sum_{n=0}^{N} p(m, n); \tag{4}$$

$$L_{av} = \sum_{n=1}^{N} n \sum_{m=0}^{S} p(m, n); \tag{5}$$

As the inventory item reserved for the servicing demand could not perish the average perishing intensity is calculated as follows:

$$\Gamma_{av} = \gamma \left(\sum_{m=1}^{S} (mp(m, 0) + \sum_{n=1}^{N} (m-1)p(m, n)) \right) \tag{6}$$

Replenishment order is placed either every time after servicing the demands that require the inventory item or after the item perishing. Consequently, the average reorder rate is calculated as follows:

$$RR = \sum_{m=1}^{S} (m\gamma p(m, 0) + ((m-1)\gamma + \mu_2\sigma_2)(1 - p(m, 0))) \tag{7}$$

As it is noted above, the balking occurs if at the moment of demand arrival the waiting hall (queue) is full. Therefore, the average loss rate of demands due to balking RL_b is given by:

$$RL_b = \lambda \left(\sum_{m=0}^{S} p(m, N) + \phi_2 \sum_{n=0}^{N-1} p(0, n) \right) \tag{8}$$

The reneging occurs only in the case of zero inventory level. Therefore, the average loss rate of the demands due to reneging RL_r is given by:

$$RL_r = \tau \sum_{n=1}^{N} p(0, n) \tag{9}$$

Analytic solution for the system could not be found due to the complexity of Q-matrix. The known numerical methods of linear algebra are only applicable for the Markov Chains of the moderate dimensions and become useless for the chains of larger dimension.

Therefore, approximate method [18] is used in this work that allows to perform asymptotic analysis of the performance measures of the given system for the large sizes of the inventory level and waiting hall for demands.

This method could be effectively applied for the models that work under large load; in other words, it is assumed that demand arrival intensity is far larger than the product perishing and replenishment rate, that is, $\lambda \gg max\{\gamma, \nu\}$. It should be noted that, this assumption is hold in many real PQIS. Moreover, as it was stated above, $\mu_1 \gg \mu_2$.

Taking into account the above conditions, let's consider the following split of the initial state space (1):

$$E = \bigcup_{m=0}^{S} E_m, \ E_{m_1} \bigcap E_{m_2} = 0, \ m_1 \neq m_2 \tag{10}$$

where $E_m = \{(m,n) : n = 0,1,\ldots,N\}$, $m = 0,1,\ldots,S$.

Additionally, we conclude that the transition intensities inside a row are far larger than the transition intensities between the rows. Further, based on the split (10) of the initial state space (1), the following merge function is defined:

$$U((m,n)) = \langle m \rangle$$

where $\langle m \rangle$ is merged state that consists of all the states E_m, $m = 0,1,\ldots,S$. Let's denote $\Omega = \{\langle m \rangle : m = 0,1,\ldots,S\}$.

Approximate values of steady state probabilities $\tilde{p}(m,n)$, $(m,n) \in E$ of the current model are defined as follows (see [18]):

$$\tilde{p}(m,n) = \rho_m(n)\pi(\langle m \rangle) \tag{11}$$

where $\rho_m(n)$ - is the probability of state (m,n) inside the merged model with the state space E_m and $\pi(\langle m \rangle)$ - is the probability of a merged state $\langle m \rangle \in \Omega$.

Steady-state probabilities of the split and merged models are calculated as follows.

In the all states (m,n) within the split model with the state space E_m the first component is a constant, therefore, all the states of such class is determined only by the second component. Consequently, in the analysis of the split models with the state space E_m the state $(m,n) \in E_m$ could only be specified with the second component, so for the convenience, while studying the split models with the state space E_m its states (m,n) are simply denoted by n, $n = 0,1,\ldots,N$.

It is concluded from the formulas (2) that the state probabilities within all the split models with the state space E_m, $m = 1,2,\ldots,S$ are the same as in classical model $M/M/1/N$ with load $a = \lambda/\mu_1\sigma_1$, i.e.:

$$\rho_m(n) = a^n \frac{1-a}{1-a^{N+1}}, \ m = 1,2,\ldots,S \tag{12}$$

Remark 1. As the quantities $\rho_m(n)$ do not depend on the index m, $m = 1, 2, \ldots, S$ below these indexes are omitted.

It is concluded from the formula (3) that the state probabilities within the merged model with the state space E_0 are the same as in Erlang model $M/M/N/0$ with load $b = \lambda\phi_1/\tau$, that is:

$$\rho_0(n) = \frac{\theta(n)}{\sum\limits_{n=0}^{N} \theta(j)}, \quad n = 0, 1, \ldots, N \tag{13}$$

Further, the following notation is accepted: $\theta(j) = \dfrac{b_j}{j!}$

Let's denote the transition intensity from the merged state $\langle m_1 \rangle$ to another merged state $\langle m_2 \rangle$ with $q(\langle m_1 \rangle, \langle m_2 \rangle)$, $\langle m_1 \rangle, \langle m_2 \rangle \in \Omega$. According to [18] these parameters are defined as follows:

$$q(\langle m_1 \rangle, \langle m_2 \rangle) = \sum\limits_{\substack{(m_1, n_1) \in E_{m_1}, \\ (m_2, n_2) \in E_{m_2}}} q((m_1, n_1), (m_2, n_2)) p(m_1, n_1) \tag{14}$$

Taking into account (2), (3) and (12), (13), (14) after some transformations we found that the given intensities are calculated as follows:

$$q(\langle m_1 \rangle, \langle m_2 \rangle) = \begin{cases} \Lambda(m_1), & \text{if } m_2 = m_1 - 1 \\ (S - m_1)\nu, & \text{if } m_2 = m_1 + 1 \\ 0, & \text{otherwise} \end{cases} \tag{15}$$

where $\Lambda(m_1) = m_1 \gamma \rho_0 + (1 - \rho(0))(\mu_2 \sigma_2 + (m_1 - 1)\gamma)$, $m_1 = 1, 2, \ldots, S$
Then from (15) we get:

$$\pi(\langle m \rangle) = \frac{S! \nu^m}{(S - m)!} \frac{\pi(0)}{\prod\limits_{i=1}^{m} \Lambda(i)}, \quad m = 1, 2, \ldots, S \tag{16}$$

where $\pi(0) = \left(\sum\limits_{m=0}^{S} \frac{S! \nu^m}{(S - m)!} \frac{1}{\prod\limits_{i=1}^{m} \Lambda(i)} \right)^{-1}$

Remark 2. We assume that $\prod\limits_{i=m}^{n} x_i = 1$, if $m > n$

Afterwards, taking into account (12), (13), (14), (15), (16) from (11) approximate joint distributions $\tilde{p}(m, n)$, $(m, n) \in E$ of inventory level and number of demands in the system are found. Using these probabilities after some transformations from (4), (5), (6), (7), (8), (9) following formulas are obtained for the calculation of performance measures:

$$S_{av} \approx \sum_{m=1}^{S} m\pi(\langle m \rangle)$$

$$\Gamma_{av} \approx \gamma \sum_{m=1}^{S} \pi(\langle m \rangle)(m\rho(0) + (m-1)(1-\rho(0)))$$

$$RR \approx \sum_{m=1}^{S} (m\gamma\rho(0)\pi(m) + (1-\rho(0))\pi(\langle m \rangle))((m-1)\gamma + \mu_2\sigma_2))$$

$$RL_b \approx \lambda(\rho(N)(1 - \pi(\langle 0 \rangle)) + \pi(\langle 0 \rangle)(\rho_0(N) + \phi_2(1 - \rho_0(N))))$$

$$RL_r \approx \tau\pi(\langle 0 \rangle) \sum_{n=1}^{N} n\rho_0(n)$$

$$L_{av} \approx \pi(\langle 0 \rangle) \sum_{n=1}^{N} n\rho_0(n) + (1 - \pi(0)) \sum_{n=1}^{N} n\rho(n)$$

4   Numerical Results

Due to the limitations to the volume of the work, only accuracy of the steady-state probabilities of the initial 2-D MC and performance measures is considered. It should be noted that, the evaluation of the accuracy of given formulas analytically is impossible. Therefore, comparative analysis of the obtained numerical results is used. The accuracy of the approximate values are evaluated by using following norms:

Maximum absolute value of differences:

$$\|N\|_1 = \max_{n \in E} |p(n) - \tilde{p}(n)| \tag{17}$$

Cosine similarity:

$$\|N\|_2 = \frac{\sum\limits_{n \in E} p(n)\tilde{p}(n)}{\sqrt{\sum\limits_{n \in E} (p(n))^2} \sqrt{\sum\limits_{n \in E} (\tilde{p}(n))^2}} \tag{18}$$

Jaccard coefficient [22]:

$$\|N\|_3 = \frac{\sum\limits_{n \in E} \min\{p(n), \tilde{p}(n)\}}{\sum\limits_{n \in E} \max\{p(n), \tilde{p}(n)\}} \tag{19}$$

Results of the comparative analysis of the steady-state probabilities for the exact and approximate methods are given in Table 1. The initial parameters of

Table 1. Estimation of accuracy of the steady-state probabilities versus various norms

Values of parameters			Norms		
S	N	λ	$\|N\|_1$	$\|N\|_2$	$\|N\|_3$
10	10	60	0.009167	0.999201	0.930769
	30	60	0.009354	0.999193	0.931855
	50	60	0.00966	0.99913	0.928858
	70	60	0.010289	0.998972	0.914683
20	10	60	0.006392	0.999313	0.932304
	30	60	0.006392	0.999313	0.932337
	50	60	0.006393	0.999312	0.932285
	70	60	0.006393	0.999312	0.932011
	90	60	0.006393	0.999311	0.931672
	110	60	0.006394	0.999311	0.931521
30	10	60	0.005361	0.999334	0.932366
	30	60	0.005361	0.999334	0.932366
	50	60	0.005361	0.999334	0.932366
	70	60	0.005361	0.999334	0.932361
	90	60	0.005361	0.999334	0.932355
	110	60	0.005361	0.999334	0.932352
40	10	40	0.006973	0.998495	0.900238
		60	0.00465	0.999344	0.932367
	30	40	0.006973	0.998495	0.900238
		60	0.00465	0.999344	0.932367
	50	40	0.006973	0.998495	0.900237
		60	0.00465	0.999344	0.932367
	70	40	0.006973	0.998495	0.900237
		60	0.00465	0.999344	0.932367
	90	40	0.006973	0.998495	0.900237
		60	0.00465	0.999344	0.932367
	110	40	0.006973	0.998495	0.900237
		60	0.00465	0.999344	0.932367
	120	60	0.00465	0.999344	0.932367
50	10	40	0.006206	0.998509	0.900238
		60	0.00413	0.999351	0.932367
	30	40	0.006206	0.998509	0.900238
		60	0.00413	0.999351	0.932367
	50	40	0.006206	0.998509	0.900238
		60	0.00413	0.999351	0.932367
	70	40	0.006206	0.998509	0.900238
		60	0.00413	0.999351	0.932367
	90	40	0.006206	0.998509	0.900238
		60	0.00413	0.999351	0.932367
	110	40	0.006206	0.998509	0.900238
		60	0.00413	0.999351	0.932367

the system are assumed as follows: $\mu_1 = 15, \mu_2 = 3, \gamma = 2, \nu = 1, \tau = 0.5, \sigma_1 = 0.3, \phi_1 = 0.6$

The exact values of steady-state probabilities are calculated from corresponding balance equations using MATLAB package. Solving time of balance equations depends on its dimension and takes several hours for $S \times N > 5000$ (e.g.: for $S = 50$ and $N = 100$ with quad core CPU Core i7 2.40 Ghz and 8 GB RAM at least 3–4 h are required). It should be noted that, in the same PC approximately 3–4 s are needed for the calculation of performance measures for $S = 100$ and $N = 500$ while using the approximate method.

It is obvious from the Table 1 that the higher the arrival intensity is, the better accuracy of the calculated steady state probabilities of the model with respect to all norms is acquired, that is, with the increase of the arrival intensity the norm (17) is approaching 0, while the norms (18) and (19) are approaching 1. It is clear from split scheme (10) that with the increase of arrival intensity, the transition intensities between the state classes $E_m, m = 1, 2, \ldots, S$ decrease; the smaller intensities between the classes of states of split model we have, the more accurate state probabilities of initial model we get. For the above initial data the analysis of accuracy of the system performance measures was performed as well (see Tables 2 and 3). It should be noted that, the performance measures (4), (6), (7), (8), (9) are almost the same when using exact and approximate approaches (see Table 2). Only the small errors (less than 5%) are observed while calculating measure (5) and this is acceptable in engineering calculations (see Table 3).

Table 2. Estimation of accuracy of accuracy of the performance measures (4), (6) and (7) for $N \in [20, 120], \lambda \in [20, 60]$; EV - Exact Value, AV - Approximate Value

S	Sav		Гav		RR	
	EV	AV	EV	AV	EV	AV
10	3.300632	3.300632	4.639206	4.639206	111	111
20	6.633345	6.633345	11.26737	11.26737	422	422
30	9.966667	9.966667	17.933346	17.933346	933	933
40	13.3	13.3	24.6	24.6	1,644.00	1,644.00
50	16.633333	16.633333	31.266667	31.266667	2,555.00	2,555.00

Remark 3. Zero values in Table 3 are not exactly equal to zero and are obtained after rounding the numbers to the 6th order precision.

Now let's consider the problem of choosing the most optimal server. Let it is possible to choose the server from the predefined collection, where with the increase of the service rate the cost associated with the corresponding server increases as well. It is required to choose such server that will minimize the long-run expected total cost (TC).

$$TC = c_h S_{av} + c_r RR + c_b RL_b + c_g RL_r + c_p \Gamma_{av} + c_w L_{av} + c_s P_b \qquad (20)$$

Table 3. Estimation of accuracy of performance measures (5), (8) and (9)

Values of parameters			RL_b		RL_r		L_{av}	
S	N	λ	EV	AV	EV	AV	EV	AV
10	10	60	53.434375	55.491991	0.090831	0.093377	9.873783	9.917505
	30	40	33.251864	35.313561	0.273342	0.271806	29.782387	29.850142
		60	53.248114	55.307103	0.277091	0.278264	29.864775	29.907871
	50	40	33.069064	35.168018	0.456141	0.417349	49.75709	49.761818
		60	53.063085	55.128522	0.462121	0.456846	49.851439	49.885624
	70	60	52.87945	54.974741	0.645755	0.610626	69.832213	69.813776
20	10	60	53.400653	55.499856	0.001592	0.001674	9.876329	9.918894
	30	40	33.397364	35.496658	0.004881	0.004873	29.802009	29.872825
		60	53.3973	55.496542	0.004944	0.004989	29.876169	29.918721
	50	40	33.394035	35.494048	0.00821	0.007482	49.801639	49.871242
		60	53.393956	55.49334	0.008288	0.00819	49.875961	49.918322
	70	40	33.390713	35.493373	0.011532	0.008158	69.80117	69.865791
		60	53.39062	55.490583	0.011625	0.010947	69.875698	69.917034
	90	40	33.387395	35.493368	0.01485	0.008162	89.800598	89.858999
		60	53.38729	55.489352	0.014954	0.012178	89.875373	89.912694
	110	40	33.38408	35.493368	0.018164	0.008162	109.79992	109.8522
		60	53.383966	55.489287	0.018278	0.012243	109.87498	109.90602
30	10	60	53.400012	55.499997	0.000028	0.00003	9.876403	9.918918
	30	60	53.399953	55.499939	0.000087	0.000088	29.8764	29.918915
	50	40	33.399895	35.499895	0.000145	0.000132	49.802382	49.873204
		60	53.399893	55.499882	0.000146	0.000145	49.876396	49.918908
	70	40	33.399835	35.499883	0.000204	0.000144	69.802374	69.873108
		60	53.399834	55.499833	0.000206	0.000194	69.876392	69.918886
	90	40	33.399776	35.499883	0.000264	0.000144	89.802366	89.872988
		60	53.399775	55.499812	0.000265	0.000215	89.876387	89.918809
	110	40	33.399717	35.499883	0.000323	0.000144	109.80236	109.87287
		60	53.399715	55.49981	0.000324	0.000217	109.87638	109.91869
40	10	60	53.4	55.5	0	0.000001	9.876404	9.918919
	30	60	53.399999	55.499999	0.000002	0.000002	29.876404	29.918919
	50	60	53.399998	55.499998	0.000003	0.000003	49.876404	49.918919
	70	60	53.399997	55.499997	0.000004	0.000003	69.876404	69.918918
	90	60	53.399996	55.499997	0.000005	0.000004	89.876404	89.918917
	110	60	53.399995	55.499997	0.000006	0.000004	109.8764	109.91892
	120	40	33.399994	35.499998	0.000006	0.000003	119.80239	119.87323
		60	53.399994	55.499997	0.000006	0.000004	119.8764	119.91891
50	10	60	53.4	55.5	0	0	9.876404	9.918919
	30	60	53.4	55.5	0	0	29.876404	29.918919
	50	60	53.4	55.5	0	0	49.876404	49.918919
	70	60	53.4	55.5	0	0	69.876404	69.918919
	90	60	53.4	55.5	0	0	89.876404	89.918919
	110	60	53.4	55.5	0	0	109.8764	109.91892

where P_b is probability that server is busy, i.e. $P_b = \sum_{m=1}^{S} \sum_{n=1}^{N} p(m,n)$. Here c_h is inventory carrying cost per unit item, c_r is setup cost per order, c_b is balking cost per customer, c_g is reneging cost per customer per unit time, c_p is the perishing cost per item per unit time, c_w is waiting time cost of a customer per unit time.

We assume that there are four possible options to choose the server: (1) $\mu_1 = 5, \mu_2 = 1$; (2) $\mu_1 = 8, \mu_2 = 2$; (3) $\mu_1 = 10, \mu_2 = 4$; (4) $\mu_1 = 15, \mu_2 = 5$. The values of the coefficients c_s when choosing the option $k, k = 1, 2, 3, 4$ are designated as $c_s^{(k)}$ and defined as: $c_s^{(1)} = 1, c_s^{(2)} = 2, c_s^{(3)} = 3, c_s^{(4)} = 4$. The values of other parameters in the (20) are constants: $c_h = 1, c_r = 0.1, c_b = 3, c_g = 2, c_p = 2, c_w = 1$.

Table 4. Results of solution of the problem (20) for $N = 150$

S	λ			
	5	10	15	30
10	4	4	4	4
50	4	2	2	2
80	1	1	1	1

Some results of the problem solution for the above data are given in Table 4., where the numbers 1, 2 and 4 indicate the index of the optimal server selection option. It is obvious from the Table 4. that if the inventory level is increasing the optimal option is to choose the server with the lesser service rate, and, vice versa, for the smaller values of the inventory level the optimal option will be the servers with the greater service rates.

5   Conclusion

PQIS model with perishable inventory and positive service time is studied in this paper. It is assumed that some demands do not acquire the item after service completion. When inventory level is zero, demands join or leave the system according to Bernoulli trial. Demands are impatient in the queue when the inventory level is zero. Order lead time, as well as, item perishing time are random variables with the exponential distributions and finite mean. Inventory replenishment policy belongs to $(S - 1, S)$ class. Exact and approximate formulas are given for calculation of steady-state probabilities of the given 2-D MC being the mathematical model of the studied system. Exact method is based on the solving of balance equations and is suitable for the moderate values of inventory level and length of the waiting hall for queuing the demands. Approximate approach is based on the state phase merging algorithms of 2-D Markov Chains and it is applicable for the systems of any dimension. High accuracy of the given formulas are shown using numerical experiments. Finally, the optimization problem of choosing optimal server for cost minimization is solved.

References

1. Sigman, K., Simchi-Levi, D.: Light traffic heuristic for an M/G/1 queue with limited inventory. Ann. Oper. Res. **40**, 371–380 (1992)
2. Melikov, A.Z., Molchanov, A.A.: Stock optimization in transport/storage systems. Cybernetics **27**(3), 484–487 (1992)
3. Krishnamoorthy, A., Lakshmy, B., Manikandan, R.: A survey on inventory models with positive service time. OPSEARCH **48**(2), 153–169 (2011)
4. Krishnamoorthy, A., Manikandan, R., Lakshmy, B.: Revisit to queueing-inventory system with positive service time. Ann. Oper. Res. **233**, 221–236 (2015)
5. Krishnamoorthy, A., Manikandan, R., Shajin, D.: Analysis of a multi-server queueing-inventory system. Adv. Oper. Res. (Hindawi Publ. Corp.) **2015**, 16 (2015). Article ID: 747328
6. Baron, O., Berman, O., Perry, D.: Continuous review inventory models for perishable items ordered in batches. Math. Methods Oper. Res. **72**, 217–247 (2010)
7. Lawrence, A.S., Sivakumar, B., Arivarignan, G.: A perishable inventory system with service facility and finite source. Appl. Math. Model. **37**, 4771–4786 (2013)
8. Goyal, S., Giri, B.: Recent trends in modeling of deteriorating inventory. Eur. J. Oper. Res. **134**, 1–16 (2011)
9. Karaesmen, I., Scheller-Wolf, A., Deniz, B.: Managing perishable and aging inventories: review and future research directions. In: Kempf, K., Keskinocak, P., Uzsoy, R. (eds.) Planning Production and Inventories in the Extended Enterprise. ISOR, vol. 151, pp. 393–436. Springer, Boston (2011). doi:10.1007/978-1-4419-6485-4_15
10. Nahmias, S.: Perishable Inventory Theory. Springer, Heidelberg (2011)
11. Sivakumar, B., Arivarignan, G.: A perishable inventory system with service facilities and negative customers. Adv. Model. Optim. **7**, 193–210 (2006)
12. Manuel, P., Sivakumar, B., Arivarignan, G.: A perishable inventory system with service facilities, MAP arrivals and PH-service times. J. Syst. Sci. Syst. Eng. **16**, 62–73 (2007)
13. Manuel, P., Sivakumar, B., Arivarignan, G.: A perishable inventory system with service facilities and retrial customers. Comput. Ind. Eng. **54**, 484–501 (2008)
14. Amirthakodi, M., Radhamami, V., Sivakumar, B.: A perishable inventory system with service facility and feedback customers. Ann. Oper. Res. **233**, 25–55 (2015)
15. Hamadi, H.M., Sangeetha, N., Sivakumar, B.: Optimal control of service parameter for a perishable inventory system maintained at service facility with impatient customers. Ann. Oper. Res. **233**, 3–23 (2015)
16. Berman, O., Sapna, K.P.: Optimal service rate of service facility with perishable inventory items. Nav. Res. Logist. **49**, 464–482 (2002)
17. Jajaraman, B., Sivakumar, B., Arivarignan, G.: A perishable inventory system with postponed demands and multiple server vacations. Model. Simul. Eng. (Hindawi Publ. Corp.) **2012**, 17 (2012). Article ID: 620960
18. Melikov, A.Z., Ponomarenko, L.A., Shahmaliyev, M.: Models of perishable queueing-inventory system with repeated customers. J. Autom. Inf. Sci. **48**(2), 22–38 (2016)
19. Kalpakam, S., Sapna, K.P.: (S-1, S) perishable systems with stochastic lead times. Math. Comput. Model. **21**(6), 95–104 (1995)
20. Kranenburg, A.A., van Houtum, G.J.: Cost optimization in the (S-1, S) lost sales inventory model with multiple demand classes. Oper. Res. Lett. **35**, 493–502 (2007)

21. Isotupa, K.P.S.: Cost analysis of an (S-1, S) inventory system with two demand classes and rationing. Ann. Oper. Res. **233**, 411–421 (2015)
22. Jaccard, P.: Etude de la distribution florale dans une portion des Alpes et du Jura. Bull. de la Soc. Vaud. des Sci. Nat. **37**, 547–579 (1901). (in French)

Comparative Analysis of Methods of Residual and Elapsed Service Time in the Study of the Closed Retrial Queuing System M/GI/1//N with Collision of the Customers and Unreliable Server

Anatoly Nazarov[1,2], Janos Sztrik[3]([⊠]), and Anna Kvach[1]

[1] National Research Tomsk State University,
36 Lenina Avenue, Tomsk 634050, Russia
kvach_as@mail.ru, nazarov.tsu@gmail.com
[2] Department of Applied Probability and Informatics,
Peoples' Friendship University of Russia,
Miklukho-Maklaya Street 6, Moscow 117198, Russia
[3] University of Debrecen, Debrecen, Hungary
sztrik.janos@inf.unideb.hu

Abstract. The aim of the present paper is to investigate a finite-source M/GI/1 retrial queuing system with collision of the customers where the server is subject to random breakdowns and repairs depending on whether it is idle or busy. The method of elapsed service time and the method of residual service time are considered using asymptotic approach under the condition of unlimited growing number of sources. It is proved, as it was expected, that basic characteristics of the system, such as the stationary probability distribution of the server states and the asymptotic average of the normalized number of customers in the system are the same and do not depend on the applied method.

Keywords: Finite-source queuing system · Closed queuing systems · Retrial queue · Collision · Server breakdowns and repairs · Unreliable server · Asymptotic analysis · Method of residual service time · Method of elapsed service time

1 Introduction

Retrial queues have been widely used to model many problems arising in telephone switching systems, telecommunication networks, computer networks and computer systems, call centers, wireless communication systems, etc.

In many practical situations it is important to take into account the fact that the rate of generation of new primary calls decreases as the number of customers in the system increases. This can be done with the help of finite-source, or quasi-random input models. Moreover, usually in the study of various

© Springer International Publishing AG 2017
A. Dudin et al. (Eds.): ITMM 2017, CCIS 800, pp. 97–110, 2017.
DOI: 10.1007/978-3-319-68069-9_8

queuing systems, servers are assumed to be absolutely reliable. But in practice it is necessary to take into account the possibility of failure and repair of the server. Finite-source retrial queues with unreliable server have been investigated in, for example [1,2,9,10]. Recent results on retrial queues with collisions can be found in, for example [3,6,8].

The aim of the present paper is to investigate such systems which has the above mention properties, that is finite-source, retrial, collision, and non-reliability of the server. The introduced model is a generalization of the systems treated in [4,5,7]. Two methods are considered using asymptotic approach under the condition of unlimited growing number of sources. It is proved, as it was expected, that basic characteristics of the system, such as the stationary probability distribution of the server states and the asymptotic average of the normalized number of customers in the system are the same and do not depend on the applied method.

The rest of the paper is organized as follows. In Sect. 2 the description of the model is given, the corresponding two-dimensional non-Markov process is defined. In Sects. 3 and 4 the residual service time method and the elapsed service time method are considered by using asymptotic analysis, respectively. Section 5 is devoted to the comparison of the offered methods. Finally, the paper ends with a Conclusion.

2   Model Description and Notations

Let us consider a closed retrial queuing system of type M/GI/1//N with collision of the customers and unreliable server (Fig. 1). The number of sources is N and each of them can generate a primary request during an exponentially distributed time with rate λ/N. A source cannot generate a new call until end of the successful service of this customer. If a primary customer finds the server idle, he enters into service immediately, in which the required service time has

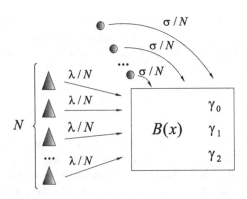

Fig. 1. Closed retrial queuing system M/GI/1//N with collision of the customers and unreliable server

a probability distribution function $B(x)$. Let us denote its hazard rate function by $\mu(y) = B'(y)(1 - B(y))^{-1}$ and Laplace -Stieltjes transform by $B^*(y)$, respectively. If the server is busy, an arriving (primary or repeated) customer involves into collision with customer under service and they both moves into the orbit. The retrial time of requests are exponentially distributed with rate σ/N. We assume that the server is unreliable, that is its lifetime is supposed to be exponentially distributed with failure rate γ_0 if the server is idle and with rate γ_1 if it is busy. When the server breaks down, it is immediately sent for repair and the recovery time is assumed to be exponentially distributed with rate γ_2. We deal with the case when the server is down all sources continue generation of customers and send it to the orbit, similarly customers may retry from the orbit to the server but all arriving customers immediately go into the orbit. Furthermore, in this unreliable model we suppose the interrupted request goes to the orbit immediately and its next service is independent of the interrupted one. All random variables involved in the model construction are assumed to be independent of each other.

Let $i(t)$ be the number of customers in the system at time t, that is, the total number of customers in orbit and in service. Similarly, let $k(t)$ be the server state at time t, that is

$$k(t) = \begin{cases} 0, & \text{if the server is idle,} \\ 1, & \text{if the server is busy,} \\ 2, & \text{if the server is down (under repair).} \end{cases}$$

Thus, we will investigate the process $\{k(t), i(t)\}$, which is not a Markov-process. To be a Markov one we will use method of supplementary variable, namely, we will consider two variants: the residual service time method and the elapsed service time method, and then we will compare them.

3   Method of Residual Service Time

Let us denote by $z(t)$ the random process, equal to the residual service time, that is time interval from the moment t until the end of successful service of the customer.

Thus, we will investigate the Markov process $\{k(t), i(t), z(t)\}$, which has a variable number of components, depending on the server state, since the component $z(t)$ is determined only in those moments when $k(t) = 1$.

Let us define the stationary probabilities as follows:

$$P_0(i) = P\{k(t) = 0, i(t) = i\},$$
$$P_1(i, z) = P\{k(t) = 1, i(t) = i, z(t) < z\},$$
$$P_2(i) = P\{k(t) = 2, i(t) = i\}.$$

To get $P_0(i)$, $P_1(i, z)$ and $P_2(i)$ the following system of Kolmogorov equations can be derived

$$\frac{\partial P_1(1,0)}{\partial z} - [\lambda + \gamma_0] P_0(0) + \gamma_2 P_2(0) = 0,$$

$$\frac{\partial P_1(1,z)}{\partial z} - \frac{\partial P_1(1,0)}{\partial z} - \left[\lambda\frac{N-1}{N} + \gamma_1\right] P_1(1,z)$$

$$+ \lambda B(z)P_0(0) + \frac{\sigma}{N}B(z)P_0(1) = 0,$$

$$- [\lambda + \gamma_2] P_2(0) + \gamma_0 P_0(0) = 0,$$

$$\frac{\partial P_1(i+1,0)}{\partial z} - \left[\lambda\frac{N-i}{N} + \gamma_0 + \frac{i}{N}\sigma\right] P_0(i) + \gamma_2 P_2(i)$$

$$+ \lambda\frac{N-i+1}{N}P_1(i-1) + \frac{i-1}{N}\sigma P_1(i) = 0,$$

$$\frac{\partial P_1(i,z)}{\partial z} - \frac{\partial P_1(i,0)}{\partial z} - \left[\lambda\frac{N-i}{N} + \gamma_1 + \frac{i-1}{N}\sigma\right] P_1(i,z)$$

$$+ \lambda\frac{N-i+1}{N}P_0(i-1)B(z) + \frac{i}{N}\sigma P_0(i)B(z) = 0,$$

$$- \left[\lambda\frac{N-i}{N} + \gamma_2\right] P_2(i) + \gamma_0 P_0(i) + \gamma_1 P_1(i)$$

$$+ \lambda\frac{N-i+1}{N}P_2(i-1) = 0.$$

(1)

Let us introduce the partial characteristic functions

$$H_k(u) = \sum_{i=0}^{N} e^{jui} P_k(i), \; k = 0, 2 \qquad H_1(u,z) = \sum_{i=1}^{N} e^{jui} P_1(i,z),$$

where $j = \sqrt{-1}$ is imaginary unit, then system (1) can be rewritten as

$$e^{-ju}\frac{\partial H_1(u,0)}{\partial z} + j\frac{(\sigma-\lambda)}{N}\frac{dH_0(u)}{du} + j\frac{(\lambda e^{ju}-\sigma)}{N}\frac{dH_1(u)}{du}$$

$$- [\lambda + \gamma_0] H_0(u) + \left[\lambda e^{ju} - \frac{\sigma}{N}\right] H_1(u) + \gamma_2 H_2(u) = 0,$$

$$\frac{\partial H_1(u,z)}{\partial z} - \frac{\partial H_1(u,0)}{\partial z} + j\frac{(\lambda e^{ju}-\sigma)}{N}B(z)\frac{dH_0(u)}{du}$$

(2)

$$+ j\frac{(\sigma-\lambda)}{N}\frac{\partial H_1(u,z)}{\partial u} + \lambda e^{ju}B(z)H_0(u) - \left[\lambda + \gamma_1 - \frac{\sigma}{N}\right] H_1(u,z) = 0,$$

$$j\frac{\lambda(e^{ju}-1)}{N}\frac{dH_2(u)}{du} + \gamma_0 H_0(u) + \gamma_1 H_1(u) + [\lambda(e^{ju}-1) - \gamma_2] H_2(u) = 0.$$

Summarizing the equations of the system (2) and executing limiting transition under condition $z \to \infty$ we obtain equation in the form

$$-e^{-ju}\frac{\partial H_1(u,0)}{\partial z} + j\frac{\lambda}{N}\left[H_0'(u) + H_1'(u) + H_2'(u)\right]$$
$$+ \lambda\left[H_0(u) + H_1(u) + H_2(u)\right] = 0 \ . \quad (3)$$

The solution of systems (2) and (3) for finite values N causes certain difficulties therefore we will find solution under condition of unlimited growing number of sources, that is $N \to \infty$.

3.1   Asymptotic Analysis

Theorem 1. *Let $i(t)$ be number of customers in a closed retrial queuing system $M/GI/1//N$ with the collisions of customers and unreliable server, then*

$$\lim_{N\to\infty} E\exp\left\{jw\frac{i(t)}{N}\right\} = \exp\left\{jw\kappa\right\}, \quad (4)$$

where value of parameter κ is the positive solution of the equation

$$(1 - \kappa)\lambda - \delta(\kappa)\left[R_0(\kappa) - R_1(\kappa)\right] + \gamma_1 R_1(\kappa) = 0, \quad (5)$$

here $\delta(\kappa)$ is

$$\delta(\kappa) = (1 - \kappa)\lambda + \sigma\kappa, \quad (6)$$

and the stationary distributions of probabilities $R_k(\kappa)$ of the service state k are determined as follows

$$R_0(\kappa) = \left\{\frac{\gamma_0 + \gamma_2}{\gamma_2} + \frac{\gamma_1 + \gamma_2}{\gamma_2} \cdot \frac{\delta(\kappa)}{\delta(\kappa) + \gamma_1}\left[1 - B^*(\delta(\kappa) + \gamma_1)\right]\right\}^{-1},$$

$$R_1(\kappa) = R_0(\kappa)\frac{\delta(\kappa)}{\delta(\kappa) + \gamma_1} \cdot \left[1 - B^*(\delta(\kappa) + \gamma_1)\right], \quad (7)$$

$$R_2(\kappa) = \frac{1}{\gamma_2}\left[\gamma_0 R_0(\kappa) + \gamma_1 R_1(\kappa)\right].$$

Proof. Denoting $\frac{1}{N} = \varepsilon$ and executing the following replacements in system (2)

$$u = \varepsilon w, \qquad H_k(u) = F_k(w, \varepsilon)\,, k = 0, 2; \qquad H_1(u, z) = F_1(w, z, \varepsilon),$$

we can write systems (2) and (3) in the form:

$$e^{-j\varepsilon w}\frac{\partial F_1(w,0,\varepsilon)}{\partial z} + j(\sigma - \lambda)\frac{\partial F_0(w,\varepsilon)}{\partial w} + j\left(\lambda e^{j\varepsilon w} - \sigma\right)\frac{\partial F_1(w,\varepsilon)}{\partial w}$$

$$- (\lambda + \gamma_0)F_0(w,\varepsilon) + \left[\lambda e^{j\varepsilon w} - \varepsilon\sigma\right]F_1(w,\varepsilon) + \gamma_2 F_2(w,\varepsilon) = 0,$$

$$\frac{\partial F_1(w,z,\varepsilon)}{\partial z} - \frac{\partial F_1(w,0,\varepsilon)}{\partial z} + j\left(\lambda e^{j\varepsilon w} - \sigma\right)B(z)\frac{\partial F_0(w,\varepsilon)}{\partial w}$$

$$+ j(\sigma - \lambda)\frac{\partial F_1(w,z,\varepsilon)}{\partial w} + \lambda e^{j\varepsilon w}B(z)F_0(w,\varepsilon)$$

$$- [\lambda + \gamma_1 - \varepsilon\sigma]F_1(w,z,\varepsilon) = 0, \quad (8)$$

$$j\lambda\left(e^{j\varepsilon w} - 1\right)\frac{\partial F_2(w,\varepsilon)}{\partial w} + \gamma_0 F_0(w,\varepsilon) + \gamma_1 F_1(w,\varepsilon)$$

$$+ \left[\lambda\left(e^{j\varepsilon w} - 1\right) - \gamma_2\right]F_2(w,\varepsilon) = 0,$$

$$-e^{-j\varepsilon w}\frac{\partial F_1(w,0,\varepsilon)}{\partial z} + j\lambda\frac{\partial}{\partial w}\left[F_0(w,\varepsilon) + F_1(w,\varepsilon) + F_2(w,\varepsilon)\right]$$

$$+ \lambda\left[F_0(w,\varepsilon) + F_1(w,\varepsilon) + F_2(w,\varepsilon)\right] = 0 .$$

Carrying out limiting transition under conditions $\varepsilon \to 0$, denoting $\lim_{\varepsilon\to 0} F_k(w,\varepsilon) = F_k(w)$, $k = 0,2$; $\lim_{\varepsilon\to 0} F_1(w,z,\varepsilon) = F_1(w,z)$ system (8) can be rewritten as

$$\frac{\partial F_1(w,0)}{\partial z} + j(\sigma - \lambda)\frac{dF_0(w)}{dw} + j(\lambda - \sigma)\frac{dF_1(w)}{dw} - (\lambda + \gamma_0)F_0(w)$$

$$+ \lambda F_1(w) + \gamma_2 F_2(w) = 0,$$

$$\frac{\partial F_1(w,z)}{\partial z} - \frac{\partial F_1(w,0)}{\partial z} + j(\lambda - \sigma)B(z)\frac{dF_0(w)}{dw} + j(\sigma - \lambda)\frac{\partial F_1(w,z)}{\partial w}$$

$$+ \lambda B(z)F_0(w) - [\lambda + \gamma_1]F_1(w,z) = 0, \quad (9)$$

$$\gamma_0 F_0(w) + \gamma_1 F_1(w) - \gamma_2 F_2(w) = 0,$$

$$-\frac{\partial F_1(w,0)}{\partial z} + j\lambda\frac{d}{dw}\left[F_0(w) + F_1(w) + F_2(w)\right]$$

$$+ \lambda\left[F_0(w) + F_1(w) + F_2(w)\right] = 0 .$$

Let us write the solution of system (9) in product-form

$$F_k(w) = R_k\Phi(w), \quad k = 0,2; \quad F_1(w,z) = R_1(z)\Phi(w), \quad (10)$$

where R_0, $R_1(z)$, R_2 are the limiting probability distributions of the server state k under conditions $N \to \infty$ and $\Phi(w)$ is limiting characteristic function of the stationary distribution of random process $\frac{i(t)}{N}$. Substituting this solution into (9), we obtain

$$R_1'(0) + j(\sigma - \lambda)[R_0 - R_1]\frac{\partial \Phi(w)/\partial w}{\Phi(w)} - (\lambda + \gamma_0)R_0 + \lambda R_1 + \gamma_2 R_2 = 0,$$

$$R_1'(z) - R_1'(0) + j(\sigma - \lambda)[R_1(z) - R_0 B(z)]\frac{\partial \Phi(w)/\partial w}{\Phi(w)} + \lambda B(z)R_0$$

$$- [\lambda + \gamma_1]R_1(z) = 0, \qquad (11)$$

$$\gamma_0 R_0 + \gamma_1 R_1 - \gamma_2 R_2 = 0,$$

$$j\lambda \frac{\partial \Phi(w)/\partial w}{\Phi(w)} + \lambda - R_1'(0) = 0.$$

The above relations allows to write down this function in the following form

$$\Phi(w) = \exp(jw\kappa),$$

which coincides with equality (4). Using notation (6) and taking into account that $j\dfrac{\partial \Phi(w)/\partial w}{\Phi(w)} = -\kappa$, system (11) can be rewritten as

$$R_1'(0) - \delta(\kappa)[R_0 - R_1] - \gamma_0 R_0 + \gamma_2 R_2 = 0,$$

$$R_1'(z) = R_1'(0) + [\delta(\kappa) + \gamma_1]R_1(z) - \delta(\kappa)R_0 B(z)0,$$

$$\gamma_0 R_0 + \gamma_1 R_1 - \gamma_2 R_2 = 0, \qquad (12)$$

$$R_1'(0) - \lambda(1 - \kappa) = 0.$$

Let us consider the second equation of the system (12) in more details. It can be proved that the solution of this equation has the form

$$R_1(z) = e^{[\delta(\kappa)+\gamma_1]z}\int_0^z e^{-[\delta(\kappa)+\gamma_1]x}\left\{R_1'(0) - \delta(\kappa)R_0 B(x)\right\}dx. \qquad (13)$$

Executing the limiting transition at $z \to \infty$ and taking into account that the first factor of the right hand side of (13) in a limiting condition tends to infinity, we can conclude that the second factor will be equal to zero, that is

$$\int_0^\infty e^{-[\delta(\kappa)+\gamma_1]x}\left\{R_1'(0) - \delta(\kappa)R_0 B(x)\right\}dx = 0.$$

Performing simple transformations, we will obtain

$$R'_1(0) = \delta(\kappa)R_0 B^*(\delta(\kappa) + \gamma_1). \tag{14}$$

Now, let us add the first and third equations of system (12) and, taking into account the received equality (14), the system (12) can be rewritten in the form

$$R'_1(0) - \delta(\kappa)R_0 + [\delta(\kappa) + \gamma_1] R_1 = 0,$$

$$R'_1(0) = \delta(\kappa)R_0 B^*(\delta(\kappa) + \gamma_1),$$

$$\gamma_0 R_0 + \gamma_1 R_1 - \gamma_2 R_2 = 0, \tag{15}$$

$$R'_1(0) - \lambda(1 - \kappa) = 0.$$

From the first three equations of system (15) and the normalization condition it is not difficult to obtain expressions for R_k, which coincides with (7) and, finally, equality (5) obviously follows from the first and fourth equations of system (15).
    Theorem is proved.   □

4   Method of Elapsed Service Time

Let us denote by $y(t)$ the supplementary random process, equal to the elapsed service time of the customer till the moment t.
    It is obvious that $\{k(t), i(t), y(t)\}$ is Markov process. Let us note, $y(t)$ is defined only in those moments when the server is busy, that is, when $k(t) = 1$.
    Define the stationary probabilities as

$$p_0(i) = P\{k(t) = 0, i(t) = i\},$$

$$p_1(i, y) = \frac{\partial P\{k(t) = 1, i(t) = i, y(t) < y\}}{\partial y},$$

$$p_2(i) = P\{k(t) = 2, i(t) = i\}.$$

To determine $p_0(i)$, $p_1(i, y)$ and $p_2(i)$ the following system of Kolmogorov equations can be written

$$-\left[\lambda\frac{N-i}{N} + \frac{i}{N}\sigma + \gamma_0\right]p_0(i) + \int_0^\infty p_1(i+1, y)\mu(y)dy$$

$$+\lambda\frac{N-i+1}{N}p_1(i-1) + \frac{i-1}{N}\sigma p_1(i) + \gamma_2 p_2(i) = 0,$$

$$\frac{\partial p_1(i, y)}{\partial y} = -\left[\lambda\frac{N-i}{N} + \frac{i-1}{N}\sigma + \mu(y) + \gamma_1\right]p_1(i, y),$$

$$-\left[\lambda\frac{N-i}{N} + \gamma_2\right]p_2(i) + \lambda\frac{N-i+1}{N}p_2(i-1) + \gamma_0 p_0(i) + \gamma_1 p_1(i) = 0,$$

(16)

with boundary condition

$$p_1(i, 0) = \lambda\frac{N-i+1}{N}p_0(i-1) + \frac{i}{N}\sigma p_0(i).$$

(17)

Introducing the partial characteristic functions

$$H_k(u) = \sum_{i=0}^N e^{jui}p_k(i), \quad k = 0, 2; \qquad H_1(u, y) = \sum_{i=1}^N e^{jui}p_1(i, y),$$

system (16) and Eq. (17) we will rewrite in the form

$$-(\lambda + \gamma_0)H_0(u) + \left[\lambda e^{ju} - \frac{\sigma}{N}\right]H_1(u) + e^{-ju}\int_0^\infty H_1(u, y)\mu(y)dy$$

$$+\gamma_2 H_2(u) + j\frac{(\sigma - \lambda)}{N}\frac{dH_0(u)}{du} + j\frac{(\lambda e^{ju} - \sigma)}{N}\frac{dH_1(u)}{du} = 0,$$

$$\frac{\partial H_1(u, y)}{\partial y} = \left[\frac{\sigma}{N} - \lambda - \mu(y) - \gamma_1\right]H_1(u, y) - j\frac{(\lambda - \sigma)}{N}\frac{\partial H_1(u, y)}{\partial u},$$

$$\gamma_0 H_0(u) + \gamma_1 H_1(u) + \left[\lambda(e^{ju} - 1) - \gamma_2\right]H_2(u)$$

(18)

$$+ j\frac{\lambda(e^{ju} - 1)}{N}\frac{dH_2(u)}{du} = 0,$$

$$H_1(u, 0) = \lambda e^{ju}H_0(u) + j\frac{(\lambda e^{ju} - \sigma)}{N}\frac{dH_0(u)}{du}.$$

4.1 Asymptotic Analysis

By using asymptotic methods for the first order solution to (18) we obtain

Theorem 2. *Let $i(t)$ be number of customers in a closed retrial queuing system $M/GI/1//N$ with the collisions of customers and unreliable server, then*

$$\lim_{N \to \infty} \mathsf{E} \exp\left\{ jw\frac{i(t)}{N} \right\} = \exp\left\{ jw\kappa \right\}, \tag{19}$$

where value of parameter κ is the positive solution of the equation

$$(1 - \kappa)\lambda - \delta(\kappa)\left[R_0(\kappa) - R_1(\kappa)\right] + \gamma_1 R_1(\kappa) = 0, \tag{20}$$

here $\delta(\kappa)$ is

$$\delta(\kappa) = (1 - \kappa)\lambda + \sigma\kappa, \tag{21}$$

and the stationary distributions of probabilities $R_k(\kappa)$ of the service state k are defined as follows

$$R_0(\kappa) = \left\{ \frac{\gamma_0 + \gamma_2}{\gamma_2} + \frac{\gamma_1 + \gamma_2}{\gamma_2} \cdot \frac{\delta(\kappa)}{\delta(\kappa) + \gamma_1} \left[1 - B^*(\delta(\kappa) + \gamma_1)\right] \right\}^{-1},$$

$$R_1(\kappa) = R_0(\kappa)\frac{\delta(\kappa)}{\delta(\kappa) + \gamma_1} \cdot \left[1 - B^*(\delta(\kappa) + \gamma_1)\right], \tag{22}$$

$$R_2(\kappa) = \frac{1}{\gamma_2}\left[\gamma_0 R_0(\kappa) + \gamma_1 R_1(\kappa)\right].$$

Proof. Denoting $\dfrac{1}{N} = \varepsilon$, in system (18) let us introduce the following substitutions

$$u = \varepsilon w, \qquad H_k(u) = F_k(w, \varepsilon), \, k = 0, 2; \qquad H_1(u, y) = F_1(w, y, \varepsilon),$$

then we will receive system of the equations

$$-(\lambda + \gamma_0)F_0(w, \varepsilon) + \left[\lambda e^{jew} - \varepsilon\sigma\right]F_1(w, \varepsilon) + e^{-jew}\int_0^\infty F_1(w, y, \varepsilon)\mu(y)dy$$

$$+ \gamma_2 F_2(w, \varepsilon) + j(\sigma - \lambda)\frac{\partial F_0(w, \varepsilon)}{\partial w} + j(\lambda e^{jew} - \sigma)\frac{\partial F_1(w, \varepsilon)}{\partial w} = 0,$$

$$\frac{\partial F_1(w, y, \varepsilon)}{\partial y} = \left[\varepsilon\sigma - \lambda - \mu(y) - \gamma_1\right]F_1(w, y, \varepsilon) - j(\lambda - \sigma)\frac{\partial F_1(w, y, \varepsilon)}{\partial w}, \tag{23}$$

$$\gamma_0 F_0(w, \varepsilon) + \gamma_1 F_1(w, \varepsilon) + \left[\lambda(e^{jew} - 1) - \gamma_2\right]F_2(w, \varepsilon)$$

$$+ j\lambda(e^{jew} - 1)\frac{\partial F_2(w, \varepsilon)}{\partial w} = 0,$$

$$F_1(w, 0, \varepsilon) = \lambda e^{jew}F_0(w, \varepsilon) + j(\lambda e^{jew} - \sigma)\frac{\partial F_0(w, \varepsilon)}{\partial w}.$$

Taking the limiting transition under conditions $\varepsilon \rightarrow 0$ let us denote $\lim_{\varepsilon \to 0} F_k(w, \varepsilon) = F_k(w)$, $k = 0, 2$; $\lim_{\varepsilon \to 0} F_1(w, y, \varepsilon) = F_1(w, y)$. Then system (23) can be rewritten as

$$-(\lambda + \gamma_0)F_0(w) + \lambda F_1(w) + \gamma_2 F_2(w) + \int_0^\infty F_1(w, y)\mu(y)dy$$

$$+ j(\lambda - \sigma)\left[\frac{dF_1(w)}{dw} - \frac{dF_0(w)}{dw}\right] = 0 ,$$

$$\frac{\partial F_1(w, y)}{\partial y} = -\left[\lambda + \mu(y) + \gamma_1\right] F_1(w, y) - j(\lambda - \sigma)\frac{\partial F_1(w, y)}{\partial w}, \qquad (24)$$

$$\gamma_0 F_0(w) + \gamma_1 F_1(w) - \gamma_2 F_2(w) = 0,$$

$$F_1(w, 0) = \lambda F_0(w) + j(\lambda - \sigma)\frac{dF_0(w)}{dw}.$$

The solution of the system (24) can be written in product-form

$$F_k(w) = R_k \Psi(w), \ k = 0, 2; \qquad F_1(w, y) = R_1(y)\Psi(w). \qquad (25)$$

Substituting this solution into (24) we will receive

$$\int_0^\infty R_1(y)\mu(y)dy - \lambda (R_0 - R_1) - \gamma_0 R_0 + \gamma_2 R_2$$

$$+ j(\lambda - \sigma)(R_1 - R_0)\frac{\partial \Psi(w)/\partial w}{\Psi(w)} = 0,$$

$$R_1'(y) = -\left[\lambda + \mu(y) + \gamma_1\right] R_1(y) - j(\lambda - \sigma)R_1(y)\frac{\partial \Psi(w)/\partial w}{\Psi(w)}, \qquad (26)$$

$$\gamma_0 R_0 + \gamma_1 R_1 - \gamma_2 R_2 = 0,$$

$$R_1(0) = \lambda R_0 + j(\lambda - \sigma)R_0\frac{\partial \Psi(w)/\partial w}{\Psi(w)}.$$

from which it follows, that function $\Psi(w)$ has the form

$$\Psi(w) = \exp\left(jw\kappa\right), \qquad (27)$$

coinciding with equality (19). Using the notation (21) the system (26) can be rewritten as

$$\int_0^\infty R_1(y)\mu(y)dy = \delta(\kappa)(R_0 - R_1) + \gamma_0 R_0 - \gamma_2 R_2,$$

$$R_1'(y) = -\left[\delta(\kappa) + \mu(y) + \gamma_1\right] R_1(y), \tag{28}$$

$$\gamma_0 R_0 + \gamma_1 R_1 - \gamma_2 R_2 = 0,$$

$$R_1(0) = \delta(\kappa) R_0.$$

Let us consider the second equation of system (28) in more details. It is not difficult to obtain a solution of this equation, taking the fourth equality of system (28) as the initial condition, and as a result we get

$$R_1(y) = \delta(\kappa) R_0 \left[1 - B(y)\right] e^{-[\delta(\kappa)+\gamma_1]y}. \tag{29}$$

To find R_1 integrate equality (29) with respect to y from 0 to ∞ and receive an expression in the form

$$R_1 = R_0 \frac{\delta(\kappa)}{\delta(\kappa) + \gamma_1} \cdot \left[1 - B^*(\delta(\kappa) + \gamma_1)\right]. \tag{30}$$

Expression for R_2 obviously follows from the third equation of system (28)

$$R_2 = \frac{1}{\gamma_2} \left[\gamma_0 R_0 + \gamma_1 R_1\right], \tag{31}$$

and, finally, from equalities (30) and (31), keeping in mind the normalization condition for R_0 we have

$$R_0 = \left\{ \frac{\gamma_0 + \gamma_2}{\gamma_2} + \frac{\gamma_1 + \gamma_2}{\gamma_2} \cdot \frac{\delta(\kappa)}{\delta(\kappa) + \gamma_1} \left[1 - B^*(\delta(\kappa) + \gamma_1)\right] \right\}^{-1}.$$

Thus, we have determined R_0, R_1 and R_2 that coincides with equalities (22).

Let us return to system (24). Integrating the second equation of the system with respect to y from 0 to ∞, adding it with other equations of system (24), substituting the decomposition (25) and taking into account the explicit form (27) of the function $\Psi(w)$, we obtain an equation in the form

$$\int_0^\infty R_1(y)\mu(y)dy = \lambda(1 - \kappa). \tag{32}$$

From (32) and the first and third equations of the system (28) it is obviously follows Eq. (20) for κ.

Theorem is proved.   □

5 Comparison of the Methods of Residual and Elapsed Time

At a research of the closed retrial queuing system $M/GI/1//N$ with collision of customers and unreliable server by asymptotic analysis for a Markovization of process $\{k(t), i(t)\}$ two methods were considered: the method of elapsed service time and the method of residual service time. From the Theorems 1 and 2 it follows, as it was expected, that the basic characteristics of the system, such as the stationary probability distribution R_k of the server states k and the asymptotic average κ of the normalized number of customers in the system are the same and do not depend on the method of investigation. Of course, it should be so, since only the proofs are different.

Let us note that the use of the elapsed service time method is necessary for a further research of number of transitions of a customer into the orbit, and also for a further research of the sojourn time of a customer in the orbit.

The residual service time method is used for finding the probability distribution of the number of customers in the system and also it is necessary at a further research of the mean sojourn time of a customer under service.

6 Conclusion

In this paper, a finite-source retrial queuing system $M/GI/1$ with collisions of customers and unreliable server was considered. Two methods of an supplementary variable was presented: method of elapsed service time and method of residual service time. The research of system has been conducted by an asymptotic analysis under condition of unlimited growing number of sources. As a result of the investigation the first order approximations of the basic characteristics of the system, such as a stationary probability distribution of the server states and the asymptotic average of the normalized number of customers in the system was obtained. It was shown, as it was expected, that specified characteristics are the same and do not depend on a type of applied method of a supplementary variable. In addition, advantages for using each of the considered methods were given and the necessity of their application for further researches of the system was indicated.

Acknowledgments. The publication was financially supported by the Ministry of Education and Science of the Russian Federation (Agreement number 02.a03.21.0008) and by Peoples Friendship University of Russia (RUDN University).

References

1. Almási, B., Roszik, J., Sztrik, J.: Homogeneous finite-source retrial queues with server subject to breakdowns and repairs. Math. Comput. Model. **42**(5–6), 673–682 (2005)
2. Dragieva, V.I.: Number of retrials in a finite source retrial queue with unreliable server. Asia Pac. J. Oper. Res. **31**(2), 23 (2014)
3. Kim, J.S.: Retrial queueing system with collision and impatience. Commun. Korean Math. Soc. **25**(4), 647–653 (2010)
4. Kvach, A., Nazarov, A.: The research of a closed RQ-system $M/GI/1//N$ with collision of the customers in the condition of an unlimited increasing number of sources. In: Probability Theory, Random Processes, Mathematical Statistics and Applications: Materials of the International Scientific Conference Devoted to the 80th Anniversary of Professor Gennady Medvedev, Doctor of Physical and Mathematical Sciences, pp. 65–70 (2015). (in Russian)
5. Nazarov, A., Sztrik, J., Kvach, A., Bérczes, T.: Asymptotic analysis of finite-source $M/M/1$ retrial queueing system with collisions and server subject to breakdowns and repairs. Ann. Oper. Res. (2017, submitted)
6. Nazarov, A., Sudyko, E.: Method of asymptotic semi-invariants for studying a mathematical model of a random access network. Probl. Inf. Transm. **46**(1), 86–102 (2010)
7. Nazarov, A., Kvach, A., Yampolsky, V.: Asymptotic analysis of closed markov retrial queuing system with collision. In: Dudin, A., Nazarov, A., Yakupov, R., Gortsev, A. (eds.) ITMM 2014. CCIS, vol. 487, pp. 334–341. Springer, Cham (2014). doi:10.1007/978-3-319-13671-4_38
8. Peng, Y., Liu, Z., Wu, J.: An $M/G/1$ retrial G-queue with preemptive resume priority and collisions subject to the server breakdowns and delayed repairs. J. Appl. Math. Comput. **44**(1–2), 187–213 (2014)
9. Sztrik, J.: Tool supported performance modelling of finite-source retrial queues with breakdowns. Publ. Math. **66**(1–2), 197–211 (2005)
10. Zhang, F., Wang, J.: Performance analysis of the retrial queues with finite number of sources and service interruptions. J. Korean Stat. Soc. **42**(1), 117–131 (2013)

The Renewal-Based Asymptotics and Accelerated Estimation of a System with Random Volume Customers

Evsey Morozov[1,2] and Lyubov Potakhina[1(✉)]

[1] Institute of Applied Mathematical Research,
Karelian Research Centre, Petrozavodsk, Russia
emorozov@karelia.ru, lpotahina@gmail.com
[2] Petrozavodsk State University, Petrozavodsk, Russia

Abstract. We consider a single-server system in which each customer is described by its service time and a random volume. The total volume of customers accepted by the system is upper bounded by a finite constant (system capacity) M. We give renewal-based approximations for a number of important stationary parameters of the system, in particular, the mean lost volume. For a large M, the loss is typically a rare event, and Crude Monte-Carlo method is time-consuming to obtain accurate estimate of the loss probability in an acceptable simulation time. We apply splitting method to speed-up estimation of the parameters by simulation. In particular, we focus on heavy load. We perform simulations for different values of capacity, different volume size distributions, including heavy- and light-tailed distributions, and also for different values of traffic intensity.

Keywords: Queueing system · Random volume customer · Finite capacity · Accelerated simulation · Splitting · Heavy-tailed volume

1 Introduction

Some important problems related to the high performance computer and communication systems can be described by the models in which customers have both random service time and random volume [1,2]. At that, in the most important cases, the buffer space (volume capacity) for the summary accumulated volume is finite. In [3] an analogy between the lost customer volume and a covering interval in the associated renewal process has been proposed. This approach uses the so-called *inspection paradox* and, as simulation confirms, in some cases leads to a useful approximation of the stationary parameters of lost volume. When the volume capacity is large enough (or system is low loaded), a customer loss becomes a rare event. In this case Crude Monte-Carlo method turns out to be ineffective, requiring a huge simulation time for an accurate estimation. Moreover, the relative error increases unlimitedly, as the loss becomes rarer. To overcome this problem, in this work we apply the so-called *splitting technique*

© Springer International Publishing AG 2017
A. Dudin et al. (Eds.): ITMM 2017, CCIS 800, pp. 111–121, 2017.
DOI: 10.1007/978-3-319-68069-9_9

to increase the occurrence of these rare events in an acceptable simulation time. The splitting method is based on the idea to generate a few stochastic copies of the basic underlying (Markov) process upon hitting the predefined thresholds. This multiplication increases the number of losses but must be compensated in the final expression of the loss probability estimator. We simulate this model for different values of the system capacity, different customer volume distributions and different system regimes (including heavy and light traffic). The obtained estimates (based on the accelerated simulation) are compared with the renewal-based approximation proposed in [3].

To the best of our knowledge, the application of the speed-up simulation technique to verify the accuracy of asymptotic renewal-based relations for lost/accepted volumes is performed for the first time. This is the key contribution of the paper. Moreover we present a numerical analysis of the covariance function between successive lost volumes, and it is also a contribution of this work. Besides, we detect and discuss some interesting results related to the behavior of the sample mean of the lost/accepted volumes depending on the capacity, traffic intensity and the volume size distribution.

The paper is organized as follows. In Sect. 2, we describe the model, while the splitting method is described in brief in Sect. 3. Section 4 contains description of experiments, simulation results and illuminating discussions.

2   Model Description and Analogy with Renewal Theory

We consider a general single-server $GI/G/1$-type queueing system, where each customer is described by both service time and a random volume. It is assumed that the service times $\{S_n, n \geq 1\}$, are independent identically distributed (i.i.d.) with generic element S, and the volumes $\{v_n, n \geq 1\}$ are i.i.d. as well, with generic element v. The two-dimensional sequence $\{S_n, v_n\}$ is assumed to be i.i.d., but, for a given n, a dependence between S_n and v_n is allowed. The arrival instants $\{t_n\}$ form the i.i.d. (renewal) sequence of the interarrival times $\tau_n = t_{n+1} - t_n, n \geq 0$ $(t_0 := 0)$ with rate $\lambda := 1/\mathsf{E}\tau \in (0, \infty)$ and generic element τ. Define $\rho = \lambda \mathsf{E}S$, the traffic intensity of the system. Denote $V(t)$ the *accumulated volume*, which is the sum of the volumes of all customers being in the system at instant t. It is assumed that the buffer for the *number of customers* waiting in the queue is infinite, while the summary accumulated volume in the system is *upper bounded* by a finite constant (capacity) M. Thus, in this system, customer n is *lost* if and only if $V(t_n^-) + v_n > M$.

Now we consider an important parameter, $\mathsf{E}V_M$, the *mean stationary lost volume*. Denote $R(t)$ the set of numbers of the rejected customers in interval $[0, t]$. Then $\mathsf{E}V_M$ is defined as the limit

$$\mathsf{E}V_M = \lim_{t \to \infty} \frac{\sum_{i \in R(t)} v_i}{|R(t)|}, \tag{1}$$

when exists, where $|R(t)|$ is the capacity of $R(t)$, and we explicitly show in the notation a dependence of this quantity on the capacity M. In turn, a loss can be treated as a crossing of the level M by a renewal process generated by the i.i.d. volumes $\{v_k\}$ [3]. Define random sums $Z_k = v_1 + \cdots + v_k$, $k \geq 1$, so $0 \leq Z_1 < Z_2 < \ldots$. As a result, the lost volume can be interpreted as a renewal interval covering the *"time instant"* M in this renewal process [4,5]. (A difference with exact renewal process caused by a dynamics of customers is discussed in [3].) This analogy of the accumulated volume with a time scale of the renewal process is widely used below. Let F be the distribution function of customer volume. Then we can deduce distribution of the lost volume V_t, provided the "capacity equals t", using an analogy with renewal interval covering "instant t" [5]. Further, applying the total probability formula, we have

$$
P(V_t \leq x) = \sum_{k=1}^{\infty} \int_{t-x}^{t} P(t - x < v \leq x) P(Z_k \in du)
$$

$$
= \int_0^t \mathbb{I}\{x > t - u\}[F(x) - F(t - u)]dH(u), \qquad (2)
$$

where $H(t)$ is the *renewal function*, defined as $H(0) = 0$,

$$
H(t) := E N_t, \quad N_t := \sum_{n=1}^{\infty} \mathbb{I}\{Z_n \leq t\}, \quad t \geq 0,
$$

and $\mathbb{I}$ denotes indicator function. That is $H(t)$ is the mean number of the renewal in interval $(0, t]$, where renewal intervals are represented by the customer volumes. A key observation is that, under this interpretation, the mean of the lost volume V_t from (2) coincides with $E V_t$ from (1), if we take $M = t$. An analogy with renewal theory makes it promising to study the asymptotic of V_M, as the capacity $M \to \infty$, to apply this result to a large but finite M. Indeed it follows from the *key renewal theorem* [4] that if g is a real bounded function and volume distribution F is *non-lattice* with $Ev < \infty$, then,

$$
\int_0^t g(t - u)dH(u) \to \frac{1}{Ev} \int_0^{\infty} g(u)du, \quad t \to \infty. \qquad (3)
$$

Applying (3) to (2), we obtain

$$
\lim_{t \to \infty} P(V_t \leq x) = \frac{1}{Ev} \int_0^x [F(x) - F(u)]du = \frac{1}{Ev} \int_0^x u F(du), \ x \geq 0, \qquad (4)
$$

where integration by parts is applied at the last step. This limiting distribution is well-known and called *integrated-tail distribution*. These results indicate that calculation of the lost volume distribution by formula (4) could lead to an accurate approximation, provided the capacity M is large enough. By (4), we obtain the following representation of the p-th moment of the stationary lost volume V_M, when fixed M is large and $\rho < 1$ [3]:

$$
E V_M^p = \frac{E v^{p+1}}{Ev} + o(1), \ p > 0, \qquad (5)
$$

where $o(1) \to 0$ as $M \to \infty$. In particular, it gives the following form of the two first moments of the stationary lost volume:

$$EV_M = \frac{Ev^2}{Ev} + o(1), \quad EV_M^2 = \frac{Ev^3}{Ev} + o(1). \tag{6}$$

In this work, we verify by simulation the accuracy of the approximation based on (5) and depending on the capacity M, distribution of volume v and traffic intensity ρ.

Now we turn to studying another important QoS parameter describing the model under consideration. Denote $A(t)$ the number of arrivals in interval $[0, t]$. Then the quantity

$$Q_{loss} := \lim_{t \to \infty} \frac{\sum_{i \in R(t)} v_i}{\sum_{k=1}^{A(t)} v_k},$$

when exists, is the limiting fraction of the lost volume in interval $[0, t]$. A closely related quantity is the stationary loss probability, P_{loss}, which is defined as

$$P_{loss} = \lim_{t \to \infty} \frac{|R(t)|}{A(t)},$$

when the limit exists. Note that if the loss happens *independently of the volume*, for instance, when $M = \infty$ but the number of waiting places is limited, then the summands in the numerator become i.i.d. Then, by the strong law of large numbers, we obtain the equality $Q_{loss} = P_{loss}$, because

$$Q_{loss} = \lim_{t \to \infty} \frac{\sum_{i \in R(t)} v_i}{\sum_{i=1}^{A(t)} v_i} = \lim_{t \to \infty} \frac{\sum_{i \in R(t)} v_i}{|R(t)|} \frac{A(t)}{\sum_{i=1}^{A(t)} v_i} \frac{|R(t)|}{A(t)} = P_{loss}. \tag{7}$$

The same equality holds for an important case, when $S_n = c\, v_n$ and $c > 0$ is a constant. The latter assumption is justified for a wide class of telecommunication models and expresses the proportionality between the service time and the volume of a given customer. However (7) is not true if the capacity $M < \infty$. More exactly, the following inequality

$$Q_{loss} \geq P_{loss} \tag{8}$$

has been established in the paper [6]. An intuition behind inequality (8) is that the bigger volume is lost with a bigger probability, implying $EV_M \geq Ev$. That is, the rejected volumes are *atypically large*, and this in turn implies (8).

Also we note that inequality (8) can be strictly proved by another method, if customer volume distribution F belongs to the class of *New-Worse-Than-Used distributions*, see [3]. In this case the tail distribution $\bar{F} := 1 - F$ satisfies the inequality $\bar{F}(y + x) \geq \bar{F}(y)\bar{F}(x)$.

In [3], the following renewal-based approximation of Q_{loss} has been proposed, provided the capacity M is large enough,

$$Q_{loss} \approx \frac{Ev^2}{[Ev]^2} P_{loss}, \tag{9}$$

which is based on asymptotic (6) and in turn implies (8). Note that result (9) is based on the assumption that $Q(t)$ consists of the *i.i.d. summands*, which are stochastically equivalent to stationary *covering interval* in the renewal process generated by $\{v_n\}$, given by (4). The accuracy of approximation (9) has been verified for large M in [3].

3 Splitting Method

An important purpose of this research is to study the estimates of the lost volume, when the loss is a *rare event*. In this case the Crude Monte-Carlo simulation turns out to be inefficient, because it requires unacceptable large simulation time to obtain the estimate with a given accuracy. In simulations below, we apply the *splitting technique* to reduce simulation time when the capacity M is large enough [7,8]. The splitting method is based on the idea to copy of the basic process upon reaching a given state, to make a rare event more frequent. In our setting, we assume $V(0) = 0$ and consider (equidistant) thresholds $0 < x_1 < x_2 < \cdots < x_K < M$. When the process $V(t)$ crosses the threshold x_i for the first time, we start R_i independent paths of process at the state $V(x_i)$, $i = 1, \ldots, x_K$. It is important to stress, that in general, the basic process must be Markovian, otherwise, the new paths are not stochastic copies. In our simulation, when the volume is not exponential, we indeed ignore the remaining service times at the arrival instants. In this case the *main* component $V(t)$ is not a Markov process, and the new trajectories (after the splitting) in general are not stochastic copies. We note that it may affect the accuracy of the estimation.

Figure 1 shows the dynamics of the accumulated volume process, $V(t)$, $t \geq 0$, when the splitting is used. The "circles" denote the instances at which the new paths start, upon reaching given thresholds.

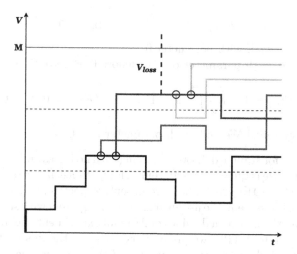

Fig. 1. An illustration of the splitting method

To realize simulation procedure, an important question arises: how to choose the thresholds x_i and multipliers R_i, $i = 1, \ldots, K$ in an "optimal" way. In general, these problems remain open, however in the next section, we will apply the following scheme, which has been successfully used for variance reduction in the estimation of a rare event related to queue size in the standard $M/M/1$ queue [9, 10]:

$$K = -\log \mathsf{P}_{loss}/2; \quad R_i = e^2, \, i = 1, \ldots, K; \quad R_0 = 1. \qquad (10)$$

Note that the loss probability P_{loss} is a priori unknown, and it is possible to use the upper bound (8) to choose a suitable number of thresholds K, when the estimate of the quantity Q_{loss} is more available. Otherwise, we may take K arbitrary.

4   Simulation Results

In this section, we present numerical results for different volume distributions and different M. We estimate the mean lost volume, the mean accepted volume and variance of the lost volume, when the system is highly or low loaded. Then we compare simulation results with theoretical results (6). Moreover, the correlation between two adjacent lost volumes is calculated as well. Simulations have been carried by means of the *system R* [11] and high performance cluster of Karelian Research Centre [12].

We consider $M/M/1$-type system with exponential interarrival times with parameter $\lambda \in [0.7, 2]$ with step 0.05, and exponential service times with parameter $\mu = 1$. These parameters give traffic intensity in the range $\rho \in [0.7, 2]$, covering both low and heavy (high) load. Also we consider the following volume size distributions:

1. Case 1: light-tailed Weibull, with parameter $i = 2$ (denoted "Weibull(2)" below):
$$F(x) = 1 - e^{-x^i}, \quad x \geq 0, \, i > 0;$$

2. Case 2: exponential, with parameter 0.5;
3. Case 3: Pareto, with parameter $\alpha = 4$ (denoted "Pareto(4)" below):
$$F(x) = 1 - \left(\frac{1}{x}\right)^{\alpha}, \, x \geq 1, \alpha > 0 \quad (F(x) = 0, \, x \leq 1),$$

4. Case 4: heavy-tailed Weibull, with parameter $i = 0.5$.

We recall that for standard $M/M/1$ system (with no losses and $\rho < 1$) the mean stationary queue size is $\mathsf{E}\nu = \rho/(1-\rho)$. Hence, for instance, if $\rho = 0.9$, then $\mathsf{E}\nu = 9$, and the values $M = 10\,\mathsf{E}\nu$, $20\,\mathsf{E}\nu$, $30\,\mathsf{E}\nu$ represent, respectively, small, medium and large capacity of the system. We use an analogy of our system with standard system $M/M/1$ to apply the mentioned values of M in all experiments below.

To calculate K from (10), we preliminary estimate the loss probability P_{loss} by Crude Monte-Carlo as a ratio the number of the lost customers and the summary number of arrivals. As we mentioned above, this estimate is typically hardly available and is not enough accurate, but can indicate (as well as the estimate of Q_{loss}) a suitable value K. We put $V(0) = 0$, and when $V(t)$ reaches a threshold x_i, new R_i paths of the process $V(t)$ start at the state $V(x_i)$. If a trajectory crosses, say thresholds x_i and x_{i+1} at once, we produce $R_i R_{i+1}$ paths, etc.

In general, the summary number of paths equals $\prod_{i=1}^{N} R_i$ and is big, if the target probability P_{loss} is small. As we mentioned above it leads to an unacceptable large simulation time. To reduce the required number of paths (and hence, simulation time), we use the following modification of the basic approach. If a trajectory, generated after crossing the threshold x_i, falls below the threshold $x_j(< x_i)$ such that $i - j > \lfloor N/2 \rfloor$, then we ignore this trajectory. (This approach is a modification of the so-called RESTART method [13].)

Now we describe the obtained simulation results.

Case 1: $v \sim Weibull(2)$, $M = 30\,Ev$

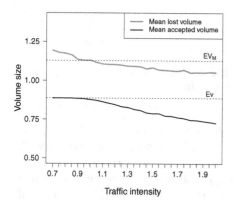
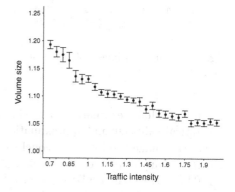

Fig. 2. The sample means of the lost and accepted volumes

Fig. 3. Confidence interval for the mean lost volume

Figures 2, 3, 4 and 5 correspond to *Case 1*, light-tailed Weibull volume. (We denote $v \sim F$ if random variable v has distribution F.) In particular, Fig. 2 shows estimates of the mean lost volume $\hat{E}V_M$ and the mean accepted volume $\hat{E}V_a$ for $M = 30\,Ev$. (The results for $M = 10\,Ev$ and $M = 20\,Ev$ are similar.) The theoretical mean Ev and the approximation of EV_M based on (6) are presented as well. We can summarize our observations as follows.

When $\rho \in [0.7, 0.9]$, the accepted volume estimate $\hat{E}V_a$ is very close to the mean volume Ev, since losses are rather rare and by this reason almost do not affect the final estimate. Note that the estimate $\hat{E}V_M$ is bigger than EV_M given by approximation (6), because typically bigger volumes are lost in this case. When $\rho \in (0.9, 1)$, the estimation is agreed with EV_M. When $\rho \in [1, 2]$, a considerable part of the customers are lost, and $\hat{E}V_M$ approaches the theoretical mean volume Ev. We suggest that it is because the size of the lost volume ceases to play a role when $\rho > 1$ and the system is permanently being in the saturated regime.

Also the accepted volume $\hat{E}V_a$ decreases because, when the system is highly loaded, the customers with small volumes are mainly accepted.

Figure 3 shows 90% confidence intervals for the mean lost volume.

Case 1: $v \sim Weibull(2)$

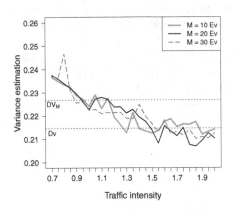

Fig. 4. Variance estimate of the lost volume

Fig. 5. Correlation between adjacent lost volumes

Figure 4 shows estimation of the lost volume variance $\hat{D}V_M$ for different M. Theoretical value Dv and approximation of DV_M based on (6) are given as well.

We can make the following conclusions. If $\rho \in [0.7, 0.9]$, the estimate $\hat{D}V_M$ is bigger than DV_M because atypically large volumes are mainly lost. When $\rho \in (0.9, 1)$ the estimate $\hat{D}V_M$ is agreed with theoretical value DV_M. Finally, if $\rho \in [1, 2]$, then $\hat{D}V_M$ is close to Dv because the losses become frequent and, by this reason, are less dependent on the volume size.

Figure 5 demonstrates the correlation between the adjacent lost volumes for different M. It is seen that when ρ increases, a dependence between the lost volumes disappears (correlation approaches zero). We note that it is agreed with the behavior of $\hat{D}V_M$ for $\rho > 1$, mentioned above. We note that the results for *Case 2* (exponential volume sizes) are quite similar to the Case 1 and, by this reason, we omit them.

Figures 6, 7, 8 and 9 correspond to *Case 3*, heavy-tailed Pareto volume. In particular, Fig. 6 shows estimates of the mean lost volume $\hat{E}V_M$ and the mean accepted volume $\hat{E}V_a$ for $M = 30\,Ev$. (The results for $M = 10\,Ev$ and $M = 20\,Ev$ are similar, if $\rho \in [0.9, 2]$, but the bursts are smaller for $\rho \in [0.7, 0.9)$.) In general, results are similar to Cases 1, 2, but the bursts are now much bigger. This can be explained by the following property of the i.i.d. heavy-tailed $\{X_i\}$ [14]:

$$P(X_1 + X_2 + \cdots + X_n > x) \sim P(\max\{X_1, X_2, \ldots X_n\} > x), \quad x \to \infty.$$

(We write $a \sim b$ if $a/b \to 1$.) Hence, the sum is likely to get large because of one of the summand gets large.

Case 3: $v \sim Pareto(4)$, $M = 30Ev$

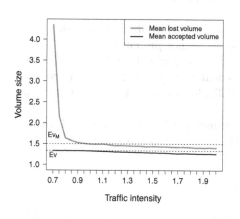

Fig. 6. The sample means of the lost and accepted volumes

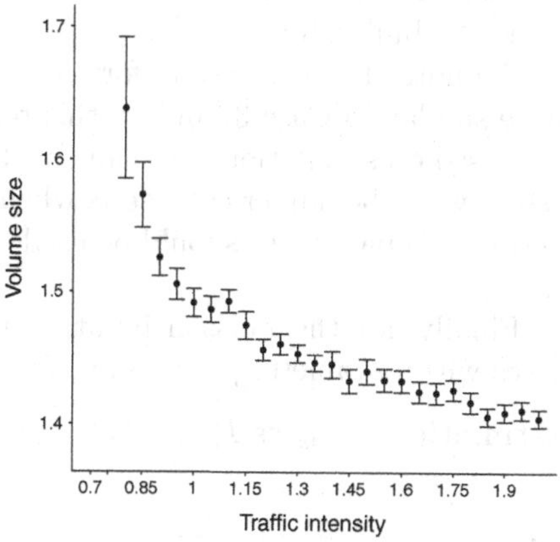

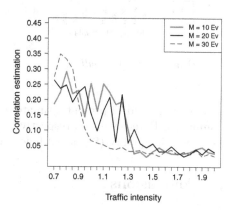

Fig. 7. Confidence interval for the mean lost volume

Figure 7 gives 90% confidence intervals for the mean lost volume when $\rho \geq 0.8$. Moreover, we found that confidence interval for $\rho = 0.7$ is $[4.08, 4.61]$, while for $\rho = 0.75$ it is $[2.05, 2.25]$.

Case 3: $v \sim Pareto(4)$

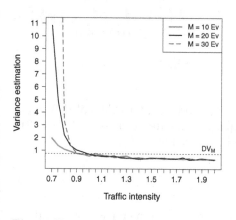

Fig. 8. Variance estimate of the lost volume

Fig. 9. Correlation between adjacent lost volumes

Figure 8 shows the estimate of the lost volume variance, $\hat{D}V_M$, for different M. Also, we note that for $M = 30\,Ev$, $\hat{D}V_M = 127.46$ if $\rho = 0.7$, and $\hat{D}V_M = 25.41$ if $\rho = 0.75$. Note that larger bursts can be explained by heavy-tailed volume distribution. Figure 9 depicts the estimate of correlation between the adjacent lost volumes for different M. It is seen that, as ρ increases, the dependence between the lost volumes decreases, however, more slowly than for the light-tailed Weibull volume.

We note that the results for *Case 4* (heavy-tailed Weibull volume sizes) are quite similar to Case 3, and by this reason we do not present it.

Based on simulations, we can conclude that approximation (6) is highly consistent with the numerical results when ρ is near 1 but $\rho < 1$. However, in general the approximation (6) should be used carefully, for instance, for heavy-tailed volumes.

Finally, for the Poisson input with parameter $\lambda = 0.9$, exponential service times with parameter $\mu = 1$ and $M = 20\,\mathrm{Ev}$, we illustrate inequality (8).

Estimation Q_{loss} vs P_{loss}

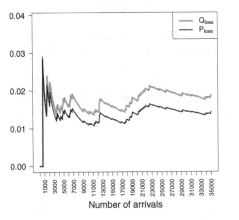
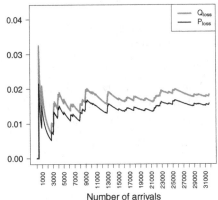

Fig. 10. $v \sim Weibull(2)$ **Fig. 11.** $v \sim Pareto(4)$

Figures 10 and 11 present estimates of Q_{loss} and P_{loss} for light-tailed Weibull volume and Pareto volume, respectively (parameters are taken as in Cases 1, 3). As we see, inequality (8) indeed holds and is agreed with (9).

5   Conclusions

We consider a single-server system in which each customer has both random service time and random volume. It is assumed that the summary accumulated volume is upper bounded by a finite constant (capacity) M. We consider a renewal-based approximation of the lost volumes and compare it with the numerical result for highly and low loaded system for different M and different volume distributions. In particular, heavy-tailed Pareto volumes and light-tailed Weibull volumes are considered. We also estimate correlation function between two adjacent lost volumes. For a low loaded system and large M, when a loss is a rare event, we apply splitting technique to accelerate estimation by simulation. The results detect the range of the parameters (in particular, the value M and traffic intensity ρ) where renewal-based approximation can be effectively used in QoS analysis of the system with random volume customers.

Acknowledgements. Research is supported by Russian Foundation for Basic Research, projects 15-07-02341, 15-07-02354, 15-07-02360.

References

1. Morozov, E., Potakhina, L., Tikhonenko, O.: Regenerative analysis of a system with a random volume of customers. In: Dudin, A., Gortsev, A., Nazarov, A., Yakupov, R. (eds.) ITMM 2016. CCIS, vol. 638, pp. 261–272. Springer, Cham (2016). doi:10.1007/978-3-319-44615-8_23
2. Tikhonenko, O.M.: Queuing system with processor sharing and limited resources. Autom. Remote Control **71**(5), 803–815 (2010)
3. Morozov, E., Nekrasova, R., Potakhina, L., Tikhonenko, O.: Asymptotic analysis of queueing systems with finite buffer space. In: Kwiecień, A., Gaj, P., Stera, P. (eds.) CN 2014. CCIS, vol. 431, pp. 223–232. Springer, Cham (2014). doi:10.1007/978-3-319-07941-7_23
4. Asmussen, S.: Applied Probability and Queues, 2nd edn. Springer, New York (2003). doi:10.1007/b97236
5. Feller, W.: An Introduction to Probability Theory and Its Applications. Wiley, New York (1971)
6. Tikhonenko, O.: Determination of loss characteristics in queueing systems with demands of random space requirement. In: Dudin, A., Nazarov, A., Yakupov, R. (eds.) ITMM 2015. CCIS, vol. 564, pp. 209–215. Springer, Cham (2015). doi:10.1007/978-3-319-25861-4_18
7. Botev, Z.I., Kroese, D.P.: Efficient Monte Carlo simulation via the generalized splitting method. Stat. Comput. **22**(1), 1–16 (2012)
8. Rubinstein, R.Y., Ridder, A., Vaisman, R.: Fast Sequential Monte Carlo Methods for Counting and Optimization. Wiley, New York (2014)
9. Borodina, A.V.: Regenerative modification of the splitting method for overload probability estimation in queueing systems. Ph.D. thesis, The Petrozavodsk State University (2008). (in Russian)
10. Garvels, M.: The splitting method in rare event simulation. Ph.D. thesis, The University of Twente, The Netherlands (2000)
11. R Foundation for Statistical Computing, Vienna, Austria. http://www.R-project.org/. ISBN 3-900051-07-0
12. Center for collective use of Karelian Research Centre of Russian Academy of Science. http://cluster.krc.karelia.ru/
13. Garvels, M., Kroese, D.: A comparison of RESTART implementations. In: Proceedings of the 1998 Winter Simulation Conference, pp. 601–608 (1998)
14. Asmussen, S., Klppelberg, C., Sigman, C.: Sampling at subexponential times, with queueing applications. Stoch. Process. Appl. **79**(2), 265–286 (1999)

Research of Heterogeneous Queueing System $SM|M^{(n)}|\infty$

Ekaterina Pankratova[1(✉)], Mais Farkhadov[1], and Erol Gelenbe[2]

[1] V. A. Trapeznikov Institute of Control Sciences of Russian Academy of Sciences,
Moscow, Russia
pankate@sibmail.com, mais@ipu.ru
[2] Imperial College, London EEE Bldg Rm 1009, South Kensington Campus,
London, England
e.gelenbe@imperial.ac.uk

Abstract. One of the modifications of the mathematical models used to describe processes in multi-service communication networks and telecommunication systems is the queueing system with heterogeneous servers. As a rule, for simulation of such processes the system with non-Poisson input flows is used. We consider the queuing system with infinite number of servers of n different types and exponential service time. Incoming flow is a Semi Markovian Process (SM-flow). Investigation of n-dimensional stochastic process characterizing the number of occupied servers of different types is performed using the initial moments method.

Keywords: Queueing system · Incoming sm-flow · Heterogeneous servers · Method of initial moments

1 Introduction

Systems with heterogeneous servers [4,5,10,11,13,14,18] and non-Poisson incoming flows [12,20,21] are suitable to simulate the functioning of real information systems. Such systems include queueing systems with non-ordinary Poisson incoming flows and exponential service time [2,8,9]; systems with parallel functioning blocks [3,6,7,15,19]. These papers deal with different configurations of parallel-service systems: single-line queueing systems with finite and infinite buffer, priority maintenance, impatient applications and a common ordinary incoming flow; queueing systems with two or more service blocs with a finite number of servers and a common final queue. Mathematical models of inhomogeneous infinite-linear systems with different types of servicing devices allow taking into account the heterogeneity of incoming applications requiring different maintenance time, which more adequately describes real information systems [16,17]. In this paper we study a heterogeneous queueing system with SM-incoming flow and exponential service time.

© Springer International Publishing AG 2017
A. Dudin et al. (Eds.): ITMM 2017, CCIS 800, pp. 122–132, 2017.
DOI: 10.1007/978-3-319-68069-9_10

2   Statement of the Problem

Consider the queuing system with infinite number of servers of n different types and exponential service time. Incoming flow is a Semi Markovian Process (SM-flow) which given by matrix $\mathbf{A}(x)$ consisting of elements $A_{k_1 k_2}(x)(k_1 = 0, \ldots, N, k_2 = 0, \ldots, N)$.

$$A_{k_1 k_2}(x) = F(k_2, x; k_1) = P\left\{\xi(k+1) = k_2, \tau(k+1) < x | \xi(k) = k_1\right\}, \quad (1)$$

where $\xi(k)$ — the Markov chain with discrete time and the transition probability matrix $\mathbf{P}$, $\tau(k)$ — non-Markov process for which

$$F(x) = P\left\{\tau(k) < x\right\} = \sum_{i=0}^{N} A_i(x) r(i), \quad (2)$$

$r(i)$ — stationary probability distribution of the Markov chain $\xi(k)$.

At the time of occurrence of the event in this stream only one customer flows in the system. The type of incoming customer is defined as i-type with probability p_i $(i = 1, \ldots, n)$. It goes to the appropriate device type, where its' service is performed during a random time having an exponential distribution function with parameter μ_i $(i = 1, \ldots, n)$ corresponding to the type of the customer.

Set the problem of exploring of n-dimensional stochastic process $\{l_1(t), \ldots, l_n(t)\}$ describing the number of occupied units of i-type at time t. Incoming flow is not Poisson, hence the n-dimensional process $\{l_1(t), \ldots, l_n(t)\}$ is non-Markov. Consider a $(n+2)$-dimensional Markov process $\{s(t), z(t), l_1(t), \ldots, l_n(t)\}$, here $z(t)$ — the time from t until the occurrence of the following event of SM-flow, $s(t)$ — the process is defined as follows

$$s(t) = \xi(k+1) \text{ if } t_k < t < t_{k+1}, \ t_k = \sum_{i=1}^{k} \tau(i).$$

For the joint probability distribution

$$P(s, z, l_1, \ldots, l_n, t) = P\{s(t) = s, z(t) < z, l_1(t) = l_1, \ldots, l_n(t) = l_n\}$$

we can write

$$P(s, z - \Delta t, l_1, \ldots, l_n, t + \Delta t)$$

$$= [P(s, z, l_1, \ldots, l_n, t) - P(s, \Delta t, l_1, \ldots, l_n, t)] \prod_{i=1}^{n} (1 - l_i \mu_i)$$

$$+ \sum_{\nu=1}^{K} P(\nu, \Delta t, l_1 - 1, \ldots, l_n, t) A_{\nu s}(z) p_1 + \ldots \quad (3)$$

$$+ \sum_{\nu=1}^{K} P(\nu, \Delta t, l_1, \ldots, l_n - 1, t) A_{\nu s}(z) p_n + P(s, z, l_1 + 1, \ldots, l_n, t)(l_1 + 1)\mu_1 \Delta t + \ldots$$

$$+ P(s, z, l_1, \ldots, l_n + 1, t)(l_n + 1)\mu_n \Delta t + o(\Delta t), \ s = 1, \ldots, K.$$

System of Kolmogorov differential equations for the probability distribution $P\{s, z, l_1, \ldots, l_n, t\}$ is the following:

$$\frac{\partial P(s, z, l_1, \ldots, l_n, t)}{\partial t} = \frac{\partial P(s, z, l_1, \ldots, l_n, t)}{\partial z} - \frac{\partial P(s, 0, l_1, \ldots, l_n, t)}{\partial z}$$

$$- \sum_{i=1}^{n} l_i \mu_i P(s, z, l_1, \ldots, l_n, t) \tag{4}$$

$$+ p_1 \sum_{\nu=1}^{K} \frac{\partial P(\nu, 0, l_1 - 1, \ldots, l_n, t)}{\partial z} A_{\nu s} + \ldots + p_n \sum_{\nu=1}^{K} \frac{\partial P(\nu, 0, l_1, \ldots, l_n - 1, t)}{\partial z} A_{\nu s}$$

$$+ \mu_1 (l_1 + 1) P(s, z, l_1 + 1, \ldots, l_n, t) + \ldots + \mu_n (l_n + 1) P(s, z, l_1, \ldots, l_n + 1, t),$$

$$s = 1, \ldots, K.$$

We will find the solution of the system (4) during stationary operation of the system. Denote $\lim_{t \to \infty} P(s, z, l_1, \ldots, l_n, t) = \Pi(s, z, l_1, \ldots, l_n)$, $s = 1, \ldots, K$.

Then the equation (4) takes the form

$$\frac{\partial \Pi(s, z, l_1, \ldots, l_n)}{\partial z} - \frac{\partial \Pi(s, 0, l_1, \ldots, l_n)}{\partial z} - \sum_{i=1}^{n} l_i \mu_i \Pi(s, z, l_1, \ldots, l_n)$$

$$+ p_1 \sum_{\nu=1}^{K} \frac{\partial \Pi(\nu, 0, l_1 - 1, \ldots, l_n)}{\partial z} A_{\nu s} + \ldots$$

$$+ p_n \sum_{\nu=1}^{K} \frac{\partial \Pi(\nu, 0, l_1, \ldots, l_n - 1)}{\partial z} A_{\nu s} \tag{5}$$

$$+ \mu_1 (l_1 + 1) \Pi(s, z, l_1 + 1, \ldots, l_n) + \ldots$$

$$+ \mu_n (l_n + 1) \Pi(s, z, l_1, \ldots, l_n + 1) = 0$$

$$s = 1, \ldots, K.$$

Introduce partial characteristic functions [1]:

$$H(s, z, u_1, \ldots, u_n) = \sum_{l_1=0}^{\infty} \ldots \sum_{l_n=0}^{\infty} e^{j u_1 l_1} \times \ldots \times e^{j u_n l_n} \Pi(s, z, l_1, \ldots, l_n),$$

where $s = 1, \ldots, K$, $j = \sqrt{-1}$ — imaginary unit.

In view of

$$\frac{\partial H(s, z, u_1, \ldots, u_n)}{\partial u_i} = j \sum_{l_1=1}^{\infty} \ldots \sum_{l_n=1}^{\infty} l_i e^{j u_1 l_1} \times \ldots \times e^{j u_n l_n} \Pi(s, z, l_1, \ldots, l_n),$$

$$i = 1, \ldots, n, \ s = 1, \ldots, K,$$

and using (4) write the system of differential equations for partial characteristic functions $H(s, z, u_1, \ldots, u_n)$

$$\frac{\partial H(s, z, u_1, \ldots, u_n)}{\partial z} - \frac{\partial H(s, 0, u_1, \ldots, u_n)}{\partial z} \tag{6}$$

$$+ j \sum_{i=1}^{n} \mu_i (1 - e^{ju_i}) \frac{\partial H(s, z, u_1, \ldots, u_n)}{\partial u_i} + \sum_{i=1}^{n} p_i e^{ju_i} \sum_{\nu=1}^{K} \frac{\partial H(\nu, 0, u_1, \ldots, u_n)}{\partial z} A_{\nu s}(z) = 0,$$

$$s = 1, \ldots, K,$$

which we rewrite in the form of the vector-matrix equation

$$\frac{\partial \mathbf{H}(z, u_1, \ldots, u_n)}{\partial z} + j \sum_{i=1}^{n} \mu_i (1 - e^{-ju_i}) \frac{\partial \mathbf{H}(z, u_1, \ldots, u_n)}{\partial u_i}$$

$$+ \frac{\partial \mathbf{H}(0, u_1, \ldots, u_n)}{\partial z} \left(\sum_{i=1}^{n} p_i e^{ju_i} \mathbf{A}(z) - \mathbf{I} \right) = 0, \tag{7}$$

$\mathbf{H}(z, u_1, \ldots, u_n) = [H(1, z, u_1, \ldots, u_n), H(2, z, u_1, \ldots, u_n), \ldots, H(K, u_1, \ldots, u_n)]$ — row vector consisting of characteristic functions of the random process $\{s(t), z, (t)l_1(t), \ldots, l_n(t)\}$ for each state of the process $s(t)$,

$$\frac{\partial \mathbf{H}(0, u_1, \ldots, u_n)}{\partial z} = \frac{\partial \mathbf{H}(z, u_1, \ldots, u_n)}{\partial z} \bigg|_{z=0}. \tag{8}$$

The solution $\mathbf{H}(z, u_1, \ldots, u_n)$ of system (7) satisfies condition

$$\mathbf{H}(z, 0, \ldots, 0) = \mathbf{r}(z)$$

and determines the characteristic function of the number of occupied servers in the stationary mode for the system SM|M$^{(n)}$|∞ by the equality

$$Me^{j \sum_{i=1}^{n} u_i l_i(t)} = \mathbf{H}(\infty, u_1, \ldots, u_n)\mathbf{e}. \tag{9}$$

$\mathbf{r}(z)$ — stationary probability distribution of a two-dimensional stochastic process $\{s(t), z(t)\}$, which has the form

$$\mathbf{r}(z) = \kappa_1 \mathbf{r} \int_0^z (\mathbf{P} - \mathbf{A}(x)) \, dx, \tag{10}$$

where $\mathbf{r}$ — stationary probability distribution of the Markov chain $\xi(k)$, $k = 1, \ldots, K$, $\kappa_1 = \frac{1}{\mathbf{r}\mathbf{A}\mathbf{e}}$, $\mathbf{A} = \int_0^{\infty} (\mathbf{P} - \mathbf{A}(x)) \, dx$.

The equation (7) will be considered as the basis for further research.

3 The Main Probabilistic Characteristics for System $SM|M^{(n)}|\infty$

Theorem 1. *For the initial moments of number of employed devices of each type for the steady-state functioning of the heterogeneous system $SM|M^{(n)}|\infty$ the following statements are true:*

Statement 1

The average value of number employed devices of the i-th type $fm_i(i = 1,\ldots,n)$ in the heterogeneous system $SM|M^{(n)}|\infty$ has the form:

$$fm_i = \frac{p_i}{\mu_i}\lambda, \qquad (11)$$

where $\lambda = \mathbf{r}'(0)\mathbf{e}$, $\mathbf{e} = [1,\ldots,1]^T$ — a unit column vector.

Statement 2

Initial moments of the second order of number of employed devices of the i-th type sm_i $(i = 1,\ldots,n)$ in the heterogeneous system $SM|M^{(n)}|\infty$ has the form:

$$sm_i = \frac{p_i}{\mu_i}\left[\lambda + p_i\mathbf{r}'(0)\mathbf{A}^*(\mu_i)(\mathbf{I} - \mathbf{A}^*(\mu_i))^{-1}\mathbf{e}\right],$$

where $\mathbf{A}^*(\alpha) = \int\limits_0^\infty e^{-\alpha z}d\mathbf{A}(z)$.

Statement 3

Correlation moment of number of employed devices of the i-th and g-th types cm_{ig} $(i = 1,\ldots,n,\ g = 1,\ldots,n,\ i \neq g)$ in the heterogeneous system $SM|M^{(n)}|\infty$ has the form:

$$cm_{ig} = \frac{p_i p_g}{\mu_i + \mu_g}\mathbf{r}'(0)\left[\mathbf{A}^*(\mu_i)\left(\mathbf{I} - \mathbf{A}^*(\mu_i)\right)^{-1} + \mathbf{A}^*(\mu_g)\left(\mathbf{I} - \mathbf{A}^*(\mu_g)\right)^{-1}\right]\mathbf{e}.$$
$$(12)$$

Proof. Denote:

- $\mathbf{fm}_i(z) = [fm_i(1,z), fm_i(2,z),\ldots,fm_i(K,z)]$ — row-vector of conditional mathematical expectations of number employed devices of i-th type ($i = 1,\ldots,n$);
- $\mathbf{sm}_i(z) = [sm_i(1,z), sm_i(2,z),\ldots,sm_i(K,z)]$ — row-vector of conditional moments of the second order of number employed devices of i-th type ($i = 1,\ldots,n$);
- $\mathbf{cm}_{ig}(z) = [cm_{ig}(1,z), cm_{ig}(2,z),\ldots,cm_{ig}(K,z)]$ — row-vector of correlation moments of number employed devices of i-th and g-th types ($i = 1,\ldots,n,\ g = 1,\ldots,n,\ i \neq g$).

We use the following properties of the characteristic function:

$$\left.\frac{\partial \mathbf{H}(z, u_1,\ldots,u_n)}{\partial u_i}\right|_{u_1=0,\ldots,u_n=0} = j\mathbf{fm}_i(z),$$

$$\left.\frac{\partial^2 \mathbf{H}(z, u_1, \ldots, u_n)}{\partial u_i^2}\right|_{u_1=0,\ldots,u_n=0} = j^2 \mathbf{sm}_i(z), \tag{13}$$

$$\left.\frac{\partial^2 \mathbf{H}(z, u_1, \ldots, u_n)}{\partial u_i u_g}\right|_{u_1=0,\ldots,u_n=0} = j^2 \mathbf{cm}_{ig}(z),$$

$$i = 1, \ldots, n, \; g = 1, \ldots, n, \; i \neq g.$$

Initial moments of the first order.

The average number of occupied devices of each type in the system is determined as follows:

$$fm_i = \mathbf{fm}_i(\infty)\mathbf{e}, \; i = 1, \ldots, n, \; \mathbf{e} = [1, \ldots, 1]^T. \tag{14}$$

Differentiate equation (7) with respect to u_i, $i = 1, \ldots, n$.

$$\left.\frac{\partial^2 \mathbf{H}(z, u_1, \ldots, u_n)}{\partial u_i \partial z}\right|_{u_1=0,\ldots,u_n=0} + j^2 \mu_i e^{-ju_i} \left.\frac{\partial \mathbf{H}(z, u_1, \ldots, u_n)}{\partial u_i}\right|_{u_1=0,\ldots,u_n=0}$$

$$+ j \sum_{\nu=1}^{n} \mu_\nu (1 - e^{-ju_\nu}) \left.\frac{\partial^2 \mathbf{H}(z, u_1, \ldots, u_n)}{\partial u_i \partial u_\nu}\right|_{u_1=0,\ldots,u_n=0} \tag{15}$$

$$+ \left.\frac{\partial^2 \mathbf{H}(0, u_1, \ldots, u_n)}{\partial u_i \partial z} \left(\sum_{i=1}^{n} p_i e^{ju_i} \mathbf{A}(z) - \mathbf{I}\right)\right|_{u_1=0,\ldots,u_n=0}$$

$$+ j \left.\frac{\partial \mathbf{H}(0, u_1, \ldots, u_n)}{\partial z} p_i e^{ju_i} \mathbf{A}(z)\right|_{u_1=0,\ldots,u_n=0} = 0, \; i = 1, \ldots, n,$$

taking into account (13) we obtain

$$\mathbf{fm}_i'(z) - \mu_i \mathbf{fm}_i(z) + \mathbf{fm}_i'(0)(\mathbf{A}(z) - \mathbf{I}) + p_i \mathbf{r}'(0)\mathbf{A}(z) = 0, \; i = 1, \ldots, n. \tag{16}$$

This equation will be solved by the conversation of Laplace-Stieltjes, denoting

$$\mathbf{\Phi}_i(\alpha) = \int_0^{\infty} e^{-\alpha z} d\mathbf{fm}_i(z), \; i = 1, \ldots, n, \; \mathbf{A}^*(\alpha) = \int_0^{\infty} e^{-\alpha z} d\mathbf{A}(z). \tag{17}$$

Completing the conversation of Laplace-Stieltjes in (16), we obtain the equality

$$(\mu_i - \alpha)\mathbf{\Phi}_i(\alpha) = \mathbf{fm}_i'(0)(\mathbf{A}^*(\alpha) - \mathbf{I}) + \mathbf{r}'(0)p_i\mathbf{A}^*(\alpha), \; i = 1, \ldots, n, \tag{18}$$

putting in which $\alpha = \mu_i$, $i = 1, \ldots, n$, we find the form of the vector $\mathbf{fm}_i'(0)$

$$\mathbf{fm}_i'(0) = p_i \mathbf{r}'(0)\mathbf{A}^*(\mu_i)(\mathbf{I} - \mathbf{A}^*(\mu_i))^{-1}. \tag{19}$$

Substituting the expression (19) in the (18) we obtain

$$\mathbf{\Phi}_i(\alpha) = \frac{1}{\mu_i - \alpha}\left\{\mathbf{fm}_i'(0)(\mathbf{A}^*(\alpha) - \mathbf{I}) + p_i \mathbf{r}'(0)\mathbf{A}^*(\alpha)\right\}, \; i = 1, \ldots, n. \tag{20}$$

Since $\mathbf{fm}_i(\infty) = \mathbf{\Phi}_i(0)$ and $\mathbf{A}^*(\infty) = \mathbf{P}$ then putting $\alpha = 0$ in (20) we obtain

$$\mathbf{\Phi}_i(0) = \mathbf{fm}_i(\infty) = \frac{1}{\mu_i}\left\{\mathbf{fm}_i'(0)(\mathbf{P} - \mathbf{I}) + p_i\mathbf{r}'(0)\mathbf{P}\right\}, \; i = 1,\ldots,n. \qquad (21)$$

Thus we have the following expression for the average value of number employed devices of the i-th type fm_i, $(i = 1,\ldots,n)$:

$$fm_i = \mathbf{fm}_i(\infty)\mathbf{e} = \frac{p_i}{\mu_i}\mathbf{r}'(0)\mathbf{e} = \frac{p_i}{\mu_i}\lambda, \; i = 1,\ldots,n, \; \mathbf{e} = [1,\ldots,1]^T.$$

Initial moments of the second order.

To find the second-order moment of the number of employed devices, we differentiate with respect to u_i, $i = 1,\ldots,n$ the equality (15).

$$\left.\frac{\partial^3 \mathbf{H}(z, u_1, \ldots, u_n)}{\partial u_i^2 \partial z}\right|_{u_1=0,\ldots,u_n=0} + j\mu_i e^{-ju_i}\left.\frac{\partial \mathbf{H}(z, u_1, \ldots, u_n)}{\partial u_i}\right|_{u_1=0,\ldots,u_n=0}$$

$$+ 2j^2\mu_i e^{-ju_i}\left.\frac{\partial^2 \mathbf{H}(z, u_1, \ldots, u_n)}{\partial u_i^2}\right|_{u_1=0,\ldots,u_n=0}$$

$$+ j\sum_{\nu=1}^{n}\mu_\nu(1 - e^{-ju_\nu})\left.\frac{\partial^3 \mathbf{H}(z, u_1, \ldots, u_n)}{\partial u_i^2 \partial u_\nu}\right|_{u_1=0,\ldots,u_n=0} \qquad (22)$$

$$+ \frac{\partial^3 \mathbf{H}(0, u_1, \ldots, u_n)}{\partial u_i^2 \partial z}\left.\left(\sum_{i=1}^{n}p_i e^{ju_i}\mathbf{A}(z) - \mathbf{I}\right)\right|_{u_1=0,\ldots,u_n=0}$$

$$+ 2j\frac{\partial^2 \mathbf{H}(0, u_1, \ldots, u_n)}{\partial u_i \partial z}p_i e^{ju_i}\left.\mathbf{A}(z)\right|_{u_1=0,\ldots,u_n=0}$$

$$+ j^2\frac{\partial \mathbf{H}(0, u_1, \ldots, u_n)}{\partial z}p_i e^{ju_i}\left.\mathbf{A}(z)\right|_{u_1=0,\ldots,u_n=0} = 0, \; i = 1,\ldots,n,$$

taking into account (13), we obtain the differential equation to find $\mathbf{sm}_i(z)$, $i = 1,\ldots,n$

$$\mathbf{sm}_i'(z) + \mu_i\mathbf{fm}_i(z) - 2\mu_i\mathbf{sm}_i(z) + \mathbf{sm}_i'(0)(\mathbf{A}(z) - \mathbf{I})$$
$$+ p_i\left\{\mathbf{fm}_i'(0) + \mathbf{r}'(0)\right\}\mathbf{A}(z) = 0, \; i = 1,\ldots,n. \qquad (23)$$

We will solve equation (23) using the conversation of Laplace-Stiltjes. Denote

$$\mathbf{\Psi}_i(\alpha) = \int_0^\infty e^{-\alpha z}d\mathbf{sm}_i(z), \; i = 1,\ldots,n, \qquad (24)$$

then the equation (23) takes the form

$$(2\mu_i - \alpha)\mathbf{\Psi}_i(\alpha) = \mu_i\mathbf{\Phi}_i(\alpha) + \mathbf{sm}_i'(0)(\mathbf{A}^*(\alpha) - \mathbf{I})$$
$$+ p_i\left\{2\mathbf{fm}_i'(0) + \mathbf{r}'(0)\right\}\mathbf{A}^*(\alpha), \; i = 1,\ldots,n, \qquad (25)$$

$\mathbf{A}^*(\alpha)$ is determined by the expression (17).

Let $\alpha = 2\mu_i$ in (25), we obtain the system of differential equations for $\mathbf{sm}'_i(0)$, $i = 1, \ldots, n$

$$
\begin{aligned}
\mathbf{sm}'_i(0) = [\mu_i \mathbf{\Phi}_i(2\mu_i) \\
+ p_i \left\{ 2\mathbf{fm}'_i(0) + \mathbf{r}'(0) \right\} \mathbf{A}^*(2\mu_i)] \left(\mathbf{I} - \mathbf{A}^*(2\mu_i) \right)^{-1}, \; i = 1, \ldots, n.
\end{aligned}
\tag{26}
$$

It follows from (25) that

$$
\begin{aligned}
\mathbf{\Psi}_i(\alpha) = \frac{1}{2\mu_i - \alpha} [\mu_i \mathbf{\Phi}_i(\alpha) + \mathbf{sm}'_i(0)(\mathbf{A}^*(\alpha) - \mathbf{I}) \\
+ p_i \left\{ 2\mathbf{fm}'_i(0) + \mathbf{r}'(0) \right\} \mathbf{A}^*(\alpha)], \; i = 1, \ldots, n,
\end{aligned}
\tag{27}
$$

and taking into account that

$$
\begin{aligned}
\mathbf{sm}_i(\infty) = \mathbf{\Psi}_i(0) = \frac{1}{2\mu_i} [\mu_i \mathbf{fm}_i(\infty) \\
+ \mathbf{sm}'_i(0) \left(\mathbf{P} - \mathbf{I} \right) + p_i \left\{ 2\mathbf{fm}'_i(0) + \mathbf{r}'(0) \right\} \mathbf{P}], i = 1, \ldots, n.
\end{aligned}
\tag{28}
$$

we can write

$$
\begin{aligned}
sm_i = \mathbf{sm}_i(\infty)\mathbf{e} = \frac{1}{2\mu_i} \mu_i \mathbf{fm}_i(\infty)\mathbf{e} + \frac{p_i}{2\mu_i} \left\{ 2\mathbf{fm}'_i(0) + \mathbf{r}'(0) \right\} \mathbf{P}\mathbf{e} \\
= \frac{p_i}{\mu_i} \left(\mathbf{fm}'_i(0)\mathbf{e} + \lambda \right), \; i = 1, \ldots, n.
\end{aligned}
\tag{29}
$$

Thus, taking into account (19) we have expression for initial moment of the second order

$$
sm_i = \frac{p_i}{\mu_i} \left[\lambda + p_i \mathbf{r}'(0) \mathbf{A}^*(\mu_i)(\mathbf{I} - \mathbf{A}^*(\mu_i))^{-1} \mathbf{e} \right], \; i = 1, \ldots, n.
$$

Correlation moment.

Differentiate the equality (15) respect to u_g, $g = 1, \ldots, n, g \neq i$.

$$
\left. \frac{\partial^3 \mathbf{H}(z, u_1, \ldots, u_n)}{\partial u_i \partial u_g \partial z} \right|_{u_1=0,\ldots,u_n=0} + j\mu_i e^{-ju_i} \left. \frac{\partial^2 \mathbf{H}(z, u_1, \ldots, u_n)}{\partial u_i \partial u_g} \right|_{u_1=0,\ldots,u_n=0}
$$

$$
+ j^2 \mu_g e^{-ju_g} \left. \frac{\partial^2 \mathbf{H}(z, u_1, \ldots, u_n)}{\partial u_i \partial u_g} \right|_{u_1=0,\ldots,u_n=0}
$$

$$
+ j \sum_{\nu=1}^{n} \mu_\nu (1 - e^{-ju_\nu}) \left. \frac{\partial^3 \mathbf{H}(z, u_1, \ldots, u_n)}{\partial u_i \partial u_\nu \partial u_g} \right|_{u_1=0,\ldots,u_n=0}
\tag{30}
$$

$$
+ \frac{\partial^3 \mathbf{H}(0, u_1, \ldots, u_n)}{\partial u_i \partial u_g \partial z} \left. \left(\sum_{i=1}^{n} p_i e^{ju_i} \mathbf{A}(z) - \mathbf{I} \right) \right|_{u_1=0,\ldots,u_n=0}
$$

$$
+ \frac{\partial^2 \mathbf{H}(0, u_1, \ldots, u_n)}{\partial u_i \partial z} jp_g e^{ju_g} \mathbf{A}(z) \Big|_{u_1=0,\ldots,u_n=0}
$$

$$+ j\frac{\partial^2 \mathbf{H}(0, u_1, \ldots, u_n)}{\partial u_g \partial z} p_i e^{j u_i} \mathbf{A}(z)\bigg|_{u_1=0,\ldots,u_n=0} = 0,$$

$$i = 1, \ldots, n, \ g = 1, \ldots, n, g \neq i,$$

taking into account (13):

$$\mathbf{cm}'_{ig}(z) - (\mu_i + \mu_g)\mathbf{cm}_{ig}(z) + \mathbf{cm}'_{ig}(0)\left(\mathbf{A}(z) - \mathbf{I}\right)$$
$$+ \left\{ p_g \mathbf{fm}'_i(0) + p_i \mathbf{fm}'_g(0) \right\} \mathbf{A}(z) = 0, \ i = 1, \ldots, n, \ g = 1, \ldots, n, g \neq i. \tag{31}$$

We will solve equation (31) using the conversation of Laplace-Stiltjes. Denote

$$\mathbf{\Theta}_{ig}(\alpha) = \int_0^\infty e^{-\alpha z} d\mathbf{cm}_{ig}(z), \ i = 1, \ldots, n, g = 1, \ldots, n, g \neq i, \tag{32}$$

then the equation (31) takes the form

$$(\mu_i + \mu_g - \alpha)\mathbf{\Theta}_{ig}(\alpha) = \mathbf{cm}'_{ig}(0)\left(\mathbf{A}^*(\alpha) - \mathbf{I}\right)$$
$$+ \left\{ p_g \mathbf{fm}'_i(0) + p_i \mathbf{fm}'_g(0) \right\} \mathbf{A}^*(\alpha) = 0, \tag{33}$$
$$i = 1, \ldots, n, \ g = 1, \ldots, n, g \neq i,$$

$\mathbf{A}^*(\alpha)$ is determined by the expression (17).

Put $\alpha = \mu_i + \mu_g$ in (33), we obtain the system of differential equations for $\mathbf{cm}'_{ig}(0)$, $i = 1, \ldots, n, \ g = 1, \ldots, n, g \neq i$

$$\mathbf{cm}'_{ig}(0) = \left\{ p_g \mathbf{fm}'_i(0) + p_i \mathbf{fm}'_g(0) \right\} \mathbf{A}^*(\mu_i + \mu_g)\left(\mathbf{I} - \mathbf{A}^*(\mu_i + \mu_g)\right)^{-1} = 0,$$

$$i = 1, \ldots, n, \ g = 1, \ldots, n, g \neq i. \tag{34}$$

Since $\mathbf{cm}_{ig}(\infty) = \mathbf{\Theta}_{ig}(0)$, it follows from (33) that the expression for the correlation moment cm_{ig} is as follows

$$cm_{ig} = \mathbf{cm}_{ig}(\infty)\mathbf{e} = \frac{1}{\mu_i + \mu_g}\left[\mathbf{cm}'_{ig}(0)\left(\mathbf{P} - \mathbf{I}\right)\right.$$
$$+ \left.\left\{ p_g \mathbf{fm}'_i(0) + p_i \mathbf{fm}'_g(0) \right\}\mathbf{P}\right]\mathbf{e}$$
$$= \frac{p_i p_g}{\mu_i + \mu_g}\mathbf{r}'(0)\left[\mathbf{A}^*(\mu_i)\left(\mathbf{I} - \mathbf{A}^*(\mu_i)\right)^{-1} + \mathbf{A}^*(\mu_g)\left(\mathbf{I} - \mathbf{A}^*(\mu_g)\right)^{-1}\right]\mathbf{e}, \tag{35}$$

$$i = 1, \ldots, n, \ g = 1, \ldots, n, g \neq i. \qquad \square$$

We can write the expression for finding the variance of the number of occupied servers of each types in the heterogeneous system $SM|M^{(n)}|\infty$

$$Var_i = sm_i - [fm_i]^2, \ i = 1, \ldots, n,$$

$$Var_i = \frac{p_i}{\mu_i}\lambda + \frac{p_i^2}{\mu_i}\mathbf{r}'(0)\mathbf{A}^*(\mu_i)\left(\mathbf{I} - \mathbf{A}^*(\mu_i)\right)^{-1}\mathbf{e}, \ i = 1, \ldots, n. \tag{36}$$

Now, using the obtained expressions for the main probabilistic characteristics, we can write the equality for the correlation coefficient r_{ig} of the number of different types devices employed in system $SM|M^{(n)}|\infty$

$$r_{ig} = \frac{cov_{ig}}{\sqrt{Var_i Var_g}} = \frac{cm_{ig} - fm_i fm_g}{\sqrt{Var_i Var_g}}, \ i = 1, \ldots, n, \ g = 1, \ldots, n, g \neq i. \tag{37}$$

4   Conclusion

In this paper we construct and investigate a mathematical model as a queueing system with the Semi Markovian incoming flow and heterogeneous service. The main probabilistic characteristics are found for the system under investigation, namely, the initial moments of the first and the second order of the number of employed devices of different type. Furthermore, we found an expression for the correlation coefficient between the number of different types devices employed. The resulting correlation coefficient indicates that the processes of change in the number of employed devices of different type in the system are dependent. Therefore, we can conclude that this infinitely linear queuing system with n types of servers can not be considered as a set of n separate systems with only one type of servers.

In the future it is planned to apply the asymptotic methods of investigation for finding moments of a higher order and for studying the functioning of the system under different special conditions. This may include the development of methods for investigating heterogeneous systems, for example, in the asymptotic condition of: high intensity of the incoming flow or an equivalent increase in the service time on devices of different type or extremely rare changes in special flow states (MMPP, MAP, SM).

There is great interest in the studying of various modifications of heterogeneous queueing systems: heterogeneous queueing systems with returns, with different volumes of applications of special incoming flows, and many others.

References

1. Artalejo, J.R., Gómes-Coral, A.: Retrial Queueing systems: A Computational Approach. Springer, Heidelberg (2008). 318 p
2. Chechelnitsky, A.A., Kucherenko, O.V.: Stationary characteristics of parallel queueing systems with two-dimensional input flow. Collection of Scientific Articles, Minsk, vol. 2, pp. 262–268 (2009). (in Russian)
3. Down, D.G., Wu, R.: Multi-layered round robin routing for parallel servers. Queueing Syst. **53**(4), 177–188 (2006)
4. Efrosinin, D., Farhadov, M., Kudubaeva, S.: Performance analysis and monotone control of a tandem queueing system. In: Vishnevsky, V., Kozyrev, D., Larionov, A. (eds.) DCCN 2013. CCIS, vol. 279, pp. 241–255. Springer, Cham (2014). doi:10.1007/978-3-319-05209-0_21
5. Efrosinin, D., Sztrik, J.: Performance analysis of a two-server heterogeneous retrial queue with threshold policy. Qual. Technol. Quant. Manage. **8**(3), 211–236 (2011)
6. Iravani, S.M.R., Luangkesorn, K.L., Simchi-Levi, D.: A general decomposition algorithm for parallel queues with correlated arrivals. Queueing Syst. **47**(4), 313–344 (2004)
7. Kargahi, M., Movaghar, A.: Utility accrual dynamic routing in real-time parallel systems. IEEE Trans. Parallel Distrib Syst. (TDPS) **21**(12), 1822–1835 (2010)
8. Lebedev, E.: On Asymptotic enlargement problem for stochastic networks. In: Mathematicals Methods and Optimization of Telecommunication Networks: Queues Flows, Systems, Networks, Minsk, pp. 108–109 (2011)

9. Lebedev, E., Chechelnitsky, A.: Limit diffusions for multi-channel networks with interdependent inputs. In: Modern Probabilistic Methods for Analysis and Optimization of Information and Telecommunication Networks, Minsk, pp. 133–136 (2011)

10. Lisovskaya, E., Moiseeva, S.: Study of the queuing systems $M|GI|N|\infty$. In: Dudin, A., Nazarov, A., Yakupov, R. (eds.) ITMM 2015. CCIS, vol. 564, pp. 175–184. Springer, Cham (2015). doi:10.1007/978-3-319-25861-4_15

11. Lisovskaya, E., Moiseeva, S., Pagano, M.: The total capacity of customers in the infinite-server queue with MMPP arrivals. In: Vishnevskiy, V.M., Samouylov, K.E., Kozyrev, D.V. (eds.) DCCN 2016. CCIS, vol. 678, pp. 110–120. Springer, Cham (2016). doi:10.1007/978-3-319-51917-3_11

12. Lozhkhovsky, A.G., Kaptu, V.A., Verbanov, O.V., Kolchar, V.M.: A mathematical model of packet traffic. Bul. Tomsk Polytech. Univ. **9**, 113–119 (2011). TPU, Tomsk (in Russian)

13. Moiseev, A.: Asymptotic analysis of the queueing network $SM - (GI/\infty)^K$. In: Dudin, A., Nazarov, A., Yakupov, R. (eds.) ITMM 2015. CCIS, vol. 564, pp. 73–84. Springer, Cham (2015). doi:10.1007/978-3-319-25861-4_7

14. Moiseeva, S., Zadiranova, L.: Feedback in infinite-server queuing systems. In: Vishnevsky, V., Kozyrev, D. (eds.) DCCN 2015. CCIS, vol. 601, pp. 370–377. Springer, Cham (2016). doi:10.1007/978-3-319-30843-2_38

15. Movaghar, A.: Analysis of a dynamic assignment of impatient customers to parallel queues. Queueing Syst. **67**(3), 251–273 (2011)

16. Pankratova, E., Moiseeva, S.: Queueing system with renewal arrival process and two types of customers. In: IEEE International Congress on Ultra Modern Telecommunications and Control Systems (ICUMT), pp. 514–517. IEEE, St. Petersburg (2015)

17. Pankratova, E., Moiseeva, S.: Queueing system $GI|GI|\infty$ with n types of customers. In: Dudin, A., Nazarov, A., Yakupov, R. (eds.) ITMM 2015. CCIS, vol. 564, pp. 216–225. Springer, Cham (2015). doi:10.1007/978-3-319-25861-4_19

18. Rykov, V., Efrosinin, D.: On the slow server problem. Autom. Remote Control **70**(12), 2013–2023 (2009)

19. Sheu, R.S., Ziedins, I.: Asymptotically optimal control of parallel tandem queues with loss. Queueing Syst. **65**(3), 211–227 (2010)

20. Taqqu, M.S., Leland, W.E., Willinger, W., Wilson, D.V.: Self-similarity in high-speed packet traffic: analysis and modeling of ethernet traffic measurements. Stat. Sci. **10**, 67–85 (1995)

21. Willinger, W.: The discovery of self-similar traffic. In: Haring, G., Lindemann, C., Reiser, M. (eds.) Performance Evaluation: Origins and Directions. LNCS, vol. 1769, pp. 513–527. Springer, Heidelberg (2000). doi:10.1007/3-540-46506-5_24

On Steady-State Analysis of $[M|M|m|m + n]$ -Type Retrial Queueing Systems

Eugene Lebedev, Igor Makushenko, Hanna Livinska$^{(\boxtimes)}$, and Iryna Usar

Applied Statistics Department, Taras Shevchenko National University of Kyiv,
Volodymyrska str., 64, Kyiv 01601, Ukraine
leb@unicyb.kiev.ua, {iamak,usar69}@ukr.net, livinskaav@gmail.com,
http://applstat.univ.kiev.ua

Abstract. In this paper we introduce a bivariate Markov process $Q(t) = (Q_1(t)), Q_2(t)) \in \{0, 1, ..., m + n\} \times Z_+$. The process $Q(t), t \geq 0$, can be seen as the joint process of the number of servers and waiting positions occupied and the number of customers in the orbit of a $[M|M|m|m + n]$ -type retrial queueing system. For the truncated model of $Q(t)$ stationary probabilities are written in explicit vector-matrix form. The result obtained is used for stationary distribution calculation in the model of $Q(t)$ with the infinite orbit and for construction of explicit formulas for stationary probabilities of a $[M|M|1|1 + 1]$ -model.

Keywords: Retrial queueing system · Truncated model · Explicit formulas · Stationary distribution

1 Introduction

One of the queueing theory important topics is the theory of retrial queues (or queues with returning customers, repeated attempts, etc.). Retrial queues arise naturally in our daily activities, in phone systems and computer networks, in the field of data transmission systems. They are widely used in designing of computer networks, in studying of stochastic information processing networks, modern mobile communication systems, etc. (see, for example, [5,6]). This explains the fact that over the past decades the theory of retrial queues has been developed widely. Review of retrial queue literature could be found in [1,4,5].

In all the retrial queueing models considered so far, the underlying assumption has been that the model have some servers and a customer who finds all the servers busy upon arrival joins a group of unsatisfied customers called "orbit" and repeats his request after some random time. Such customers become sources of retrial calls and generate a secondary input flow of customers.

Unfortunately, explicit formulas of stationary probabilities for the most types of retrial systems were obtained only in simplest cases ([2,3,7], etc.). Note also, that some sort of a recurrent algorithm for $[M|M|m|m + n]$ queues with constant retrial rate is presented in [8]. In [9] for multi-server retrial queue with the rate of a repeated flow independent of the number of retrial sources,

© Springer International Publishing AG 2017
A. Dudin et al. (Eds.): ITMM 2017, CCIS 800, pp. 133–146, 2017.
DOI: 10.1007/978-3-319-68069-9_11

an explicit vector-matrix representation of the stationary distribution was obtained. This representation allows to write down stationary probabilities via the model parameters in closed form and to derive explicit formulas for main performance measures.

In many practical situations, queueing system have in addition some places for waiting (queue). So, if a customer arrives and all the servers are busy, he/she can occupy a waiting place. But if all the waiting places are occupied as well, he/she leaves the service area, joins the orbit and retries to get service after some time.

For example, let us imagine a person arriving into a shopping center who would like to take some money out from a cash machine. He/she can decide to do it immediately if there are no people or few people near the cash machine, or decide to do some shopping first and repeat their attempt to get money later. This is the concept of multiserver retrial queue under our consideration. Such systems are considered in works [8, 10].

So, in this paper we consider a multiserver retrial queueing system with the finite number of servers, a finite length queue and an infinite orbit. Such a system can be denoted by a symbol $[M|M|m|m + n]$ (see, for example, [8, 10]), where m is the number of servers, d is the number of waiting places in the queue. We also use a symbol $[M|M|m|m + n]^{(N)}$ for defining the correspondent truncated model with the size N orbit. It is assumed that the primary input flow of customers from outside has a rate λ_j dependent on the number of retrial sources j, $j = 0, 1, ...$, and that service times are exponentially distributed with a constant rate μ. The investigative techniques is similar to approach in [9] and it uses an approximation of the initial model with infinite orbit by means of the truncated one and the direct passage to the limit.

For this aim, a class of bivariate migration processes is introduced to describe the service process of retrial systems with queue. Its first component is associated with the number of customers in the working (service) area, that means the customers being under service and in the queue. The second component is the number of customers in the orbit. Type of the system is chosen by means of controlled migration parameters. This paper is organized as follows. In Sect. 2, we give a brief description of the model as a bivariate continuous-time Markov chain. Some basic assumptions and the model are described. Conditions of existence of steady-state regime for service process in the basic model are given. In Sect. 3, we analyze the steady-state distribution of a truncated model. Stationary probabilities for the system with finite orbit is obtained as vector-matrix formulas. In Sect. 4, we deal with the consequences of result obtained for the case of one server and one place in the queue. Here we present some numerical results as well. Finally, the conclusions and some suggestions for future research are given in Sect. 5.

2   Description of Service Process

Let us define the main model under consideration as a bivariate continuous-time Markov chain $Q(t) = (Q_1(t), Q_2(t))'$, with the set of states $S = \{0, 1, ..., m + n\} \times \{0, 1, ...\}$. This chain $Q(t)$ is defined with its infinitesimal rates $q_{\alpha\beta}$, $\alpha, \beta \in S$, $\alpha \neq \beta$, determined by the system of the following relations:

– if $\alpha = (i,j)$ and $i < m+n$:

$$q_{\alpha\beta} = \begin{cases} \lambda_j, & \text{if } \beta = (i+1,j), \\ \min\{i,m\} \cdot \mu, & \text{if } \beta = (i-1,j), \\ j\nu, & \text{if } \beta = (i+1,j-1); \end{cases}$$

– if $\alpha = (m+n,j)$:

$$q_{\alpha\beta} = \begin{cases} \lambda_j, & \text{if } \beta = (m+n,j+1), \\ m\mu, & \text{if } \beta = (m+n-1,j). \end{cases}$$

By agreement, the rates of those transitions that go beyond the region S are equal zero.

We will simulate the service process of customers in the system with a queue and an infinite number of retrial sources by the migration process $Q(t)$. Thus we construct a typical model of $[M|M|m|m+n]$ multiserver retrial system with queue and an infinite orbit, where m is the number of servers, n is the queue size. The rate λ_j of the input flow depends on the number of retrial sources j. Every retrial source generates Poisson flow with the rate ν. Service times at the each of m servers are independent exponentially distributed random values with the rate μ. So, in terms of the system, $Q_1(t), t \geq 0$, is the number of customers in working area, i.e. the number of occupied servers and occupied waiting places in the queue at the instant t, when $Q_2(t), t \geq 0$, is the number of retrial sources at the instant t. In the paper we propose an effective approach to finding stationary distribution for the bivariate service process $Q(t)$.

At first, let us find out the conditions of steady-state regime existence for $Q(t)$.

Lemma 1. *If* $\lambda = \overline{\lim}_{j\to\infty} \lambda_j < \infty$ *and* $\lambda/m\mu < 1$, *then the Markov chain* $Q(t)$ *is ergodic and its limit distribution coincides with the unique stationary distribution.*

Proof. As Lyapunov test functions, we consider functions of the following form:

$$\phi(i,j) = ai + j, \quad (i,j) \in S,$$

where the parameter a will be determined later.

For these test functions, the mean transfer

$$y_{ij} = \sum_{(i',j') \neq (i,j)} q_{(i,j)(i',j')} \left(\phi(i',j') - \phi(i,j) \right)$$

will be

$$y_{ij} = \begin{cases} \lambda_j a - \min\{i,m\}\mu + j\nu(a-1), & 0 \leq i \leq m+n-1, \\ \lambda_j - m\mu a, & i = m+n. \end{cases}$$

When $\lambda/m\mu < 1$, there exists such a value $\varepsilon > 0$ for any $a \in (\lambda/m\mu, 1)$, that $y_{ij} < -\varepsilon$ for any $(i,j) \in S$ except of a finite number of states $(i,j) \in S$. Thus, for the test functions $\phi(i,j)$, $(i,j) \in S$, the conditions of Tweedy's theorem are satisfied ([7], p. 97).

The Lemma is proved.

To construct computational algorithms and to obtain explicit formulas for the system $[M|M|m|m+n]$, we consider the corresponding truncated model $[M|M|m|m+n]^{(N)}$, wherein the number of retrial sources N is finite. In the last model provided that all places in the working area and in the orbit are occupied, an arrived customer is lost and does not receive service in the system. If we find the stationary distribution of such a system and proceed to the limit under the conditions of Lemma 1 as $N \to \infty$, then the stationary distribution for the system $[M|M|m|m+n]$ can be obtained as well (see [7], Sect. 2.4).

3   Stationary Probabilities for the System with Finite Orbit

Denote a service process for the multiserver system with a queue and a finite orbit $[M|M|m|m+n]^{(N)}$ by $Q^{(N)}(t) = \left(Q_1^{(N)}(t), Q_2^{(N)}(t)\right)'$.

The process $Q^{(N)}(t) = \left(Q_1^{(N)}(t), Q_2^{(N)}(t)\right)'$ takes values in a finite set of states $S^{(N)} = \{0, 1, ..., m+n\} \times \{0, 1, ..., N\}$, and there exists a steady-state regime for it. We denote its stationary distribution by $\pi_{ij}^{(N)}$, $(i,j) \in S$.

Since the rate of the input flow depends in arbitrary way on the number of customers in the orbit, then using generating function method is impossible. In our case we apply the theorem on the equality of probability flows in steady-state regime ([11], Sect. 2).

To formulate the main result, let us introduce the following notations. Let $A(j)$, $j = 0, 1, ..., N-1$ be a tridiagonal matrix of the form:

$$A(j) = \begin{pmatrix} a_0^{(0)} & a_0^{(+)} & 0 & 0 & \cdots & 0 & 0 & 0 \\ a_1^{(-)} & a_1^{(0)} & a_0^{(+)} & 0 & \cdots & 0 & 0 & 0 \\ 0 & a_2^{(-)} & a_2^{(0)} & a_2^{(+)} & \cdots & 0 & 0 & 0 \\ \cdots & \cdots & \cdots & \cdots & \cdots & \cdots & \cdots & \cdots \\ 0 & 0 & 0 & 0 & \cdots & a_{m+n-2}^{(-)} & a_{m+n-2}^{(0)} & a_{m+n-2}^{(+)} \\ 0 & 0 & 0 & 0 & \cdots & 0 & a_{m+n-1}^{(-)} & a_{m+n-1}^{(0)} \end{pmatrix}$$

where

$$a_i^{(0)} = \begin{cases} \lambda_j + i\mu + j\nu, & i = 0, 1, ..., m, \\ \lambda_j + m\mu + j\nu, & i = m+1, m+2, ..., m+n-1, \end{cases}$$

$$a_i^{(+)} = -\lambda_j, \quad i = 0, 1, ..., m+n-2,$$

$$a_i^{(-)} = \begin{cases} -i\mu, & i = 0, 1, ..., m, \\ -m\mu, & i = m+1, m+2, ..., m+n-1, \end{cases}$$

$$B = \begin{pmatrix} 0\,1\,0 \ldots 0 \\ 0\,0\,1 \ldots 0 \\ \ldots \\ 0\,0\,0 \ldots 1 \\ 0\,0\,0 \ldots 0 \end{pmatrix}, \quad C = \begin{pmatrix} 0\,0 \ldots 0\,1 \\ 0\,0 \ldots 0\,1 \\ \ldots \\ 0\,0 \ldots 0\,1 \\ 0\,0 \ldots 0\,1 \end{pmatrix},$$

are matrices of size $(m+n) \times (m+n)$.

By $D(N)$ we will denote the following tridiagonal matrix of size $(m+n-1) \times (m+n-1)$:

$$D(N) = \begin{pmatrix} \mu & 0 & \ldots & 0 & 0 & 0 & \ldots & 0 & 0 \\ -(N\nu+\lambda_N) & 2\mu & \ldots & 0 & 0 & 0 & \ldots & 0 & 0 \\ -N\nu & -(N\nu+\lambda_N) & \ldots & 0 & 0 & 0 & \ldots & 0 & 0 \\ \ldots & \ldots & \ldots & \ldots & \ldots & \ldots & \ldots & \ldots & \ldots \\ -N\nu & -N\nu & \ldots & -(N\nu+\lambda_N) & m\mu & 0 & \ldots & 0 & 0 \\ -N\nu & -N\nu & \ldots & -N\nu & -(N\nu+\lambda_N) & m\mu & \ldots & 0 & 0 \\ \ldots & \ldots & \ldots & \ldots & \ldots & \ldots & \ldots & \ldots & \ldots \\ -N\nu & -N\nu & \ldots & -N\nu & -N\nu & -N\nu & \ldots & -(N\nu+\lambda_N) & m\mu \end{pmatrix}.$$

We need also the following vectors:

$$\pi'^{(N)}(j) = \left(\pi_{0j}^{(N)}, \pi_{1j}^{(N)}, ..., \pi_{m+n-1j}^{(N)} \right),$$

$$G'^{(N)}(j) = \frac{\pi'^{(N)}(j)}{\pi_{0N}^{(N)}} = \left(G_{0j}^{(N)}, G_{1j}^{(N)}, ..., G_{m+n-1j}^{(N)} \right).$$

Denote an $(m+n-1)$ -dimensional vector composed of units by $\bar{1}(m+n-1)$, an $(m+n-1)$ -dimensional vector with i-th entry equal one and other entries equal zero by $e_i(m+n-1)$. By $\bar{1}$, e_i we will denote similar vectors with dimension $m+n$.

Theorem 1. *If $\lambda_i > 0$, $j = 0, 1, ..., N$, then stationary probabilities $\pi_{ij}^{(N)}$, $(i,j) \in S^{(N)}$ can be written as follows:*

$$\left(\pi_{1N}^{(N)}, \pi_{2N}^{(N)}, ..., \pi_{m+n-1N}^{(N)} \right)'$$
$$= \pi_{0N}^{(N)} D^{-1}(N) \left(N\nu \bar{1}(m+n-1) + \lambda_N e_1(m+n-1) \right), \quad (1)$$

$$\pi_{m+nN}^{(N)} = \frac{\pi_{0N}^{(N)}}{m\mu} G'^{(N)}(N) \left(N\nu \bar{1} + \lambda_N e_{m+n} \right),$$

$$\pi_j'^{(N)} = \frac{\pi_{0N}^{(N)} N! \nu^{N-j}}{j!} G'^{(N)}(N) T(N-1) \times ... \times T(j), \, j = 0, 1, ..., N-1, \quad (2)$$

$$\pi_{m+n\,j}^{(N)} = \frac{\pi_{0N}^{(N)} N! \nu^{N-j}}{\lambda_j j!} G'^{(N)}(N) T(N-1) \times \ldots \times T(j+1)\overline{1}, \qquad (3)$$

$$j = 0, 1, \ldots, N-1,$$

where

$$\pi_{0N}^{(N)} = \left\{ G'^{(N)}(N) \left(\overline{1} + N! \sum_{j=0}^{N-1} \frac{\nu^{N-j}}{j!} T(N-1) \times \ldots \times T(j+1) \right. \right.$$

$$\left. \left. \times \left[T(j) + \frac{1}{\lambda_j} \right] \overline{1} + \frac{1}{m\mu} (N\nu\overline{1} + \lambda_N e_{m+n}) \right) \right\}^{-1}, \qquad (4)$$

$$G^{(N)}(N) = \left(\frac{1}{D^{-1}(N) \left(N\nu\overline{1}(m+n-1) + \lambda_N e_1(m+n-1) \right)} \right), \qquad (5)$$

$$T(j) = \left[B + \frac{m\mu}{\lambda_j} C \right] A^{-1}(j), \ j = 0, 1, \ldots, N-1.$$

Proof. For convenience let us denote by $\pi_{ij}^{(N)} = \widetilde{\pi}_{ij}$, $G_{ij}^{(N)} = \widetilde{G}_{ij}$, $(i,j) \in S^{(N)}$. For every $k = 0, 1, \ldots, m+n-1$ we shall divide $S^{(N)}$ into two subsets $E_k = \{(0,N), (1,N), \ldots, (k,N)\}$ and $\overline{E}_k = S^{(N)} \backslash E_k$. By virtue of the equality of the probability flows through the closed contour in the steady-state regime ([11], Sect. 2), we have:

$$\begin{cases} \sum_{i=0}^{k-1} N\nu\widetilde{\pi}_{iN} + (N\nu + \lambda_N)\widetilde{\pi}_{kN} = (k+1)\mu\widetilde{\pi}_{k+1\,N}, \\ \qquad\qquad\qquad\qquad k = 0, 1, \ldots, m-1, \\ \sum_{i=0}^{k-1} N\nu\widetilde{\pi}_{iN} + (N\nu + \lambda_N)\widetilde{\pi}_{kN} = m\mu\widetilde{\pi}_{k+1\,N}, \\ \qquad\qquad\qquad\qquad k = m, \ldots, m+n-1. \end{cases} \qquad (6)$$

For $\widetilde{G}_{ij} = \widetilde{\pi}_{ij}/\widetilde{\pi}_{0N}$, $(i,j) \in S^{(N)}$, the first $(m+n-1)$ equations from system (6) have the following form:

$$\begin{cases} \mu G_{1N} = N\nu + \lambda_N, \\ -(N\nu + \lambda_N)G_{1N} + 2\mu G_{2N} = N\nu, \\ -N\nu G_{1N} - (N\nu + \lambda_N)G_{2N} + 3\mu G_{3N} = N\nu, \\ \ldots \\ -N\nu G_{1N} - \ldots - N\nu G_{m-2\,N} - (N\nu + \lambda_N)G_{m-1\,N} + m\mu G_{mN} = N\nu, \\ -N\nu G_{1N} - \ldots - N\nu G_{m-1\,N} - (N\nu + \lambda_N)G_{mN} + m\mu G_{m+1\,N} = N\nu, \\ \ldots \\ -N\nu G_{1N} - \ldots - N\nu G_{m+n-3\,N} - (N\nu + \lambda_N)G_{m+n-2\,N} + m\mu G_{m+n-1\,N} = N\nu. \end{cases} \qquad (7)$$

With respect to $G_{1N}, G_{2N}, \ldots, G_{m+n-1\,N}$, the solution of (7) is:

$$\begin{pmatrix} \widetilde{G}_{1N} \\ \ldots \\ \widetilde{G}_{m+n-1\,N} \end{pmatrix} = D^{-1}(N) \left(N\nu\overline{1}(m+n-1) + \lambda_N e_1(m+n-1) \right),$$

which yields (5). From (6) for $k = m + n - 1$ we obtain:

$$\widetilde{G}_{m+nN} = \frac{1}{m\mu}\widetilde{G}'(N)(N\nu\overline{1} + \lambda_N e_{m+n}). \tag{8}$$

Now we have to obtain $\widetilde{G}_{m+nj}$ under assumption, that $j = 0, 1, ..., N - 1$. Divide $S^{(N)}$ into two subsets $S_j^{(N)} = \{(\alpha, \beta) \in S^{(N)} : \beta \leq j\}$ and $\overline{S}_j^{(N)} = S^{(N)} \setminus S_j^{(N)}$. Using again the equality of the probabilities flows through the closed contour, we have:

$$\lambda_j \widetilde{\pi}_{m+nj} = (j+1)\nu\widetilde{\pi}_{0j+1} + ... + (j+1)\nu\widetilde{\pi}_{m+n-1j+1},$$

or

$$\lambda_j \widetilde{G}_{m+nj} = (j+1)\nu\widetilde{G}_{0j+1} + ... + (j+1)\nu\widetilde{G}_{m+n-1j+1},$$

whence it follows that

$$\widetilde{G}_{m+nj} = \frac{(j+1)\nu}{\lambda_j}\widetilde{G}'(j+1)\overline{1}, \ j = 0, 1, ..., N - 1. \tag{9}$$

Consider now $(m+n) \times N$ closed contours, which contain one point (i, j) from the domain $\widetilde{S}^{(N)} = \{0, 1, ..., m + n - 1\} \times \{0, 1, ..., N - 1\}$. The corresponding equations for G_{ij}, $(i, j) \in \widetilde{S}^{(N)}$, have the form:

$$(\lambda_j + j\nu)\widetilde{G}_{0j} = \mu\widetilde{G}_{1j}, \ i = 0, \tag{10}$$

$$\begin{cases} (\lambda_j + i\mu + j\nu)\widetilde{G}_{ij} = (j+1)\nu\widetilde{G}_{i-1j+1} + \lambda_j\widetilde{G}_{i-1j} + (i+1)\mu\widetilde{G}_{i+1j}, \\ \qquad\qquad\qquad\qquad\qquad\qquad\qquad\qquad\qquad i = 1, 2, ..., m - 1, \\ (\lambda_j + m\mu + j\nu)\widetilde{G}_{ij} = (j+1)\nu\widetilde{G}_{i-1j+1} + \lambda_j\widetilde{G}_{i-1j} + m\mu\widetilde{G}_{i+1j}, \\ \qquad\qquad\qquad\qquad\qquad\qquad\qquad\qquad\qquad i = m, m+1, ..., m+n-2. \end{cases} \tag{11}$$

When $i = m + n - 1$, taking into account (9), we obtain:

$$(\lambda_j + m\mu + j\nu)\widetilde{G}_{m+n-1j}$$
$$= (j+1)\nu\widetilde{G}_{m+n-2j+1} + \lambda_j\widetilde{G}_{m+n-2j} + \frac{(j+1)\nu m\mu}{\lambda_j}\widetilde{G}'(j+1)\overline{1}. \tag{12}$$

System $(10) - (12)$ can be represented in a vector-matrix form:

$$\widetilde{G}'(j) = (j+1)\nu\widetilde{G}'(j+1)\left[B + \frac{m\mu}{\lambda_j}C\right]A^{-1}(j)$$
$$= (j+1)\nu\widetilde{G}'(j+1)T(j), \ j = 0, 1, ..., N - 1. \tag{13}$$

The solution of recurrent relation (13) is the sequence of vectors

$$G'(j) = \frac{N!\nu^{N-j}}{j!}\widetilde{G}'(N)T(N-1) \times ... \times T(j), \ j = 0, 1, ..., N - 1. \tag{14}$$

Substituting the right-hand side of (14) in (9), we get

$$\widetilde{G}_{m+nj} = \frac{N!\nu^{N-j}}{j!\lambda_j}\widetilde{G}'(N)T(N-1)\times \dots \times T(j+1)\bar{I}, \quad j=0,1,...,N-1. \quad (15)$$

The normalization condition for stationary probabilities $\widetilde{\pi}_{ij}$, $(i,j)\in S$, which looks like $\sum_{i=0}^{m+n}\sum_{j=0}^{N}\widetilde{\pi}_{ij} = 1$, can be rewritten as follows:

$$\widetilde{\pi}_{0N}\left(\sum_{i=0}^{m+n-1}G_{iN} + G_{m+nN} + \sum_{j=0}^{N-1}G_{m+nj} + \sum_{i=0}^{m+n-1}\sum_{j=0}^{N-1}G_{ij}\right) = 1.$$

That allows us to find $\widetilde{\pi}_{0N}$. Thus we obtain formula (4). Relations (1) − (3) are a direct consequence of (8), (14), (15).

Theorem is proved.

Obviously, these formulas are an effective recurrent procedure for computing the stationary distribution.

4   Case of One Server and One Waiting Place

Let us apply the obtained result for systems $[M|M|1|1+1]^{(N)}$ and $[M|M|1|1+1]$. In this case, we can carry out more detailed analysis and derive explicit formulas.

Denote by

$$A_i(j) = \begin{cases} \prod_{k=i}^{j-1}\dfrac{(1+\rho_k)\mu+\lambda_{k+1}+(k+1)\nu}{\rho_k[(\lambda_k+k\nu)^2+k\nu\mu]}, & i < j, \\ 1, & i = j, \end{cases}$$

where $\rho_k = \lambda_k/\mu$ is system load with primary customers when the orbit $Q_2(t) = k$.

Corollary 1. *If* $\lambda_j > 0$, $j = 0,1,...,N$, *then for any* N *stationary probabilities of the* $[M|M|1|1+1]^{(N)}$*-retrial system have the following form:*

$$\pi_{0j}^{(N)} = \pi_{0N}^{(N)}\frac{N!\nu^{N-j}A_j(N)}{j!}, \quad \pi_{1j}^{(N)} = \pi_{0N}^{(N)}\frac{N!\nu^{N-j}(\lambda_j+j\nu)A_j(N)}{j!\mu}, \quad (16)$$

$$\pi_{2j}^{(N)} = \pi_{0N}^{(N)}\frac{N!\nu^{N-j}(\mu+\lambda_{j+1}+(j+1)\nu)A_{j+1}(N)}{j!\lambda_j\mu}, \quad j=0,1,...,N-1, \quad (17)$$

$$\pi_{1N}^{(N)} = \pi_{0N}^{(N)}\frac{\lambda_N+N\nu}{\mu}, \quad \pi_{2N}^{(N)} = \pi_{0N}^{(N)}\frac{(\lambda_N+N\nu)^2+N\nu\mu}{\mu^2},$$

$$\pi_{0N}^{(N)} = \left\{ 1 + \frac{1}{\mu}(\lambda_N + N\nu) + N! \sum_{j=0}^{N-1} \frac{\nu^{N-j}(\mu + \lambda_j + j\nu)A_j(N)}{j!\mu} \right.$$

$$\left. + N! \sum_{j=0}^{N-1} \frac{\nu^{N-j}(\mu + \lambda_{j+1} + (j+1)\nu)A_{j+1}(N)}{j!\lambda_j\mu} + \frac{1}{\mu^2}\left((\lambda_N + N\nu)^2 + N\nu\mu\right) \right\}^{-1}.$$

Proof. For the $[M|M|1|1+1]^{(N)}$- retrial system we have

$$B = \begin{pmatrix} 0 & 1 \\ 0 & 0 \end{pmatrix}, \quad C = \begin{pmatrix} 0 & 1 \\ 0 & 1 \end{pmatrix},$$

$$A(j) = \begin{pmatrix} \lambda_j + j\nu & -\lambda_j \\ -\mu & \lambda_j + \mu + j\nu \end{pmatrix}.$$

Since

$$\begin{vmatrix} \lambda_j + j\nu & -\lambda_j \\ -\mu & \lambda_j + \mu + j\nu \end{vmatrix} = (\lambda_j + j\nu)(\lambda_j + \mu + j\nu) - \lambda_j\mu = (\lambda_j + j\nu)^2 + \mu j\nu,$$

then

$$A^{-1}(j) = \frac{1}{(\lambda_j + j\nu)^2 + \mu j\nu} \begin{pmatrix} \lambda_j + \mu + j\nu & \lambda_j \\ \mu & \lambda_j + j\nu \end{pmatrix},$$

and matrix $T(j)$ has the form:

$$T(j) = \left[B + \frac{\mu}{\lambda_j} C \right] A^{-1}(j) = \left[B + \rho_j^{-1} C \right] A^{-1}(j)$$

$$= \frac{1}{\rho_j[(\lambda_j + j\nu)^2 + \mu j\nu]} \begin{pmatrix} (1 + \rho_j)\mu & (1 + \rho_j)(\lambda_j + j\nu) \\ \mu & \lambda_j + j\nu \end{pmatrix},$$

$$j = 0, 1, ..., N - 1.$$

Let

$$D(j) = \begin{pmatrix} (1 + \rho_j)\mu & (1 + \rho_j)(\lambda_j + j\nu) \\ \mu & \lambda_j + j\nu \end{pmatrix}, \quad j = 0, 1, ...$$

It is easy to see that

$$D(j+1)D(j)$$

$$= ((1 + \rho_j)\mu + \lambda_{j+1} + (j+1)\nu) \begin{pmatrix} (1 + \rho_j)\mu & (1 + \rho_j)(\lambda_j + j\nu) \\ \mu & \lambda_j + j\nu \end{pmatrix}.$$

Which implies

$$D(m)...D(j+1)D(j)$$

$$= \prod_{k=j}^{m-1} ((1 + \rho_k)\mu + \lambda_{k+1} + (k+1)\nu) \begin{pmatrix} (1 + \rho_m)\mu & (1 + \rho_m)(\lambda_j + j\nu) \\ \mu & \lambda_j + j\nu \end{pmatrix}$$

for $m > j$, and therefore

$$T(N-1)...T(j+1)T(j) =$$

$$= \frac{1}{(1+\rho_{N-1})\mu + \lambda_N + N\nu} \prod_{k=j}^{N-1} \frac{(1+\rho_k)\mu + \lambda_{k+1} + (k+1)\nu}{\rho_k[(\lambda_k + k\nu)^2 + k\mu\nu]}$$

$$\times \begin{pmatrix} (1+\rho_{N-1})\mu & (1+\rho_{N-1})(\lambda_j + j\nu) \\ \mu & \lambda_j + j\nu \end{pmatrix}$$

$$= \frac{A_j(N)}{(1+\rho_{N-1})\mu + \lambda_N + N\nu} \begin{pmatrix} (1+\rho_{N-1})\mu & (1+\rho_{N-1})(\lambda_j + j\nu) \\ \mu & \lambda_j + j\nu \end{pmatrix}. \quad (18)$$

From the first two equations of system (5) we have $G^{(N)'}(N) = (G_{0N}^{(N)}, G_{1N}^{(N)}) = (1, 1/\mu \cdot (\lambda_N + N\nu))$. It is not difficult to verify that

$$G^{(N)'}(N) \begin{pmatrix} (1+\rho_{N-1})\mu & (1+\rho_{N-1})(\lambda_j + j\nu) \\ \mu & \lambda_j + j\nu \end{pmatrix}$$

$$= ((1+\rho_{N-1})\mu + \lambda_N + N\nu)(1, \frac{1}{\mu}(\lambda_j + j\nu)).$$

And thus

$$G^{(N)'}(N)T(N-1)...T(j+1)T(j) = A_j(N)(1, \frac{1}{\mu}(\lambda_j + j\nu)), \quad (19)$$

$$G_{2j}^{(N)} = \frac{N!\nu^{N-j}}{j!\lambda_j} G^{(N)'}(N)T(N-1)...T(j+1)\bar{1}$$

$$\frac{N!\nu^{N-j}}{j!\lambda_j} A_{j+1}(N)(1, \frac{1}{\mu}(\lambda_{j+1} + (j+1)\nu))\bar{1}. \quad (20)$$

Now from (18) we obtain:

$$\pi_{0N}^{(N)} = \left\{ G^{(N)'}(N) \left(\bar{1} + N! \sum_{j=0}^{N-1} \frac{\nu^{N-j}}{j!} T(N-1) \times ... \times \right. \right.$$

$$\times T(j+1)[T(j)\bar{1} + \frac{1}{\lambda_j}\bar{1}] + \frac{1}{m\mu}(N\nu\bar{1} + \lambda_N e_{m+n}) \Bigg) \Bigg\}^{-1}$$

$$= \left\{ 1 + \frac{1}{\mu}(\lambda_N + N\nu) + N! \sum_{j=0}^{N-1} \frac{\nu^{N-j}(1 + \frac{1}{\mu}(\lambda_j + j\nu))A_j(N)}{j!} \right.$$

$$+ N! \sum_{j=0}^{N-1} \frac{\nu^{N-j}(1 + \frac{1}{\mu}(\lambda_{j+1} + (j+1)\nu))A_{j+1}(N)}{j!\lambda_j} + \frac{1}{\mu}\left(\frac{1}{\mu}(\lambda_N + N\nu)^2 + N\nu \right) \Bigg\}^{-1}.$$

Relations (16), (17) can be deduced from (1)–(3) and (18)–(20). The corollary is proved.

We now consider the system $[M|M|1|1+1]$. Lemma 1 yields that this system has a steady-state regime provided that $\overline{\lim}_{j\to\infty}[\lambda_j/\mu] < 1$. This condition is assumed to be satisfied.

Denote

$$R_j = \lim_{N\to\infty} \pi_{0N}^{(N)} N! \nu^N A_j(N).$$

It is easy to show that

$$R_j = \left\{ \sum_{i=0}^{j} \frac{(\mu + \lambda_i + i\nu)A_i(j)}{\nu^i \mu i!} + \sum_{i=j+1}^{\infty} \frac{\mu + \lambda_i + i\nu}{A_j(i)\nu^i \mu i!} \right.$$
$$\left. + \sum_{i=0}^{j-1} \frac{(\mu + \lambda_{i+1} + (i+1)\nu)A_{i+1}(j)}{\nu^i \lambda_i \mu i!} + \sum_{i=j}^{\infty} \frac{\mu + \lambda_{i+1} + (i+1)\nu}{A_j(i+1)\nu^i \lambda_i \mu i!} \right\}^{-1}. \quad (21)$$

Indeed, using the representation for $\pi_{0N}^{(N)}$ from Corollary 1, we obtain the follows:

$$\lim_{N\to\infty} \pi_{0N}^{(N)} N! \nu^N A_j(N)$$

$$= \left\{ \lim_{N\to\infty} \sum_{i=0}^{N-1} \frac{(\mu + \lambda_i + i\nu)A_i(N)}{\nu^i \mu i! A_j(N)} \right.$$
$$\left. + \lim_{N\to\infty} \sum_{i=0}^{N-1} \frac{(\mu + \lambda_{i+1} + (i+1)\nu)A_{i+1}(N)}{\nu^i \lambda_i \mu i! A_j(N)} \right\}^{-1}. \quad (22)$$

It is easy to see that when $i, j < N$, then

$$\frac{A_i(N)}{A_j(N)} = \begin{cases} A_i(j), & \text{if } i < j, \\ A_i(i) = 1, & \text{if } i = j, \\ A_j^{-1}(i), & \text{if } i > j. \end{cases}$$

Accordingly, we can write:

$$\lim_{N\to\infty} \sum_{i=0}^{N-1} \frac{(\mu + \lambda_i + i\nu)A_i(N)}{\nu^i \mu i! A_j(N)}$$

$$= \sum_{i=0}^{j} \frac{(\mu + \lambda_i + i\nu)A_i(j)}{\nu^i \mu i!} + \lim_{N\to\infty} \sum_{i=j+1}^{N} \frac{\mu + \lambda_i + i\nu}{\nu^i \mu i! A_j(i)}$$

$$= \sum_{i=0}^{j} \frac{(\mu + \lambda_i + i\nu)A_i(j)}{\nu^i \mu i!} + \sum_{i=j+1}^{\infty} \frac{\mu + \lambda_i + i\nu}{\nu^i \mu i! A_j(i)}.$$

Similarly, it can be shown that

$$\lim_{N\to\infty} \sum_{i=0}^{N-1} \frac{(\mu + \lambda_{i+1} + (i+1)\nu)A_{i+1}(N)}{\nu^i \lambda_i \mu i! A_j(N)}$$

$$= \sum_{i=0}^{j-1} \frac{(\mu + \lambda_{i+1} + (i+1)\nu)A_{i+1}(j)}{\nu^i \lambda_i \mu i!}$$

$$+ \sum_{i=j}^{\infty} \frac{\mu + \lambda_{i+1} + (i+1)\nu}{\nu^i \lambda_i \mu i! A_j(i+1)}.$$

From these relations and (22) it follows that (21) takes place.

Corollary 2. *If $\lambda_j > 0$, $j = 0, 1, ...$, and $\overline{\lim}_{j\to\infty}(\lambda_j/\mu) < 1$ for the $[M|M|1|1+1]$- retrial system, then for the system there exist stationary probabilities:*

$$\pi_{0j} = \frac{R_j}{\nu^j j!}, \quad \pi_{1j} = \frac{(\lambda_j + j\nu)R_j}{\nu^j \mu j!},$$

$$\pi_{2j} = \frac{(\mu + \lambda_{j+1} + (j+1)\nu)R_j}{\nu^j \lambda_j \mu j!} \cdot \frac{\rho_j[(\lambda_j + j\nu)^2 + j\nu\mu]}{(1+\rho_j)\mu + \lambda_{j+1} + (j+1)\nu},$$

where

$$R_j = \left\{ \sum_{i=0}^{j} \frac{(\mu + \lambda_i + i\nu)A_i(j)}{\nu^i \mu i!} + \sum_{i=j+1}^{\infty} \frac{\mu + \lambda_i + i\nu}{\nu^i \mu i! A_j(i)} \right.$$

$$\left. + \sum_{i=0}^{j-1} \frac{(\mu + \lambda_{i+1} + (i+1)\nu)A_{i+1}(j)}{\nu^i \lambda_i \mu i!} + \sum_{i=j}^{\infty} \frac{\mu + \lambda_{i+1} + (i+1)\nu}{\nu^i \lambda_i \mu i! A_j(i+1)} \right\}^{-1}.$$

Let us consider an example of a system $[M|M|1|1+1]^{(20)}$. Let $\lambda_j = 1.1$, $j = 0, 1, ..., 20$, $\mu = 1$, $\nu = 0, 1$.

The program, based on the formulas obtained, gives us the following values of the stationary distribution shown in Table 1.

From here blocking probability is calculated:

$$\pi_2^{(20)} = \sum_{j=0}^{20} \pi_{2j}^{(20)} = 0.34658565.$$

Table 1. Stationary probabilities for the system $[M|M|1|1+1]^{(20)}$ when $\lambda_j = 1.1$, $j = 0, 1, ..., 20$, $\mu = 1$, $\nu = 0, 1$.

j	$\pi_{0j}^{(20)}$	$\pi_{1j}^{(20)}$	$\pi_{2j}^{(20)}$
0	0.00734806	0.00808286	0.00355646
1	0.0177823	0.0213388	0.01105
2	0.0264239	0.034351	0.020315
3	0.0310369	0.0434516	0.0287473
4	0.031622	0.047433	0.0345834
5	0.0292629	0.0468206	0.0371954
6	0.0252561	0.0429354	0.0368362
7	0.0206734	0.0372121	0.0342339
8	0.0162316	0.0308401	0.0302457
9	0.0123223	0.0246446	0.0256405
10	0.00909825	0.0191063	0.0210012
11	0.00656287	0.0144383	0.0167072
12	0.00464089	0.010674	0.0129628
13	0.00322603	0.00774247	0.00984138
14	0.00220929	0.00552322	0.00733058
15	0.00149327	0.00388249	0.00536908
16	0.000997633	0.00269361	0.0038738
17	0.000659625	0.00184695	0.00275753
18	0.000432092	0.00125307	0.00193919
19	0.000280672	0.000842016	0.00134874
20	0.00018093	0.00056087	0.00105029

5   Conclusion

In this paper a multiserver retrial queueing system of the $[M|M|m+n]$ -type with a queue is considered. A class of bivariate migration processes is introduced to describe the service process of the system.

For finding the stationary distribution of the service process in the system, a two-stage approach is offered. At the first stage, explicit formulas for the stationary distribution in the vector-matrix form are derived for a truncated system with a finite orbit. The next step is to approximate the stationary probabilities for a system with an infinite orbit by the probabilities obtained in the first stage.

As an example, stationary probabilities as well as blocking probability are calculated for a retrial system of $[M|M|1|1+1]^{(20)}$ -type.

Note that the input flow rate for the models considered has variable nature. This feature allows us to use the results obtained for formulation and solution of various problems on the system optimal control.

References

1. Anisimov, V.V., Artalejo, J.R.: Analysis of Markov multiserver retrial queues with negative arrivals. Queuing Syst. **39**, 157–182 (2001)
2. Artalejo, J.R.: Stationary analysis of the characteristics of the M/M/2 queue with constant repeated attempts. Opsearch **33**, 83–95 (1996)
3. Artalejo, J.R., Gómez-Corral, A., Neuts, M.F.: Analysis of multiserver queues with constant retrial rate. Eur. J. Oper. Res. **135**, 569–581 (2001)
4. Artalejo, J.R., Falin, G.I.: Standard and retrial queueing systems: a comparative analysis. Rev. Mat. Complut. **15**, 101–129 (2002)
5. Artalejo, J.R., Gómez-Corral, A.: Retrial Queueing Systems. Springer, Heidelberg (2008)
6. Avrachenkov, K., Yechiali, U.: Retrial networks with finite buffers and their application to internet data traffic. Probab. Eng. Inf. Sci. **22**, 519–536 (2008)
7. Falin, G.I., Templeton, J.G.C.: Retrial Queues. Chapman and Hall, London (1997)
8. Gómez-Corral, A., Ramalhoto, M.F.: The stationary distribution of a Markovian process arising in the theory of multiserver retrial queueing systems. Math. Comput. Model. **30**, 141–158 (1999)
9. Lebedev, E., Ponomarov, V.: Steady-state analysis of M/M/c/c-type retrial queueing systems with constant retrial rate. TOP **24**, 693–704 (2016)
10. Ramalhoto, M.F., Gómez-Corral, A.: Some decomposition formulae for M/M/r/r+d queues with constant retrial rate. Stoch. Models **14**, 123–145 (1998)
11. Walrand, J.: An Introduction to Queueing Networks. Prentice Hall, Englewood Cliffs (1988)

Analysis of Queueing Tandem with Feedback by the Method of Limiting Decomposition

Maria Shklennik[1(✉)], Svetlana Moiseeva[1], and Alexander Moiseev[1,2]

[1] Tomsk State University, Tomsk, Russia
shklennikm@yandex.ru, smoiseeva@mail.ru, moiseev.tsu@gmail.com
[2] Peoples' Friendship University of Russia (RUDN University), Moscow, Russia

Abstract. The paper presents the study of a two-stage infinite-server queueing system with feedback. The service time at each stage is given by an arbitrary distribution function. The method of limiting decomposition is used for the study. As a result of the research, stationary distributions of the number of customers at each stage of the system are found. The obtained analytical results are compared with the asymptotic ones which were obtained in previous papers.

Keywords: Infinite-server queueing tandem · Method of limiting decomposition · Feedback

1 Introduction

Multi-stage queueing systems are models in which customers' service is performed sequentially stage by stage. In these systems, a customer arrives at the first stage. After the service at this stage is complete, the customer moves to the second stage, and so on, until it completes the service at the last stage of the system. Then it leaves the system. Two-stage tandems is a subclass of multi-stage queueing systems and they are most often considered in scientific literature on the queueing theory. E.g., such systems were considered in the articles [1–3].

Queueing systems with feedback are mathematical models for many real systems, in which a customer needs to return to the system again to get an additional service [4,5]. In the paper [6], infinite-server systems with feedback and various types of arrival processes were studied.

Infinite-server queueing systems are used in cases when the probability of losing a customer can be neglected. An important property of such systems is that customers in the system do not depend on each other, which allows to analytically solve problems for such systems. In [7,8] it is proved that the number of customers in the $M/M/\infty$ system has the Poisson distribution. In 1958 B.A. Sevastyanov solved the Erlang problem for the systems with an arbitrary distribution function of the service times in the system $M/G/N$ [9]. He showed that the distribution of the number of customers in the system converges to a Poisson distribution if $N \to \infty$. In 1969, L. Takács [10] showed that the number

© Springer International Publishing AG 2017
A. Dudin et al. (Eds.): ITMM 2017, CCIS 800, pp. 147–157, 2017.
DOI: 10.1007/978-3-319-68069-9_12

of customers in the $M/G/\infty$ system has a Poisson distribution in the steady-state regime and it depends on the average rate of arrivals and the average time of service. In many cases, non-Poisson arrivals can be approximated by the Poisson process, for example, if we have a large number of independent input flows [11].

An asymptotic approximation of the investigated processes in queueing systems with an infinite number of servers is of interest for research also. In the papers [12,13], it was proved that if the intensity of the arrivals in the system $GI/G/\infty$ tends to infinity, then the process describing the number of customers in the system converges to the Gaussian process. In the article [14], this theorem was supplemented, and it was proved that the Gaussian approximation is the Ornstein-Uhlenbeck diffusion process if and only if the service time is exponential. Similar studies for infinite-server queueing networks were carried out in [15,16].

For models with an infinite number of servers with arbitrary service times, as well as multi-stage queueing systems, a few number of analytical results were obtained. One of the methods for studying such systems is the method of limiting decomposition, which was proposed to study the $M/G/\infty$ system in [17]. In our paper, the method of limiting decomposition is used to analyze a tandem of queues with an infinite number of servers and a feedback which is possible at any stage of the tandem.

Briefly about the content of the paper. In the Sect. 2, the model under study and the problem are formulated. The main result of the article is presented in the Sect. 3. This is the probability distribution of the number of customers at the stages of the system, which is obtained by the method of limiting decomposition. In the Sect. 4, we present a numerical example and comparison of the analytical results of this article with the asymptotic ones obtained earlier for more general models.

2   Mathematical Model

We consider an infinite-server queuing tandem $M/G/\infty \to G/\infty$ with feedback (Fig. 1). The arrival process is a stationary Poisson process with the rate equals to λ. Service times at the first stage are independent and identically distributed (i.i.d.) with an arbitrary distribution function $B_1(x)$. After a completion of the service at the first stage, the customer may return back to the first stage for a new service with the probability r_{11} or it may move to the second stage with the probability r_{12} or it may leave the system with the probability $(1 - r_{11} - r_{12})$. Service times at the second stage are i.i.d. with an arbitrary distribution function $B_2(x)$. When the service at the second stage is completed, the customer may return to the first stage with the probability r_{21} or it may get a new service at the second stage with the probability r_{22} or it may leave the system with the probability $(1 - r_{21} - r_{22})$.

Such models may describe the set of servers where one group of servers can handle user's requests and perform some actions similar to these handling,

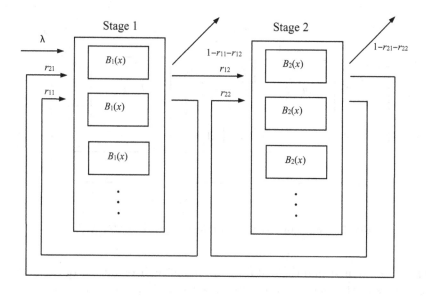

Fig. 1. Queueing tandem $M/G/\infty \to G/\infty$ with feedback.

and another group of servers perform another type of work and users can not directly send requests to these servers.

The problem is to determine the steady-state probability distribution of the number of customers at the stages of the system.

3   Method of Limiting Decomposition

To study the queueing system with an infinite number of lines, an arbitrary service time distribution function and Poisson arrivals, we use the method of limiting decomposition [17]. It is known that the division of the Poisson point process performed according to the binomial scheme gives independent Poisson processes as a result. Proceeding from this, we divide the arrival process in the considered tandem into N independent processes according to a polynomial scheme with identical probabilities. As the rate of the original arrivals was equal to λ, then the intensity of each generated Poisson process will be equal to λ/N. After that we construct a single-line tandem for each of these arrival processes to serve their customers. The considered single-line two-stage queueing tandem $M/G/1 \to G/1$ (Fig. 2) is a system with loses, that is, customers are not servicing if they arrive during a period when at least one stage is busy (they are lost). In the article [18], it was proved that the total probability characteristics of the independent one-line systems constructed in this way coincide with the corresponding characteristics of the original infinite-server system if N tends to ∞.

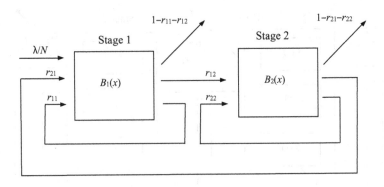

Fig. 2. Single-line queueing tandem $M/G/1 \to G/1$ with loses.

Therefore, we first find the stationary probability distributions of the number of customers at the stages of the constructed single-line tandems.

Let the state of the single-line tandem be a random process $k(t)$ with the following values:

$$
k(t) = \begin{cases} 0, & \text{if the system is free,} \\ 1, & \text{if the first stage is busy,} \\ 2, & \text{if the second stage is busy.} \end{cases}
$$

The process $k(t)$ is not Markovian. Let $z(t)$ be a length of the interval from the time moment t to the end of the current service (either at the first or second stage). If both servers are free at the time moment t, then the component $z(t)$ is not defined.

The random process $\{k(t), z(t)\}$ is Markovian. We denote the probability distribution of the process as $P_0(t) = \mathrm{P}\{k(t) = k\}$, $P_1(z,t) = \mathrm{P}\{k(t) = 1, z(t) < z\}$, $P_2(z,t) = \mathrm{P}\{k(t) = 2, z(t) < z\}$.

Using the total probability formula, we can write the system of equations:

$$
P_0(t + \Delta t) = \left(1 - \frac{\lambda}{N}\Delta t\right)P_0(t) + (1 - r_{11} - r_{12})P_1(\Delta t, t)
$$
$$
+ (1 - r_{21} - r_{22})P_2(\Delta t, t) + o(\Delta t),
$$

$$
P_1(z, t + \Delta t) = P_1(z + \Delta t, t) - P_1(\Delta t, t) + \frac{\lambda}{N}\Delta t B_1(z)P_0(t)
$$
$$
+ r_{11}B_1(z)P_1(\Delta t, t) + r_{21}B_1(z)P_2(\Delta t, t) + o(\Delta t),
$$

$$
P_2(z, t + \Delta t) = P_2(z + \Delta t, t) - P_2(\Delta t, t) + r_{12}B_2(z)P_1(\Delta t, t)
$$
$$
+ r_{22}B_2(z)P_2(\Delta t, t) + o(\Delta t).
$$

Hence we obtain the following system of Kolmogorov differential equations:

$$
\frac{\partial P_0(t)}{\partial t} = (1 - r_{11} - r_{12})\frac{\partial P_1(0,t)}{\partial z} + (1 - r_{11} - r_{12})\frac{\partial P_1(0,t)}{\partial z} - \frac{\lambda}{N}P_0(t), \quad (1)
$$

$$\frac{\partial P_1(z,t)}{\partial t} = \frac{\partial P_1(z,t)}{\partial z} + [r_{11}B_1(z) - 1]\frac{\partial P_1(0,t)}{\partial z}$$

$$+ r_{21}B_1(z)\frac{\partial P_2(0,t)}{\partial z} + \frac{\lambda}{N}P_0(t)B_1(z), \tag{2}$$

$$\frac{\partial P_2(z,t)}{\partial t} = \frac{\partial P_2(z,t)}{\partial z} + [r_{22}B_2(z) - 1]\frac{\partial P_2(0,t)}{\partial z} + r_{12}B_2(z)\frac{\partial P_1(0,t)}{\partial z}. \tag{3}$$

We denote the probability distribution in the stationary regime as $P_0(t) = \Pi_0$, $P_1(z,t) = \Pi_1(z)$, $P_2(z,t) = \Pi_2(z)$. Then the system (1)–(3) can be rewritten in the form

$$\frac{\lambda}{N}\Pi_0 = (1 - r_{11} - r_{12})\frac{\partial \Pi_1(0)}{\partial z} + (1 - r_{11} - r_{12})\frac{\partial \Pi_1(0)}{\partial z}, \tag{4}$$

$$\frac{\partial \Pi_1(z)}{\partial z} = [1 - r_{11}B_1(z)]\frac{\partial \Pi_1(0)}{\partial z} - r_{21}B_1(z)\frac{\partial \Pi_2(0)}{\partial z} - \frac{\lambda}{N}\Pi_0 B_1(z), \tag{5}$$

$$\frac{\partial \Pi_2(z)}{\partial z} = [1 - r_{22}B_2(z)]\frac{\partial \Pi_2(0)}{\partial z} - r_{12}B_2(z)\frac{\partial \Pi_1(0)}{\partial z}. \tag{6}$$

Letting $z \to \infty$ in system (4)–(6), we derive the following system of three equations

$$\frac{\lambda}{N}\Pi_0 = (1 - r_{11} - r_{12})\frac{\partial \Pi_1(0)}{\partial z} + (1 - r_{11} - r_{12})\frac{\partial \Pi_1(0)}{\partial z}, \tag{7}$$

$$(1 - r_{11})\frac{\partial \Pi_1(0)}{\partial z} - r_{21}\frac{\partial \Pi_2(0)}{\partial z} = \frac{\lambda}{N}\Pi_0, \tag{8}$$

$$- r_{12}\frac{\partial \Pi_1(0)}{\partial z} + (1 - r_{22})\frac{\partial \Pi_2(0)}{\partial z} = 0. \tag{9}$$

In this system, it is easy to see that Eq. (7) can be represented as a linear combination of Eqs. (8) and (9). Therefore, we will not use it for further derivations.

We can express the constants $\frac{\partial \Pi_1(0)}{\partial z}$ and $\frac{\partial \Pi_2(0)}{\partial z}$ from the system of equations (8)–(9):

$$\frac{\partial \Pi_1(0)}{\partial z} = \frac{1 - r_{22}}{(1 - r_{11})(1 - r_{22}) - r_{12}r_{21}} \cdot \frac{\lambda}{N}\Pi_0, \tag{10}$$

$$\frac{\partial \Pi_2(0)}{\partial z} = \frac{r_{12}}{(1 - r_{11})(1 - r_{22}) - r_{12}r_{21}} \cdot \frac{\lambda}{N}\Pi_0. \tag{11}$$

Substituting expressions (10), (11) into Eqs. (5) and (6), we derive equations for $\Pi_1(z)$ and $\Pi_2(z)$ as follows:

$$\frac{\partial \Pi_1(z)}{\partial z} = \frac{1 - r_{22}}{(1 - r_{11})(1 - r_{22}) - r_{12}r_{21}} \cdot \frac{\lambda}{N}\Pi_0[1 - B_1(z)],$$

$$\frac{\partial \Pi_2(z)}{\partial z} = \frac{r_{12}}{(1 - r_{11})(1 - r_{22}) - r_{12}r_{21}} \cdot \frac{\lambda}{N}\Pi_0[1 - B_2(z)].$$

Hence, we obtain

$$\Pi_1(z) = \frac{1 - r_{22}}{(1 - r_{11})(1 - r_{22}) - r_{12}r_{21}} \cdot \frac{\lambda}{N} \Pi_0 \int_0^z [1 - B_1(z)]dz, \qquad (12)$$

$$\Pi_2(z) = \frac{r_{12}}{(1 - r_{11})(1 - r_{22}) - r_{12}r_{21}} \cdot \frac{\lambda}{N} \Pi_0 \int_0^z [1 - B_2(z)]dz. \qquad (13)$$

If we make a limit transition $z \to \infty$ in expressions (12), (13), then we obtain

$$\Pi_1 = \frac{1 - r_{22}}{(1 - r_{11})(1 - r_{22}) - r_{12}r_{21}} \cdot \frac{\lambda}{N} \Pi_0 b_1,$$

$$\Pi_2 = \frac{r_{12}}{(1 - r_{11})(1 - r_{22}) - r_{12}r_{21}} \cdot \frac{\lambda}{N} \Pi_0 b_2,$$

where $b_k = \int_0^\infty [1 - B_k(z)]\, dz$ is the average service time at the k-th stage of the system ($k = 1, 2$).

Using the normalization condition $\Pi_0 + \Pi_1 + \Pi_2 = 1$, we can write the following expression for Π_0:

$$\Pi_0 = \frac{N\left[(1 - r_{11})(1 - r_{22}) - r_{12}r_{21}\right]}{N\left[(1 - r_{11})(1 - r_{22}) - r_{12}r_{21}\right] + \lambda\left[(1 - r_{22})b_1 + r_{12}b_2\right]}. \qquad (14)$$

Then the expressions for Π_1 and Π_2 take the form

$$\Pi_1 = \frac{\lambda(1 - r_{22})b_1}{N\left[(1 - r_{11})(1 - r_{22}) - r_{12}r_{21}\right] + \lambda\left[(1 - r_{22})b_1 + r_{12}b_2\right]}, \qquad (15)$$

$$\Pi_2 = \frac{\lambda r_{12}b_2}{N\left[(1 - r_{11})(1 - r_{22}) - r_{12}r_{21}\right] + \lambda\left[(1 - r_{22})b_1 + r_{12}b_2\right]}. \qquad (16)$$

Thus, the expressions (14)–(16) determine the probability characteristics of the state of a single-line tandem. Now let us return to the original infinite-server two-stage queueing system.

Let $i_1(t)$ be the number of customers at the first stage of the system and $i_2(t)$ be the number of customers at the second stage. Denote the probabilities $P_1(i,t) = \mathrm{P}\{i_1(t) = i\}$, $P_2(i,t) = \mathrm{P}\{i_2(t) = i\}$, and let $P_1(i) := P_1(i,t)$, $P_2(i) := P_2(i,t)$ be the probabilities in a stationary regime. Then, using the Bernoulli formula, if we let $N \to \infty$, we can derive the following:

$$P_1(i) = \lim_{N\to\infty} C_N^{i_1} \Pi_1^{i_1} (1-\Pi_1)^{N-i_1} =$$

$$\lim_{N\to\infty} \left[\frac{N!}{i_1!(N-i_1)!} \cdot \left(\frac{\lambda(1-r_{22})b_1}{N\left((1-r_{11})(1-r_{22})-r_{12}r_{21}\right) + \lambda\left((1-r_{22})b_1 + r_{12}b_2\right)} \right)^{i_1} \cdot \right.$$

$$\left. \left(1 - \frac{\lambda(1-r_{22})b_1}{N\left((1-r_{11})(1-r_{22})-r_{12}r_{21}\right) + \lambda\left((1-r_{22})b_1 + r_{12}b_2\right)} \right)^{N-i_1} \right] =$$

$$\lim_{N\to\infty} \left[\frac{(\lambda(1-r_{22})b_1)^{i_1}}{i_1!} \cdot \frac{N\cdot(N-1)\cdots(N-i_1+1)}{\left(N\left((1-r_{11})(1-r_{22})-r_{12}r_{21}\right) + \lambda\left((1-r_{22})b_1 + r_{12}b_2\right)\right)^{i_1}} \cdot \right.$$

$$\left. \left(1 - \frac{\lambda(1-r_{22})b_1}{N\left((1-r_{11})(1-r_{22})-r_{12}r_{21}\right) + \lambda\left((1-r_{22})b_1 + r_{12}b_2\right)} \right)^{N-i_1} \right] =$$

$$\left(\frac{\lambda(1-r_{22})b_1}{(1-r_{11})(1-r_{22})-r_{12}r_{21}} \right)^{i_1} \cdot \frac{e^{-\frac{\lambda(1-r_{22})b_1}{(1-r_{11})(1-r_{22})-r_{12}r_{21}}}}{i_1!}$$

Thus, we obtain

$$P_1(i) = \frac{\gamma_1^i}{i!} \cdot e^{-\gamma_1}, \tag{17}$$

where

$$\gamma_1 = \frac{\lambda(1-r_{22})b_1}{(1-r_{11})(1-r_{22})-r_{12}r_{21}}. \tag{18}$$

Similarly, one can derive

$$P_2(i) = \frac{\gamma_2^i}{i!} \cdot e^{-\gamma_2}, \tag{19}$$

where

$$\gamma_2 = \frac{\lambda r_{12}b_2}{(1-r_{11})(1-r_{22})-r_{12}r_{21}}. \tag{20}$$

Thus, the marginal probability distributions of the number of customers at the first and second stages of the system are Poisson distributions with the parameters γ_1 and γ_2 respectively.

4   Numerical Example

Let us consider an example of applying the obtained results (17)–(20), and also compare these results with the asymptotic ones obtained earlier in [15, 19].

Consider a two-stage queueing tandem with feedback, the configuration of which is described in the Sect. 2. Let the parameters of the tandem have the following values: $r_{11} = 0.2$, $r_{12} = 0.4$, $r_{21} = 0.6$, $r_{22} = 0.2$. The service time at the first stage has gamma distribution with a shape parameter equal to 0.25 and an inverse scale parameter equal to 0.5. At the second stage, the service time has gamma distribution with a shape parameter equal to 0.25 and inverse scale parameter equal to 0.25. The arrival process is a stationary Poisson process with the intensity $\lambda = K$, where $K > 0$ is the high-intensity parameter of the arrivals as described in [20]. We will change value of this parameter in a series of experiments.

For this system, according to the results (17)–(20), we obtain that the marginal probability distributions of the number of customers at each stage of the system are Poisson distributions with parameter 1.

In the paper [15], it was obtained that under a condition of a high rate of arrivals, these distributions can be approximated by Gaussian distributions with means and variances equal to 1.

In the article [19], a more accurate expression for approximating the multidimensional characteristic function of the number of customers at the nodes of the queueing network was obtained for the similar asymptotic condition (it is called as *third-order asymptotic* or *third-order approximation*). From that expression, we can write the following general form of third-order approximations for the characteristic functions of the marginal stationary probability distributions of the number of customers at the stages of the tandem under study:

$$h(u) = \exp\left\{ juc_1 + \frac{(ju)^2}{2}c_1 + \frac{(ju)^3}{6}c_3 \right\}, \qquad (21)$$

where $c_1 = K$, $c_3 = K + 6K\theta$. In our example, $\theta \approx 0.17644$ for the first stage, and $\theta \approx 0.03024$ for the second stage of the system. To obtain the distribution law from expression (21), we should make the inverse Fourier transformation of its right-hand side and perform a normalization of obtained values.

The graphs at the Fig. 3 show a comparison of the probability distribution laws for the number of customers at the second stage of the system for different values of the arrivals intensity K. The graphs are constructed on the base of the analytical results (19), the Gaussian approximation [15], and the third-order approximation (21) [19].

Table 1 presents the values of the Kolmogorov distance

$$d_k = \max_x \left| \sum_{i=0}^{x} [P_2(i) - G_k(i)] \right|, \quad k = 2, 3,$$

between the Poisson distribution function $P_2(i)$ with the parameter (20) and the distribution functions for the Gaussian approximation $G_2(i)$ and the third-order approximation $G_3(i)$.

Table 1. The Kolmogorov distance for the Gaussian approximation d_2 and third-order approximation d_3.

K	1	2	3	4	5	6
d_2	0.109	0.070	0.050	0.040	0.033	0.029
d_3	0.026	0.020	0.018	0.016	0.012	0.009

If we choose $d_k \leq 0.03$ as a permissible error of the asymptotic result, then we can conclude that the Gaussian approximation is applicable for values $K \geq 6$,

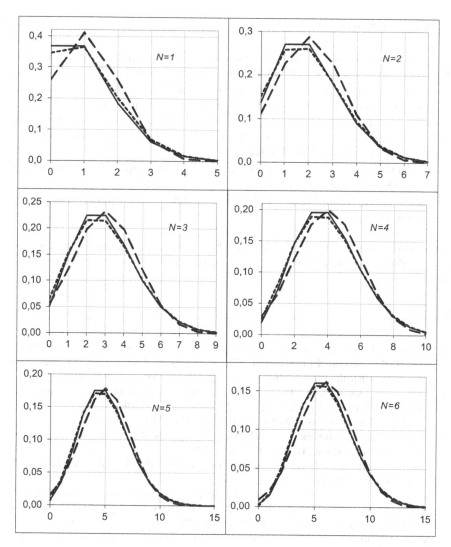

Fig. 3. Distribution laws of the number of customers at the second stage of the tandem for different values of the arrivals intensity K, built on the basis of analytical results (solid line), Gaussian approximation (dashed line) and third-order approximation (dotted line).

and the third-order approximation is applicable for values $K \geq 1$. This result is approximately twice a quality (low boundary of applicability) of the results obtained in papers [15,19]. This effect is probably due to the fact that we are considering a system with a Poisson arrival process, but those papers consider models with renewal arrival process.

5   Conclusions

In this paper, we consider a two-stage tandem of queues with an infinite number of servers and a possibility of repeated service at each stage. The service times has an arbitrary distributions. The arrival process is a stationary Poisson point process. The study is carried out by the method of limiting decomposition. It is shown that the marginal steady-state probability distributions of the number of customers at each stage are Poisson distributions. Parameters of these distributions are obtained in the paper.

Also in the paper, the obtained analytical result is compared with the asymptotic ones that was obtained earlier in the papers [15,19].

Future research can be devoted to the analysis of customers' flows in the considered model.

Acknowledgments. This work was partially supported (A. Moiseev) by the Ministry of Education and Science of the Russian Federation [Agreement number 02.a03.21.0008].

References

1. Boxma, O.J.: $M/G/\infty$ Tandem Queues. Stoch. Proc. Apps. **18**, 153–164 (1984)
2. Ahn, H.-S., Duenyas, I., Lewis, M.E.: Optimal control of a two-stage tandem queuing system with flexible servers. Prob. Eng. Inf. Sci. **16**, 453–469 (2002)
3. Moiseev, A., Nazarov, A.: Tandem of infinite-server queues with Markovian arrival process. In: Vishnevsky, V., Kozyrev, D. (eds.) DCCN 2015. CCIS, vol. 601, pp. 323–333. Springer, Cham (2016). doi:10.1007/978-3-319-30843-2_34
4. Coffman, E.G., Kleinrock, L.: Feedback queueing models for time-shared systems. J. Ass. Comput. Mach. **15**(4), 549–576 (1968)
5. Foley, R.D., Disney, R.L.: Queues with delayed feedback. Adv. Appl. Probab. **15**(1), 162–182 (1983)
6. Moiseeva, S., Zadiranova, L.: Feedback in infinite-server queuing systems. In: Vishnevsky, V., Kozyrev, D. (eds.) DCCN 2015. CCIS, vol. 601, pp. 370–377. Springer, Cham (2016). doi:10.1007/978-3-319-30843-2_38
7. König, D., Stoyan, D.: Methoden der Bedienungstheorie. Akademie-Verlag, Berlin (1976)
8. Kleinrock, L.: Queueing Systems: Volume 1, Theory, vol. 1. Wiley Interscience, New York (1975)
9. Sevastyanov, B.A.: An ergodic theorem for Markov processes and its application to telephone systems with fefusals. Theor. Prob. Appl. **2**, 104–112 (1957)
10. Takács, L.: On Erlang's formula. Annal. Math. Stat. **40**, 71–78 (1969)
11. Grigelionis, B.: Asymptotic expansions in the compound Poisson limit theorem. Acta Applicandae Math. **58**(1), 125–134 (1999)
12. Iglehart, D.L.: Limit diffusion approximations for the many server queue and repairman problem. J. Appl. Prob. **2**, 429–441 (1965)
13. Borovkov, A.A.: On limit laws for service processes in multi-channel systems. Siberian Math. J. **8**, 746–763 (1967)
14. Whitt, W.: On the heavy-traffic limit theorem for $GI/G/\infty$ queues. Adv. Appl. Prob. **14**, 171–190 (1982)

15. Nazarov, A.A., Moiseev, A.N.: Analysis of an open Non-Markovian $GI - (GI|\infty)^K$ queueing network with high-rate renewal arrival process. Prob. Inf. Trans. **49**(2), 167–178 (2013)
16. Moiseev, A., Nazarov, A.: Queueing network $MAP - (GI - \infty)^K$ with high-rate arrivals. Europ. J. Oper. Res. **254**, 161–168 (2016)
17. Nazarov, A.A., Moiseeva, S.P., Morozova, A.S.: Analysis of infinite-server queueing system with feedback by the method of limiting decomposition (in Russian). Comput. Technol. **13**(55), 88–92 (2008)
18. Nazarov, A.A., Terpugov, A.F.: Queueing Theory - Tutorial. NTL, Tomsk (2010). (in Russian)
19. Nazarov, A.A., Moiseev, A.N.: Distributed system of processing of data of physical experiments. Rus. Phys. J. **57**(7), 984–990 (2014)
20. Moiseev, A., Nazarov, A.: Investigation of high intensive general flow. In: 4th International Conference Problems of Cybernetics and Informatics (PCI 2012), pp. 161–163. IEEE, Baku (2012)

Combination of Queueing Systems of Different Types with Common Buffer: A Theoretical Treatment

Oleg Tikhonenko[1]([✉]) and Marcin Ziółkowski[2]

[1] Faculty of Mathematics and Natural Sciences, College of Sciences,
Cardinal Stefan Wyszyński University in Warsaw,
Ul. Wóycickiego 1/3, 01-938 Warsaw, Poland
o.tikhonenko@uksw.edu.pl
[2] Institute of Mathematics and Computer Sciences,
Jan Długosz University in Częstochowa,
Al. Armii Krajowej 13/15, 42-200 Częstochowa, Poland
marionesta5@vp.pl

Abstract. In the paper, we investigate a combination of two following different queueing systems connected via common limited buffer space: (1) the system of $M/G/1$-type, in which service time does not depend on demand volume; (2) the processor-sharing system, in which demand length arbitrarily depends on its volume. For such combination, we determine the steady-state loss probability and distribution of number of demands present in each system of the combination.

Keywords: Queueing system · Demand volume · Total demands capacity · Buffer space capacity · Processor-sharing system

1 Introduction

We consider queueing systems with demands of random space requirement. It means that (1) each demand is characterized by some non-negative demand space requirement ore demand volume ζ; (2) the total sum $\sigma(t)$ of volumes of all demands present in the system at an arbitrary time instant t is limited by some constant value V, which is named the buffer space capacity of the system; (3) we also assume that service time ξ of the demand and its volume can be dependent.

Such systems have been used to model and solve the various practical problems occurring in the design of computer and communicating systems.

The joint distribution of the random variables ζ and ξ we characterize by the joint distribution function $F(x,t) = \mathsf{P}\{\zeta < x, \xi < t\}$. The buffer space is occupied by the demand at the epoch it arrives to the system and is released entirely at the epoch it completes service. The random process $\sigma(t)$ is called the total demands capacity. The limitation of the buffer capacity V leads to additional losses of demands. A demand having volume x, which arrives at the

© Springer International Publishing AG 2017
A. Dudin et al. (Eds.): ITMM 2017, CCIS 800, pp. 158–167, 2017.
DOI: 10.1007/978-3-319-68069-9_13

epoch τ when there are idle servers or waiting positions, is admitted to the system if $\sigma(\tau - 0) + x \leq V$. Otherwise (if $\sigma(\tau - 0) + x > V$) the demand is lost.

Queueing systems of different types with limited buffer capacity were analyzed, for example, in [1–8]. For these systems, the stationary demands number distribution and loss probability were determined. In particular, in [8] the combination of two different systems of the same $(M/G/1)$ type were investigated under assumption that service time does not depends on the demand volume for each of the systems.

In present paper, we investigate combinations of two following systems connected via common limited (by V) buffer: (1) $M/G/1/\infty$-type system, (2) $M/G/1 - PS$-type system. Such models obviously can be used in computer and communicating networks designing. A statement allowing determination of the steady-state demands number distribution and loss probability for each of the systems connected via common buffer will be presented.

The paper is organized as follows. Section 2 contains a description of the queueing model and necessary notations. In Sect. 3, we define a Markovian process describing the evolution of the system and the functions characterizing the system behavior. In Sect. 4, we build a system of partial differential equations for the transient system characteristics and give their steady-state solution. In this section, we also determine the steady-state queue-size distribution for the system under consideration. Section 5 presents conclusions and final remarks.

2   The Model and Notation

By a demand length for the second (processor sharing) system in the considered combination we mean the amount of work required to serve it, that is, the time of demand sojourn in this system at hand, provided that there are no other demands in the system during this time [9]. By the demand remaining length for this system we mean the amount of work required to complete its service after some time instant, that is, the remaining time of demand sojourn, provided that there are no other demands in the system during this time.

We shall use the following notation $(i = 1, 2)$: a_i – the rate of demand arrival process for ith system, ζ_i – the demand space requirement for ith system, ξ_1 – the first system demand service time; ξ_2 – the second system demand length; $L_i(x)$ – the distribution function of ζ_i random variable, $B_1(t)$ – the distribution function of ξ_1 random variable that is assumed to be independent of the demand space requirement ζ_1.

We assume that, for the second system, the demand length ξ_2 can be dependent on its space requirement, and we denote the joint distribution function $F(x, t) = \mathsf{P}\{\zeta_2 < x, \xi_2 < t\}$. So, we have evidently that $L_2(x) = F(x, \infty)$, $B_2(t) = F(\infty, t)$, where $B_2(t)$ is the distribution function of the demand length for the second system. Let $\eta_i(t)$ be the number of demands present in ith system at time moment t; $\zeta_j^{(i)}(t)$ – the space requirement of jth demand present in ith system at the moment t, $j = \overline{1, \eta_i(t)}$; $\xi_*^{(1)}(t)$ – the remaining service time of the demand being on service in the first system at the moment t; $\xi_{1*}^{(2)}, \ldots, \xi_{\eta_2(t)*}^{(2)}$ — the remaining lengths of demands in the second one at this moment.

We agree to assume that the demands present in the second system at an arbitrary time instant are numerated randomly, that is, if at the time instant t there are k demands in this system, then any of the possible $k!$ numerations can be used with the same probability $1/k!$.

It is clear that $\sigma(t) = \sum\limits_{i=1}^{2} \sum\limits_{j=1}^{\eta_i(t)} \varsigma_j^{(i)}(t)$.

3   Random Process and Functions Describing the System Behavior

The combination of queues under consideration can be described by the following Markovian random process:

$$\left(\eta_i(t), \varsigma_j^{(i)}(t), j = \overline{1, \eta_i(t)}, i = 1, 2; \xi_*^{(1)}(t); \xi_{l*}^{(2)}(t), l = \overline{1, \eta_2(t)} \right) \tag{1}$$

It is clear that the components $\varsigma_1^{(i)}(t), \xi_*^{(1)}(t)$ are absent in (1), if $\eta_1(t) = 0$; $\varsigma_2^{(i)}(t), \xi_{l*}^{(2)}(t)$ are absent, if $\eta_2(t) = 0$.

Let us introduce the vector $Z_k = (z_1, \ldots, z_k)$. We also shall use the notations $(Z_k, y) = (z_1, \ldots, z_k, y)$, $Z_k^j = (z_1, \ldots, z_{j-1}, z_{j+1}, \ldots z_k)$.

The process (1) is characterized by the functions with the following probability sense:

$$P_0(t) = \mathsf{P}\{\eta_i(t) = 0, i = 1, 2\} = \mathsf{P}\{\sigma(t) = 0\} \tag{2}$$

$$G(k, 0, x, y, t) = \mathsf{P}\{\eta_1(t) = k, \eta_2(t) = 0; \sigma(t) < x; \xi_*^{(1)}(t) < y\}, \ k = 1, 2, \ldots \tag{3}$$

$$G(0, k, x, Z_k, t) = \mathsf{P}\{\eta_1(t) = 0, \eta_2(t) = k; \sigma(t) < x; \xi_{j*}^{(2)}(t) < z_j, j = \overline{1, k}\}, \ k = 1, 2, \ldots \tag{4}$$

$$\begin{aligned} G(k_1, k_2, x, y, Z_{k_2}, t) = \mathsf{P}\{\eta_1(t) = k_1, \eta_2(t) = k_2; \\ \sigma(t) < x; \xi_*^{(1)}(t) < y, \xi_{j*}^{(2)}(t) < z_j, j = \overline{1, k_2}\}, \\ k_1, k_2 = 1, 2, \ldots. \end{aligned} \tag{5}$$

We also introduce the functions

$$\begin{aligned} W(k_1, k_2, y, Z_{k_2}, t) = \mathsf{P}\{\eta_1(t) = k_1, \eta_2(t) = k_2; \xi_*^{(1)}(t) < y, \xi_{j*}^{(2)}(t) < z_j, j = \overline{1, k_2}\} \\ = G(k_1, k_2, V, y, Z_{k_2}, t), \ k_1, k_2 = 1, 2, \ldots. \end{aligned} \tag{6}$$

The functions $W(k, 0, y, t)$ and $W(0, k, Z_k, t)$ can be introduced by similar way. The demands number distribution is defined by the following functions:

$$P(k_1, k_2, t) = \mathsf{P}\{\eta_i(t) = k_i, i = 1, 2\} = W(k_1, k_2, \infty, \infty_{k_2}, t), \ k_1, k_2 = 1, 2, \ldots, \tag{7}$$

where $\infty_k = \underbrace{(\infty, \ldots, \infty)}_{k}$.

We can define the functions $P(k, 0, t)$, $P(0, k, t)$ analogously.

We are interested in steady-state demands number distribution and loss probability. Therefore, let us write out the stationary analogies of the functions (2)–(5):

$$p_0 = \lim_{t \to \infty} P_0(t) \tag{8}$$

$$g(k, 0, x, y) = \lim_{t \to \infty} G(k, 0, x, y, t), \ k = 1, 2, \ldots \tag{9}$$

$$g(0, k, x, Z_k) = \lim_{t \to \infty} G(0, k, x, Z_k, t), \ k = 1, 2, \ldots \tag{10}$$

$$g(k_1, k_2, x, y, Z_{k_2}) = \lim_{t \to \infty} G(k_1, k_2, x, y, Z_{k_2}, t), \ k_1, k_2 = 1, 2, \ldots \tag{11}$$

Now we can define stationary analogies of (6) and (7) functions:

$$w(k_1, k_2, y, Z_{k_2}) = g(k_1, k_2, V, y, Z_{k_2})), \ k_1, k_2 = 1, 2, \ldots \tag{12}$$

$$p(k_1, k_2) = w(k_1, k_2, \infty, \infty_{k_2}), \ k_1, k_2 = 1, 2, \ldots \tag{13}$$

The functions $w(k, 0; y)$, $w(0, k, Z_k)$, $p(k, 0)$, $p(0, k)$ can be defined analogously.

The functions $p(k_1, k_2)$ define the steady-state demands number distribution in the combination of systems under consideration. We can determine loss probabilities for each system of the combination using the functions (8)–(13). Let $P_L^{(1)}$ be a loss probability for the first system. Let us define the following functions:

$$r_k^{(1)}(y) = w(k, 0; y) + \sum_{k_2=1}^{\infty} w(k, k_2; y, \infty_{k_2}), \ k = 1, 2, \ldots$$

Then, the loss probability can be determined from the following equilibrium equation:

$$a_1 \left(1 - P_L^{(1)} \right) = \sum_{k=1}^{\infty} \frac{\partial r_k^{(1)}(y)}{\partial y} \bigg|_{y=0},$$

whereas we get

$$P_L^{(1)} = 1 - \frac{1}{a_1} \sum_{k=1}^{\infty} \frac{\partial r_k^{(1)}(y)}{\partial y} \bigg|_{y=0}. \tag{14}$$

For the second system, let us define the functions

$$r_k^{(2)}(Z_k) = w(0, k; Z_k) + \sum_{k_1=1}^{\infty} w(k_1, k; \infty, Z_k), \ k = 1, 2, \ldots$$

So, we obtain the following equilibrium equation:

$$a_2 \left(1 - P_L^{(2)} \right) = \sum_{k=1}^{\infty} \frac{\partial r_k^{(2)}((\infty_{k-1}, y))}{\partial y} \bigg|_{y=0},$$

whereas

$$P_L^{(2)} = 1 - \frac{1}{a_2} \sum_{k=1}^{\infty} \frac{\partial r_k^{(2)}((\infty_{k-1}, y))}{\partial y} \bigg|_{y=0}. \tag{15}$$

4   The Main Statement

Further we shall use the following (convenient for our aims) notation for Stieltjes convolution (see [8]):

1. $\displaystyle\int_0^x f_1(x-u)\mathrm{d}f_2(u) = f_1 * f_2(x) = f_2 * f_1(x).$

2. $\displaystyle\int_{u=0}^x f_1(y_1,...,y_{i-1},x-u,y_{i+1},...,y_k)\mathrm{d}_u f_2(z_1,...,z_{j-1},u,z_{j+1},...,z_l)$
$= f_1(y_1,...,y_{i-1},*,y_{i+1},...,y_k) * f_2(z_1,...,z_{j-1},*,z_{j+1},...,z_l)(x)$
$= f_2(z_1,...,z_{j-1},*,z_{j+1},...,z_l) * f_1(y_1,...,y_{i-1},*,y_{i+1},...,y_k)(x).$

3. $\displaystyle\int_0^x f_1(y_1,...,y_{i-1},x-u,y_{i+1},...,y_k)\mathrm{d}f_2(u)$
$= f_1(y_1,...,y_{i-1},*,y_{i+1},...,y_k) * f_2(x) = f_2 * f_1(y_1,...,y_{i-1},*,y_{i+1},...,y_k)(x).$

Similar notations are used for a convolution of more than two functions.

By supplementary variables method [10], partial differential equations for the functions (2)–(6) can be written out. From these equations we obtain the following ones for steady-state functions (8)–(12):

$$0 = -(a_1 L_1(V) + a_2 L_2(V))p_0 + \frac{\partial w(1,0,y)}{\partial y}\bigg|_{y=0} + \frac{\partial w(0,1,z)}{\partial z}\bigg|_{z=0} \tag{16}$$

$$-\frac{\partial w(1,0,y)}{\partial y} + \frac{\partial w(1,0,y)}{\partial y}\bigg|_{y=0} = a_1 p_0 L_1(V)B_1(y) - a_1 g(1,0,*,y) * L_1(V)$$
$$-a_2 g(1,0,*,y) * L_2(V) + \frac{\partial w(2,0,u)}{\partial u}\bigg|_{u=0} B_1(y) + \frac{\partial w(1,1,y,u)}{\partial u}\bigg|_{u=0} \tag{17}$$

$$-\frac{\partial w(0,1,z)}{\partial z} + \frac{\partial w(0,1,z)}{\partial z}\bigg|_{z=0} = a_2 p_0 F(V,z) - a_1 g(0,1,*,z) * L_1(V)$$
$$-a_2 g(0,1,*,z) * L_2(V) + \frac{\partial w(1,1,u,z)}{\partial u}\bigg|_{u=0} + \frac{\partial w(0,2,(z,u))}{\partial u}\bigg|_{u=0}; \tag{18}$$

$$-\frac{\partial w(k,0,y)}{\partial y} + \frac{\partial w(k,0,y)}{\partial y}\bigg|_{y=0} = a_1 g(k-1,0,*,y) * L_1(V)$$
$$-a_1 g(k,0,*,y) * L_1(V) - a_2 g(k,0,*,y) * L_2(V)$$
$$+\frac{\partial w(k+1,0,u)}{\partial u}\bigg|_{u=0} B_1(y) + \frac{\partial w(k,1,y,u)}{\partial u}\bigg|_{u=0}, \quad k=2,3,\ldots \tag{19}$$

$$-\frac{1}{k}\sum_{j=1}^k \left[\frac{\partial w(0,k,Z_k)}{\partial z_j} - \frac{\partial w(0,k,Z_k)}{\partial z_j}\bigg|_{z_j=0}\right]$$
$$= \frac{a_2}{k}\sum_{j=1}^k g(0,k-1,*,Z_k^j) * F(*,z_j)(V)$$
$$-a_1 g(0,k,*,Z_k) * L_1(V) - a_2 g(0,k,*,Z_k) * L_2(V)$$
$$+\frac{\partial w(1,k,u,Z_k)}{\partial u}\bigg|_{u=0} + \frac{\partial w(0,k+1,(Z_k,u))}{\partial u}\bigg|_{u=0}, \quad k=2,3,\ldots \tag{20}$$

$$-\frac{\partial w(1,1,y,z)}{\partial y} + \frac{\partial w(1,1,y,z)}{\partial y}\bigg|_{y=0} - \frac{\partial w(1,1,y,z)}{\partial z} + \frac{\partial w(1,1,y,z)}{\partial z}\bigg|_{z=0}$$
$$= a_1 g(0,1,*,z) * L_1(V)B_1(y) + a_2 g(1,0,*,y) * F(*,z)(V)$$
$$-a_1 g(1,1,*,y,z) * L_1(V) - a_2 g(1,1,*,y,z) * L_2(V)$$
$$+\frac{\partial w(2,1,u,z))}{\partial u}\bigg|_{u=0} B_1(y) + \frac{\partial w(1,2,y,(z,u))}{\partial u}\bigg|_{u=0}; \tag{21}$$

$$-\frac{\partial w(k_1,k_2,y,Z_{k_2})}{\partial y} + \frac{\partial w(k_1,k_2,y,Z_{k_2})}{\partial y}\bigg|_{y=0} - \frac{1}{k_2}\sum_{j=1}^{k_2}\left[\frac{\partial w(k_1,k_2,y,Z_{k_2})}{\partial z_j} - \frac{\partial w(k_1,k_2,y,Z_k)}{\partial z_j}\bigg|_{z_j=0}\right]$$
$$= a_1 g(k_1-1,k_2,*,y,Z_{k_2}) * L_1(V) + \frac{a_2}{k_2}\sum_{j=1}^{k_2} g(k_1,k_2-1,*,y,Z_{k_2}^j) * F(*,z_j)(V)$$
$$-a_1 g(k_1,k_2,*,y,Z_{k_2}) * L_1(V) - a_2 g(k_1,k_2,*,y,Z_{k_2}) * L_2(V)$$
$$+\frac{\partial w(k_1+1,k_2,u,Z_{k_2})}{\partial u}\bigg|_{u=0} B_1(y) + \frac{\partial w(k_1,k_2+1,y,(Z_{k_2},u))}{\partial u}\bigg|_{u=0}, \tag{22}$$
$$k_1, k_2 = 1, 2, \ldots$$

The following evident boundary conditions take place in steady state:

$$a_1 L_1(V)p_0 = \frac{\partial w(1,0,y)}{\partial y}\bigg|_{y=0}; \quad a_2 L_2(V)p_0 = \frac{\partial w(0,1,z)}{\partial z}\bigg|_{z=0}; \tag{23}$$

$$a_1 g(k,0,*,y) * L_1(V) = \frac{\partial w(k+1,0,y)}{\partial y}\bigg|_{y=0} B_1(y), \quad k = 1,2,\ldots; \tag{24}$$

$$a_2 g(k,0,*,y) * L_2(V) = \frac{\partial w(k,1,y,z)}{\partial z}\bigg|_{z=0}, \quad k = 1,2,\ldots; \tag{25}$$

$$a_2 g(0,k,*,Z_k) * L_2(V) = \frac{\partial w(0,k+1,(Z_k,u))}{\partial u}\bigg|_{u=0}, \quad k = 1,2,\ldots; \tag{26}$$

$$a_1 g(0,k,*,Z_k) * L_1(V) = \frac{\partial w(1,k,y,Z_k)}{\partial y}\bigg|_{y=0}, \quad k = 1,2,\ldots; \tag{27}$$

$$a_1 g(k_1,k_2,*,y,Z_{k_2}) * L_1(V) = \frac{\partial w(k_1+1,k_2,u,Z_{k_2})}{\partial u}\bigg|_{u=0} B_1(y),$$
$$k_1, k_2 = 1,2,\ldots; \tag{28}$$

$$a_2 g(k_1,k_2,*,y,Z_{k_2}) * L_2(V) = \frac{\partial w(k_1,k_2+1,y,(Z_{k_2},u))}{\partial u}\bigg|_{u=0}, $$
$$k_1, k_2 = 1,2,\ldots. \tag{29}$$

Denote by η_i, σ the stationary number of demands in ith system ($i = 1,2$) and total demands capacity in the combination of systems, respectively. Introduce also the notation $p_0^{(i)} = \mathsf{P}\{\eta_i = 0\}$.

Introduce the following notation for the first system of the combination:

$$g_k^{(1)}(x,y) = \mathsf{P}\{\eta_1 = k, \sigma < x, \xi_*^{(1)} < y\},$$

$$w_k^{(1)}(y) = g_k^{(1)}(V,y), \tag{30}$$

$$v_k^{(1)}(x) = \mathsf{P}\{\eta_1 = k, \sigma < x\} = \lim_{y\to\infty} g_k^{(1)}(x,y) = g_k^{(1)}(x,\infty), \tag{31}$$

$$p_k^{(1)} = \mathsf{P}\{\eta_1 = k\} = v_k^{(1)}(V). \tag{32}$$

Let us write out equations for an independent $M/G/1/(\infty, V)$ system. For example, they follow from the Eqs. (16)–(29), if $a_2 = 0$. So, for a separate first queue of the combination with limited (by V) buffer space capacity we obtain the following equations:

$$0 = -a_1 L_1(V) p_0^{(1)} + \left.\frac{\partial w_1^{(1)}(y)}{\partial y}\right|_{y=0}; \tag{33}$$

$$
\begin{aligned}
-\frac{\partial w_1^{(1)}(y)}{\partial y} &+ \left.\frac{\partial w_1^{(1)}(y)}{\partial y}\right|_{y=0} = a_1 p_0^{(1)} L_1(V) B_1(y) \\
-a_1 g_1^{(1)}(*, y) * L_1(V) &+ \left.\frac{\partial w_2^{(1)}(y)}{\partial y}\right|_{y=0} B_1(y);
\end{aligned}
\tag{34}
$$

$$
\begin{aligned}
-\frac{\partial w_k^{(1)}(y)}{\partial y} &+ \left.\frac{\partial w_k^{(1)}(y)}{\partial y}\right|_{y=0} = a_1 g_{k-1}^{(1)}(*, y) * L_1(V) \\
-a_1 g_k^{(1)}(*, y) * L_1(V) &+ \left.\frac{\partial w_{k+1}^{(1)}(y)}{\partial y}\right|_{y=0} B_1(y), \ k = 2, 3, \ldots;
\end{aligned}
\tag{35}
$$

$$a_1 g_k^{(1)}(*, y) * L_1(V) = \left.\frac{\partial w_{k+1}^{(1)}(y)}{\partial y}\right|_{y=0} B_1(y), \ k = 1, 2, \ldots. \tag{36}$$

For the second system, we introduce the notation:

$$g_k^{(2)}(x, Z_k) = \mathsf{P}\{\eta_2 = k, \sigma < x; \xi_{j*}^{(2)} < z_j, j = \overline{1, k}\},$$

$$w_k^{(2)}(Z_k) = g_k^{(2)}(V, Z_k), \tag{37}$$

$$v_k^{(2)}(x) = \mathsf{P}\{\eta_2 = k, \sigma < x\} = g_k^{(2)}(x, \infty_k), \tag{38}$$

$$p_k^{(2)} = \mathsf{P}\{\eta_2 = k\} = v_k^{(2)}(V). \tag{39}$$

Then, for a separate second system of the combination, we can write out the following equations (which follow from the Eqs. (16)–(29), when $a_1 = 0$):

$$0 = -a_2 L_2(V) p_0^{(2)} + \left.\frac{\partial w_1^{(2)}(z)}{\partial z}\right|_{z=0}; \tag{40}$$

$$
\begin{aligned}
-\frac{\partial w_1^{(2)}(z)}{\partial z} &+ \left.\frac{\partial w_1^{(2)}(z)}{\partial z}\right|_{z=0} = a_2 p_0^{(2)} F(V, z) \\
-a_2 g_1^{(2)}(*, z) * L_2(V) &+ \left.\frac{\partial w_2^{(2)}((z,u))}{\partial u}\right|_{u=0};
\end{aligned}
\tag{41}
$$

$$
\begin{aligned}
-\frac{1}{k}\sum_{j=1}^k \left[\frac{\partial w_k^{(2)}(Z_k)}{\partial z_j} - \left.\frac{\partial w_k^{(2)}(Z_k)}{\partial z_j}\right|_{z_j=0}\right] &= \frac{a_2}{k}\sum_{j=1}^k g_{k-1}^{(2)}(*, Z_k^j) * F(*, z_j)(V) \\
-a_2 g_k^{(2)}(*, Z_k) * L_2(V) &+ \left.\frac{\partial w_{k+1}^{(2)}((Z_k, u))}{\partial u}\right|_{u=0}, \ k = 2, 3, \ldots;
\end{aligned}
\tag{42}
$$

$$a_2 g_k^{(2)}(*, Z_k) * L_2(V) = \frac{\partial w_{k+1}^{(2)}((Z_k, u))}{\partial u}\bigg|_{u=0}, \quad k = 1, 2, \ldots. \tag{43}$$

The following statement takes place.

Theorem. *Let the number $p_0^{(1)}$ and the functions $g_k^{(1)}(x, y)$ satisfy the Eqs. (33)–(36) and the normalization condition*

$$p_0^{(1)} + \sum_{k=1}^{\infty} g_k^{(1)}(V, \infty) = 1;$$

and the number $p_0^{(2)}$ and the functions $g_k^{(2)}(x, Z_k)$ satisfy the Eqs. (40)–(43) and the normalization condition

$$p_0^{(2)} + \sum_{k=1}^{\infty} g_k^{(2)}(V, \infty_k) = 1.$$

Then the functions

$$g(k, 0, x, y) = \frac{p_0}{p_0^{(1)}} g_k^{(1)}(x, y), \quad g(0, k, x, Z_k) = \frac{p_0}{p_0^{(2)}} g_k^{(2)}(x, Z_k), \quad k = 1, 2, \ldots,$$

$$g(k_1, k_2, x, y, Z_{k_2}) = \frac{p_0}{p_0^{(1)} p_0^{(2)}} g_{k_1}^{(1)}(*, y) * g_{k_2}^{(2)}(*, Z_{k_2})(x), \quad k_1, k_2 = 1, 2, \ldots$$

satisfy the Eqs. (16)–(29) (the number p_0 can be determined from the normalization condition).

The theorem can be proved by direct substitution of $g(k_1, k_2, x, y, z)$ functions into Eqs. (16)–(29).

Corollary. *For the functions $w(k_1, k_2, y, Z_{k_2})$ we get:*

$$w(k, 0, y) = \frac{p_0}{p_0^{(1)}} w_k^{(1)}(y), \quad w(0, k, Z_k) = \frac{p_0}{p_0^{(2)}} w_k^{(2)}(Z_k), \quad k = 1, 2, \ldots,$$

where the functions $w_k^{(1)}(y)$, $w(0, k, Z_k)$ are determined by relations (30) and (36), respectively,

$$w(k_1, k_2, y, Z_{k_2}) = \frac{p_0}{p_0^{(1)} p_0^{(2)}} g_{k_1}^{(1)}(*, y) * g_{k_2}^{(2)}(*, Z_{k_2})(V), \quad k_1, k_2 = 1, 2, \ldots.$$

The determination of the steady-state demands number distribution in the system under consideration has the following form:

$$p(k, 0) = \frac{p_0}{p_0^{(1)}} p_k^{(1)}, \quad p(0, k) = \frac{p_0}{p_0^{(2)}} p_k^{(2)}, \quad k = 1, 2, \ldots, \tag{44}$$

where $p_k^{(1)}$ and $p_k^{(2)}$ are determined by relations (32) and (39), respectively,

$$p(k_1, k_2) = \frac{p_0}{p_0^{(1)} p_0^{(2)}} v_{k_1}^{(1)} * v_{k_2}^{(2)}(V), \quad k_1, k_2 = 1, 2, \ldots, \tag{45}$$

where $v_{k_1}^{(1)}(x)$ and $v_{k_2}^{(2)}(x)$ are determined by relations (31) and (38), respectively.

Let, for example, the first system of the combination is $M/M/1/(\infty, V)$ (see e.g. [2]) with exponentially distributed demand capacity: $L_1(x) = 1 - e^{-f_1 x}$, $f_1 > 0$. Let μ_1 be the parameter of service time of the system, $\rho_1 = a_1/\mu_1$. Then, we have:

$$p_0^{(1)} = \begin{cases} \dfrac{1 - \rho_1}{1 - \rho_1 e^{-(1-\rho_1)f_1 V}}, & \text{if } \rho_1 \neq 1, \\ (1 + f_1 V)^{-1}, & \text{if } \rho_1 = 1; \end{cases}$$

$$v_k^{(1)}(x) = p_0^{(1)} \rho_1^k \left[1 - e^{-f_1 x} \sum_{i=0}^{k-1} \frac{(f_1 x)^i}{i!} \right], \quad k = 1, 2, \ldots;$$

$$p_k^{(1)} = p_0^{(1)} \rho_1^k \left[1 - e^{-f_1 V} \sum_{i=0}^{k-1} \frac{(f_1 V)^i}{i!} \right], \quad k = 1, 2, \ldots.$$

Assume that, in the second system, demand capacity has an exponential distribution with parameter f_2, $f_2 > 0$, demand length is proportional to its capacity: $\xi_2 = c\zeta_2$, $c > 0$, $\rho_2 = a_2 c/f_2$. Then, we have (see [7]):

$$p_0^{(2)} = \begin{cases} \dfrac{1 - \rho_2}{1 - \sqrt{\rho_2} e^{-f_2 V} \left[\sinh(\sqrt{\rho_2} f_2 V) + \sqrt{\rho_2} \cosh(\sqrt{\rho_2} f_2 V) \right]}, & \text{if } \rho_2 \neq 1, \\ \dfrac{4}{3 + 2 f_2 V + e^{-2 f_2 V}}, & \text{if } \rho_2 = 1; \end{cases}$$

$$v_k^{(2)}(x) = p_0^{(2)} \rho_2^k \left[1 - e^{-f_2 x} \sum_{i=0}^{2k-1} \frac{(f_2 x)^i}{i!} \right], \quad k = 1, 2, \ldots;$$

$$p_k^{(2)} = p_0^{(2)} \rho_2^k \left[1 - e^{-f_2 V} \sum_{i=0}^{2k-1} \frac{(f_2 V)^i}{i!} \right], \quad k = 1, 2, \ldots.$$

Now, we can calculate demands number distribution for the combination of these systems using relations (44) and (45).

5 Conclusions

In the paper, we investigate combinations of two different queueing systems ($M/G/1$-type with independent service time and demand volume and a processor-sharing with dependent ones) connected via common buffer of limited space capacity. We determine the loss probability and demands number distribution for each system of the combination.

We show that the formulas for characteristics of demands number distributions of the combination have the form of Stieltjs convolution of the characteristics of separate systems.

The formulas obtained in the paper are not generally convenient for precise calculation, but the calculation is possible in some special cases. In other cases we can use the numeric inversion of Laplace transform [11,12] to approximate calculations of Stieltjs convolutions.

References

1. Tikhonenko, O.M.: Queueing Models in Computer Systems. Universitetskoe, Minsk (1990, in Russian)
2. Tikhonenko, O.: Computer Systems Probability Analysis. Akademicka Oficyna Wydawnicza EXIT, Warsaw (2006, in Polish)
3. Alexandrov, A.M., Katz, B.A.: Non-homogeneous Demands Flows Service. Izvestiya AN SSSR. Tekhnicheskaya Kibernetika, No. 2, 47–53 (1973, in Russian)
4. Tikhonenko, O.M.: Destricted capacity queueing systems: determination of their characteristics. Autom. Remote Control 58(6), 969–972 (1997)
5. Tikhonenko, O.M., Klimovich, K.G.: Queuing systems for random-length arrivals with limited cumulative volume. Probl. Inf. Transm. 37(1), 70–79 (2001)
6. Tikhonenko, O.M.: Generalized erlang problem for service systems with finite total capacity. Probl. Inf. Transm. 41(3), 243–253 (2005)
7. Tikhonenko, O.M.: Queuing systems with processor sharing and limited resources. Autom. Remote Control 71(5), 803–815 (2010)
8. Tikhonenko, O.: Queueing systems with common buffer: a theoretical treatment. In: Kwiecień, A., Gaj, P., Stera, P. (eds.) CN 2011. CCIS, vol. 160, pp. 61–69. Springer, Heidelberg (2011). doi:10.1007/978-3-642-21771-5_8
9. Yashkov, S.F., Yashkova, A.S.: Processor sharing: a survey of the mathematical theory. Autom. Remote Control 68(9), 1662–1731 (2007)
10. Bocharov, P.P., D'Apice, C., Pechinkin A.V., Salerno, S.: Queueing Theory. VSP, Utrecht-Boston (2004)
11. Gaver, D.P.: Observing stochastic processes, and approximate transform inversion. Oper. Res. 14(3), 444–459 (1966)
12. Stehfest, H.: Algorithm 368: numeric inversion of laplace transform. Commun. ACM 13(1), 47–49 (1970)

Minimization of Packet Loss Probability
in Network with Fractal Traffic

Vladimir N. Zadorozhnyi and Tatiana R. Zakharenkova[✉]

Omsk State Technical University, Omsk, Russia
zwn2015@yandex.ru, ZakharenkovaTatiana@gmail.com

Abstract. Methods for radical reduction of packet loss probability in telecommunication networks with fractal traffic are developed. We investigate ways of preventing the losses within the framework of queueing theory; relevant simulation experiments are carried out. It is determined that strategy for the channel number increase in the network nodes has principally higher efficiency than that for the buffer increasing and/or channel performance increasing. Approximation methods for loss probability in the nodes of multiserver queueing system without buffers are investigated. The paper offers to approximate the loss probability in the node with n channels by steady-state probability in the state n of relating infinite-server queueing systems. We develop an analytical-statistical technique of optimal channel distribution over the nodes in networks with fractal traffic which is based on such approximation. The example of the method application is provided. The developed method could be used by engineers designing the telecommunication networks.

Keywords: Telecommunication networks · Analytical-statistical techniques · Fractal traffic · Queueing theory

1 Introduction

It is known that the traffic of modern telecommunication networks has a fractal (self-similar) structure [1]. Random variables describing such traffic are given by asymptotically power-law distributions (we will call them power-law distributions) [2]. The properties of power-law distributions generate specific difficulties that arise while measuring traffic [3] and designing network devices.

In designing network, devices at a system level are presented in the form of queueing systems [4,5]. We will call fractal systems the $GI/GI/n/m$ class systems in which intervals of requests arrival and/or their service time belong to power-law distributions and have finite mathematical expectation (m.e.) and infinite dispersion. The load coefficient ρ of the examined systems does not exceed one: $\rho = \lambda b/n \leq 1$ where $b < \infty$ is the average request service time, $\lambda = 1/a$ is the intensity of arriving request flow, $a < \infty$ is the average time between the request arrivals, n is a number of channels in the system. We shall call the $GI/GI/n/m$ systems set only by exponential-tailed distributions (and also by

© Springer International Publishing AG 2017
A. Dudin et al. (Eds.): ITMM 2017, CCIS 800, pp. 168–183, 2017.
DOI: 10.1007/978-3-319-68069-9_14

distributions with tails positive only in finite intervals) the classical systems. Correspondingly, a queueing network will be called fractal if there is power-law distribution with infinite dispersion among distributions describing the network.

For instance, the queueing systems $Pa/M/n/m$, $M/Pa/n/m$ and $Pa/Pa/n/m$ are fractal when Pa distribution has finite m.e. and infinite dispersion. Here the Pa symbol corresponds to Pareto distribution with cumulative distribution function (d.f.)

$$F(t) = 1-(K/t)^{\alpha}, \quad \alpha > 0, \quad K > 0, \quad t \geq K,$$

where α is a shape parameter, K is the smallest value of a random variable (r.v.) and simultaneously, a scale parameter. We denote Pareto distribution with K, α as $Pa(K, \alpha)$. The range of α values, typical for the fractal traffic, belongs to the interval $1 < \alpha \leq 2$. At such α $Pa(K, \alpha)$ distribution has m.e. equal to $\alpha K/(\alpha - 1)$, and infinite dispersion. In a general case, fractal systems are calculated through simulation [6, 7].

In simulation of fractal systems, there are significant difficulties caused by hidden defect of power-law r.v. generators - the moments shifting [8]. For correct realization of power-law distributions, it is necessary either to use random number generators (RNG) with infinite number of digit positions or to develop special RNGs. In [9] the correct realization problem for power-law distributions in simulation has been solved in general terms by constructing ARAND algorithm (Accurate RAND), efficiently applying random numbers, resulting from any standard well-tested RNG. The simulation, the results of which are to be used below, has been implemented with the ARAND algorithm and "Mersene twister" RNG. At the same time, we used widely known classical methods providing the necessary accuracy of simulation results, adapted to fractal systems modeling [10]. Taking the aforesaid into consideration, the simulation conducted in this research is to be called a high-precision one.

The present investigation aims to develop methods for radical reduction of packet loss probability in telecommunication networks with fractal traffic.

2   Problem Statement

The following notations will be used for queueing networks description: $A(t)$ is d.f. for intervals of requests arrival, a and σ^2 are m.e. and arrival intervals dispersion, $\lambda = 1/a$ is the intensity of arriving flow, $B(t)$ is d.f. of service time with m.e. b. Parameter α in the Pareto distribution is in the range defined by $1 < \alpha \leq 2$.

Let us consider the $M/Pa/1$ fractal system as an illustrative example. According to the Pollaczek-Khinchine formula [4], average queue length L in this system is infinite at any $\rho > 0$:

$$L = \frac{\lambda^2 b^{(2)}}{2(1 - \rho)} = \infty,$$

since the second moment $b^{(2)}$ of the service time, distributed herein by the Pareto distribution at $1 < \alpha \leq 2$, is infinite. This example clarifies why in the case of a finite buffer (i.e., in the $M/Pa/1/m$ system) the reduction of the request loss probability due to the sufficiently large m and/or the channel operation speed increase (load coefficient decrease) is inefficient. In [11] this is justified by a more complete and detailed analysis.

Therefore, the need of finding more efficient ways of combating requests losses in fractal systems and networks arises.

The theoretical problem considered in this paper is to investigate the efficiency of loss probability reduction in fractal networks by increasing the number of channels in their nodes.

The development of an efficient technique for loss minimization using the optimal channel distribution over the nodes of fractal networks is the applied problem. The creation and use of sufficiently accurate expressions describing dependence of loss probability on the channels number in fractal system are the key elements of the developed technique.

3   Classical Infinite-Server Queueing Systems

In classical infinite-serverl system $GI/GI/\infty$ both $A(t)$, $B(t)$ distributions have finite dispersion. With the rise of a load $R = \lambda b$, the probability distribution p_k of the occupied channel number k converges to the Gaussian distribution $N(\bar{k}, \sigma_k)$ [12], i.e., $p_k \to g_k$,

$$g_k = \frac{1}{\sqrt{2\pi}\sigma_k}\exp\left[-\frac{(k - \bar{k})^2}{2\sigma_k^2}\right],\tag{1}$$

where $\bar{k} = \lambda b, \sigma_k^2 = \lambda b + \kappa\beta$,

$\kappa = \lambda^3(\sigma^2 - a^2), \beta = \int\limits_0^\infty [1 - B(\tau)]^2 d\tau$.

In practice there is a problem of choosing the smallest channels number n which, in the case of a request buffer absence, i.e., for $m = 0$, could provide a low request loss probability not exceeding the given value Q. In other words, it is necessary to find the smallest n, at which a multiserver system $GI/GI/n/0$ with the same $A(t)$ and $B(t)$ (i.e., a $GI/GI/n/0$ system that corresponds to initial infinite-server $GI/GI/\infty$ system) will have loss probability not exceeding Q. This problem will be called the problem of finding $n(Q)$, meaning that Q is a sufficiently low probability, for instance, $Q = 10^{-4}$ or $Q = 10^{-15}$.

When distribution p_k of the initial $GI/GI/\infty$ system is known, the problem of finding $n(Q)$ can be reformulated and solved as the one of finding the smallest n which satisfies the $P(k \geq n) = \sum_{k=n}^{\infty} p_k \leq Q$ condition, i.e., the $1 - P(k < n) \leq Q$ condition. Considering $k, n, P(k < n)$ as continuous quantities we can find such n that $1 - P(k < n) = Q$, i.e., we can obtain n for the required low Q by solving the equation

$$1 - F(n) = Q,\tag{2}$$

by defining d.f. $F(n)$ of r.v. k as $P(k < n)$. A solution n should be rounded up to an integer.

Problem (2) of finding $n(Q)$ can be solved by using approximation (1) of distribution p_k [12] in case of large load $R = \lambda b$ and small loss probability.

Let us consider, for instance, the classical infinite-server $\Gamma_1/\Gamma_2/\infty$ system where the gamma distribution [13] Γ_1 has parameters $\alpha_1 = 1/3$, $\beta_1 = 2/3$, and the gamma distribution Γ_2 has parameters $\alpha_2 = 1/3$, $\beta_2 = 1/30$. The m.e. of requests arrival intervals here is equal to $\lambda^{-1} = a = \alpha_1/\beta_1 = 1/2$ and the dispersion is equal to $\sigma^2 = \alpha_1/\beta_1^2 = 3/4$. The average service time in this system is $b = \alpha_2/\beta_2 = 10$.

Further, using formula (1), we obtain the parameters $\bar{k}$ and σ_k of the Gaussian approximation g_k for distribution p_k:

$$\bar{k} = \lambda b = 20, \kappa = \lambda^3(\sigma^2 - a^2) = 4, \beta = 2.86826...,$$
$$\sigma^2{}_k = \lambda b + \kappa\beta = 31.47304..., \sigma_k = 5.61008....$$

For solving problem (2), we will use the derived Gaussian approximation $g_k = N(\bar{k}, \sigma_k) = N(20, 5.61008...)$ concurrently with the actual distribution p_k which is obtained by means of the high-precision simulation.

One might see that graphs of the actual distribution p_k and its Gaussian approximation $g_k = N(20, 5.61008...)$ are visually almost superimposed on one another [14]. However, besides, the relative error in the form of p_k/g_k ratio is several orders of magnitude for the small p_k and increases with k growth.

Therefore, tails of the actual distribution p_k and its Gaussian approximation g_k vary by orders of magnitude (Fig. 1).

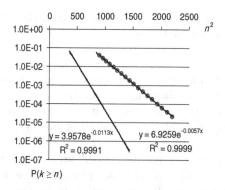

Fig. 1. Tail $Q = 1 - F(n) = P(k \geq n)$ of the distribution $F(n)$, calculated on the basis of probabilities p_k (marked line) and its Gaussian approximation g_k

We will compare the results of the problem (2) solution based on the distributions p_k, g_k and their associated regression equation (see Fig. 1).

Regression equation for the actual distribution p_k can be written as $1 - F(n) \approx 6.9259e^{-0.057n^2}$. According to this formula, Eq. (2) takes the

$Q \approx 6.9259e^{-0.057n^2}$ form, and its solution $n(Q)$ is expressed by the following simple approximate formula

$$n(Q) \approx \sqrt{-175.4 \ln(Q) + 339.5} \,. \tag{3}$$

A similar solution based on the Gaussian approximation g_k (see Fig. 1) is given by

$$n(Q) \approx \sqrt{-88.5 \ln(Q) + 121.7} \,. \tag{4}$$

Figure 2 compares solutions (3) and (4). The comparison shows that solution based on the Gaussian approximation leads to (in this case) a significant understatement of the channel number compared to the needed one. The use of this solution in practice would result in loss probability exceeding the admissible limit value Q by orders of magnitude.

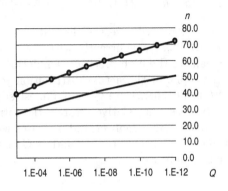

Fig. 2. Comparison of the problem (2) solution based on the distribution p_k (marked line) and its Gaussian approximation g_k

Generally, the performed examination of a specific classical multiserver system allows one to draw the following conclusions.

Firstly (this is the main conclusion), it is possible to effectively ensure a low loss probability by increasing the number of channels even without using a request buffer.

Secondly, in the general way, one should prefer simulation to asymptotic approximations to solve problem (2) which provides a low loss probability.

Thirdly, the reduction of a loss probability by several orders of magnitude is achieved via the relatively small increase of channel number redundancy (see Fig. 2 where the average number of the occupied channels equals $\lambda b = 20$).

In addition to that, relatively high sensitivity of problem (2) solution towards the errors of the used approximations, which has emerged in the investigation, results in the need to analyse errors of such approximations. In the method under consideration, the state probabilities of an infinite-server system are such approximations for loss probabilities of relating multiserver system.

4   Loss Probabilities Approximation by State Probabilities of Infinite-Server Systems

Let us consider the infinite-server $M/GI/\infty$ system. The Poisson distribution describes its state probabilities

$$p_k = \frac{(\lambda b)^k}{k!} e^{-(\lambda b)} = \frac{R^k}{k!} e^{-R}. \tag{5}$$

Request loss probabilities in the relating multiserver $M/GI/n/0$ system is defined by the Erlang loss formula

$$p_{loss}(n) = \frac{(\lambda b)^n}{n!} \left(\sum_{i=0}^{n} \frac{(\lambda b)^i}{i!} \right)^{-1} = \frac{R^n}{n!} \left(\sum_{i=0}^{n} \frac{R^i}{i!} \right)^{-1}. \tag{6}$$

Formulae (5) and (6) hold for any d.f. $B(t)$ with a finite m.e.

If the arriving request flow is not Poisson, one can use approximations to estimate loss probabilities $p_{loss}(n)$. The papers [12,15] give the expressions for approximation of $p_{loss}(n)$ in the $GI/GI/n/0$ systems with the finite first and second moments of $A(t)$ and $B(t)$ distributions. In case of the fractal system $GI/GI/n/0$, the state probabilities $p_k = p_n$ or the tail $P(k \geq n))$, which are obtained by simulation, of the relating infinite-server system can be used to approximate $p_{loss}(n)$ [16]. This approximation allows one to significantly (by several orders of magnitude) accelerate the process of finding the optimal distribution of the channel number over the fractal networks nodes, as will be indicated below.

We now show that the approximation is consistent. The ratio of approximation $p_k = p_n$ (5) to probability $p_{loss}(n)$ (6) when $n \to \infty$ converges to one in systems with the Poisson arrival flow

$$\lim_{n \to \infty} \frac{p_n}{p_{loss}(n)} = \lim_{n \to \infty} \frac{\frac{R^n}{n!} e^{-R}}{\frac{R^n}{n!} \left(\sum_{i=0}^{n} \frac{R^i}{i!} \right)^{-1}} = \frac{e^{-R} e^{R}}{1} = 1. \tag{7}$$

The ratio of tail $P(k \geq n)$ to $p_{loss}(n)$ when $n \to \infty$ also converges to one:

$$\lim_{n \to \infty} \frac{P(k \geq n)}{p_{loss}(n)} = \lim_{n \to \infty} \frac{\left(\sum_{i=n}^{\infty} \frac{R^i}{i!} \right) e^{-R}}{\frac{R^n}{n!} \left(\sum_{i=0}^{n} \frac{R^i}{i!} \right)^{-1}} = \frac{(1+0)e^{-R}}{e^{-R}} = 1 \tag{8}$$

since summands of the sum $\sum_{i=n}^{\infty} \frac{R^i}{i!} = \frac{R^n}{n!} + \frac{R^{n+1}}{(n+1)!} + \dots$ in (8) are decreasing too fast at any finite $R > 0$ with n growth, the sum converges to its first summand. For large finite n we have $\sum_{i=n}^{\infty} \frac{R^i}{i!} = \frac{R^n}{n!}(1 + \varepsilon)$, where $\varepsilon \to 0$ for $n \to \infty$.

By comparing relative errors $\delta_1 = p_n/p_{loss}(n)$ and $\delta_2 = P(k \geq n)/p_{loss}(n)$ of approximations p_n and $P(k \geq n)$, it can be seen that the state probabilities p_n converge to loss probabilities $p_{loss}(n)$ faster than tails $P(k \geq n)$. Figure 3

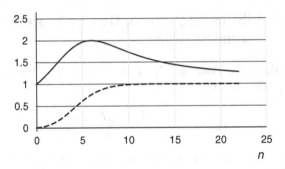

Fig. 3. Relative errors δ_1 and δ_2 (dashed and solid lines, respectively) of approximations p_n and $P(k \geq n)$

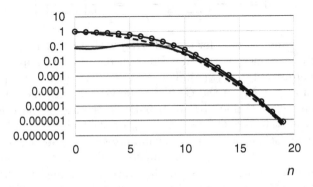

Fig. 4. Loss probabilities $p_{loss}(n)$ (dashed line), state probabilities p_n (solid line) and tails $P(k \geq n)$ (marked line) for the $Pa_1/Pa_2/n/0$, $Pa_1/Pa_2/\infty$ systems with their parameters $\alpha_1 = 1.6, K_1 = 0.1$ (for Pa_1) and $\alpha_2 = 1.5, K_2 = 0.445$ (for Pa_2)

illustrates the relative errors for the $M/Pa/n/0$ and $M/Pa/\infty$ systems with parameters $\lambda = 1.154, \alpha = 1.3$ and $K = 1$ calculated by (5) and (6).

In [16] a number of simulation experiments show that for general fractal systems the approximations p_n and $P(k \geq n)$ also may be used, and that p_n approximations are more accurate. Figure 4 shows the results of one such experiment.

The approximations p_n, as it may be noted, are simple. Estimation of the $GI/GI/\infty$ system using simulation gives us direct estimators of all steady p_n (as a ratio of associated cumulative occupancy times in the states n to all simulation times of steady-state process). To obtain the tails, the relative sums are to be calculated through p_n.

Results of the performed comparison of two approximations are taken into account in the developed below method for fractal networks optimization. Problem (2) of choosing the smallest channels number $n(Q)$ providing the loss probability not exceeding Q for multiserver systems is stated above in a traditional form, while now it is formulated as the problem of solving the equation

$$p(n) = Q, \tag{9}$$

where the probability $p_n = p(n)$ of state n in the corresponding infinite-server system is considered as a continuous function of continuous n due to the applied approximate expressions. The obtained solution of problem (9) should be rounded up to the nearest integer value.

5 Fractal Infinite-Server Systems

When in the $GI/GI/\infty$ system only d.f. $B(t)$ is fractal, one may use the Gaussian approximation g_k of state probabilities p_k, since parameter β can be finite in this case. Suppose, for example, $B(t) = 1 - \left(\frac{K}{t}\right)^{\alpha}, 1 < \alpha \leq 2$. Then, according to (1),

$$\beta = \int_0^\infty [1 - B(\tau)]^2 d\tau = \int_0^\infty \left(\frac{K}{\tau}\right)^{2\alpha} d\tau = \frac{2\alpha K}{2\alpha - 1},$$

and certainly for any $\alpha > 0.5$ (that holds automatically since $\alpha > 1$ always). Further, it is easy to define the other parameters of the Gaussian approximation (1) for this system:

$$b = \frac{K\alpha}{\alpha - 1}, \bar{k} = \lambda b, \kappa = \lambda^3(\sigma^2 - a^2), \sigma_k^2 = \lambda b + \kappa \frac{2\alpha K}{2\alpha - 1}. \tag{10}$$

Thus if $A(t)$ has finite dispersion σ^2, then in the $GI/Pa/\infty$ fractal system the distribution p_k for the number k of the occupied channels will converge to the Gaussian distribution $N(\bar{k}, \sigma_k)$ with parameters (10) with load $R = \lambda b$ growth.

As an example, let us consider the $\Gamma/Pa/\infty$ system with the gamma distribution which has parameters $\alpha_1 = 2, \beta_1 = 2$ (its m.e. $a = 1$ and dispersion $\sigma^2 = 0.5$) and with distribution $Pa(2, 1.25)$ (its dispersion is infinite and m.e. $b = 10$). Here the average number of the occupied channels is $\bar{k} = \lambda b = (1/a)b = 10$ and according to (10) parameter σ_k^2 of the Gaussian approximation equals $25/3 \approx 8.3333$. Figure 5 compares the Gaussian approximation g_k (dashed line) and the actual distribution p_k (solid line), calculated with simulation. In the simulation 100 million requests were generated. As it is seen, at $\lambda b = 10$ the Gaussian approximation g_k already agrees well with the actual distribution of r.v. k.

If we solve problem (2) of finding $n(Q)$ for this $\Gamma/Pa/\infty$ system using a Gaussian approximation of distribution p_k, it will result in unacceptable errors as in the case of the considered above $\Gamma_1/\Gamma_2/\infty$ classical system. As for a classical system, an actual distribution p_n, obtained by simulation, allows solving the problem of finding $n(Q)$ for the fractal systems under study with great accuracy. Figure 6 displays dependences of probabilities p_n and tail $1 - F(n)$ on n^2, obtained by simulation, that characterize the investigated $\Gamma/Pa/\infty$ system.

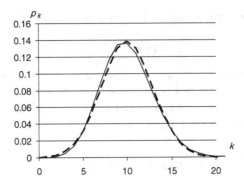

Fig. 5. Distributions p_k and g_k in the $GI/Pa/\infty$ system at $\lambda b = 10$

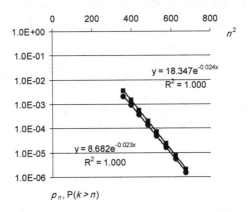

Fig. 6. Probabilities p_n (round markers) and their regression equation (at the bottom). Tails $P(k \geq n)$ (square markers) and their regression equation (at the top). Quantity n^2 is denoted by letter x in the equations

Trend line equation obtained for p_n in Fig. 6

$$p_n = 6.682e^{-0.023n^2} \tag{11}$$

provides the following solution of problem (9) for the investigated $\Gamma/Pa/\infty$ system:

$$n = \sqrt{-43.48 \ln p + 82.58} \ . \tag{12}$$

NB. The efficiency of a strategy to increase the number of channels can be seen from the following example in comparison with the strategies which increase buffer capacity and performance for a single channel. Let the arriving requests flow be served by a single channel with the performance equal to the total one of 30 channels (providing, according to (11), the loss probability $p_{loss} \approx 1 \cdot 10^{-8}$). In addition to that, suppose this one-channel system has a buffer, sufficient to store $m = 10\,000$ requests. Hence, we are discussing the $\Gamma/Pa/1/m$ system with the same arrival flow and Pareto $Pa(2/30, 1.25)$ service time "compressed" by a factor of 30. Simulation of this system shows that the request loss probability is equal to $p_{loss} \approx \mathbf{0.007}$ despite the large buffer capacity m and the low load coefficient $\rho = 1/3$. Therefore, in general the battle with losses by increasing a buffer capacity m and/or a single channel efficiency has proven to be almost inefficient for fractal traffic.

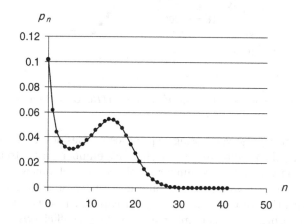

Fig. 7. Dependence of p_n on n in the investigated $Pa/Pa/\infty$ system

Figure 7 shows the dependence of state probabilities $p_n = p(n)$ for the $Pa/Pa/\infty$ system, in which intervals of requests arrival are distributed by the Pareto $Pa(1/5, 1.25)$ law, and the service time is by the $Pa(10/3, 1.5)$ law with a lighter tail. In this system $\lambda = 1, b = 10$. More than 100 million requests ran through the system in the simulation.

The trend line for the low state probabilities p_n of this system is presented in Fig. 8, which is given by the equation

$$p_n = 24.294e^{-0.0133n^2}.$$

Using it, we find a solution for problem (9) in the following form: $n = \sqrt{-75.19 \ln Q + 239.87}$. According to this, to provide, for example, the loss probability $Q = 10^6$, it is enough to set 35.8, i.e., 36 channels in the system.

High-precision simulation experiments with various fractal systems show that dependence of p_n and, consequently, p_{loss} on n with growing n in every such system at sufficient (around $\lambda b = 10$) load is well described by the following formula

$$p_n \sim c_0 e^{-Cn^2}, \tag{13}$$

where c_0, are some constants determined independently for each given system. Law (18) allows one to recommend increasing the number of channels in a system

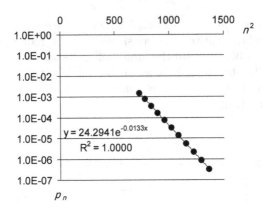

Fig. 8. Trend for dependence of p_n on n^2 in the $Pa/Pa/\infty$ system (see also Fig. 7)

as the efficient strategy for combating request losses. Besides, even for extremely low values of p_{loss} the redundancy of the channels number proves to be relatively low in comparison with average number λb of occupied channels.

Nevertheless, it is possible to combine a strategy of increasing the channels number with that of increasing the buffer capacity in networks with fractal traffic. After choosing channels number n, providing a sufficiently low loss probability p_{loss}, a request buffer can be added to the system.

Empirically obtained result (13) has all necessary characteristics of a universal law which holds for both classical and fractal systems. This law correlated well with theoretical result (1) as well, that was proven in [12] for classic systems. Therefore, taking into consideration similarity relation between dependence (13) and Gaussian distribution tails, (13) can be rewritten in a more theoretically correct form. With n growth

$$p_{loss} \sim p_n \sim c_0 e^{-C(n-\lambda b)^2}, \tag{14}$$

Consequently, one may suppose that approximations (13) and (14) will hold for the queueing network nodes as well (both fractal and classical ones).

6   Fractal Networks with Multiserver Nodes

Simulation experiments with various networks having multiserver nodes without buffers demonstrate that approximations (13) and (14) hold true for such network nodes with great accuracy. For instance, Fig. 9 gives state probabilities distributions for each of four nodes in a multiserver network, obtained by modification (sophistication) of the four-nodal network described in [12].

After modification, the routing matrix of the network

$$M = \begin{pmatrix} 0.1 & 0.4 & 0.4 & 0 \\ 0 & 0.4 & 0.1 & 0.2 \\ 0.1 & 0 & 0.2 & 0.1 \\ 0.3 & 0.2 & 0.5 & 0 \end{pmatrix}$$

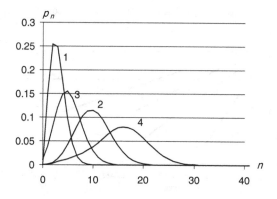

Fig. 9. State probabilities distribution for the network nodes

stays unchanged. The arriving requests flow and the service time in the nodes are changed and specified as follows.

From the outside, four flows enter the network. A regular flow with intensity 2 enters the first node from the outside. A Poisson flow enters the second node with intensity 2. A flow with Pareto $Pa(0.2, 1.25)$ distributed intervals of request arrivals enters the third node. And the fourth node has a flow with arrival intervals distributed by the $Pa(1/15, 1.5)$ law. Requests service time in the first node is deterministic and equal to 0.5. In the second one, the service time is exponentially distributed with the average 1, in the third one - by the $Pa(1/6, 1.5)$ law, and in the fourth - by the $Pa(0.4, 1.25)$ law.

Figure 10 depicts the dependences of probabilities p_{loss} on $x = (n - \lambda b)^2$, obtained by the simulation, as a relation (14), for all four nodes in the form of corresponding trend lines and trend equations. Trend line equations in the bottom of the figure are enumerated (according to the lines drawings) in the order of numbers 1, 3, 2, 4 of the corresponding network nodes. Initial dependences obtained in simulation experiment are shown in Fig. 10 as the lines almost coinciding with the trend lines.

Similar results are obtained when using approximations (13).

The experiment results support the hypothesis for the correctness of approximations (13) and (14) not only in isolated systems but in the networks nodes.

Approximations (13) and (14) can be used as a basis to develop various methods of structural optimization for fractal systems and networks to guarantee low loss probability. The most important feature of laws (13) and (14) is that the increase of the channel number in the nodes at their relatively small redundancy leads to drastic reduction of a loss probability. Under conditions of fractal traffic, it distinguishes the strategy of the channel number increase from that of the buffers capacity and/or channel performance increase.

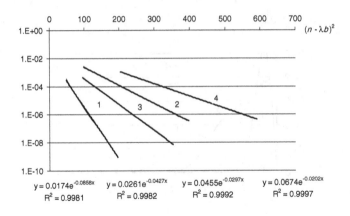

Fig. 10. Trend lines for dependences of p_n on n_2 for the network nodes

A rather simple and quick methods of optimal channels distribution over the fractal network nodes is proposed as a practical application for the results of (13) and (14).

7   Optimization of Channel Distribution over the Nodes

Suppose a fractal network routing matrix, d.f. $B_i(t)$ of service time in nodes i ($i = 1, ..., M$) and arriving requests flows are given. There are no buffers in the network nodes. It is required to distribute N channels ($N \gg M$) over the network nodes in order to minimize the sum of loss probabilities in the nodes.

In practice, whatever number of the channels we have, it is always finite. Moreover, the efficiency of the channels usage depends on the way these channels are distributed over nodes.

Considering approximations (13) and (14), formally, the problem of optimal channel distribution can be rewritten as follows:

$$\sum_{i=1}^{M} c_{0i} e^{-C_i n_i^2} \rightarrow \min, \tag{15}$$

or in a form of

$$\sum_{i=1}^{M} c_{0i} e^{-C_i (n_i - \lambda_i b_i)^2} \rightarrow \min, \tag{16}$$

in both cases, limitations are used

$$\sum_{i=1}^{M} n_i = N, \quad n_i > 0, \quad i = 1, ..., M.$$

In both cases coefficients c_{0i} and C_i are calculated by simulation, just as it was demonstrated above. Intensities λ_i of the entering flows are easily calculated by the network routing matrix through the given intensities of the entering flows (for example, by constructing and solving intensity balance equations).

In (15) or (16) the solution of optimal channels distribution problem can be obtained with any known numerical methods. For instance, we can consider variables n_i as continuous quantities, look for their optimal values by any known gradient method, then we can round these values accordingly.

Let us consider the problem of optimal distribution of 100 channels over the nodes of the network described in Sect. 6.

Using its simulation data shown in Fig. 10, we write the optimization problem in the form of (9):

$$0.0174e^{-0.0858(n_1-3.039)^2} + 0.0455e^{-0.0297(n_2-10.08)^2}$$
$$+ 0.0261e^{-0.0427(n_3-5.225)^2} + 0.0674e^{-0.0202(n_4-15.715)^2} \to \min, \qquad (17)$$

$$\sum_{i=1}^{4} n_i = 100, \quad n_i > 0, \quad i = 1, ..., 4. \qquad (18)$$

Solving problem (17) and (18) with the help of Excel add-in program Solver, we get $n_1 = 13.51, n_2 = 28.3, n_3 = 20.12, n_4 = 38.05$ or, after rounding:

$$n_1 = 13, n_2 = 29, n_3 = 20, n_4 = 38. \qquad (19)$$

In this case, there are six possible rounding procedures that preserve Eq. (18). The correct rounding is the one that provides the smallest value when substituted in target function (17).

Channels distribution (19) is tested by simulation. At such distribution, cumulative failure probability in the nodes is $\mathbf{1.34 \cdot 10^{-6}}$. All "neighboring" distributions (in which one of the four solution (19) coordinates is decreased by one, and the other is increased by one) are characterized by a worse cumulative failure probability in the nodes than that of distribution (19).

It should be noted that the values of target function (17), using approximation, differ significantly from the corresponding values obtained in the simulation. The results of problem (17) and (18) solution, however, are rather accurate.

The problem in the form of (15) gives the same solution (19).

If loss probability approximation in the form of tails $P(p_k \geq p_n)$ is used, the result is a less accurate solution $n_1 = 13, n_2 = 28, n_3 = 20, n_4 = 39$, at which the simulation provides cumulative loss probability $\mathbf{1.85 \cdot 10^{-6}}$.

Let us compare the optimal distribution (19) and a uniform channels distribution over the nodes and channels distribution providing similar coefficients ρ_i of the nodes load. With the help of simulation, having set the optimal channels distribution as 25 channels per each node, we get cumulative loss probability in the nodes $\mathbf{2.4 \cdot 10^{-3}}$. If the channels are distributed in such a way that nodes load coefficients are equal (then $n_1 = 9, n_2 = 30, n_3 = 16, n_4 = 45$ and all $\rho_i = 0.33$), the cumulative loss probability will be $2.6 \cdot 10^{-4}$. In both cases the results are worse than optimal ones by several orders of magnitude, and this indicates the practical significance of the proposed optimization method.

When designing a telecommunication network after optimal channels distribution over the nodes, the requests buffer can be added to every node. In this way, the loss probability can be reduced almost to zero.

For example, after adding buffers with size $m_i = 100$ to the nodes of the newly optimized network, there were no queues longer than 7 at 10 mln requests passing the network multiple times. It is obvious that requests loss is almost eliminated in the resulting network.

8   Conclusion

The main research results are relations (13) and (14), that not only give the key to solving the loss problem in the networks with fractal traffic but allow solving the problems of their optimization and analysis. As a rule, approximations (13) and (14) can be used when loading multiserver nodes $\lambda b = 10$ and more.

In combating the request losses in fractal networks, the strategy of increasing the channels number has the principal advantages over those of increasing the buffer capacity and the channels performance.

The proposed herein approximate method for channels distribution optimization over the nodes of a queueing fractal network is rather simple, effective and can be directly used by designers of networks.

The architecture, proposed in the article for the networks with fractal traffic and oriented at multiserver nodes, allows one to combat the messages loss efficiently and it is characterized by low redundancy of cumulative performance of hardware facilities. Moreover, such architecture exhibits high immunity of the solutions towards the second moments of the arrival intervals and the service time, for it guarantees low losses even at their infinite values.

References

1. Leland, W.E., Taqqu, M.S., Willinger, W., Wilson, D.V.: On the self-similar nature of ethernet traffic. ACM/SIGCOMM Comput. Commun. Rev. 23, 146–155 (1993)
2. Crovella, M.E., Taqqu, M., Bestavros, A.: Heavy tailed-probability distributions in the world wide web. IEEE/ACM Trans. Netw. 5(6), 835–846 (1997)
3. Czachórski, T., Domańska, J., Pagano, M.: On stochastic models of internet traffic. In: Dudin, A., Nazarov, A., Yakupov, R. (eds.) ITMM 2015. CCIS, vol. 564, pp. 289–303. Springer, Cham (2015). doi:10.1007/978-3-319-25861-4_25
4. Kleinrock, L.: Queueing Systems: Computer Applications, vol. 2. Wiley Interscience, New York (1976). 576 pages
5. Zwart, A.P.: Queueing Systems with Heavy Tails. Eindhoven University of Technology (2001). 227 pages
6. Asmussen, S., Binswanger, K., Hojgaard, B.: Rare events simulation for heavy-tailed distributions. Bernoulli 6(2), 303–322 (2000)

7. Boots, N.K., Shahabuddin, P.: Simulating GI/GI/1 queues and insurance risk processes with subexponential distributions. In: Proceedings of the 2000 Winter Simulation Conference, pp. 656–665 (2000). Unpublished manuscript, Free University, Amsterdam. Shortened version
8. Zadorozhnyi, V.: Fractal queues simulation peculiarities. In: Dudin, A., Nazarov, A., Yakupov, R. (eds.) ITMM 2015. CCIS, vol. 564, pp. 415–432. Springer, Cham (2015). doi:10.1007/978-3-319-25861-4_35
9. Zadorozhnyi, V.N.: Peculiarities and methods of fractal queues simulation. In: 2016 International Siberian Conference on Control and Communications (SIBCON), Fundamental Problems of Communications, Moscow, Russia, 12–14 May 2016
10. Zadorozhnyi, V.N., Zakharenkova, T.R.: Methods of simulation queueing systems with heavy tails. In: Dudin, A., Gortsev, A., Nazarov, A., Yakupov, R. (eds.) ITMM 2016. CCIS, vol. 638, pp. 382–396. Springer, Cham (2016). doi:10.1007/978-3-319-44615-8_33
11. Zadorozhnyi, V.N. Simulation modeling of fractal queues. In: Dynamics of Systems, Mechanisms and Machines (Dynamics), pp. 1–4 (2014). doi:10.1109/Dynamics.2014.7005703
12. Moiseev, A.N., Nazarov, A.A.: Beskonechnolinejnye sistemy i seti-massovogo obsluzhivaniya [Infinite-linear queueing systems and networks]. NTL Publ., Tomsk (2015). 240 pages
13. Korn, G.A., Korn, T.M.: Mathematical Handbook for Scientists and Engineers: Definitions, Theorems, and Formulas for Reference and Review. General Publishing Company (2000). 1151 pages
14. Zadorozhnyi, V.N., Zakharenkova, T.R.: Optimization of channel distribution over nodes in networks with fractal traffic. In: 2016 Dynamics of Systems, Mechanisms and Machines, Dynamics, Omsk, Russia, 14–16 November (2016). doi:10.1109/Dynamics.2016.7819112
15. Li, A., Whitt, W.: Approximate blocking probabilities in loss models with independence and distribution assumptions relaxed. Perform. Eval. **80**, 82–101 (2014)
16. Zakharenkova, T.R.: On loss probability in fractal multiserver queueing systems. Omsk Sci. Bull. **3**(153), 110–114 (2017)

Optimization of Pipelining and Data Processing

Pavel Mikheev[(✉)], Anastasiya Pichugina, Sergey Suschenko,
and Roman Tkachev

National Research Tomsk State University, Lenina Str., 36, 634050 Tomsk, Russia
doka.patrick@gmail.com , ssp.inf.tsu@gmail.com

Abstract. The method of optimal partitioning of subscriber messages into protocol data units by the transport layer according to the criterion of delays in the multi-hop transmission path is proposed. The terms of the appropriateness of the fragmentation of messages into packets during its transmission over multi-hop virtual channel are obtained. Analytical dependences for the optimal packet size from the structure of network traffic and settings of the virtual connections are obtained.

Keywords: Transport connection · Delay · The multiplex packet · Anon-uniform path · The pipelininig effect

1 Introduction

The most important indicator of efficiency of functioning of the network packet switching is the transmission time of user data between the communicating subscribers [1,2]. Functions for the delivery of message flow to the user and compensation of overhead in the transmission of packets that may occur in the communication network are performed by the transport layer protocol [2]. The basis of a reliable transport protocol is the principle of decision feedback. The delay in subscriber traffic in a virtual connection depends to a large extent on the characteristics of the individual links of the connecting path, the length of the data transmission path, the size of the user messages, the intensity of the network streams and the protocol parameters, among which the most important is the packet size, which actually determines the power of the pipeline effect [3–6]. It should also be noted that the connecting path of the virtual channel in the packet switching network is used in by many interacting subscribers. This leads to the fact that the load on various parts of the data path along which the virtual connection goes can be significantly different. Then the effective bandwidth of individual links for the traffic of this virtual connection will be reduced by the corresponding parts of "external" flows, as a result of which the time of packet transfer over inter-node connections even of a uniform virtual channel can be substantially different [6–8]. Simulation of the transport connection and analysis of its operational characteristics under various loading conditions is performed in [3–8]. A wide range of studies [9–13] are aimed at optimizing protocol parameters by various criteria and adapting protocol parameters to the

© Springer International Publishing AG 2017
A. Dudin et al. (Eds.): ITMM 2017, CCIS 800, pp. 184–193, 2017.
DOI: 10.1007/978-3-319-68069-9_15

changing network load, the level of losses, the activity of interacting subscribers. A key indicator of the quality of service for network subscribers is the message delivery time, which is determined by the pipelining effect. This indicator is also very important for pipelined implementation of the instruction processing [14,15]. The development of the results of [9–15] is to optimize the size of the protocol data units when sending the subscription message via the transport connection and the structure of the data transmission path. The most important tool of the analysis of processes of data transmission and processing in a random environment are Queuing systems [16,17]. A Queuing system in continuous time allow to investigate the operating characteristics of inhomogeneous systems of data transmission and processing, but ignores their essentially discrete nature. A Queuing system with discrete time adequately describes such processes, but does not allow analysis of heterogeneous systems of data transmission and processing. In addition, models based on queueing systems do not take into account the pipeline effect, which is manifested in the transmission and processing of data streams. However, in the particular case of a deterministic environment, data transmission and processing can be obtained the analytical results with the conveying effect of the discreteness and heterogeneity of the processes of data transmission and processing. In the proposed method of this work the partitioning the subscriber messages into packets of optimum size and the conditions the feasibility of converting transmission path and pipelined processor to a uniform appearance.

2   The Virtual Connection Model

Consider an nonuniform virtual connection consisting of D links of data transmission. Define the time of the message transmission from the N packets, according to the deterministic virtual connection in the data transfer phase. We believe that the flow control procedure carried by the virtual connection provides end-to-end confirmation of the delivery of individual messages, and each virtual connection node can simultaneously perform data reception and transmission, however, packet transmission can be started only after its reception is completed. All message packets have the same length, except the last one, which can contain the remainder of dividing the message into fragments and can be smaller. We believe that there is no competing traffic and there are no packet queues at the switching nodes to the output communication channels. Then the delay of the subscriber's message in the data transmission path will be [10]:

$$T(D, N) = (N - 1)\tau_m + \sum_{d=1}^{D} \tau_d, \quad \tau_m = \max_{d=\overline{1,D}} \tau_d, \tag{1}$$

where τ_d, $d = \overline{1, D}$ — packet delay in d sector of the hops.

3   The Optimal Partitioning of the Message into the Packets

For further analysis, we express explicitly the time of packet transmission in inter-node connection through the parameters of data transmission link. Suppose that the transmission rate and time of node packet processing is independent of the size of the package. In fact, the assumption of rate is true only for absolutely reliable inter-node communication channels included in the virtual connection. Then the packet delay at the d link of the transmission path taking into account the previously introduced notation we can write as: $\tau_d = \frac{L}{C_d} + t_d$. Here t_d, $d = \overline{1, D}$ the packet processing time in the receiver node of the d data link. Substituting this relation in (1) and taking into account that $L = \frac{B}{N} + H$, where B is the size of the transmitted message, we get:

$$T(D, N) = (N - 1)\left[\frac{B/N + H}{C_m} + t_m\right] + \sum_{d=1}^{D}\left[\frac{B/N + H}{C_d} + t_d\right]; \qquad (2)$$

$$\frac{B/N + H}{C_m} + t_m = \max_{d=\overline{1,D}}\left(\frac{B/N + H}{C_d} + t_d\right).$$

Obviously, when transmitting a message in the form of a sequence of packets, it is possible to reduce the time of its delivery significantly over a virtual connection in comparison to its transmission by one packet. This gain is due to the pipelining effect [10], as a result of which the different parts of the message are simultaneously in the transmission state at different parts of the path. On the one hand, the number of packets in the sequence should be increased in order to enhance the pipelining effect and thereby reduce the message delivery time. On the other hand, sequence growth leads to an increase in the volume of the transferred service information and the processing time of packets by nodes. Hence it follows that the dependence (2) is unimodal from the argument N. Using (2), we determine the benefit in time from the transmission of a message over a virtual connection of length D by a sequence of $N > 1$ packets in comparison with its delivery as a whole:

$$\Delta(D, N) = T(D, 1) - T(D, N) = (N - 1)\left\{\frac{B}{N}\sum_{\substack{d=1, \\ d \neq m}}^{D}\frac{1}{C_d} - \frac{H}{C_m} - t_m\right\}. \qquad (3)$$

For uniform virtual connection, $C_d = C$, $t_d = t$, $d = \overline{1, D}$ the benefit (3) is converted to the form:

$$\Delta(D, N) = \frac{N - 1}{C}\left\{\frac{B}{N}(D - 1) - H - Ct\right\}. \qquad (4)$$

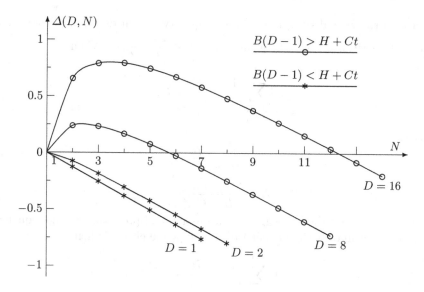

Fig. 1. The dependence of the benefit from argument N

The relations (3) and (4) define unimodal of the argument $N \geq 1$ functions (Fig. 1) with asymptotes

$$\Delta(D, N) = -(N-1)\left(\frac{H}{C_m} + t_m\right) + B\sum_{\substack{d=1, \\ d\neq m}}^{D} \frac{1}{C_d} \quad \text{and}$$

$$\Delta(D, N) = -(N-1)\left(\frac{1}{C} + t\right) + \frac{B(D-1)}{C}$$

accordingly. It can be seen from (3) and (4) that it is expedient to split/divide the message into packets only for long $D > 1$ virtual connections and if condition

$$B\sum_{\substack{d=1, \\ d\neq m}}^{D} \frac{1}{C_d} > \frac{H}{C_m} + t_m$$

is fulfilled, and benefit (3) is positive for partitions that satisfy the inequality

$$1 < N < \frac{BC_m \sum_{\substack{d=1, \\ d\neq m}}^{D} \frac{1}{C_d}}{H + C_m t_m}.$$

For a uniform virtual connection, this inequality, which determines the set of expedient partitions, takes the form of $1 < N < \dfrac{B(D-1)}{H+Ct}$. When

$$B \sum_{\substack{d=1,\\ d \neq m}}^{D} \frac{1}{C_d} > \frac{H}{C_m} + t_m$$

splitting

$$N \geq \frac{BC_m \sum_{\substack{d=1,\\ d \neq m}}^{D} \frac{1}{C_d}}{H + C_m t_m}$$

lead to the fact that the losses from transmission and processing of the sequence of packets prevail over the benefit from the pipeline effect. When

$$B \sum_{\substack{d=1,\\ d \neq m}}^{D} \frac{1}{C_d} < \frac{H}{C_m} + t_m$$

splitting $N > 1$ increase the negative effect of exceeding the overhead on user information.

On virtual connections of a single length, there is no pipelining effect, and $N > 1$ partitions lead to an increase in the multiplex packet delay due to an increase in the amount of overhead transfer of the service information and the node packet processing (Fig. 1).

From the size of the subscriber message B the benefit (3) and (4) has a linear dependence (Fig. 2). When transferring over the uniform virtual connection, the benefit (4) also grows linearly with the path length, and the values $\Delta(D, N)$ are positive for $D > 1 + \frac{N(H+Ct)}{B}$.

From (3) we find that the partition

$$N_0 = \sqrt{\frac{B}{H/C_m + t_m} \sum_{\substack{d=1,\\ d \neq m}}^{D} \frac{1}{C_d}}, \quad N_0 \geq 1$$

maximizes the benefit (3). Substituting the relation for N in (3) we obtain:

$$\Delta(D, N_0) = B \sum_{\substack{d=1,\\ d \neq m}}^{D} \frac{1}{C_d} + \frac{H}{C_m} + T_m - 2 \sqrt{B \sum_{\substack{d=1,\\ d \neq m}}^{D} d\frac{1}{C_d} \left(\frac{H}{C_m} + t_m \right)}.$$

Hence it is easy to see that the optimal benefit is equal to twice the difference between the arithmetic mean and geometric mean values

$$B \sum_{\substack{d=1,\\ d \neq m}}^{D} \frac{1}{C_d} \quad \text{and} \quad \frac{H}{C_m} + t_m,$$

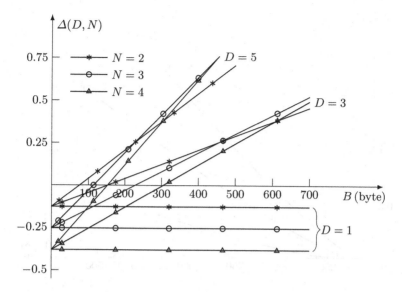

Fig. 2. The dependence of the benefit from the message size

which correspond respectively to the transmission time of the information part of the message as a single unit over a virtual connection without a narrow link and the overhead of a narrow link in the form of the time for transmitting the service part of the packet and the processing time of the packet. For a uniform virtual connection, the optimal benefit is:

$$\Delta(D, N_0) = \left\{ \sqrt{B(D-1)} - \sqrt{H + Ct} \right\}^2$$

Since $N \geq 1$, we can conclude that the area of definition $\Delta(D, N_0)$ for the uniform virtual connection is the length of the paths that satisfy the inequality $D \geq \frac{1+H+Ct}{B}$. Figure 3 shows the dependence $\Delta(D, N_0)$ from B.

Knowing the optimal relation of splitting N, it is easy to determine the packet size L_0, that minimizes the delivery time of a message over the virtual connection:

$$L_0 = H + \frac{B}{N_0} = H + \sqrt{\frac{B(H/C_m + t_m)}{\displaystyle\sum_{\substack{d=1, \\ d \neq m}}^{D} 1/C_d}}.$$

On the uniform connection the expression for the optimal packet size is simplified: $L_0 = H + \sqrt{\frac{B(H+Ct)}{D-1}}$.

4   Select the Size of the Package

In order to apply the obtained ratios of the optimal packet length in practice, it is necessary to take into account the sheer size of the messages transmitted via

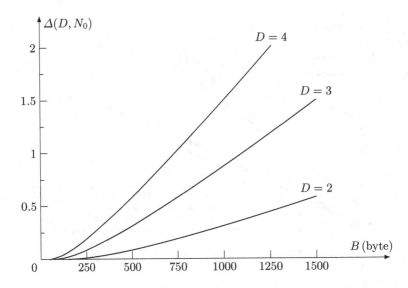

Fig. 3. The dependence of the benefit from the message size

the virtual network connections. In addition, since the largest packet size for a static selection should usually have a single value for the entire network, it should be determined from the maximum benefit condition for virtual connections of all possible lengths. Thus, the obtained dependencies must be generalized to the case of an integral criterion.

Assume that the transmission network is set to all possible path lengths D_j, $j = \overline{1, J}$ and the distribution of intensities (parts) of network traffic on the transmission paths of the data α_j, $j = \overline{1, J}$ which satisfies the normalization condition $\sum_{j=1}^{J} \alpha_j = 1$ where J is the number of different information flows. Let also for each information flow a continuous distribution of message lengths is given $f_j(B)$. Consider, as an objective function, the average benefit, which is a natural generalization of criterion (3) to the entire data network:

$$\bar{\Delta}(N_1, N_2, \ldots, N_J) = \sum_{j=1}^{J} \alpha_j \int_0^\infty \Delta(D_j, N_j) df_j(B)$$

$$= \sum_{j=1}^{J} \alpha_j (N_j - 1) \left\{ \frac{\bar{B}_j}{N_j} \sum_{\substack{d=1, \\ d \neq m_j}}^{D_j} \frac{1}{C_{jd}} - \left(\frac{H}{C_{m_j}} + t_{m_j} \right) \right\}, \quad (5)$$

where $\bar{B}_j = \int_0^\infty B df_j(B)$ the average length of messages in j information stream. Since each virtual connection has its own partition coefficient, in this relation it is more convenient to go directly to the required length of the frame L. Substituting $L_0 = H + \frac{\bar{B}_j}{N_j}$ from (5) we obtain:

$$\bar{\Delta}(L) = \sum_{j=1}^{J} \alpha_j \bar{B}_j \sum_{\substack{d=1, \\ d \neq m_j}}^{D_j} \frac{1}{C_{jd}} - (L-H) \sum_{j=1}^{J} \alpha_j \sum_{\substack{d=1, \\ d \neq m_j}}^{D_j} \frac{1}{C_{jd}}$$

$$- \frac{1}{L-H} \sum_{j=1}^{J} \alpha_j \bar{B}_j \left(\frac{H}{C_{m_j}} + t_{m_j} \right) + \sum_{j=1}^{J} \alpha_j \left(\frac{H}{C_{m_j}} + t_{m_j} \right).$$

Hence we determine the optimal value L_0:

$$L_0 = H + \sqrt{\frac{\sum\limits_{j=1}^{J} \alpha_j \bar{B}_j \left(\dfrac{H}{C_{m_j}} + t_{m_j} \right)}{\sum\limits_{j=1}^{J} \alpha_j \sum\limits_{\substack{d=1, \\ d \neq m_j}}^{D_j} \dfrac{1}{C_{jd}}}}. \tag{6}$$

Here the parameters C_{m_j} and t_{m_j} which determine the narrow link of the j virtual connection, are found from condition

$$\frac{L}{C_{m_j}} + t_{m_j} = \max_{d=1,D_j} \left(\frac{L}{C_{jd}} + t_{jd} \right). \tag{7}$$

In the case of the uniform network the dependence for the optimal L_0 is transformed to the form:

$$L_0 = H + \sqrt{\frac{\bar{B}(H + Ct)}{\bar{D} - 1}},$$

where $\bar{B} = \sum_{j=1}^{J} \alpha_j \bar{B}_j$ the average size of messages transmitted over the network $\bar{D} = \sum_{j=1}^{J} \alpha_j D_j$ — average length of network transmission paths.

It is not difficult to see that condition (7) uniquely determines the narrow links of virtual connections only when the packet processing time is zero at the switching nodes, in general, the parametric dependence of condition (7) on L does not allow unambiguous definition of bottlenecks before calculating the optimal frame size. For the case when packet processing time in nodes can not be neglected, it is possible to propose a procedure for consistent calculation of the optimal frame size. According to this procedure, the optimal value of L is calculated iteratively. As initial value L for definition of narrow links by a condition (7) it is possible to accept $L = H$. Using the parameters found in this way C_{m_j}, t_{m_j}, $j = \overline{1,J}$, the frame size is calculated from (6), which is used to determine the narrow links in the next step. The stop criteria for stopping the iterative process is the match of the frame size or the set of indices of narrow links m_j, $j = \overline{1,J}$ in two consecutive iterations.

5   The Conditions of Feasibility of Unification of the Non-uniform Phases in the Pipeline

One of the most important conditions for achieving minimum latency on the line as in (1) is to eliminate the most time-consuming stages of processing (transmission) data through their pipeline. Most often, this approach of bringing the individual phases of the pipeline to a uniform duration of stages is used in the processor of the data processing or telecommunication systems to eliminate low-speed, geographically distributed network sections. In this case, the narrow phases of the pipeline are broken (if possible) into sub-phases of the minimal complexity of the input of the original or desired conveyor, which leads to an increase in its length. Let us analyze the conditions under which such a partition of complex phases into sub-phases reduces the processing time of the data stream. Consider fully ununiformed pipeline, which should lead to a uniform with the duration of the phases equal to $\tau \leq \tau_d$, $d = \overline{1, D}$. We assume that each phase of an ununiformed pipeline has a duration $\tau_d = l_d$, $d = \overline{1, D}$, where $l_d \geq 1$—is an integer. Then every d phase of the source pipeline should be pipelined in the form of stages of the same duration τ. The normalized delay in the original ununiformed pipeline length is

$$t_n(D, N) = \frac{T(D, N)}{\tau} = (N - 1)l_m + \sum_{d=1}^{D} L_d, \quad l_m = \max_{d=\overline{1,D}} l_d,$$

and in the uniformed—with the resulting number of stages, equal $\sum_{d=1}^{D} l_d$, will take the form

$$t_0\left(\sum_{d=1}^{D} l_d, N\right) = \frac{T\left(\sum_{d=1}^{D} l_d, N\right)}{\tau} = N - 1 + \sum_{d=1}^{D} l_d.$$

We determine normalized to the complexity of the uniformed phase τ the benefit from the unification of pipelining N applications in the form of a difference of the processing times of the original ununiformed pipeline and extended uniformed pipeline:

$$\nabla(D, N) = t_n(D, N) - t_0\left(\sum_{d=1}^{D} l_d, N\right) = (N - 1)(l_m - 1).$$

Hence it follows that the positive values of the benefit are invariant to the complexity l_d of all stages of the pipeline, except the most labor-intensive and possible when $N > 1$ and $l_m > 1$.

6   Conclusion

The paper proposes the method of optimal partitioning of subscriber messages into protocol data units by the transport layer according to the criterion of delays in the multi-hop transmission path. Analytical dependences for the optimal packet size from the structure of network traffic and settings of the virtual connections are obtained. The terms of the appropriateness of the fragmentation of messages into packets during its transmission over multi-hop virtual channel are obtained. The direction of the further development of research on the unification of the pipeline should be distinguished by the task of analyzing the delay in conditions of rebooting of the pipeline with repeated transmissions of distorted data in networks or incorrect branch prediction processed by the processor of instruction stream.

References

1. Fall, K., Stevens, R.: TCP/IP Illustrated: The Protocols. Addison-Wesley Professional Computing Series, vol. 1, 2nd edn. Addison-Wesley, Reading (2012). 1017 pages
2. Boguslavskii, L.B.: Upravlenie potokami dannykh v setyakh EVM (Controlling Data Flows in Computer Networks). Energoatomizdat, Moscow (1984). 168 pages (in Russian)
3. Boguslavskij, L.B., Gelenbe, E.: Analytical models of data link control procedures in packet-switching computer networks. Autom. Remote Control. 41(7), 1033–1042 (1980)
4. Gelenbe, E., Labetoulle, J., Pujolle, G.: Performance evaluation of the HDLC protocol. Comput. Netw. 2(4/5), 409–415 (1978)
5. Kokshenev, V.V., Suschenko, S.P.: Analysis of the asynchronous performance management procedures link transmission data. Comput. Technol. 15(5), 61–65 (2008)
6. Kokshenev, V.V., Mikheev, P.A., Suschenko, S.P.: Comparative analysis of the performance of selective and group repeat transmission models in a transport protocol. Autom. Remote Control. 78(2), 247–261 (2017)
7. Kokshenev, V.V., Mikheev, P.A., Suschenko, S.P.: Transport protocol selective acknowledgements analysis in loaded transmission data path. Tomsk State Univ. J. Control Comput. Sci. 3(24), 78–94 (2014). (in Russian)
8. Suschenko, S.P.: The influence of buffer overfilling on the speed of synchronous data-transmission control procedures. Autom. Remote Control. 60(10), 1460–1468 (1999)
9. Suschenko, S.P.: Method of rational choice of packet length of packet switching network. Autom. Comput. Eng. 3, 24–28 (1984)
10. Suschenko, S.P.: Parametric optimization of a packet switching network. Autom. Comput. Eng. (2), 43–49 (1985)
11. Suschenko, S.P.: Analysis of end-to-end message delay in a multi-tier virtual channel. Autom. Comput. Technol. (3), 52–64 (1989)
12. Suschenko, S.P.: Influence of the duration of a through time-out on the delay of data in a virtual channel. Autom. Comput. Technol. (6), 36–40 (1991)
13. Suschenko, S.P.: Analysis of the influence of the duration of a through time-out on the operational characteristics of a virtual channel. Autom. Comput. Technol. (4), 43–66 (1995)
14. Patterson, D.A., Hennessy, J.L.: Computer Organization and Design. Elsevier, Amsterdam (2014). 793 pages
15. Gorshenin, A.K., Zamkovets, S.V., Zakharov, V.N.: Parallelism in microprocessors. Syst. Means Inform. 24(1), 46–60 (2014)
16. Kleinrock, L.: Queueing Systems: Theory, vol. 1. Wiley, New York (1975)
17. Kleinrock, L.: Queueing Systems, vol. 2. Wiley, New York (1976)

Performance Simulation of Non-reliable Servers in Finite-Source Cognitive Radio Networks with Collision

Hamza Nemouchi and Janos Sztrik$^{(\boxtimes)}$

Faculty of Informatics, University of Debrecen, Debrecen, Hungary
nemouchih@gmail.com, sztrik.janos@inf.unideb.hu

Abstract. This paper introduces a finite-source retrial queueing system which models cognitive radio networks. We assume two non-independent frequency bands servicing two classes of users: Primary Users (PUs) and Secondary Users (SUs). A service unit with a priority queue and another service unit with an orbit are assigned to the PUs and SUs, respectively.

In this work, we focus on the non-reliability of the servers and the collisions at the secondary servers. The primary and secondary servers are subject to random breakdowns and repairs. A collision is introduced at the retrial part of the cognitive radio network. This conflict invokes the interruption of a servicing packet when a new arriving call requests the server unit.

The novelty of the investigation is the non-reliability of servers and the inclusion of conflicts at the secondary server.

By the use of simulation, we analyze the effect of the non-reliability of the servers on the mean response time of the secondary users.

Keywords: Finite source queuing systems · Simulation · Cognitive Radio Networks · Performance and reliability measures · Collision · Non-reliable servers

1 Introduction

Cognitive radio (CR) has emerged as a promising technology to realize dynamic spectrum access and increase the efficiency of a largely under utilized spectrum. As it was defined in [1,2], the cognitive radio network (CRN) is a network made up of CRs by extending the radio link features to network layer function and above. By means of CRs cooperation, the network is able to sense its environment, learn from the history, and accordingly decide the best spectrum settings.

In other words, cognitive radio allows efficient use of the available spectrum by defining two types of users in wireless networks: licensed and unlicensed users. An unlicensed user (also called secondary user (SU)) can use the spectrum if it is not being used at that time by licensed users (also called primary user (PU)). When the licensed user appears to use the spectrum, the unlicensed user must find another spectrum to use. see for example [3–5].

© Springer International Publishing AG 2017
A. Dudin et al. (Eds.): ITMM 2017, CCIS 800, pp. 194–203, 2017.
DOI: 10.1007/978-3-319-68069-9_16

In this paper we introduce a finite-source queueing model with two (non independent) frequency channels. The cognitive radio architecture consists of two main networks: The Primary Channel Service (PCS) and Secondary Channel Service (SCS). The PCS refers to the existing network, wherein the primary users (PUs) have got a licensed frequency which does not suffer from overloading. The SCS does not have a license to operate in a licensed frequency. Hence, SCS is designed to work with PCS to provide the capability to utilize or share the unused spectrum in an opportunistic way. The secondary users have got also a frequency band but it suffers from overloading.

In our environment the band of the PUs is modelled by a queue where the requests has preemptive priority over the SUs requests. The band of the SUs is described by a retrial queue: if the band is free when the request arrives then it is transmitted. Otherwise, the request goes to the orbit if both bands are busy. The primary server unit and the secondary server unit are not reliable and are assumed to be subject to breakdown and repair. Also, the retrial part of the cognitive radio network suffered from collision at the secondary server unit, which means that the arriving packets involve into collision with the servicing packets [6,7].

Hence, it should be noted that the novelty of this model is the introduction of the non-reliability of the servers with conflict (collision), and by using simulation, we analyze the effect of the request generation, retrial, service, failure and repair rate of the primary and secondary users on the mean response time of the secondary users.

2   System Model

Figure 1 illustrates a finite source queueing system which is used to model the considered cognitive radio network. The queueing system contains two interconnected, not independent sub-systems. The first part is for the requests of the PUs. The number of sources is denoted by N_1. These sources generate high priority requests with an exponentially distributed inter-request times with the parameter λ_1. The generated requests are sent to a single server unit (Primary Channel Service - PCS). The service times are supposed to be also exponentially distributed with the parameter μ_1.

The second part is for the requests of the SUs. There are N_2 sources, the inter-request times and service times of the single server unit (Secondary Channel Service - SCS) are assumed to be exponentially distributed random variables with rate λ_2 and μ_2, respectively.

The servers can be in three states: idle, busy and failed. For the primary server unit, if it is idle, the service of the generated high priority packet starts immediately. If the server is busy with a high priority request, the packet joins the preemptive priority queue. When the unit is engaged with a request from SUs, the service is interrupted and the interrupted low priority task is sent back to the SCS. Depending on the state of secondary channel the interrupted job is directed to either the server or the orbit. The server unit can fail during an idle

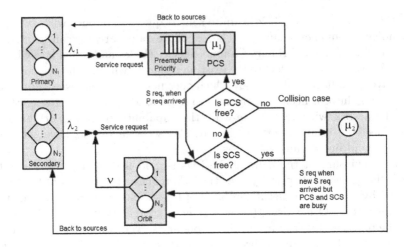

Fig. 1. A priority and a retrial queue with collision

or busy state according to an exponentially distributed time with parameter γ_1. If the server fails in busy state, the service is interrupted and the interrupted request joins the preemptive priority queue. The repair time is exponentially distributed random variable with a parameter σ_1.

In case of requests from SUs. If the SCS is idle, the service starts. If it is busy, the packet looks for the PCS. In case of an idle PCS, the service of the low priority packet begins at the high priority channel (PCS). If the PCS is busy, the packet involves into collision with the low priority servicing packet and both goes to the orbit. the same failure state can occur at the secondary server unit according to an exponentially distributed time with parameter γ_2, the repair time is exponentially distributed with the parameter σ_2. The interrupted packet also goes to the orbit. From the orbit it retries to be served after an exponentially distributed time with parameter ν. All the random variables involved in the model construction are supposed to be independent of each other.

To create a stochastic process describing the behaviour of the system, the following notations are introduced

- $k_1(t)$ is the number of high priority sources at time t,
- $k_2(t)$ is the number of low priority sources at time t,
- $q(t)$ denotes the number of high priority requests in the priority queue at time t,
- $o(t)$ is the number of requests in the orbit at time t,
- $y(t) = 0$ if there is no job in the PCS unit, $y(t) = 1$ if the PCS unit is busy with a job coming from the high priority class, $y(t) = 2$ when the PCS unit is servicing a job coming from the secondary class at time t
- $c(t) = 0$ when the SCS unit is idle and $c(t) = 1$, when the SCS is busy at time t.

It is easy to see that

$$k_1(t) = \begin{cases} N_1 - q(t), & y(t) = 0, 2, \\ N_1 - q(t) - 1, & y(t) = 1. \end{cases}$$

$$k_2(t) = \begin{cases} N_2 - o(t) - c(t), & y(t) = 0, 1, \\ N_2 - o(t) - c(t) - 1, & y(t) = 2. \end{cases}$$

Since all the random variables involved in the model construction are assumed to be exponentially distributed we could create a continuous time Markov chain with multidimensional state space. However, the main problem is the determination of its stationary distribution. Instead we prefer the stochastic simulation and in a further paper we aim to use non-exponentially distributions and to investigate the effect of distribution of specific random time on different performance measures.

The input parameters are collected in Table 1.

Table 1. List of simulation parameters

Parameter	Maximum	Value at t
Active primary sources	N_1	$k_1(t)$
Active secondary sources	N_2	$k_2(t)$
Primary generation rate		λ_1
Secondary generation rate		λ_2
Requests in priority queue	$N_1 - 1$	$q(t)$
Requests in orbit	$N_2 - 1$	$o(t)$
Primary service rate		μ_1
Secondary service rate		μ_2
Retrial rate		ν
Primary failure rate		γ_1
Secondary failure rate		γ_2
Primary repair rate		σ_1
Secondary repair rate		σ_2

3   Simulation Results

In order to estimate the mean response times of the requests, the batch mean method is used which is the most popular confidence interval technique for the output analysis of a steady-state simulation, see for example [8–10].

There are many possible combinations of the cases, we considered only the following sample results showing the effect of the non-reliability of the servers on the mean response time of the secondary users.

Table 2. Numerical values of model parameters

No	N_1	N_2	λ_1	λ_2	μ_1	μ_2	σ_1	σ_2	γ_1	γ_2	ν
Figures 2 and 3	6	6	0.6	x - axis	4	4	1	1	0.05	0.05	0.4
Figures 4 and 5	6	6	0.6	0.6	4	4	1	1	0.05	0.05	x - axis
Figures 6 and 7	6	6	0.6	0.6	x - axis	4	1	1	0.05	0.05	0.4
Figure 8	10	10	0.1	0.1	4	4	x - axis	x - axis	0.05	0.05	0.4
Figure 9	10	10	0.1	0.1	4	4	0.05	0.05	x - axis	x - axis	0.4
Figure 10	10	10	0.1	0.1	4	4	0	x - axis	0	0.05	0.4

For the easier understanding the numerical value of parameters are collected in Table 2.

Figure 2 shows the effect of the request generation rate on the mean response time of the secondary users in the two cases: Secondary server unit non-reliable and both servers non-reliable, where the Fig. 3 shows the same effect in the two cases of non-reliability with collision in the retrial part of the system. Figures show the phenomenon of having a maximum value of the mean response time which was noticed in [11]. The collision involves longer response time for the users as it was expected.

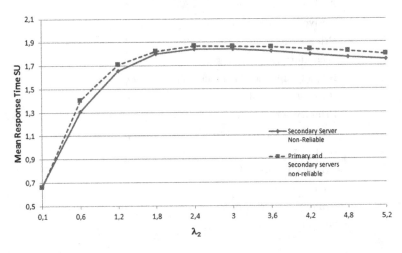

Fig. 2. The effect of servers non-reliability on the mean response time of secondary users vs λ_2

Figures 4 and 5 shows the effect of the retrial rate on the mean response time of the secondary users. In Fig. 4, the non-reliability of the primary server unit does not have any effect on the mean response time of secondary users where the retrial rate is increasing. However, the non-reliability of the primary server has an effect on the mean response time which can be shown on the Fig. 5. It means that in the cognitive radio networks, having a reliable primary server involves

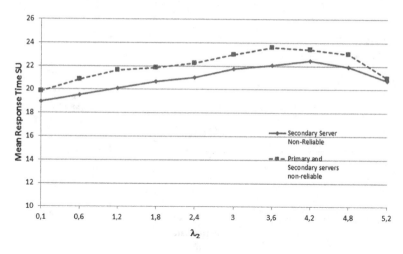

Fig. 3. The effect of servers non-reliability with collision on the mean response time of secondary users vs λ_2

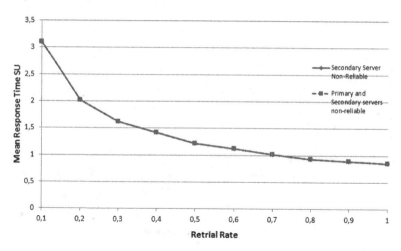

Fig. 4. The effect of servers non-reliability on the mean response time of secondary users vs the retrial rate ν

shortest mean response time of secondary users where there is a collision in the retrial part of the system.

Figures 6 and 7 illustrate the effect of the primary service rate on the mean response time of the secondary users. The non-reliability of the primary server has an effect on the mean response time of the secondary users in the case of the collision where the primary service rate is increasing. A longer response time can be seen in the case of the collision in the retrial part, as it was expected.

Figure 8 shows the effect of the non-reliability of the servers on the mean response time of the secondary users where the repair rate is increasing.

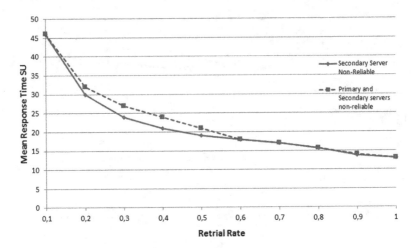

Fig. 5. The effect of servers non-reliability with collision on the mean response time of secondary users vs the retrial rate ν

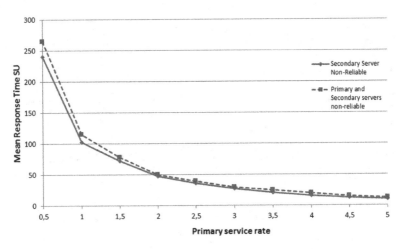

Fig. 6. The effect of servers non-reliability with collision on the mean response time of secondary users vs Primary service rate μ_1

The first case is where the primary server is non-reliable, in this case the value of the mean response time of the secondary users becomes a constant when the primary repair rate (σ_1) is higher. The second case is where the secondary server is non-reliable, in this case, the value of the mean response time of the secondary users is decreasing when the secondary repair rate (σ_2) is increasing.

Figure 9 illustrates the effect of the non-reliability of the servers on the mean response time of the secondary users where the failure rate is increasing. As it was expected, increasing the failure rate involves longer response time in the both cases (primary server non-reliable and secondary server non-reliable).

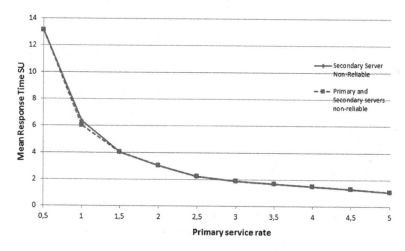

Fig. 7. The effect of servers non-reliability on the mean response time of secondary users vs Primary service rate μ_1

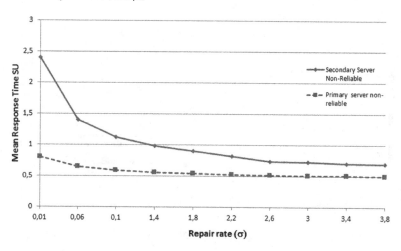

Fig. 8. The effect of servers non-reliability on the mean response time of secondary users vs the repair rate σ

The last Figure shows the effect of the collision on the mean response time of the secondary users. The conflict on a non-reliable secondary server causes a very long response time.

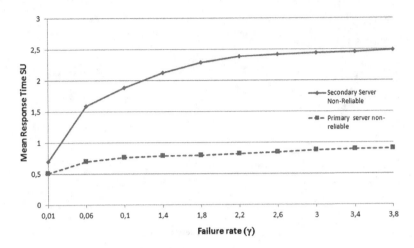

Fig. 9. The effect of servers non-reliability on the mean response time of secondary users vs the failure rate γ

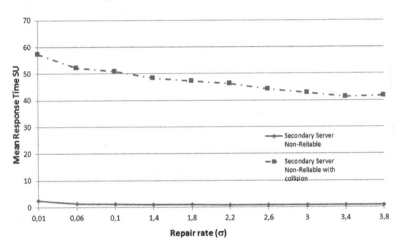

Fig. 10. The effect of servers non-reliability and collision on the mean response time of secondary users vs the repair rate σ

4   Conclusions

In this paper a finite-source retrial queueing model was proposed with two bands servicing primary and secondary users in a cognitive radio network. Primary users have preemptive priority over the secondary ones in servicing at primary channel. At the secondary channel an orbit is installed for the secondary packets finding the server busy upon arrival. The non-reliability and conflict (collision) of the servers were introduced. By using simulation, several sample examples illustrates the effect of the non-reliability of the servers and the collisions at the secondary service on the mean response time of the secondary users. This paper is the starting point of a more complex investigation where generally distributed random variables are introduced to see the effect of the distribution of the specific random variable on the main performance measures of the system.

Acknowledgments. The work of Nemouchi H. was supported by the Stipendium Hungaricum Scholarship.

References

1. Léon, O., Hernández-Serrano, J., Soriano, M.: Securing cognitive radio networks. Int. J. Commun. Syst. **23**, 633–652 (2010)
2. Ahmed, H., AlQahtani, S.A.: Performance evaluation of joint admission and eviction controls of secondary users in cognitive radio networks. Arab. J. Sci. Eng. **40**, 34469–34481 (2015)
3. Wong, E.W., Foch, C.H., Adachi, F.: Analysis of cognitive radio spectrum decision for cognitive radio networks. IEEE J. Sel. Areas Commun. **29**, 757–769 (2011)
4. Zhang, Y., Zhend, J., Chen, H.: Cognitive Radio Networks: Architectures, Protocols, and Standards. CRC Press, Boca Raton (2010)
5. Gao, S., Wang, J.: Performance analysis of a cognitive radio network based on preemptive priority and guard channels. Int. J. Comput. Math. (2014). P.GCOM-2013-0621-A
6. Kvach, A., Nazarov, A.: Sojourn time analysis of finite source Markov retrial queueing systems with collision. In: Dudin, A., Nazarov, A., Yakupov, R. (eds.) Information Technologies and Mathematical Modelling - Queueing Theory and Applications. Communications in Computer and Information Science, vol. 564, pp. 64–72. Springer, Cham (2015). doi:10.1007/978-3-319-25861-4_6
7. Nazarov, A., Kvach, A., Yampolsky, V.: Asymptotic analysis of closed Markov retrial queuing system with collision. In: Dudin, A., Nazarov, A., Yakupov, R., Gortsev, A. (eds.) ITMM 2014. CCIS, vol. 487, pp. 334–341. Springer, Cham (2014). doi:10.1007/978-3-319-13671-4_38
8. Law, A.M., Kelton, W.D.: Simulation Modeling and Analysis, 2nd edn. McGraw-Hill College, New York (1991)
9. Carlstein, E., Goldsman, D.: The use of subseries values for estimating the variance of a general statistic from a stationary sequence. Ann. Stat. **14**, 1171–1179 (1986)
10. Sztrik, J., Bérczes, T., Nemouchi, H., Melikov, A.Z.: Performance modeling of finite-source cognitive radio networks using simulation. DCCN 2016. Communications in Computer and Information Science, vol. 678, pp. 64–73. Springer, Cham (2016)
11. Sztrik, J., Roszik, J., Almási, B.: The effect of server's breakdowns on the finite-source retrial queueing systems. In: 6th International Conference on Applied Informatics, vol. 2, pp. 221–230. Eger, Hungary (2004)

Research of Mathematical Model of Insurance Company in the Form of Queueing System in a Random Environment

Diana Dammer$^{(\boxtimes)}$

Tomsk State University, Tomsk, Russia
di.dammer@yandex.ru

Abstract. The present paper is devoted to the research of the mathematical model of an insurance company in the form of the queueing system with an infinite number of servers. The arrival process of risks is regarded as a modulated Poisson arrival process. Applying the asymptotic analysis method under the condition of a high-rate arrivals, the characteristic function of the probability distribution for the two-dimensional process of the number of risks and the number of claims for insurance payments is obtained. It is shown that this probability distribution can be approximated by Gaussian distribution. These results can be applied to the estimation of functioning of the various economic systems.

Keywords: Mathematical model · Insurance company · Insurance payments · Queueing system · Characteristic function · Asymptotic analysis

1 Introduction

At present, the research and modeling of economic systems are paid a great deal of attention. These problems is usually related to research in the field of arrival processes. The results of these studies show that the classic models (for example, the Poisson ones) are not exactly modeling real arrival processes. Thus, the problem of the research models of economic systems with reference to this aspect becomes quite relevant. For example, the intensity of incoming risks into the insurance company is not a constant and it depends on the impact of external random factors such as season, state policy, probability of natural disasters, fashion, etc. On the whole, all papers focused on the research of mathematical models of insurance company include characteristics of the performance of a company with a stationary Poisson arrival process of risks. Thus, these models are reviewed in [1]. In the paper [2] the distribution of claims for insurance payments with a random value of contract duration is obtained. Applying an asymptotic analysis method, in [3] we obtain the two-dimensional probability distribution of the number of risks and the number of payments. The model with a possibility of reinsurance is investigated in [4]. Papers [5,6] cover the model with implicit advertisement and one-time insurance payment for limited

© Springer International Publishing AG 2017
A. Dudin et al. (Eds.): ITMM 2017, CCIS 800, pp. 204–214, 2017.
DOI: 10.1007/978-3-319-68069-9_17

and unlimited insurance coverage. In this paper, we consider the mathematical model of an insurance company in a random environment, when the rate of the arrival process, the rate of occurrences of the insured events and the contract duration are not constants and depend on the impact of external factors and change with time, which is undoubtedly present in real life.

2   Mathematical Model

Let us have a look at the model of an insurance company with infinite insurance coverage [7] (Fig. 1) in the form of a queueing system with an infinite number of servers. We can assume that risks (customers) coming into the company form high-rate modulated Poisson arrival process that is regulated by the random process $k(t)$ [8]. This process is a Markov chain with a continuous time that is defined by the matrix $N\mathbf{Q}$ of infinitesimal characteristics $Nq_{k\nu}$, where $k = 1,\ldots,K$, $\nu = 1,\ldots,K$ and N has a large value (we suppose that $N \to \infty$).

Let us define the matrix $N\mathbf{\Lambda}$ with elements $N\lambda_k$ on the main diagonal. Here $N\lambda_k$ — the intensity of risks coming into the company, when Markov chain is in k state, λ_k — fixed value. Thus, the Markov chain $k(t)$ state defines the state of a random environment.

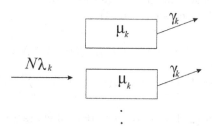

Fig. 1. Model of the insurance company in the form of queueing system with an infinite number of servers in a random environment.

After coming into the company the risk makes the insurance contract. The contract duration is the duration of serving a customer at a server. Each risk that is in the company during the contract duration generates claim for the insurance payment with intensity γ_k independently from other risks. These intensities also depend on the environment state and form a diagonal matrix $\mathbf{\Gamma}$. Requirements for insurance payments also form a random process. It is natural to assume that the claim for payment is determined by the occurrence of the insured event. The contract duration for each risk in the company is considered to be random value, exponentially distributed with a parameter μ_k, that is also dependent on the environment state. These values form the diagonal matrix $\mathbf{M}$.

Let us denote: $n(t)$ — number of claims for payments over the time interval $[0,t]$, $i(t)$ — number of insurance risks that are in the company at the moment t, $P_k(i,n,t) = \mathrm{P}\{i(t) = i, n(t) = n, k(t) = k\}$ — probability of a number of risks in the company at the moment t equals to i, a number of claims for payments at the moment t equals to n and environment is in the k state at the moment t. The problem is to obtain the expression for characteristic function of two-dimensional random process $(i(t), n(t))$.

3   Kolmogorov Equations

Let us set up a system of Kolmogorov differential equations [9] for probability distribution $P_k(i, n, t)$. Using the notation $P_k(i, n, t) = \mathrm{P}\{i(t) = i, n(t) = n, k(t) = k\}$ and applying the formula of total probability, we can write the following equations

$$P_k(i, n, t + \Delta t) = P_k(i, n, t)(1 - N\lambda_k \Delta t)(1 - i\gamma_k \Delta t)(1 - i\mu_k \Delta t)$$

$$\times (1 + Nq_{kk}\Delta t) + N\lambda_k \Delta t P_k(i - 1, n, t) + i\gamma_k \Delta t P_k(i, n - 1, t) \tag{1}$$

$$+ (i + 1)\mu_k \Delta t P_k(i + 1, n, t) + \sum_{\nu \neq k} P_\nu(i, n, t) N q_{\nu k} \Delta t + o(\Delta t) .$$

for $k = 1, \ldots, K$. After performing some transformation, we derive the following system of the Kolmogorov differential equations for the probability distribution of the two-dimensional process $(i(t), n(t))$

$$\frac{\partial P_k(i, n, t)}{\partial t} = -(N\lambda_k + i\mu_k + i\gamma)P_k(i, n, t) + N\lambda P_k(i - 1, n, t)$$

$$+ (i + 1)\mu_k P_k(i + 1, n, t) + i\gamma_k P_k(i, n - 1, t) + \sum_{\nu=1}^{K} P_\nu(i, n, t) N q_{\nu k} . \tag{2}$$

To solve the system (2) let us consider partial characteristic functions:

$$H_k(u_1, u_2, t) = \sum_{i,n=0}^{\infty} e^{ju_1 i} e^{ju_2 n} P_k(i, n, t),$$

for $k = 1, \ldots, K$, j—imaginary unit. Then, using system (2) and takint into account the properties of characteristic functions, we will obtain the first-order partial differential equation for $H_k(u_1, u_2, t)$ in the following form:

$$\frac{\partial H_k(u_1, u_2, t)}{\partial t} = N\lambda_k(e^{ju_1} - 1)H_k(u_1, u_2, t) + \sum_{\nu=1}^{K} H_\nu(u_1, u_2, t)q_{\nu k}$$

$$+ j\frac{\partial H_k(u_1, u_2, t)}{\partial u_1}(\mu_k - \mu_k e^{-ju_1} + \gamma_k - \gamma_k e^{ju_2}) . \tag{3}$$

Let us consider the vector characteristic function

$$\mathbf{H}(u_1, u_2, t) = \{H_1(u_1, u_2, t), H_2(u_1, u_2, t), \ldots, H_K(u_1, u_2, t)\} .$$

Thus, using (3) we can write the matrix differential equation for the function $\mathbf{H}(u_1, u_2, t)$

$$\frac{\partial \mathbf{H}(u_1, u_2, t)}{\partial t} = \mathbf{H}(u_1, u_2, t)[N\mathbf{\Lambda}(e^{ju_1} - 1) + N\mathbf{Q}]$$

$$+ j \frac{\partial \mathbf{H}(u_1, u_2, t)}{\partial u_1}[(1 - e^{-ju_1})\mathbf{M} + (1 - e^{ju_2})\mathbf{\Gamma}],$$

(4)

matrixes $N\mathbf{\Lambda}$, $\mathbf{M}$, $\mathbf{\Gamma}$, $N\mathbf{Q}$ are defined above.

We will solve the Eq. (4) for vector characteristic function $\mathbf{H}(u_1, u_2, t)$ using the asymptotic analysis method [10] under conditions of high-rate arrival process and extremely often changes of a random environment states ($N \to \infty$).

4   The First-Order Asymptotic Analysis

Let us make the following changes to the variables in the Eq. (4):

$$\varepsilon = \frac{1}{N}, \; u_1 = \varepsilon\omega_1, \; u_2 = \varepsilon\omega_2, \; \mathbf{H}(u_1, u_2, t) = \mathbf{F}(\omega_1, \omega_1, t, \varepsilon) . \quad (5)$$

Using these new variables we will write the equation for function $\mathbf{F}(\omega_1, \omega_1, t, \varepsilon)$:

$$\varepsilon \frac{\partial \mathbf{F}(\omega_1, \omega_2, t, \varepsilon)}{\partial t} = \mathbf{F}(\omega_1, \omega_2, t, \varepsilon)[\mathbf{\Lambda}(e^{j\omega_1\varepsilon} - 1) + \mathbf{Q}]$$

$$+ j \frac{\partial \mathbf{F}(\omega_1, \omega_2, t, \varepsilon)}{\partial \omega_1}[(1 - e^{-j\omega_1\varepsilon})\mathbf{M} + (1 - e^{j\omega_2\varepsilon})\mathbf{\Gamma}] .$$

(6)

Denote an asymptotic solution to this equation under the condition $\varepsilon \to 0$ by $\mathbf{F}(\omega_1, \omega_2, t)$:

$$\lim_{\varepsilon \to 0} \mathbf{F}(\omega_1, \omega_2, t, \varepsilon) = \mathbf{F}(\omega_1, \omega_2, t) .$$

Let us perform the asymptotic transition $\varepsilon \to 0$ in the Eq. (6). We will obtain

$$\mathbf{F}(\omega_1, \omega_2, t)\mathbf{Q} = \mathbf{0} . \quad (7)$$

Thus, the function $\mathbf{F}(\omega_1, \omega_2, t)$ is a solution for the homogeneous system of the linear algebraic Eq. (7). Solution for this system has the following form:

$$\mathbf{F}(\omega_1, \omega_2, t) = \mathbf{R}\Phi(\omega_1, \omega_2, t) , \quad (8)$$

where $\Phi(\omega_1, \omega_2, t)$ — some scalar function, $\mathbf{R}$ — a row vector of stationary probability distribution of Markov chain $k(t)$, that is defined by the equations system $\mathbf{RQ} = \mathbf{0}$ and a normalization condition $\mathbf{RE} = 1$, where $\mathbf{0}$ — a row vector with zeros and $\mathbf{E}$ — a column vector with enteries all equal to 1. To obtain function

$\Phi(\omega_1, \omega_2, t)$, we will sum up equations of the system (6). Taking into account condition $\mathbf{F}(\omega_1, \omega_2, t)\mathbf{Q} = \mathbf{0}$, we can write

$$\varepsilon \frac{\partial \mathbf{F}(\omega_1, \omega_2, t, \varepsilon)}{\partial t} \mathbf{E} = \mathbf{F}(\omega_1, \omega_2, t, \varepsilon)(e^{j\omega_1 \varepsilon} - 1)\mathbf{\Lambda E}$$

$$+ j \frac{\partial \mathbf{F}(\omega_1, \omega_2, t, \varepsilon)}{\partial \omega_1}[(1 - e^{-j\omega_1 \varepsilon})\mathbf{ME} + (1 - e^{j\omega_2 \varepsilon})\mathbf{\Gamma E}] .$$

(9)

Let us divide left and right sides of the Eq. (9) by ε and perform the asymptotic transition $\varepsilon \to 0$. We obtain the equation for $\mathbf{F}(\omega_1, \omega_2, t)$:

$$\frac{\partial \mathbf{F}(\omega_1, \omega_2, t)}{\partial t} \mathbf{E} = \mathbf{F}(\omega_1, \omega_2, t) j\omega_1 \mathbf{\Lambda E}$$

$$- \omega_1 \frac{\partial \mathbf{F}(\omega_1, \omega_2, t)}{\partial \omega_1} \mathbf{ME} + \omega_2 \frac{\partial \mathbf{F}(\omega_1, \omega_2, t)}{\partial \omega_1} \mathbf{\Gamma E} .$$

(10)

Now we can write down the equation for the unknown scalar function $\Phi(\omega_1, \omega_2, t)$ considering $\mathbf{F}(\omega_1, \omega_2, t) = \mathbf{R}\Phi(\omega_1, \omega_2, t)$ and $\mathbf{RE} = 1$ in the following form:

$$\frac{\partial \Phi(\omega_1, \omega_2, t)}{\partial t} = \Phi(\omega_1, \omega_2, t) j\omega_1 \mathbf{R\Lambda E}$$

$$- \omega_1 \frac{\partial \Phi(\omega_1, \omega_2, t)}{\partial \omega_1} \mathbf{RME} + \omega_2 \frac{\partial \Phi(\omega_1, \omega_2, t)}{\partial \omega_1} \mathbf{R\Gamma E} .$$

(11)

We have the first-order partial differential equation. Its solution is defined by solving a system of ordinary differential equations for characteristic curves [11]:

$$\frac{\partial t}{1} = \frac{\partial \Phi(\omega_1, \omega_2, t)}{\Phi(\omega_1, \omega_2, t) j\omega_1 \kappa} = \frac{\partial \omega_1}{\omega_1 \kappa_1 - \omega_2 \kappa_2} ,$$

(12)

where $\kappa = \mathbf{R\Lambda E}$, $\kappa_1 = \mathbf{RME}$, $\kappa_2 = \mathbf{R\Gamma E}$. Let us obtain first two integrals of this system. We can write following equation:

$$\frac{\partial t}{1} = \frac{\partial \Phi(\omega_1, \omega_2, t)}{\Phi(\omega_1, \omega_2, t) j\omega_1 \kappa} .$$

(13)

The solution of Eq. (13) we will write down in the following form:

$$t = \frac{1}{\kappa_1} \ln(\omega_1 \kappa_1 - \omega_2 \kappa_2) - \ln C ,$$

(14)

where C is constant. Let us denote $C_1 = C^{\kappa_1}$, then $C_1 = (\omega_1 \kappa_1 - \omega_2 \kappa_2)e^{-t\kappa_1}$. The other first integral we will obtain from the equation

$$\frac{\partial \Phi(\omega_1, \omega_2, t)}{\Phi(\omega_1, \omega_2, t) j\omega_1 \kappa} = \frac{\partial \omega_1}{\omega_1 \kappa_1 - \omega_2 \kappa_2} .$$

(15)

The solution of the Eq. (15) has the following form:

$$\Phi(\omega_1, \omega_2, t) = e^{j\frac{\kappa\omega_1}{\kappa_1}} (\omega_1\kappa_1 - \omega_2\kappa_2)^{j\frac{\kappa\omega_2\kappa_2}{\kappa_1^2}} C_2 . \tag{16}$$

Let us introduce an arbitrary differentiated function $\phi(C_1) = C_2$. Then the general solution of the equation (15) will have the following form

$$\Phi(\omega_1, \omega_2, t) = e^{j\frac{\kappa\omega_1}{\kappa_1}} (\omega_1\kappa_1 - \omega_2\kappa_2)^{j\frac{\kappa\omega_2\kappa_2}{\kappa_1^2}} \phi((\omega_1\kappa_1 - \omega_2\kappa_2)e^{-t\kappa_1}) . \tag{17}$$

Let us define the partial solution with the help of initial conditions. We have to define $\Phi(\omega_1, \omega_2, 0)$ first. Let us write down value of functions $H_k(u_1, u_2, t)$, $k = 1 \ldots K$, at the moment $t = 0$:

$$H_k(u_1, u_2, 0) = \sum_{i=0}^{\infty} \sum_{n=0}^{\infty} e^{ju_1 i} e^{ju_2 n} P_k(i, n, 0) = \sum_{i=0}^{\infty} e^{ju_1 i} P(i) ,$$

because at the initial moment of time (when the insurance company starts to work) there were no claims for insurance payments, thus $P(i, n, 0) = P(i)$ if $n = 0$ and $P(i, n, 0) = 0$ if $n > 0$. Let us denote the function $H_k(u_1, u_2, 0) = G_k(u_1)$ and the vector function $\mathbf{G}(u_1) = \{G_1(u_1), G_2(u_1), \ldots, G_K(u_1)\}$. Then we will write down the equation for $\mathbf{G}(u_1)$ in the following form:

$$\mathbf{G}(u_1)[N\mathbf{\Lambda}(e^{ju_1} - 1) + N\mathbf{Q}] + j\mathbf{G}'(u_1)\mathbf{M}(1 - e^{-ju_1}) = 0 . \tag{18}$$

We will solve Eq. (18) applying an asymptotic analysis method under similar asymptotic conditions and substitutions:

$$\varepsilon = \frac{1}{N}, \; u_1 = \varepsilon\omega_1, \; \mathbf{G}(u_1) = \mathbf{F}(\omega_1, \varepsilon), \; \varepsilon \to 0 .$$

For the function $\mathbf{F}(\omega_1, \varepsilon)$ we can write

$$\mathbf{F}(\omega_1, \varepsilon)[\mathbf{\Lambda}(e^{ju_1} - 1) + \mathbf{Q}] + j\frac{\partial \mathbf{F}(\omega_1, \varepsilon)}{\partial \omega_1}\mathbf{M}(1 - e^{-ju_1}) = 0 . \tag{19}$$

Let us denote

$$\lim_{\varepsilon \to 0} \mathbf{F}(\omega_1, \varepsilon) = \mathbf{F}(\omega_1)$$

and perform the asymptotic transition $\varepsilon \to 0$ in the Eq. (19). We will obtain $\mathbf{F}(\omega_1)\mathbf{Q} = 0$, therefore $\mathbf{F}(\omega_1) = \mathbf{R}\Psi(\omega_1)$, where scalar function $\Psi(\omega_1) = \Phi(\omega_1, \omega_2, 0)$, $\mathbf{R}$—a row vector of stationary probability distribution of the Markov chain states. To obtain this function, we will sum up equations of the system (19), then divide by ε and perform the asymptotic transition $\varepsilon \to 0$. We will obtain the equation for the unknown function $\Phi(\omega_1, \omega_2, 0) = \Psi(\omega_1)$:

$$\Psi'(\omega_1)\kappa_1 = j\Psi(\omega_1)\kappa , \tag{20}$$

where $\kappa = \mathbf{R}\mathbf{\Lambda}\mathbf{E}$, $\kappa_1 = \mathbf{R}\mathbf{M}\mathbf{E}$. As a result, we obtain the following solution of this equation under initial condition $\Psi(0) = 1$:

$$\Psi(\omega_1) = e^{j\frac{\kappa}{\kappa_1}\omega_1} .$$ (21)

Then we can write down the expression for function $\phi(C_1)$:

$$\phi(C_1) = \left[e^{-t\kappa_1}(\omega_1\kappa_1 - \omega_2\kappa_2) \right]^{-j\omega_2\frac{\kappa\kappa_2}{\kappa_1^2}} .$$ (22)

Taking into account (22), the function $\Phi(\omega_1, \omega_2, t)$ will have the following form:

$$\Phi(\omega_1, \omega_2, t) = \exp\left\{ j\omega_1\frac{\kappa}{\kappa_1} + j\omega_2\frac{\kappa\kappa_2}{\kappa_1}t \right\} .$$ (23)

Substituting this expression into (8), we obtain the expression for the function $\mathbf{F}(\omega_1, \omega_2, t)$ in the following form:

$$\mathbf{F}(\omega_1, \omega_1, t) = \mathbf{R}\exp\left\{ j\omega_1\frac{\kappa}{\kappa_1} + j\omega_2\frac{\kappa\kappa_2}{\kappa_1}t \right\} .$$ (24)

For the function $\mathbf{H}(u_1, u_2, t)$ we can write down

$$\mathbf{H}(u_1, u_2, t) = \mathbf{F}(\omega_1, \omega_2, t, \varepsilon) \approx \mathbf{F}(\omega_1, \omega_2, t) = \mathbf{R}\exp\left\{ j\omega_1\frac{\kappa}{\kappa_1} + j\omega_2\frac{\kappa\kappa_2}{\kappa_1}t \right\} .$$

Let us make in this formula substitutions that are inverse to changes (5). Using expression (8), we obtain the following expression for the vector characteristic function of the probability distribution of the two-dimensional process $(i(t),\ n(t))$

$$\mathbf{H}(u_1, u_2, t)\mathbf{E} \approx \exp\left\{ j\omega_1\frac{\kappa}{\kappa_1} + j\omega_2\frac{\kappa\kappa_2}{\kappa_1}t \right\}$$
$$= \exp\left\{ jNu_1\frac{\kappa}{\kappa_1} + jNu_2\frac{\kappa\kappa_2}{\kappa_1}t \right\} .$$ (25)

5 The Second-Order Asymptotic Analysis

Let us denote the vector function $\mathbf{H}_2(u_1, u_2, t)$ satisfying the expression:

$$\mathbf{H}(u_1, u_2, t) = \mathbf{H_2}(u_1, u_2, t)\exp\left\{ jNu_1\frac{\kappa}{\kappa_1} + jNu_2\frac{\kappa\kappa_2}{\kappa_1}t \right\} .$$ (26)

Substituting this expression in the Eq. (4), we obtain the equation for the function $\mathbf{H}_2(u_1, u_2, t)$:

$$\frac{\partial \mathbf{H_2}(u_1, u_2, t)}{\partial t} = \mathbf{H_2}(u_1, u_2, t)[N\mathbf{\Lambda}(e^{ju_1} - 1) + N\mathbf{Q}]$$
$$- \mathbf{H_2}(u_1, u_2, t)\left[jNu_2\frac{\kappa\kappa_2}{\kappa_1}\mathbf{I} + N\frac{\kappa}{\kappa_1}\left((1 - e^{-ju_1})\mathbf{M} + (1 - e^{ju_2})\mathbf{\Gamma} \right) \right]$$
$$+ j\frac{\partial \mathbf{H_2}(u_1, u_2, t)}{\partial u_1}[(1 - e^{-ju_1})\mathbf{M} + (1 - e^{ju_2})\mathbf{\Gamma}] ,$$ (27)

where $\mathbf{I}$ — a diagonal unit matrix. Let us make the changes variables:

$$\varepsilon = \frac{1}{N^2}, \ u_1 = \varepsilon\omega_1, \ u_2 = \varepsilon\omega_2, \ \mathbf{H_2}(u_1, u_2, t) = \mathbf{F_2}(\omega_1, \omega_1, t, \varepsilon) \ . \tag{28}$$

Using the new variables, we can rewrite the problem (27) in the form:

$$\varepsilon^2 \frac{\partial \mathbf{F_2}(\omega_1, \omega_2, t, \varepsilon)}{\partial t} = \mathbf{F_2}(\omega_1, \omega_2, t, \varepsilon)[\mathbf{\Lambda}(e^{j\omega_1} - 1) + \mathbf{Q}]$$

$$- \mathbf{F_2}(\omega_1, \omega_2, t, \varepsilon), \left[j\omega_2\varepsilon\frac{\kappa\kappa_2}{\kappa_1} I + \frac{\kappa}{\kappa_1}\left((1 - e^{-j\omega_1\varepsilon})\mathbf{M} + (1 - e^{j\omega_2\varepsilon})\mathbf{\Gamma} \right) \right] \tag{29}$$

$$+ j\varepsilon\frac{\partial \mathbf{F_2}(\omega_1, \omega_2, t, \varepsilon)}{\partial \omega_1}[(1 - e^{-j\omega_1\varepsilon})\mathbf{M} + (1 - e^{j\omega_2\varepsilon})\mathbf{\Gamma}] \ .$$

Let us denote

$$\lim_{\varepsilon \to 0} F_2(\omega_1, \omega_2, t, \varepsilon) = F_2(\omega_1, \omega_2, t) \ .$$

Furthermore, we will perform the asymptotic transition $\varepsilon \to 0$ in (29), then we will obtain the equation $\mathbf{F_2}(\omega_1, \omega_2, t)\mathbf{Q} = \mathbf{0}$. Thus, the function $\mathbf{F_2}(\omega_1, \omega_2, t)$ can be written in the following form:

$$\mathbf{F_2}(\omega_1, \omega_2, t) = \mathbf{R}\Phi_2(\omega_1, \omega_2, t) \ , \tag{30}$$

where $\Phi_2(\omega_1, \omega_2, t)$ — some scalar function that will be defined below.

We will find the solution $\mathbf{F_2}(\omega_1, \omega_2, t, \varepsilon)$ of the Eq. (29) in the following expansion form

$$\mathbf{F_2}(\omega_1, \omega_2, t, \varepsilon) = \Phi_2(\omega_1, \omega_2, t) \left(\mathbf{R} + j\omega_1\varepsilon\mathbf{f_1} + j\omega_2\varepsilon\mathbf{f_2} + \mathbf{O}\left(\varepsilon^2\right) \right) \ , \tag{31}$$

where $\mathbf{f_1}$, $\mathbf{f_2}$ — some row vectors, $\mathbf{O}\left(\varepsilon^2\right)$ — a row vector that consist of the infinitesimals of the order ε^2.

Substituting (30) in the Eq. (29) and taking into account $\mathbf{RQ} = \mathbf{0}$, we obtain the matrix system of the equations for the row vectors $\mathbf{f_1}$, $\mathbf{f_2}$:

$$\mathbf{f_1}\mathbf{Q} = \frac{\kappa}{\kappa_1}\mathbf{RM} - \mathbf{R\Lambda} \ , \ \mathbf{f_2}\mathbf{Q} = \frac{\kappa}{\kappa_1}\mathbf{R}\kappa_2 - \frac{\kappa}{\kappa_1}\mathbf{R\Gamma} \ . \tag{32}$$

To obtain function $\Phi_2(\omega_1, \omega_2, t)$ let us sum up all equations of the system (29). We will obtain the following equation

$$\varepsilon^2 \frac{\partial \mathbf{F_2}(\omega_1, \omega_2, t, \varepsilon)}{\partial t}\mathbf{E} = \mathbf{F_2}(\omega_1, \omega_2, t, \varepsilon)[\mathbf{\Lambda}(e^{j\omega_1\varepsilon} - 1) + \mathbf{Q}]\mathbf{E}$$

$$- \mathbf{F_2}(\omega_1, \omega_2, t, \varepsilon)\left[j\omega_2\varepsilon\frac{\kappa\kappa_2}{\kappa_1} I + \frac{\kappa}{\kappa_1}\left((1 - e^{-j\omega_1\varepsilon})\mathbf{M} + (1 - e^{j\omega_2\varepsilon})\mathbf{\Gamma} \right) \right] \mathbf{E} \tag{33}$$

$$+ j\varepsilon\frac{\partial \mathbf{F_2}(\omega_1, \omega_2, t, \varepsilon)}{\partial \omega_1}[(1 - e^{-j\omega_1\varepsilon})\mathbf{M} + (1 - e^{j\omega_2\varepsilon})\mathbf{\Gamma}]\mathbf{E} \ .$$

In the Eq. (33) let us substitute the expansion $e^{jw_1\varepsilon} = 1 + jw_1\varepsilon + \dfrac{(jw_1\varepsilon)^2}{2} + O\left(\varepsilon^3\right)$ and the expansion (31). We obtain the following equality

$$
\begin{aligned}
\varepsilon^2 \frac{\partial \Phi_2(\omega_1,\omega_2,t)}{\partial t} &= \Phi_2(\omega_1,\omega_2,t)\mathbf{R}\left[\mathbf{\Lambda}\left(j\omega_1\varepsilon - \frac{(\omega_1\varepsilon)^2}{2}\right) + \mathbf{Q} - j\omega_2\varepsilon\frac{\kappa\kappa_2}{\kappa_1}\mathbf{I}\right]\mathbf{E} \\
&\quad - \Phi_2(\omega_1,\omega_2,t)\mathbf{R}\frac{\kappa}{\kappa_1}\left[\left(j\omega_1\varepsilon + \frac{(\omega_1\varepsilon)^2}{2}\right)\mathbf{M}\mathbf{E} + \left(-j\omega_1\varepsilon + \frac{(\omega_1\varepsilon)^2}{2}\right)\mathbf{\Gamma}\mathbf{E}\right] \\
&\quad - \Phi_2(\omega_1,\omega_2,t)\varepsilon(j\omega_1\mathbf{f_1} + j\omega_2\mathbf{f_2}) \\
&\quad \times \left(j\frac{\kappa}{\kappa_1}\kappa_2\varepsilon\omega_2\mathbf{E} - \mathbf{Q}\mathbf{E} - j\omega_1\varepsilon\mathbf{\Lambda}\mathbf{E} + j\frac{\kappa}{\kappa_1}\varepsilon\omega_1\mathbf{M}\mathbf{E} - j\frac{\kappa}{\kappa_1}\varepsilon\omega_2\mathbf{\Gamma}\mathbf{E}\right) \\
&\quad + \varepsilon^2 \frac{\partial \Phi_2(\omega_1,\omega_2,t)}{\partial\omega_1}\mathbf{R}\left(\omega_1\mathbf{M}\mathbf{E} - \omega_2\mathbf{\Gamma}\mathbf{E}\right) + O\left(\varepsilon^3\right) .
\end{aligned}
$$

In the last expression let us divide left and right sides by ε^2, and after using the asymptotic transition $\varepsilon \to 0$, we obtain the equation for the function $\Phi_2(\omega_1,\omega_2,t)$:

$$
\begin{aligned}
&\frac{\partial \Phi_2(\omega_1,\omega_2,t)}{\partial t} + \frac{\partial \Phi_2(\omega_1,\omega_2,t)}{\partial\omega_1}(\omega_1\kappa_1 - \omega_2\kappa_2) = \Phi_2(\omega_1,\omega_2,t) \\
&\times \left[\omega_1^2(\mathbf{f_1}\mathbf{A_1} - \kappa) + \omega_1\omega_2(\mathbf{f_2}\mathbf{A_1} + \mathbf{f_1}\mathbf{A_2}) + \omega_2^2\left(\mathbf{f_2}\mathbf{A_2} - \frac{\kappa\kappa_2}{2\kappa_1}\right)\right],
\end{aligned} \tag{34}
$$

under the initial condition $\Phi_2(\omega_1,\omega_2,0) = \exp\left\{\dfrac{\mathbf{A_1}\mathbf{f_1} - \kappa}{2\kappa_1}\omega_1^2\right\}$ and where vectors $\mathbf{A_1}$, $\mathbf{A_2}$ are defined by expressions

$$
\mathbf{A_1} = \left(\frac{\kappa}{\kappa_1}\mathbf{M} - \mathbf{\Lambda}\right)\mathbf{E}, \quad \mathbf{A_2} = \left(\frac{\kappa\kappa_2}{\kappa_1}\mathbf{I} - \frac{\kappa}{\kappa_1}\mathbf{\Gamma}\right)\mathbf{E}. \tag{35}
$$

We will find a solution of the Eq. (34) in the following form:

$$
\Phi_2(\omega_1,\omega_2,t) = \exp\left\{-\frac{1}{2}\left(K_{11}\omega_1^2 + 2K_{12}(t)\omega_1\omega_2 + K_{22}(t)\omega_2^2\right)\right\}. \tag{36}
$$

Substituting this expression in the Eq. (34), we obtain the following system for K_{11}, $K_{12}(t)$, $K_{22}(t)$:

$$
K_{11}\kappa_1 = \kappa - \mathbf{f_1}\mathbf{A_1},
$$

$$
K_{12}'(t) + \kappa_1 K_{12}(t) - \kappa_2 K_{11}(t) = -\mathbf{f_1}\mathbf{A_2} - \mathbf{f_2}\mathbf{A_1}, \tag{37}
$$

$$
\frac{1}{2}K_{22}'(t) - \kappa_2 K_{12}(t) = \frac{\kappa\kappa_2}{2\kappa_1} - \mathbf{f_2}\mathbf{A_2},
$$

where vectors $\mathbf{f_1}$, $\mathbf{f_2}$ are defined by the system (32), $\mathbf{A_1}$, $\mathbf{A_2}$ are defined by expressions (35) and $\kappa = \mathbf{R\Lambda E}$, $\kappa_1 = \mathbf{RME}$, $\kappa_2 = \mathbf{R\Gamma E}$. Solving the system (37) under initial conditions $K_{12}(0) = 0$, $K_{22}(0) = 0$, we obtain the expressions for K_{11}, $K_{12}(t)$, $K_{22}(t)$:

$$
K_{11} = \frac{\kappa - \mathbf{f_1}\mathbf{A_1}}{\kappa_1}, \tag{38}
$$

$$K_{12}(t) = \left(1 - e^{-\kappa_1 t}\right) \left[\frac{\kappa - A_1 f_1}{\kappa_1^2} \kappa_2 - \frac{A_1 f_2 + A_2 f_1}{\kappa_1}\right] , \qquad (39)$$

$$K_{22}(t) = 2t \left[\frac{\kappa - A_1 f_1}{\kappa_1^2} \kappa_2^2 - \frac{A_1 f_2 + A_2 f_1}{\kappa_1} \kappa_2 - \left(A_2 f_2 - \frac{\kappa \kappa_2}{2 \kappa_1}\right)\right]$$
$$+ 2 \left(1 - e^{-\kappa_1 t}\right) \left[\frac{\kappa - A_1 f_1}{\kappa_1^3} \kappa_2^2 - \frac{A_1 f_2 + A_2 f_1}{\kappa_1^2} \kappa_2\right] . \qquad (40)$$

Substituting these expressions into (30), we obtain the final form of the function $\mathbf{F_2}(\omega_1, \omega_2, t)$ as following expression

$$\mathbf{F_2}(\omega_1, \omega_2, t) = \mathbf{R} \exp\left\{-\frac{1}{2}\left(K_{11}\omega_1^2 + 2K_{12}(t)\omega_1\omega_2 + K_{22}(t)\omega_2^2\right)\right\} . \qquad (41)$$

Let us make in this formula substitutions that are inverse to changes (28). Using (26), we can write the expression for the vector characteristic function $\mathbf{H}(u_1, u_2, t)$ in the following form:

$$\mathbf{H}(u_1, u_2, t) = \mathbf{R} \exp\left\{-\frac{1}{2}\left(K_{11}(Nu_1)^2 + 2K_{12}(t)N^2 u_1 u_2\right.\right.$$
$$\left.\left. + K_{22}(t)(Nu_2)^2\right) + j\frac{N\kappa}{\kappa_1}u_1 + j\frac{N\kappa\kappa_1}{\kappa_1}u_2 t\right\} . \qquad (42)$$

Thus, we have the following formula for second-order approximation $h_2(u_1, u_2, t)$ for the characteristic function $h(u_1, u_2, t) = \mathbf{H}(u_1, u_2, t)\mathbf{E}$ of the two-dimensional process $(i(t), n(t))$ under the condition that N is large enough:

$$h(u_1, u_2, t) \approx h_2(u_1, u_2, t) = \exp\left\{-\frac{1}{2}\left(K_{11}(Nu_1)^2 + 2K_{12}(t)N^2 u_1 u_2\right.\right.$$
$$\left.\left. + K_{22}(t)(Nu_2)^2\right) + j\frac{N\kappa}{\kappa_1}u_1 + j\frac{N\kappa\kappa_1}{\kappa_1}u_2 t\right\} , \qquad (43)$$

where K_{11}, $K_{12}(t)$, $K_{22}(t)$ are defined by the expressions (38), (39) and (40).

6 Conclusions

In this paper we have researched the mathematical model of the insurance company in the form of queueing system with infinite number of servers with high-rate arrival process and in a random environment. We have shown that the probability distribution of the two-dimensional process of the insurance risks and the insurance payments under the above conditions can be approximated by the two-dimensional Gaussian distribution. These results can be used to analysis the activity of insurance companies and other economic systems.

References

1. Glukhova, E.V., Zmeev, O.A., Livshits, K.I.: Mathematical models of insurance. Published by Tomsk University, Tomsk (2004). (in Russian)
2. Nazarov, A.A., Dammer, D.D.: Research of a number of requests for insurance payments in the company with arbitrary length of duration of the contract. Tomsk State Univ. J. **2**(15), 24–32 (2011). (in Russian)
3. Dammer, D.D., Nazarov, A.A.: Research of the mathematical model of the insurance company in form of the infinite queuing system by using method of asymptotic analysis. In: Proceedings of 7th Ferghan Conference Limit Theorems and its Applications, Namangan, pp. 191–196 (2015). (in Russian)
4. Dammer, D.D.: Mathematical model of insurance company with nonstationary input process of risks and taking into account reinsurance. In: Proceedings of International Conference Theory of Probability and Mathematical Statistics and their Applications, Minsk, pp. 80–86 (2010). (in Russian)
5. Dammer, D.D.: Research of mathematical model of insurance company in the form of queueing system with unlimited number of servers considering implicit advertising. In: Dudin, A., Nazarov, A., Yakupov, R. (eds.) ITMM 2015. CCIS, vol. 564, pp. 163–174. Springer, Cham (2015). doi:10.1007/978-3-319-25861-4_14
6. Dammer, D.D.: Research of the total amount of one-time insurance payments in model with limited insurance coverage. In: Proceedings of 5th International Scientific Practical Conference Mathematical and Computer Modeling in Economics, Insurance and Risk Management, pp. 51–56, Saratov (2016). (in Russian)
7. Gafurov, S.R., Gugnin, V.I., Amanov, S.N.: Business Language. Shark, Tashkent (1995). (in Russian)
8. Nazarov, A.A., Moiseev, A.N.: Queueing Systems and Networks with Unlimited Number of Servers. NTL, Tomsk (2015). (in Russian)
9. Kleinrock, L.: Queueing Systems, vol. 1. Wiley (1975)
10. Nazarov, A.A., Moiseeva, S.P.: Asymptotic Analysis in Queueing Theory. NTL, Tomsk (2006). (in Russian)
11. Elsgolts, L.E.: Differential Equations and Calculus of Variations. Science, Moscow (1969). (in Russian)

State Reduction in Analysis of a Tandem Queueing System with Correlated Arrivals

Vladimir Vishnevsky$^{(\boxtimes)}$, Andrey Larionov, Olga Semenova, and Roman Ivanov

V.A. Trapeznikov Institute of Control Sciences of Russian Academy of Sciences,
Profsoyuznaya str. 65, 117997 Moscow, Russia
vishn@inbox.ru, larioandr@gmail.com, olgasmnv@gmail.com,
iromcorp@gmail.com
http://www.ipu.ru

Abstract. Tandem queueing systems often arise in wireless networks modeling. Queueing models are very suitable for network performance evaluation but the system complexity exponential growth (or state space explosion) could make the analysis barely feasible. The paper presents a comparative study of various methods of a state space reduction for markovian arrival processes (MAP) and phase-type distributions (PH) applied to tandem queueing systems. The applied methods include nonlinear optimization, EM-algorithm and linear minimization. While most of the described algorithms are well-studied, a number of issues arises when applying them to a tandem system of a real wireless network. Particularly, it is shown that while all the algorithms could be applied to tandems with a small number of queues, bigger tandems require additional effort to get the appropriable results. Nevertheless, the results presented show that the departure MAPs reduction may help to solve the state space explosion problem.

Keywords: Queueing systems · Random process fitting · Markov chain space reduction · MAP · PH · Wireless networks modeling

1 Introduction

Wireless backbone networks play essential role in modern communication systems. One of the crucial applications of wireless networks are backhauls along the long objects like roads, railways or pipes. Such networks could be used for data transmission from surveillance cameras or sensors to data centers, as well as for providing Internet connection. IEEE 802.11 is a frequently used technology for such networks implementation due to a reasonable data transfer rate and a wide range of the available inexpensive equipment. While IEEE 802.11-based networks have many advantages, the performance of multi-hop networks could be insufficient, thus performance estimation and analysis are required.

A prospective approach for wireless networks performance analysis involves queueing systems with correlated arrivals. Such models become especially attractive when Markov random processes are used for both arrivals and service time

© Springer International Publishing AG 2017
A. Dudin et al. (Eds.): ITMM 2017, CCIS 800, pp. 215–230, 2017.
DOI: 10.1007/978-3-319-68069-9_18

216 V. Vishnevsky et al.

distribution modeling. One of the perspective models are tandem queueing networks $MAP/PH/1/N \rightarrow \cdots \rightarrow \bullet/PH/1/N$ with cross-traffic [20]. In this queueing system data transmission time is modelled with phase-type (PH) distributions and user traffic is modelled with Markovian arrival processes (MAP) [15]. Representing the user data with Markovian arrival processes allows to take into account correlated nature of real network traffic [5,13] and PH-distributions provide sufficient approximation for a complex random process describing data transmission. The application of the tandem queueing systems described above for networks with linear topology was studied in general in the previous work [20].

The tandem queueing system analysis is affected by the exponential state space growth with the number of hops increasing. State space reduction techniques could be applied to solve this problem but their usage may lead to the precision loss and needs to be analysed carefully. Another issue to solve is to find a PH-distribution approximating the specific medium access scheme precisely enough.

While a plenty of MAP fitting approaches exists, their application to the wireless networks models analysis faces several difficulties. The data transmission over wireless channels involves a number of constant intervals for channel listening or scheduling which makes service time distribution more deterministic and causes additional correlation in departure processes. Another issue relates to very small values of distributions moments and large values of MAP generator entries (to be noted, the generator itself may contain hundreds or thousands of states), leading to the relative errors growth when applying the fitting algorithms to real data and it further requires an additional effort to improve accuracy. Last but not least is the performance issue since some algorithms could take hours of processor time to converge.

The paper presents a comparative study of various methods of state space reduction for markovian arrival processes and phase-type distributions applied to tandem queueing systems. We study the application of different methods and compare their performance and accuracy. We also provide the results of applying the state reduction techniques to a wireless tandem network containing up to ten stations and show that the departure process state space reduction methods could be applied for a real network analysis.

2 Tandem Queueing System

Let us consider $MAP/PH/1/N \rightarrow \cdots \rightarrow \bullet/PH/1/N$ system as a wireless network model. This system consists of a chain of servers with PH-distributed service time and a buffer size N. Each server receives the output flow from a previous station and a cross-traffic modelling the data flow from the external users as a Markovian arrival process (MAP), see Fig. 1.

A Markovian arrival process is defined by an irreducible continuous-time Markov chain $\nu_t, t \geq 0$ with a finite state space $\{0, \ldots, W\}$. The process $\nu_t, t \geq 0$ is in state ν during exponentially distributed time with parameter $\lambda_\nu, \nu \in \overline{0, W}$. After the time expires the chain jumps from state ν to state $\tilde{\nu}$ with probability

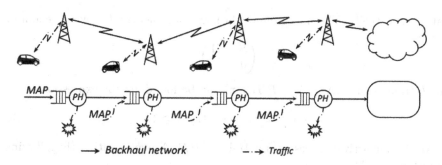

Fig. 1. A tandem network model

$p_0(\nu, \tilde{\nu})$ if the transmission is unobserved and $p_1(\nu, \tilde{\nu})$ otherwise. An observed transmission generates a message. It is also assumed that the process can not stay in the same state $\tilde{\nu} = \nu$ without message generation. Matrices D_0, D_1 are used to define the MAP:

$$(D_0)_{\nu,\nu'} = \begin{cases} -\lambda_\nu, & \text{if } \nu = \nu' \\ \lambda_\nu p_0(\nu, \nu'), & \text{otherwise} \end{cases}$$

$$(D_1)_{\nu,\nu'} = \lambda_\nu p_1(\nu, \nu').$$

The matrix $D = D_0 + D_1$ defines an infinitesimal generator of the random process $\nu_t, t \geq 0$. Its stationary probability vector $\boldsymbol{\theta}$ is obtained from the system

$$\boldsymbol{\theta} D = \mathbf{0}, \qquad \boldsymbol{\theta}\mathbf{1} = 1,$$

where $\mathbf{0}$ is a row vector of zeros and $\mathbf{1}$ is a column vector of ones. The steady-state probability vector $\boldsymbol{\pi}$ of a discrete time Markov chain embedded at arrival instants with a generator $P = (-D_0)^{-1}D_1$ can be obtained as the solution of the following linear system:

$$\boldsymbol{\pi} P = \boldsymbol{\pi}, \qquad \boldsymbol{\pi}\mathbf{1} = 1.$$

The average arrival intensity of a MAP is $\lambda = 1/\boldsymbol{\pi}(-D_0)^{-1}\mathbf{1}$. The k-th moment and lag-k correlation can be expressed as

$$m_k = k!\,\boldsymbol{\pi}(-D_0)^{-k}\mathbf{1}, \quad k \geq 1, \tag{1}$$

$$l_k = \frac{\lambda^2\boldsymbol{\pi}(-D_0)^{-1}P^k(-D_0)^{-1}\mathbf{1} - 1}{\lambda^2\boldsymbol{\pi}(-D_0)^{-2}\mathbf{1} - 1}, \quad k \geq 1. \tag{2}$$

A phase-type (PH) distribution is defined as a hitting time of the absorbing state in a continuous-time Markov chain with a single absorbing state. Formally, a random variable X is said to have PH-distribution $X \sim PH(S, \boldsymbol{\tau})$ if $\boldsymbol{\tau} \in \mathbb{R}^V$ is a probability distribution and $S \in \mathbb{R}^{V \times V}$ is a subinfinitesimal matrix defining initial states probabilities and transition rates between non-absorbing states respectively. The background Markov chain has the following generator matrix:

$$\begin{pmatrix} S & -S1 \\ 0 & 1 \end{pmatrix}$$

The k-th moment $\mathbb{E}[X^k]$, $X \sim PH(S, \tau)$ can be found via the expression

$$m_k = k!\, \tau(-S)^{-k}1, \quad k \geq 1. \tag{3}$$

Markovian arrival processes and $MAP/PH/1/N$ queues satisfy the following properties [20–22]:

1. The result of sifting a MAP with constant probability is also a MAP;
2. The composition of a finite number of MAPs is a MAP;
3. The departure process of $MAP/PH/1/N$ system is also a MAP.

Note that $MAP/PH/1/N$ queue can lose packets due to the queue overflow and the flow of lost packets is also a MAP. Taking into account these properties it can be shown that a departure process form the first server is a MAP and consequently the arrival processes to all succeeding servers are also MAPs as well as the departure processes. Thus an iterative procedure can be built to compute parameters of a queueing network [20].

However a state space of departure process is expressed as a cartesian product of the state spaces of MAP-input, PH-distribution and the queue length (the number of messages being queued and served). This fact results into an exponential state space growth also referred to as a state space explosion, making a precise analysis barely feasible for an arbitrary number of servers. To solve the state space explosion problem, the departure process of each queue can be approximated with a lower order MAP. Alternative approach is to approximate a process arriving at the queue, i.e. after the composition with cross-traffic.

Another problem considered is to find a PH-distribution adequately describing the medium access scheme operation. This problem is closely related to MAP fitting and will also be discussed further.

3   Related Work

There is a plenty of works describing various MAP and PH fitting. These studies could be divided into three areas. The first direction is the reconstruction of MAP and PH-distributions based on the known moments and lag-k correlation coefficients [7]. The second direction is to improve distributions already constructed and to choose the parameters closest (in the sense of some criterion) to the parameters of the statistical series [16]. The third one is to find MAPs and PH distributions maximizing the likelihood function based on the statistical data. These approaches are often based on the expectation-maximization (EM) algorithms [11,17,19]. We refer a reader to [2,16] for the state-of-art and open problems in this area.

Bodrog et al. [3] describe a method to find second-order matrices for MAP and phase distributions. To describe MAP with a large number of states and a specified correlation coefficient, it was suggested in [7] to build a phase-type distribution with a given number of states first, and then use it to construct MAP matrices. The inter-arrival time distribution is fitted by a PH distribution where the PH generator determines matrix D_0 of MAP and the initial probability vector π determines the steady state probability vector of the MAP embedded process. On the next stage the matrix D_1 is constructed by approximating the lag-k correlation values. Note that the system for matrix D_1 contains $2n + 1$ equations for n^2 unknowns leading to a linearly constrained non-linear optimization problem. The matching of order 3 and higher phase distributions was considered in [1,8,9].

Bobbio et al. [1] proposes a method to compose minimal order phase type distribution with first three moments and present a simple transformation from APH(n-1) (acyclic phase type distribution of order $n - 1$) to APH(n) with an additional phase. The authors also evaluate the bounds for the first three moments of APH(n).

Telek and Horvath [18] present the minimal representations of PH and MAP (Markovian, Jordan, Laplace, moments and MRP representations) and transformations between them. They construct an algorithm to optimize D_0 and D_1 matrices of MAP by means of a transformation matrix B such that matrices $B^{-1}D_0B$ and $B^{-1}D_1B$ minimize a goal function (or improve its value). The method allows improving any MAP fitted by other methods.

Casale et al. [4] propose a MAP fitting algorithm based on the first three moments and high order autocorrelations. They define a process composition method called a Kronecker Product Composition (KPC). Given J MAPs with matrices $D_0^{(j)}$ and $D_1^{(j)}$, $j = \overline{1,J}$, the composed MAP is defined as

$$D_0 = (-1)^{J-1}D_0^{(1)} \otimes \cdots \otimes D_0^{(J)}, \quad D_1 = D_1^{(1)} \otimes \cdots \otimes D_1^{(J)} \qquad (4)$$

and can be constructed of any order to fit data traces or reduce a state space of an arbitrary MAP. The algorithm consists of three steps. The first step fits the sample squared coefficient of variation and correlation coefficient to minimize the distance between sample lag-k correlations and numerical ones. On the second step, the first and third moments for each MAP $D_0^{(j)}$ and $D_1^{(j)}$, $j = \overline{1,J}$, are determined from an acceptable region and further optimized to minimize the distance between the sample joint moments and their estimated values. Based on the optimal values of the first three moments and the correlation coefficients, we can construct J MAPs $D_0^{(j)}$ and $D_1^{(j)}$, $j = \overline{1,J}$ (e.g., of the second order) and compose the final form of matrices (4).

In this paper we use three different approaches to reduce the tandem departure processes state spaces: MAP fitting as a solution of nonlinear optimization problem, EM-based approach [1,8] and a method of building a phase-type distribution with a given number of states and construction of a D_1 matrix for fitting lag-k correlation coefficients [7]. We also use G-FIT approach based on

EM procedure [19] to fit the PH distributions. These methods will be described in more details in the following sections.

4 MAP and PH Fitting

The fitting methods allow to construct a markovian arrival process or a phase-type distribution using a given trace (a set of samples) or a set of estimated metrics including moments and lag-k autocorrelation coefficients. In the tandem queueing system described above the fitting methods allow to approximate an operation of a specific communication protocol as well as to reduce the size of the departure processes (the latter case will be discussed in more details in the following section). Here we describe several fitting methods as they are; we suppose the data trace or estimated moments and lag-k autocorrelation coefficients values to be given as an input. The described methods include the expectation maximization (EM) algorithm [11,17,19], search for the MAP or PH as a solution of the nonlinear optimization problem constrained by the given moments and lag-k values, and a sequential independent fitting of the PH distribution using the trace or estimated moments values and MAP matrix D_1 using lag-k values constraints [7].

4.1 Fitting by Trace

The paper [19] describes a PH distribution fitting technique based on the EM algorithm (the authors call this algorithm G-FIT). MAP fitting using the EM algorithm is described in the papers [11,17]. While both algorithms will be used in numerical experiments, we describe briefly only G-FIT algorithm here due to the paper space limitations. The details of the algorithms could be found in the papers cited above.

G-FIT algorithm attempts to find a PH distribution fitting the given trace as a Hyper-Erlang distribution. Let X be a Hyper-Erlang random variable with M mutually independent Erlang distributions weighted with probabilities $\boldsymbol{\alpha} = (\alpha_1, \dots, \alpha_M)$, m-th chain containing r_m phases jointly forms a vector $\boldsymbol{r} = (r_1, \dots, r_M)$ and its intensities describe a vector $\boldsymbol{\lambda} = (\lambda_1, \dots, \lambda_M)$. Then the pdf of X is $f_X(x) = \sum_{m=1}^{M} \alpha_m \frac{(\lambda_m x)^{r_m-1}}{(r_m-1)!} \lambda_m e^{-\lambda_m x}$.

The parameters $(\boldsymbol{r}, \boldsymbol{\alpha}, \boldsymbol{\lambda})$ are chosen while fitting. Consider EM algorithm to maximize a log-likelihood expression. The authors first apply it for a general set of independents distributions with density functions p_m such that $p(x|\boldsymbol{\Theta}) = \sum_{m=1}^{M} \alpha_m p_m(x|\theta_m)$ where $\boldsymbol{\Theta} = (\boldsymbol{\alpha}, \boldsymbol{\theta})$ and θ_i is a parameter (or vector) of p_m.

Then the authors suggest considering an unobserved random variable Y having values in $\{1, \dots, M\}$ and specifying which component is used to generate a specific item x_k of the trace in order to simplify the a log-likelihood calculation.

Applying this idea the expected value of complete log-likelihood is

$$Q(\boldsymbol{\Theta}, \hat{\boldsymbol{\Theta}}) = \sum_{m=1}^{M} \sum_{k=1}^{K} \log(\alpha_m) q(m|x_k, \hat{\boldsymbol{\Theta}})$$
$$+ \sum_{m=1}^{M} \sum_{k=1}^{K} \log(p_m(x_k|\theta_m)) q(m|x_k, \hat{\boldsymbol{\Theta}}) \tag{5}$$

where $\hat{\boldsymbol{\Theta}}$ is an initially chosen parameter set, required to compute a conditional pdf of Y:

$$q(y_k|x_k, \hat{\boldsymbol{\Theta}}) = \frac{\hat{\alpha}_{y_k} p_{y_k}(x_k|\hat{\theta}_{y_k})}{\sum_{m=1}^{M} \hat{\alpha}_m p_m(x_k|\hat{\theta}_m)}. \tag{6}$$

Computing expression (5) for some vector $\hat{\boldsymbol{\Theta}}$ is a E-step of EM algorithm. For performing M-step (maximization), parameters $\boldsymbol{\Theta} = (\boldsymbol{\alpha}, \boldsymbol{\theta})$ maximizing $Q(\boldsymbol{\Theta}, \hat{\boldsymbol{\Theta}})$ should be found. $\boldsymbol{\alpha}$ can be found by applying Lagrange multipliers to (5); to find $\boldsymbol{\theta}$ a specific pdf required, so let $\boldsymbol{\theta} = \boldsymbol{\lambda}$ for Hyper-Erlang. Then

$$\alpha_m = \frac{1}{K} \sum_{k=1}^{K} q(m|x_k, \hat{\boldsymbol{\Theta}}), \qquad \lambda_m = \frac{r_m \cdot q(m|x_k, \hat{\boldsymbol{\Theta}})}{\sum_{k=1}^{K} q(m|x_k, \hat{\boldsymbol{\Theta}}) \cdot x_k}. \tag{7}$$

4.2 Fitting as Optimization

The MAP or PH distribution fitting may be described as a solution of the optimization problem constrained by the values of the moments and lag-k autocorrelation coefficient values. Let $\mathbf{m}_{K_m}$ be the vector of the first K_m moments of MAP, $\mathbf{l}_{K_l}$ be the vector of the first K_l lags given in (1) and (2) correspondingly; $\boldsymbol{\mu}$ and $\boldsymbol{\nu}$ be the vectors of moments and lags of a random process to fit correspondingly. Using this notation the problem of MAP fitting can be formulated via solution of a nonlinear algebraic system

$$\begin{cases} \mathbf{m}_{K_m}(D_0, D_1) = \boldsymbol{\mu}, \\ \mathbf{l}_{K_l}(D_0, D_1) = \boldsymbol{\nu}. \end{cases} \tag{8}$$

System (8) should be solved for D_0 and D_1 such that $D = D_0 + D_1$ is an infinitesimal generator and D_0 is a subgenerator. By these restrictions, the system may have no solution for some pairs $(\boldsymbol{\mu}, \boldsymbol{\nu})$ and the order N thus a MAP with such lags and moments does not exist. It should be noticed that there are no known closed form margins for the moments and lags values for MAPs and PH distributions of an arbitrary order making the problem much harder.

We suggest that approximate solution of the system can be brought to an optimization problem as follows. Define a loss function $\mathscr{L}(\cdot) = (|\cdot|)^2$ and a loss functional

$$Q(D_0, D_1) = \mathscr{L}(\mathbf{m}_{K_m}(D_0, D_1) - \boldsymbol{\mu}) + \mathscr{L}(\mathbf{l}_{K_l}(D_0, D_1) - \boldsymbol{\nu}). \tag{9}$$

Then a proper MAP is found as a solution of $(D_0, D_1) = \underset{D_0, D_1}{\arg\min} Q(D_0, D_1)$.
The fitting procedure is iterative and start with the lower possible N. The tolerance ϵ should be chosen such that $\min Q(D_0, D_1) < \epsilon$ holds. If for given N there is no solution (D_0, D_1) the order N should be incremented and the new fitting procedure starts until the criterion is satisfied or the maximum number of iteration is exhausted. Otherwise, the pair (D_0, D_1) with the lower error $\min Q$ is supposed to be a solution. Also another loss function $\mathscr{L}$ can be considered.

For PH distribution the optimization problem can be simplified as it has zero lags and less difficulty in moments computation. The loss functional for PH fitting is as follows:

$$Q(\boldsymbol{\tau}, S) = \mathscr{L}(\mathbf{m}_{K_m}(\boldsymbol{\tau}, S) - \boldsymbol{\mu}).$$

The problem described is generally nonconvex which leads to local optima solutions and require additional effort to randomize the initial vectors and look for the best solution.

4.3   MAP Fitting by Given PH

The MAP moments depend on the matrix D_0 and a steady state probability distribution of the embedded discrete process which allows to fit them independently of the lag-k autocorrelation values; the lag-k values could be used to find the appropriate matrix D_1 [10]. Suppose we have a PH$(\boldsymbol{\tau}, S)$ distribution. It is assumed that the MAP(D_0, D_1) has $D_0 = S$ and $\boldsymbol{\pi} = \boldsymbol{\tau}$ where $\boldsymbol{\pi}$ is a steady state probability distribution of the embedded Markov chain with the transition matrix $P = (-D_0)^{-1}D_1$. Combining the restrictions for D_1 to be held and considering the autocorrelation c_{corr} the authors [10] obtained a linear system:

$$D_1 \mathbf{1} = -D_0 \mathbf{1}, \quad \boldsymbol{\pi}(-D_0)^{-1}D_1 = \boldsymbol{\pi}, \quad \delta D_1 \mathbf{f} = \upsilon$$

where the values $\boldsymbol{\delta}$, $\mathbf{f}$ and υ can be derived from the lag-1 expression (2). Considering a vector $\mathbf{x} = [\mathbf{d}_1, \mathbf{d}_2, \ldots, \mathbf{d}_M]^T$ where $\mathbf{d}_i$ is the i-th column vector of D_1, the authors has transformed these three matrix equations into one that have the following form:

$$\begin{bmatrix} I & I & \ldots & I \\ \gamma & & & \\ & \gamma & & \\ & & \ddots & \\ & & & \gamma \\ f_1\delta & f_2\delta & \ldots & f_M\delta \end{bmatrix} \cdot \mathbf{x} = \begin{bmatrix} \mathbf{g} \\ \boldsymbol{\pi} \\ \upsilon \end{bmatrix}. \tag{10}$$

This linear equations should be solved for non-zero elements of $\mathbf{x}$. To find the higher-order lags the authors suggest to use an optimization procedure.

5   State Reduction

While the process fitting is the only approach for obtaining PH distributions of the service time when only the trace is known, the departure MAP processes state space could be reduce by other methods. Generally any reduction method supposes the given matrices D_0 and D_1 of a MAP to decrease the order. For PH-distribution the problem is set in the same way taking into account a vector $\boldsymbol{\pi}$ and a matrix S instead of D_0 and D_1.

To apply the fitting methods described in the previous section to MAP state space reduction, the moments and lags of the source MAP should be calculated. If the fitting method requires a trace (e.g. EM algorithm), the source MAP should be randomized to get a trace. It should be noticed that for the MAPs with a huge number of states the consistent trace may contain over a million samples. To simplify the problem and avoid the randomization, two additional techniques of state space reduction are described below including the nonlinear optimization problem solving constrained by the distance between the cumulative distribution functions and cutting the tail states of the QBD (quasi-death-and-birth) process.

5.1   Reduction as Optimization

The state reduction can be performed by solving an optimization problem. For that aim let us consider the difference of the stationary cumulative distribution function of the given MAP and some lower order MAP

$$\Delta F(t) = F(t) - F'(t) = \boldsymbol{\pi}' e^{D_0' t} \mathbf{1}' - \boldsymbol{\pi} e^{D_0 t} \mathbf{1}.$$

Taking into account that $e^{D_0 t} \approx I + \sum_{k=1}^{K} \frac{1}{k!} D_0^k t^k$ for some K and $\boldsymbol{\pi} \mathbf{1} = \boldsymbol{\pi}' \mathbf{1}' = 1$ this difference takes a form:

$$\Delta F(t) = \sum_{k=1}^{K} \frac{t^k}{k!} \left(\boldsymbol{\pi}'(D_0')^k \mathbf{1}' - \boldsymbol{\pi} D_0^k \mathbf{1} \right) = \sum_{k=1}^{K} w(k, t) \left(\boldsymbol{\pi}'(D_0')^k \mathbf{1}' - \boldsymbol{\pi} D_0^k \mathbf{1} \right), \quad (11)$$

where $w(k, a) = a^k / k!$ is a weight for the k-th power of D_0. Multiple ways to define the weights exist; here we define the weights as $\frac{T^{k+1}}{(k+1)!}$ applying $a = T$ that arises out of integrating (11) in a range $[0, T]$

$$\Delta F = \int_0^T \Delta F(t) dt = \sum_{k=1}^{K} \left[\int_0^T \frac{t^k}{k!} dt \right] \left(\boldsymbol{\pi}'(D_0')^k \mathbf{1}' - \boldsymbol{\pi} D_0^k \mathbf{1} \right)$$

Taking $D_0' = S, D_1' = \tau$ in case of PH and D_0, D_1 in case of MAP reduction, the loss functional can be expressed as

$$Q(D_0', D_1') = \mathscr{L}(\Delta F(D_0', D_1')). \quad (12)$$

5.2 State Truncation

This method is applied to servers without the memory, i.e. $MAP/PH/1$ systems when the utilization coefficient is sufficiently small. A system with a limited memory could be approximated if it has a very large capacity. The authors of [6] consider the departure MAP as a pair of block matrices and suggest to trunk the tail blocks stating with some level $N+1$ by merging the stationary probabilities into N-th state: $\pi_N^+ = \sum_{i=N+1}^{\infty} \pi_i$ and considering matrices:

$$\hat{A}_0 = A_0 + A_1,$$
$$\hat{A}_2 = \text{diag}(\pi_N)\text{diag}^{-1}(\pi_N + \pi_N^+)A_2,$$
$$\check{A}_2 = \text{diag}(\pi_N^+)\text{diag}^{-1}(\pi_N + \pi_N^+)A_2$$

to describe the reduced matrices of the departure MAP

$$D_0 = \begin{bmatrix} B_1 & B_0 & & & \\ & A_1 & A_0 & & \\ & & \ddots & \ddots & \\ & & & A_1 & A_0 \\ & & & & \hat{A}_0 \end{bmatrix}, \quad D_1 = \begin{bmatrix} O & & & \\ B_2 & & & \\ & A_2 & & \\ & & \ddots & \\ & & & \hat{A}_2 & \check{A}_2 \end{bmatrix}. \quad (13)$$

The matrices A_i, B_i for $i = 0, 1, 2$ describing the blocks of the initial departure MAP and their definition are provided in [6]. This method allows to restrict the state space growth by decreasing the effective queue length. Unfortunately, the state space continues exponential increasing along with the number of queues in the tandem since the size of the service time PH-distribution is greater than 1 which makes the method not applicable to analyse a tandem queuing systems with an arbitrary number of queues.

6   Experimental Results

In the numerical experiments we used three different methods of MAP fitting:

1. Searching for a MAP (defined by the matrices D_0 and D_1) as a solution of an optimization problem constrained by the values of the first moments and lags (this method is referred to as OPT below).
2. MAP fitting using the EM algorithm [11];
3. Successive independent fitting of D_0 matrix as a PH distribution using the moments or a trace provided and looking for D_1 as a solution of the optimization problem constrained by the lag-k correlation coefficients (INDI in the following text). The algorithm was described in [7].

Queueing system analysis framework [14] was developed in the Python 3 language using NumPy/SciPy packages. We used EM algorithms implementations from a BuTools [12] package. Simulation models were developed using OMNeT++ network simulator. All the experiments ran on a generic laptop with i7 processor and 16 GB of RAM.

While the OPT method shown good results, it often fell into a local optima and required initial solutions randomization to converge good. The key problem is a lack of easy checkable conditions on the solution existence and moments and lags values, so it was often hard or impossible to find a solution of a given order under the given constraints while the attempt to find it takes significant time. It should be noted that algorithm converged rapidly for small MAPs and PHs with up to 8 states but required lots of time when called for bigger orders (5 min and more). It was also noticed that order increasing didn't provide better results in many cases so we decided to use the algorithm with small orders. The solution error was also reduced by normalizing the moments and the D_0 and D_1 matrices consequently.

The EM algorithm provided good results but required too much time to converge. Typically, it takes up to 20 min to fit a given trace with 40000 samples using a MAP with 12 states. While it is possible to speed up the algorithm as described in [19], it didn't completely solve the problem and the algorithm still required lots of processor resources. Since the algorithm had to be applied several times, we decided to limit the search with MAPs containing up to four states. The similar problem arose during G-FIT execution while it still allowed to fit PH distributions with up to 10 states in a reasonable time. Due to the order limitation, the EM algorithm for MAP fitting provided the worst results considering moments and lags matching.

The third (INDI) approach was implemented as described in the paper [7]. We tried both nonlinear optimization and G-FIT [19] algorithm for fitting the PH distribution for D_0 matrix construction, and G-FIT provided much better overall results. To keep the problem of D_1 construction linear, we limited the constraints with lag-1 correlation. In this case the problem could be solved as a linear minimization problem $\|Ax - b\|_2 \to \min$. The key problem was that the existence of D_1 matrix was dependent on the particular D_0 and it sometimes required several D_0 fitting iterations to find an appropriate matrix to make the D_1 construction with a reasonable error possible.

First of all, the fitting algorithms were applied to fit the PH-distribution approximating data transmission intervals. To simplify the analysis, a tandem consisting of two stations was considered. The wireless channel bitrate was 5 mbps (e.g. a slow sensor network link) and an arrival traffic bitrate was 2.8 mbps. The arrival traffic was described with a MAP approximating a real network trace LBL-TCP-3 described in [10]:

$$D_0 = \begin{bmatrix} -508.11 & 0 & 0 & 0 \\ 0 & -526.82 & 0 & 0 \\ 0 & 0 & -112.88 & 0 \\ 0 & 0 & 0 & -292.87 \end{bmatrix}$$

$$D_1 = \begin{bmatrix} 281.9 & 226.06 & 0 & 0.15872 \\ 526.66 & 0.024505 & & 0.13422 \\ 0 & 0 & 82.094 & 30.79 \\ 0.056728 & 0 & 38.799 & 254.01 \end{bmatrix}$$

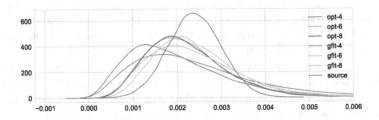

Fig. 2. Service time fitting. Lines labeled as 'opt-n' show solutions of a nonlinear optimization problem and 'gfit-n' show G-FIT [19] where 'n' is the order of PH

Table 1. Moments and lags of the approximated departure processes from the first station.

Algorithm	Order	M1	M2	M3	Std.	Lag-1	Lag-2
Original MAP	192	0.00453	0.000048	1.032e-06	7.329e-10	0.176	0.126
Nonlinear opt.	6	0.00453	0.000048	1.032e-06	7.328e-10	0.176	0.126
EM	3	0.004371	0.00005	1.248e-06	9.566e-10	0.109	0.048
G-FIT and linear opt.	8	0.00428	0.000038	6.079e-07	3.982e-10	0.176	0.075

The packets sizes were assumed to have a normal distribution truncated to positive values with an average value 12 kbit and standard deviation 3 kbit. We applied G-FIT algorithm [19] and nonlinear optimization approach with a number of moments equations equal to 3 for PH orders 4, 6 and 8. The results are shown on Fig. 2 (while more lags could allow to fit the service time distribution better, it was crucial to use a small distribution due to the state space growth appearing on the next stations). It should be also noticed that applying G-FIT for a greater number of states takes a rather long time due to combinatorial complexity of inspecting various structures of the Hyper-Erlang distributions.

The PH distribution obtained with G-FIT containing 8 states was used for the later computations. This distribution matched the mean value and had a 32% error in standard deviation. It was used to build the departure process of the first station having capacity 5, which was approximated with the EM-algorithm [11], nonlinear optimization with moments and lags constraints and the approach of independent construction of a D_0 matrix as a PH-distribution and D_1 matrix with linear constraints [7]. In the latter approach only the lag-1 correlation coefficient was constrained. The results are shown in Table 1. The last row describes a separate D_0 fitting with G-FIT and D_1 as a solution of linear minimization problem [7]. It should be noticed that EM-algorithm was used for a small MAP order equal to 3, its stop condition was reduced to 10^{-3} and the maximum number of iterations was 100. This could be the reason of the worst results shown.

The system size distribution of the second station was also investigated. Since the arrival traffic required more than a half of the modeled channel bandwidth,

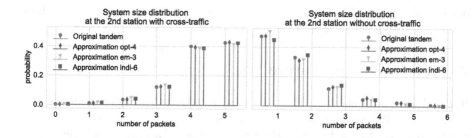

Fig. 3. Number of packets probability distributions of the second station with and without the cross-traffic and various fitting algorithms.eps

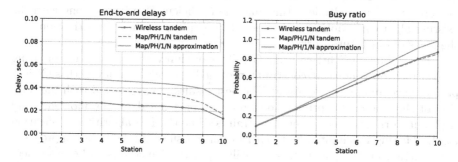

Fig. 4. End-to-end delays and busy ratios computed for a wireless network tandem model.

adding the cross-traffic caused a system overflow. All the described algorithms allowed to get sufficient approximation of the system states distributions as shown on Fig. 3.

Finally, the OPT approach was applied to fit the departure MAP processes in the model of a real wireless network containing 10 nodes and operating under the IEEE 802.11 standard. To simplify the simulation a simple DCF channel access scheme was considered and the wireless channels provided 54 mbps bitrate. Each arrival process was described with the same MAP as above. The cross-traffic arrived at each wireless station.

The measured transmission time was fitted by the first three moments with 0.05 relative error with a PH-distribution $PH(S, \tau)$:

$$S = \begin{bmatrix} -6267.56 & 1412.75 & 0.001814 & 943.60 \\ 1008.28 & -3337.85 & 0.000100 & 258.21 \\ 0.002726 & 0.0000027 & -107.766 & 2.744 \\ 1565.226 & 1563.65 & 3.9327 & -6778.49 \end{bmatrix}$$

$$\tau = \begin{bmatrix} 0.038351 & 0.961517 & 0 & 0.000132 \end{bmatrix}$$

The measured end-to-end delay and busy ratios are shown on Fig. 4. The busy ratios were approximated well but end-to-end delays approximated values were not precise. This problem could be solved with other approximation methods or fitting PH distributions and MAP arrivals with higher order processes.

7   Conclusion and Future Work

It was shown that departure MAP state space reduction provided sufficient precision while allowing to analyze the tandem queueing systems of an arbitrary length. However the fitting algorithms performance along with time limitations may lead to accuracy degradation. The nonconvex nature of the problems arising leads to local optima convergence and impossibility to find the optimal solution in many cases. To face these issues, a randomization of initial parameters should be applied to find multiple optima and more efficient algorithms along with the existing algorithms optimization should be explored. These investigations are the focus of our future work.

The combined PH fitting using G-FIT algorithm with D_1 construction using autocorrelation coefficient constraint provided a good accuracy with sufficient performance and looks promising. The best results were retrieved with the solution of the nonlinear optimization problem constrained by the moments and lag-k autocorrelation coefficient constraints while the EM algorithm application to MAP fitting was limited by the performance issues. While several approaches were studied in this paper, there is still a plenty of methods to be examined, including the KPC approach. These methods would be applied and optimized in the future works.

Finally, it should be noticed that the lag-k autocorrelation coefficients of the departure processes grow along with the number of stations in the tandem network. While the typical moments values allowed to fit the service time distribution with a good precision, it was often a problem to find a valid MAP process with the precise autocorrelation coefficients values. The solution may be found using the processes with the greater number of states, but this requires more intelligent methods of predicting the structure of the approximating MAPs due to performance limitations. These methods will also be studied in the future works.

Acknowledgements. This work has been financially supported by the Russian Science Foundation and the Department of Science and Technology (India) via grant 16-49-02021 for the joint research project by the V.A. Trapeznikov Institute of control Sciences and the CMS College Kottayam.

References

1. Bobbio, A., Horvath, A., Telek, M.: Matching three moments with minimal acyclic phase type distributions. Stochast. Models **21**, 303–326 (2005)
2. Bodrog, L., Heindl, A., Horvath, A., Horvath, G., Telek, M.: Current results and open questions on PH and MAP characterization. Numerical Methods for Structured Markov Chains (2007)

3. Bodrog, L., Heindl, A., Horvath, G., Telek, M.: A markovian canonical form of second-order matrix-exponential processes. Eur. J. Oper. Res. **190**, 459–477 (2008)
4. Casale, G., Zhang, E.Z., Smirni, E.: Trace data characterization and fitting for markov modeling. Perform. Eval. **67**, 61–79 (2010)
5. Heyman, D., Lucantoni, D.: Modelling multiple IP traffic streams with rate limits. IEEE ACM Trans. Network. **11**, 948–958 (2003)
6. Horvath, A., Horvath, G., Telek, M.: A joint moments based analysis of networks of MAP/MAP/1 queues. Perform. Eval. **67**, 759–778 (2010)
7. Horvath, G., Buchholz, P., Telek, M.: A map fitting approach with independent approximation of the inter-arrival time distribution and the lag correlation. In: Second International Conference on the Quantitative Evaluation of Systems, pp. 124–133 (2005)
8. Horvath, G., Reinecke, P., Telek, M., Wolter, K.: Heuristic representation optimization for efficient generation of PH-distributed random variates. Ann. Oper. Res. **239**, 643–665 (2016)
9. Horváth, G., Telek, M.: A canonical representation of order 3 phase type distributions. In: Wolter, K. (ed.) EPEW 2007. LNCS, vol. 4748, pp. 48–62. Springer, Heidelberg (2007). doi:10.1007/978-3-540-75211-0_5
10. Horváth, G., Buchholz, P., Telek, M.: A map fitting approach with independent approximation of the inter-arrival time distribution and the lag correlation. In: Second International Conference on the Quantitative Evaluation of Systems, pp. 124–133 (2005)
11. Horváth, G., Okamura, H.: A fast EM algorithm for fitting marked markovian arrival processes with a new special structure. In: Balsamo, M.S., Knottenbelt, W.J., Marin, A. (eds.) EPEW 2013. LNCS, vol. 8168, pp. 119–133. Springer, Heidelberg (2013). doi:10.1007/978-3-642-40725-3_10
12. Hovarth, G.: Butools: Queueing and traffic modeling related functionality for matlab, mathematica and python (2016). https://github.com/ghorvath78/butools
13. Klemm, A., Lindermann, C., Lohmann, M.: Modelling IP traffic using the batch marcovian arrival process. Perform. Eval. **54**, 149–173 (2008)
14. Larionov, A., Ivanov, R.: Pyqumo: Python queueing modeler (2017). https://github.com/larioandr/pyqumo
15. Neuts, M.: A versatile markovian point process. J. Appl. Probab. **16**, 764–779 (1979)
16. Okamura, H., Dohi, T.: PH fitting algorithm and its application to reliability engineering. J. Oper. Res. Soc. Japan **59**, 72–109 (2016)
17. Okamura, H., Dohi, T.: Faster maximum likelihood estimation algorithms for markovian arrival processes. In: IEEE Sixth International Conference on the Quantitative Evaluation of Systems (QEST 2009) (2009)
18. Telek, M., Horvath, G.: A minimal representation of markov arrival processes and a moments matching method. Perform. Eval. **64**, 1153–1168 (2007)
19. Thummler, A., Buchholz, P., Telek, M.: A novel approach for fitting probability distributions to real trace data with the EM algorithm. In: International Conference on Dependable Systems and Networks (2005)
20. Vishnevski, V., Larionov, A., Ivanov, R.: An open queueing network with a correlated input arrival process for broadband wireless network performance evaluation. In: Dudin, A., Gortsev, A., Nazarov, A., Yakupov, R. (eds.) ITMM 2016. CCIS, vol. 638, pp. 354–365. Springer, Cham (2016). doi:10.1007/978-3-319-44615-8_31

21. Klimenok, V., Dudin, A., Vishnevsky, V.: Tandem queueing system with correlated input and cross-traffic. In: Kwiecień, A., Gaj, P., Stera, P. (eds.) CN 2013. CCIS, vol. 370, pp. 416–425. Springer, Heidelberg (2013). doi:10.1007/978-3-642-38865-1_42
22. Vishnevsky, V., Dudin, A., Kozyrev, D., Larionov, A.: Methods of performance evaluation of broadband wireless networks along the long transport routes. In: Vishnevsky, V., Kozyrev, D. (eds.) DCCN 2015. CCIS, vol. 601, pp. 72–85. Springer, Cham (2016). doi:10.1007/978-3-319-30843-2_8

Analyzing of Licensed Shared Access Scheme Model with Service Bit Rate Degradation in 3GPP Network

Daria Ivanova[1], Ekaterina Karnauhova[1(✉)], Ekaterina Markova[1], and Irina Gudkova[1,2]

[1] Peoples' Friendship University of Russia (RUDN University),
6 Miklukho-Maklaya St, 117198 Moscow, Russian Federation
daria.i1996@gmail.com, ek.karnauhova@gmail.com,
{markova_ev,gudkova_ia}@rudn.university
[2] Institute of Informatics Problems, Federal Research Center
"Computer Science and Control" of the Russian Academy of Sciences,
44-2 Vavilov St, 119333 Moscow, Russian Federation

Abstract. The volume of mobile traffic is growing every year. More and more frequency resources are needed to provide users services with a required level of quality of service (QoS). One of the possible solutions to a problem of radio spectrum shortage is the sharing of spectrum between the owners and LSA licensees. Licensed shared access (LSA) framework gives the owner priority in spectrum access, to the detriment of the secondary user, LSA licensee. If the mobile operator users of both need continuous service without interruptions on the rented part of the spectrum, the rules of shared access should guarantee the possibility of simultaneous access. In this paper we simulate a queuing system and consider a scheme model of LSA framework with the limit power policy. We propose formulas for calculation of main characteristics of the model – a blocking probability and a mean bit rate. These characteristics are very important in teletraffic theory. For example, blocking probabilities help to determine the number of required channels.

Keywords: Queuing system · Licensed shared access · Limit power policy · Blocking probability · Mean bit rate

1 Introduction

Teletraffic theory is a mathematical theory, or one of branches of queueing theory. It is used for studying and designing telecommunication systems (telephony, computer networks, etc.). More generally, one can set the goal of teletraffic theory: construction of mathematical models that map real processes in information

The publication was financially supported by the Ministry of Education and Science of the Russian Federation (the Agreement number 02.a03.21.0008) and RFBR according to the research projects No. 16-37-00421 mol_a and No. 15-07-03608 a.

A. Dudin et al. (Eds.): ITMM 2017, CCIS 800, pp. 231–242, 2017.
DOI: 10.1007/978-3-319-68069-9_19

distribution systems and development of methods for assessing the quality of their functioning [11–13]. The number of user devices connected to a high-speed network as well as the volume of traffic transmitted between them is constantly increasing [1]. Consequently, an increasing amount of resources is needed to provide quality services. The problem of resource shortage could be solved by using the licensed shared access (LSA) framework [2–4]. LSA framework could improve the efficiency of resource usage and ensure the access to a spectrum which otherwise would be underused [5]. By using this framework, the spectrum is shared between the owner (so-called incumbent) and a limited number of LSA licensees (mobile operators). The LSA licensee has access to single-tenant band (the part of the spectrum, belonging only to the mobile operator) and rents the multi-tenant band (the part of the spectrum, belonging to the incumbent and the mobile operator), whereas the incumbent has access only to multi-tenant band. For interference coordination between the incumbent and the LSA licensee three policies [6] are proposed: limit power policy, shutdown policy, and ignore policy. According to the limit power policy [7], there is no interruption of service due to the incumbent accessing spectrum. It implies managing the user equipment power in uplink and eNodeB (eNB) power in downlink. According to shutdown policy [8,9], at any time, LSA spectrum could be used by incumbent or LSA licensees but not together at once. According to ignore policy LSA licensees use the shared spectrum without interference coordination.

In this paper we propose a scheme model of 3GPP wireless network within LSA framework [10]. For efficient interference coordination we consider the limit power policy, which allows us to continue the service of multi-tenant band users, even if the incumbent needs this part of the spectrum. In this case, the service of mobile operator users will not be interrupted, but the service bit rate will be reduced (degraded). At this time, the multi-tenant band goes into the so-called unavailable mode and user's requests arrived on the multi-tenant band continue their service at the degraded bit rate – minimum bit rate. After the incumbent releases the multi-tenant band, the band goes from the unavailable to the operational mode and the service bit rate for the connected mobile network users increases to the maximum value - maximum bit rate. The service on the single-tenant band is always carried out at the maximum bit rate.

This model is an improved version of the model described in [7]. One of the main disadvantages of the previous model was that after the disconnection the band was not recovered which means that even after the band was vacated by the owner, the service continued with degraded quality until all users of multi-tenant band were served. In our model this drawback is eliminated, the band goes into operational mode as soon as the owner frees it, while the quality of user service is increased to the original level.

This paper is organized as follows. In Sect. 2, we propose a mathematical model of the LSA framework with the limit power policy. In Sect. 3, we analyze main characteristics of the model: the blocking probability and the mean bit rate. Finally, we conclude the paper in Sect. 4.

2   Mathematical Model

2.1   General Assumptions and Parameters

We propose a scheme model of a single mobile network cell with LSA framework and limit power policy. We suppose that the mobile operator has access to the single-tenant band with the total capacity of C_1 bandwidth units (b.u.) and rents the multi-tenant band with the total capacity of C_2 b.u. Let the arrival rate λ be Poisson distributed and let the service time be exponentially distributed with mean μ^{-1}. Then, we denote the corresponding offered load as $\rho = \lambda/\mu$.

Each request processed on the single-tenant band is served at the maximum bit rate $d_{\max}$. Request on the multi-tenant band could be served at the maximum bit rate $d_{\max}$ or at the minimum bit rate $d_{\min}$ depending on the state of the multi-tenant band – operational or unavailable. Figure 1 shows the scheme of the model.

We assume that the multi-tenant band goes into unavailable mode with rate α and recovers into operational mode with rate β. Recovery and failure intervals follow the exponential distribution. All necessary notations are given in Table 1.

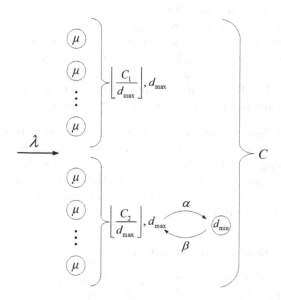

Fig. 1. The scheme of the model.

2.2   Limit Power Policy

Let us consider in more detail the limit power policy. First of all, we determine the rules for accepting requests for service.

When a new request arrives, four scenarios are possible:

Table 1. System parameters

Notation	Parameter description
C_1	Total capacity of the single-tenant band
C_2	Total capacity of the multi-tenant band
λ	Arrival rate
μ^{-1}	Mean service time
$d_{\max}$	Maximum bit rate
$d_{\min}$	Minimum bit rate
α	Rate of a transition the multi-tenant band into unavailable mode
β	Rate of a transition the multi-tenant band into operational mode
n_1	The number of single-tenant band users
n_2	The number of multi-tenant band users
s	The state of the multi-tenant band, s equals to 1 if the band is operational and s equals to 0 if the band is unavailable

- The request will be accepted for service on the single-tenant band, if the single-tenant band has not less than $d_{\max}$ free b.u.
- The request will be accepted for service on the multi-tenant band, if the single-tenant band has less than $d_{\max}$ b.u. free, the multi-tenant band is operational and has not less than $d_{\max}$ b.u. free.
- The request will be blocked, if the single-tenant band has less than $d_{\max}$ b.u. free and the multi-tenant band is unavailable or has less than $d_{\max}$ b.u. free.

Let us note if the owner does not use the frequency spectrum of the multi-tenant band, the data transfer can be carried out at the highest possible rate, which equals to $d_{\max}$, in other case the service bit rate for the mobile operator users is degraded from the maximum $d_{\max}$ to the minimum $d_{\min}$ value. When the multi-tenant band recovers, the bit rates are switched back and all users that have been degraded continue to receive service at bit rate $d_{\max}$.

2.3   System of Equilibrium Equations

The behavior of the system is defined by the Markov process $\mathbf{X}(t) = \{(N_1(t), N_2(t), S(t)), t \geq 0\}$, where $N_1(t)$ is the number of single-tenant band users, $N_2(t)$ is the number of multi-tenant band users, $S(t)$ is the state of the multi-tenant band at the moment $t \geq 0$. Let us denote $N_1 = \left\lfloor \frac{C_1}{d_{\max}} \right\rfloor$ the maximum number of single-tenant band users, $N_2 = \left\lfloor \frac{C_2}{d_{\max}} \right\rfloor$ the maximum number of multi-tenant band users. Then the system state space is the following:

$$\mathbf{X} = \{n_1 = 0, \ldots, N_1, \ n_2 = 0, \ldots, N_2, \ s = 1$$
$$\vee \ n_1 = 0, \ldots, N_1, \ n_2 = 0, \ldots, N_2, \ s = 0\}. \tag{1}$$

State space (1) could be divided into two subspaces: $\{n_1 = 0, \ldots, N_1, \ n_2 = 0, \ldots, N_2, \ s = 1\}$ if the multi-tenant band is operational and requests could be served at the maximum bit rate $d_{\max}$, and $\{n_1 = 0, \ldots, N_1, \ n_2 = 0, \ldots, N_2, \ s = 0\}$ if the multi-tenant band is unavailable and requests continue their service at the minimum bit rate $d_{\min}$. Figure 2 shows the structure of the state space, considering the two subspaces.

The corresponding Markov process $\mathbf{X}(t)$, which representing the system's states, is described by the following system of equilibrium equations

$$
\begin{aligned}
&p(n_1, n_2, s)\left[\lambda \cdot I(n_1 < N_1) + \lambda \cdot I(n_1 = N_1, \ n_2 < N_2, \ s = 1)\right. \\
&\left.+ (n_1 + n_2)\mu + \alpha \cdot I(s = 1) + \beta \cdot I(s = 0)\right] \\
&= p(n_1 + 1, n_2, s)\left[(n_1 + 1)\mu \cdot I(n_1 < N_1)\right] \\
&+ p(n_1, n_2 + 1, s)\left[(n_2 + 1)\mu \cdot I(n_2 < N_2)\right] \\
&+ p(n_1 - 1, n_2, s)\left[\lambda \cdot I(n_1 > 0)\right] \\
&+ p(n_1, n_2 - 1, 1)\left[\lambda \cdot I(n_1 = N_1, \ n_2 > 0, \ s = 1)\right] \\
&+ p(n_1, n_2, 1)\left[\alpha \cdot I(s = 0)\right] + p(n_1, n_2, 0)\left[\beta \cdot I(s = 1)\right], \quad (n_1, n_2, s) \in X,
\end{aligned}
\tag{2}
$$

where $(p(n_1, n_2, s))_{(n_1, n_2, s) \in X} = \mathbf{p}$ is the stationary probability distribution.

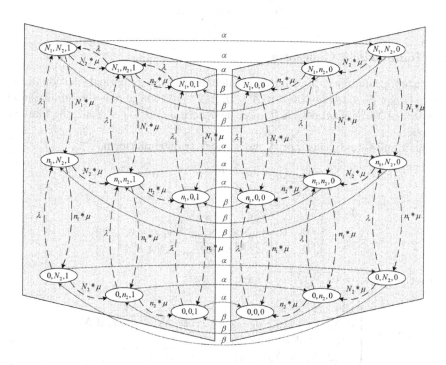

Fig. 2. The state space.

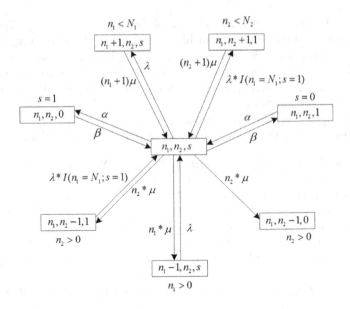

Fig. 3. Central state.

2.4 Infinitesimal Generator

The system probability distribution is calculated as the numerical solution of the system of equilibrium equations $\mathbf{p} \cdot \mathbf{A} = \mathbf{0}$, $\mathbf{p} \cdot \mathbf{1}^T = 1$, where $\mathbf{A}$ is the infinitesimal generator of Markov process $\mathbf{X}(t)$. Let us denote $n = \overline{0, N_1 + N_2}$ – the number of users.

The infinitesimal generator $\mathbf{A}$ has a block tridiagonal form

$$
\mathbf{A} = \begin{bmatrix}
\mathbf{N}_0 & \mathbf{\Lambda}_0 & \cdots & 0 & 0 \\
\mathbf{M}_1 & \mathbf{N}_1 & \cdots & 0 & 0 \\
\vdots & \vdots & \ddots & \vdots & \vdots \\
0 & 0 & \cdots & \mathbf{N}_{N_1+N_2-1} & \mathbf{\Lambda}_{N_1+N_2-1} \\
0 & 0 & \cdots & \mathbf{M}_{N_1+N_2} & \mathbf{N}_{N_1+N_2}
\end{bmatrix}.
$$

Blocks $\mathbf{\Lambda}_n$, $n = \overline{0, N_1\,N_2 - 1}$ have the sizes

$$
\dim \mathbf{\Lambda}_n = \begin{cases}
(2n+2) \times (2n+4), & n = \overline{0, N_2 - 1}, \\
(2N_2+2) \times (2N_2+2), & n = \overline{N_2, N_1 - 1}, \text{ if } N_1 > N_2, \\
(2(N_1 + N_2 - n) + 2) \\
\quad \times (2(N_1 + N_2 - n)), & n = \overline{N_1, N_1 + N_2 - 1}.
\end{cases}
$$

and the following form:

(1) $n = \overline{0, N_2 - 1}$

$$\mathbf{\Lambda}_n = \begin{bmatrix} \lambda & 0 & \cdots & 0 & 0 & 0 & 0 \\ 0 & \lambda & \cdots & 0 & 0 & 0 & 0 \\ \vdots & \vdots & \ddots & \vdots & \vdots & \vdots & \vdots \\ 0 & 0 & \cdots & \lambda & 0 & 0 & 0 \\ 0 & 0 & \cdots & 0 & \lambda & 0 & 0 \end{bmatrix},$$

(2) $n = \overline{N_2, N_1 - 1}$, if $N_1 > N_2$

$$\mathbf{\Lambda}_n = \begin{bmatrix} 0 & 0 & \cdots & 0 & 0 & 0 & 0 \\ 0 & \lambda & \cdots & 0 & 0 & 0 & 0 \\ \lambda & 0 & \cdots & 0 & 0 & 0 & 0 \\ 0 & \lambda & \cdots & 0 & 0 & 0 & 0 \\ \vdots & \vdots & \ddots & \vdots & \vdots & \vdots & \vdots \\ 0 & 0 & \cdots & \lambda & 0 & 0 & 0 \\ 0 & 0 & \cdots & 0 & \lambda & 0 & 0 \end{bmatrix},$$

(3) $n = \overline{N_1, N_1 + N_2 - 1}$

$$\mathbf{\Lambda}_n = \begin{bmatrix} 0 & 0 & \cdots & 0 & 0 \\ 0 & \lambda & \cdots & 0 & 0 \\ \lambda & 0 & \cdots & 0 & 0 \\ 0 & \lambda & \cdots & 0 & 0 \\ \vdots & \vdots & \ddots & \vdots & \vdots \\ 0 & 0 & \cdots & \lambda & 0 \\ 0 & 0 & \cdots & 0 & \lambda \end{bmatrix}.$$

Blocks $\mathbf{M}_n$, $n = \overline{1, N_1 + N_2}$ have the sizes

$$\dim \mathbf{M}_n = \begin{cases} (2n + 2) \times 2n, & n = \overline{1, N_2}, \\ (2N_2 + 2) \times (2N_2 + 2), & n = \overline{N_2 + 1, N_1}, \text{ if } N_1 > N_2, \\ (2(N_1 + N_2 - n) + 2) \\ \times (2(N_1 + N_2 - n) + 4), & n = \overline{N_1 + 1, N_1 + N_2}. \end{cases}$$

and the following form:

(1) $n = \overline{1, N_2}$

$$\mathbf{M}_n = \begin{bmatrix} n\mu & 0 & 0 & 0 & \cdots & 0 & 0 \\ 0 & n\mu & 0 & 0 & \cdots & 0 & 0 \\ \mu & 0 & (n-1)\mu & 0 & \cdots & 0 & 0 \\ 0 & \mu & 0 & (n-1)\mu & \cdots & 0 & 0 \\ 0 & 0 & 2\mu & 0 & \cdots & 0 & 0 \\ 0 & 0 & 0 & 2\mu & \cdots & 0 & 0 \\ \vdots & \vdots & \vdots & \vdots & \ddots & \vdots & \vdots \\ 0 & 0 & 0 & 0 & \cdots & \mu & 0 \\ 0 & 0 & 0 & 0 & \cdots & 0 & \mu \\ 0 & 0 & 0 & 0 & \cdots & n\mu & 0 \\ 0 & 0 & 0 & 0 & \cdots & 0 & n\mu \end{bmatrix},$$

(2) $n = \overline{N_2 + 1, N_1}$, if $N_1 > N_2$

$$\mathbf{M}_n = \begin{bmatrix} \mu & 0 & N_1\mu & 0 & \cdots & 0 & 0 \\ 0 & \mu & 0 & N_1\mu & \cdots & 0 & 0 \\ \vdots & \vdots & \vdots & \vdots & \ddots & \vdots & \vdots \\ 0 & 0 & 0 & 0 & \cdots & \mu & 0 \\ 0 & 0 & 0 & 0 & \cdots & 0 & \mu \\ 0 & 0 & 0 & 0 & \cdots & n\mu & 0 \\ 0 & 0 & 0 & 0 & \cdots & 0 & n\mu \end{bmatrix},$$

(3) $n = \overline{N_1 + 1, N_1 + N_2}$

$$\mathbf{M}_n = \begin{bmatrix} (n-N_1)\mu & 0 & N_1\mu & 0 & \cdots & 0 & 0 & 0 & 0 \\ 0 & (n-N_1)\mu & 0 & N_1\mu & \cdots & 0 & 0 & 0 & 0 \\ \vdots & \vdots & \vdots & \vdots & \ddots & \vdots & \vdots & \vdots & \vdots \\ 0 & 0 & 0 & 0 & \cdots & N_2\mu & 0 & (n-N_2)\mu & 0 \\ 0 & 0 & 0 & 0 & \cdots & 0 & N_2\mu & 0 & (n-N_2)\mu \end{bmatrix}.$$

Blocks $\mathbf{N}_n$, $n = \overline{0, N_1 + N_2}$ have the sizes

$$\dim \mathbf{N}_n = \begin{cases} (2n+2) \times (2n+2), & n = \overline{0, N_2 - 1}, \\ (2N_2 + 2) \times (2N_2 + 2), & n = \overline{N_2, N_1 + N_2 - 2}, \\ (2(N_1 + N_2 - n) + 2) & \\ \times (2(N_1 + N_2 - n) + 2), & n = \overline{N_1 + N_2 - 1, N_1 + N_2}. \end{cases}$$

and the following form:

(1) $n = \overline{0, N_1 - 1}$

$$\mathbf{N}_n = \begin{bmatrix} -(\lambda + n\mu + \beta) & \beta & \cdots & 0 & 0 \\ \alpha & -(\lambda + n\mu + \alpha) & \cdots & 0 & 0 \\ \vdots & \vdots & \ddots & \vdots & \vdots \\ 0 & 0 & \cdots & -(\lambda + n\mu + \beta) & \beta \\ 0 & 0 & \cdots & \alpha & -(\lambda + n\mu + \alpha) \end{bmatrix},$$

(2) $n = \overline{N_1, N_1 + N_2 - 1}$

$$\mathbf{N}_n = \begin{bmatrix} -(n\mu + \beta) & \beta & \cdots & 0 & 0 \\ \alpha & -(\lambda + n\mu + \alpha) & \cdots & 0 & 0 \\ \vdots & \vdots & \ddots & \vdots & \vdots \\ 0 & 0 & \cdots & -(\lambda + n\mu + \beta) & \beta \\ 0 & 0 & \cdots & \alpha & -(\lambda + n\mu + \alpha) \end{bmatrix},$$

(3) $n = N_1 + N_2$

$$\mathbf{N}_n = \begin{bmatrix} -(n\mu + \beta) & \beta \\ \alpha & -(n\mu + \alpha) \end{bmatrix}.$$

3 Numerical Analysis

3.1 Performance Measures

Having found the probability distribution $p(n_1, n_2, s)$, $(n_1, n_2, s) \in X$, one may compute performance measures of the considered scheme:

- Blocking probability

$$B = \sum_{i=0}^{N_2} p(N_1, i, 0) + p(N_1, N_2, 1); \tag{3}$$

- Mean bit rate

$$\bar{d} = \frac{\sum_{(n_1, n_2, s) \in X/(0,0,0),(0,0,1)} \frac{n_1 d_{\max} + n_2 d_{\max} \cdot I(s=1) + n_2 d_{\min} \cdot I(s=0)}{n_1 + n_2} \cdot p(n_1, n_2, s)}{\sum_{(n_1, n_2, s) \in X/(0,0,0),(0,0,1)} p(n_1, n_2, s)}; \tag{4}$$

- Mean bit rate on the multi-tenant band

$$\bar{d}(C_2) = \frac{\sum_{(n_1, n_2, s) \in X : n_2 \neq 0} (d_{\max} \cdot p(n_1, n_2, 1), + d_{\min} \cdot p(n_1, n_2, 0))}{\sum_{(n_1, n_2, s) \in X : n_2 \neq 0} p(n_1, n_2, s)}. \tag{5}$$

3.2 Numerical Example

Let us assume that users view short video in high quality at a bit rate $d_{\max} = 1$ Mbps. If a part of the frequency band has to be returned, the bit rate decreases to $d_{\min} = 0.5$ Mbps. So the users continue watching video but in a lower quality. The multi-tenant band goes into unavailable mode every hour (3600 s) or every four

Table 2. System parameters

Parameter description	Notation	Value
Total capacity of the single-tenant band	C_1	10 Mbps
Total capacity of the multi-tenant band	C_2	10 Mbps
Mean service time of one user	μ^{-1}	30 s
Mean time when multi-tenant band is available	α^{-1}	3540 s, 14340 s
Mean time when multi-tenant band is unavailable	β^{-1}	60 s
Maximum bit rate	d_{max}	1 Mbps
Minimum bit rate	d_{min}	0.5 Mbps
Offered load	ρ	$0 \div 30$

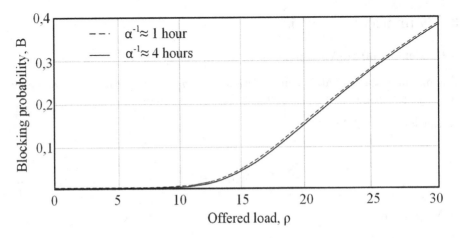

Fig. 4. Blocking probability B for different α^{-1}.

hours (14400 s) and the recovery takes around one minute. Table 2 summarizes the initial data of the example.

The figures below show the behavior of each characteristic – blocking probability B (Fig. 4), mean bit rates $\overline{d}$ and $\overline{d}(C_2)$ (Fig. 5) – for different values of α^{-1} (the mean time when the multi-tenant band is available). All figures show that the less multi-tenant band goes into unavailable mode, the better the performance metrics, namely, the blocking probability is lower, whereas the mean bit rate is higher.

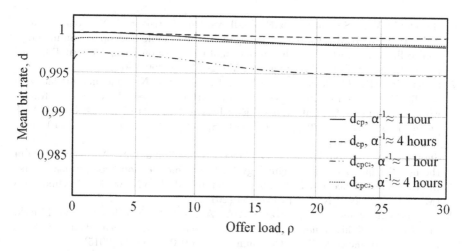

Fig. 5. Mean bit rates $\overline{d}$ and $\overline{d}(C_2)$ for different α^{-1}.

4 Conclusion

We have presented the scheme model for analyzing the simultaneous access to spectrum in 3GPP cellular network within LSA framework for intolerant to delay traffic under the limit power policy. This policy is based on the implementation of a mechanism the service bit rate degradation for the mobile operator users on multi-tenant band, if it is necessary to release the resources of this band for the owner. We have obtained the infinitesimal generator as a block tridiagonal matrix, what is required for the numerical solution of the equilibrium equations system and the calculation of the performance metrics for the considered queuing system that characterize the impact of LSA on the QoS – the blocking probability and the mean bit rate.

References

1. Cisco Visual Networking Index: Forecast and Methodology, 2015–2020 (2016)
2. ETSI: TS 103 113 Mobile broadband services in the 2300 MHz 2400 MHz band under Licensed Shared Access regime (2013)
3. Buckwitz, K., Engelberg, J., Rausch, G.: Licensed Shared Access (LSA) - regulatory background and view of Administrations. In: CROWNCOM (invited paper), pp. 413–416 (2014)
4. Ahokangas, P., Matinmikko, M., Yrjola, S., Mustonen, M., Luttinen, E., Kivimäki, A., Kemppainen, J.: Business models for mobile network operators in Licensed Shared Access (LSA). In: DYSPAN, pp. 407–412 (2014)
5. Gomez-Miguelez, I., Avdic, E., Marchetti, N., Macaluso, I., Doyle, L.E.: Cloud-RAN platform for LSA in 5G networks - tradeoff within the infrastructure. In: Communications, Control and Signal Processing, pp. 522–525 (2014)

6. Borodakiy, V.Y., Samouylov, K.E., Gudkova, I.A., Ostrikova, D.Y., Ponomarenko, A.A., Turlikov, A.M., Andreev, S.D.: Modeling unreliable LSA operation in 3GPP LTE cellular networks. In: 6th International Congress on Ultra Modern Telecommunications and Control Systems ICUMT-2014, pp. 490–496 (2014)
7. Samouylov, K., Gudkova, I., Markova, E., Yarkina, N.: Queuing model with unreliable servers for limit power policy within licensed shared access framework. In: Galinina, O., Balandin, S., Koucheryavy, Y. (eds.) NEW2AN/ruSMART -2016. LNCS, vol. 9870, pp. 404–413. Springer, Cham (2016). doi:10.1007/978-3-319-46301-8_34
8. Shorgin, S.Y., Samouylov, K.E., Gudkova, I.A., Galinina, O.S., Andreev, S.D.: On the benefits of 5G wireless technology for future mobile cloud computing. In: 1st International Science and Technology Conference Modern Networking Technologies (MoNeTec): SDN & NFV, pp. 151–154 (2014)
9. Ponomarenko-Timofeev, A., Pyattaev, A., Andreev, S., Koucheryavy, Y., Mueck, M., Karls, I.: Highly dynamic spectrum management within licensed shared access regula-tory framework. IEEE Commun. Mag. 54(3), 100–109 (2015)
10. 3GPP: TS 36.300 Evolved Universal Terrestrial Radio Access (E-UTRA) and Evolved Universal Terrestrial Radio Access Network (E-UTRAN); Overall description; Stage 2: Release 13 (2015)
11. Nunez-Queija, R.: Sojourn times in a processor sharing queue with service interruptions. Queueing Syst. 34(1), 351–386 (2000)
12. Yashkov, S.F.: Processor-sharing queues: some progress in analysis. Queueing Syst 2(1), 1–17 (1987)
13. Asmussen, S.: Applied Probability and Queues. Springer, New York (2003)

Two-Way Communication M/M/1/1 Queue with Server-Orbit Interaction and Feedback of Outgoing Retrial Calls

Velika Dragieva[1]($\boxtimes$) and Tuan Phung-Duc[2]

[1] University of Forestry, 10 Kliment Ohridsky, 1756 Sofia, Bulgaria
dragievav@yahoo.com
[2] University of Tsukuba, 1-1-1 Tennodai, Tsukuba, Ibaraki 305-8573, Japan
tuan@sk.tsukuba.ac.jp

Abstract. The paper deals with two-way communication M/M/1/1 retrial queue where the server during its idle time makes outgoing calls of two types - to the customers in orbit and to the customers outside it. Durations of these calls follow two distinct exponential distributions. After completion of the outgoing call to a customer from orbit, this customer with probability p rejoins the orbit, and with its complementary probability leaves the service area. Using generating functions approach we derive explicit and recursive formulas for the stationary system state distribution and its factorial moments.

Keywords: Two-way communication · Retrials · Server-orbit interaction · Feedback

1 Introduction

The basic characteristic of retrial queues is the behaviour of customers whose service cannot start immediately upon their arrival. These customers join a virtual waiting room, called orbit and after some time try to get service again. Retrial queues have been widely used to model diverse problems arising in telephone switching systems, telecommunication and computer networks, call centers, celular and local area networks, etc. [1,2,5,7,14,17,18]. A systematic account of the fundamental methods and the latest results, as well an classified bibliography on this topic can be found, for example in [2,11–13], and references therein.

In many real situations, especially in models with human servers, the servers in their idle time can perform some additional activity. In recent literature this additional activity is usually referred to as an outgoing call, and the models with both incoming and outgoing calls - as two-way communication queues. Queueing systems with two-way communication have been investigated in not a small number of papers. Ones of the first results on this topic are presented by Falin [9], who analyzes a single server model in which the outgoing and the incoming calls follow the same arbitrary service distribution. Later, Artalejo and

© Springer International Publishing AG 2017
A. Dudin et al. (Eds.): ITMM 2017, CCIS 800, pp. 243–255, 2017.
DOI: 10.1007/978-3-319-68069-9_20

Phung-Duc [3] extend this investigation, considering single and multiple servers retrial models with two-way communication where the service times of incoming and outgoing calls follow the exponential distribution with distinct parameters. The corresponding $M/G/1/1$ queue where the service times of incoming and outgoing calls follow two distinct arbitrary distributions is studied by Artalejo and Phung-Duc in [4], while the same model under the assumption of multiple types of outgoing calls is considered by Sakurai and Phung-Duc, [15]. The priority retrial queues with available buffers for the outgoing calls, studied in [6,10] could also be considered as two-way communication models. Deslauriers et al. [7] consider five Markovian models for blending call centers where operators not only serve incoming calls but also make calls to outside. In this article, however the retrial behavior of customers is not taken into account. Dragieva and Phung-Duc [8] consider two-way communication $M/M/1/1$ retrial queue where the server makes outgoing calls not only to customers outside the orbit (outgoing calls of type 2) but also to the customers in orbit (outgoing calls of type 1). Investigation of this model is motivated by many real situations like call centers or mobile phone where the operator can be notified about the customers in orbit, for example by registering all failed calls. The operator during his/her idle time may call to these customers to inform or to offer some different proposals. One of the main goals of the operator could be to reduce the number of customers in orbit (orbit size). In such situation it is natural to measure the operator's success or failure with a certain probability and its complementary. This motivated us to extend the model considered in [8] by introducing a feedback probability for the outgoing calls of type 1. Namely, in this paper we assume that after the service completion of an outgoing call of type 1, i.e. with a customer from the orbit, this customer returns to the orbit with probability p and with probability $(1 - p)$ leaves it, $p \in [0, 1]$. The present paper also extends [8] by assuming that both types of outgoing calls (i.e. to the orbit and to outside) follow two distinct exponential distributions.

Further on, the structure of the paper is as follows. A detailed description of the model is given in Sect. 2. Section 3 presents explicit and recursive formulas for the stationary joint distribution of the server state and orbit size. In Sect. 4 we derive explicit and recursive formulas for the partial factorial moments of this distribution and formulas for the basic macro characteristics of the system performance. Concluding remarks and some topics for future investigations are presented in Sect. 5.

2   Model Description

We consider single server retrial queue with two-way communication. This, as described in previous Section, means a queue with two flows of calls - incoming and outgoing calls. Incoming calls arrive at the system according to a Poisson process with rate λ. An incoming call is accepted for service if upon arrival it finds the server idle. Otherwise, if the server is busy, the call enters the orbit of retrial customers (calls), stays in it for an exponentially distributed time with mean $1/\mu$, and retries to get service. An arriving retrial call is accepted if the server is idle, otherwise it enters the orbit again.

On the other hand the server makes an outgoing call after some exponentially distributed idle time. There are two types of outgoing calls whose durations follow two distinct exponential distributions. The outgoing calls of the first type are directed towards the customers in the orbit, while the outgoing calls of the second type - to the customers outside it. The outgoing calls of first type are also referred to as outgoing retrial calls, while the outgoing calls of second type - as outgoing primary calls. If the server is idle it makes an outgoing retrial or primary call in an exponentially distributed time with mean $1/\alpha$ and $1/\beta$, respectively.

The service times of the incoming and outgoing retrial and primary calls are exponentially distributed with rates ν_1, ν_2 and ν_3, respectively. We accept that when the service is over the incoming calls as well as the outgoing primary calls leave the service area while, as stated in the Introduction, the outgoing retrial calls rejoin the orbit with probability p and with the complementary probability $1 - p$ leave the service area.

We assume that the arrival of incoming calls, retrial interval of incoming calls, service times of incoming and outgoing calls, and the time to make outgoing calls are mutually independent.

Let $R(t)$ denote the number of incoming calls in the orbit, and $S(t)$ - the state of the server in time t,

$$
S(t) = \begin{cases}
0, & \text{if the server is idle,} \\
1, & \text{if an incoming call is in service,} \\
2, & \text{if an outgoing call of type 1 is in service,} \\
3, & \text{if an outgoing call of type 2 is in service.}
\end{cases}
$$

It is easy to see that the process $\{(S(t), R(t)); t \geqslant 0\}$ forms a Markov chain on the state space $\{0, 1, 2, 3\} \times \mathbb{Z}_+$ where $\mathbb{Z}_+ = \{0, 1, 2, ...\}$.

In what follows we consider the system under the stability condition which will be derived later. Let

$$
\pi_{i,j} = \lim_{t \to \infty} P\left(S(t) = i, R(t) = j\right) \quad i = 0, 1, 2, 3, \ j \epsilon \mathbb{Z}_+,
$$

denote the stationary joint distribution of the server state and the orbit size.

The system of balance equations for $\{\pi_{i,j}; i = 0, 1, 2, 3, \ j \epsilon \mathbb{Z}_+\}$ is given by

$$
[\lambda + \beta + j(\alpha + \mu)] \pi_{0,j} = \nu_1 \pi_{1,j} + \nu_2 [p\pi_{2,j-1} + (1-p)\pi_{2,j}] + \nu_3 \pi_{3,j}, \quad (1)
$$

$$
(\lambda + \nu_1) \pi_{1,j} = \lambda \pi_{0,j} + (j+1)\mu\pi_{0,j+1} + \lambda \pi_{1,j-1}, \quad (2)
$$

$$
(\lambda + \nu_2) \pi_{2,j} = (j+1)\alpha\pi_{0,j+1} + \lambda \pi_{2,j-1}, \quad (3)
$$

$$
(\lambda + \nu_3) \pi_{3,j} = \beta\pi_{0,j} + \lambda \pi_{3,j-1}, \quad (4)
$$

for $j \in \mathbb{Z}_+$, where $\pi_{i,-1} = 0$ $(i = 1, 2, 3)$.

3   Stationary Distribution of the System State

In this Section, using the generating functions approach we derive explicit and recursive formulas for the joint stationary distribution of the server state and the orbit size.

3.1   Partial Generating Functions

Let $\Pi_i(z)$ denote the partial generating functions

$$\Pi_i(z) = \sum_{j=0}^{\infty} z^j \pi_{i,j}, \quad i = 0,1,2,3, \quad |z| \leqslant 1.$$

Multiplying (1)–(4) by z^j and summing up over j yields

$$(\lambda + \beta)\,\Pi_0(z) + z(\alpha + \mu)\Pi_0'(z) = \nu_1 \Pi_1(z) + \nu_2 q(z)\Pi_2(z) + \nu_3 \Pi_3(z), \quad (5)$$

$$(\lambda + \nu_1)\,\Pi_1(z) = \lambda \Pi_0(z) + \mu \Pi_0'(z) + \lambda z \Pi_1(z), \quad (6)$$

$$(\lambda + \nu_2)\,\Pi_2(z) = \alpha \Pi_0'(z) + \lambda z \Pi_2(z), \quad (7)$$

$$(\lambda + \nu_3)\,\Pi_3(z) = \beta \Pi_0(z) + \lambda z \Pi_3(z), \quad (8)$$

where

$$q(z) = 1 - (1-z)p = \bar{p} + pz,$$

$$\bar{p} = 1 - p.$$

Solving the system (5)–(8) we derive formulas for the generating functions $\Pi_i(z)$. They are presented in the next theorem.

Theorem 1. *The partial generating functions $\Pi_i(z)$ $(i = 0,1,2,3)$ are given by the following explicit expressions*

$$\Pi_0(z) = \pi_{0,0} \left(\frac{1}{1 - \tau_1 z}\right)^{\frac{D_1}{\alpha + \mu}} \left(\frac{1}{1 - \tau_2 z}\right)^{\frac{D_2}{\alpha + \mu}} \left(\frac{1}{1 - \tau_3 z}\right)^{\frac{D_3}{\alpha + \mu}}, \quad (9)$$

$$\Pi_1(z) = \left(\lambda + \mu \frac{g(z)\,(\lambda + \nu_2 - \lambda z)}{f_p(z)\,(z_3 - z)}\right) \frac{\Pi_0(z)}{\lambda + \nu_1 - \lambda z}, \quad (10)$$

$$\Pi_2(z) = \frac{\alpha g(z)}{f_p(z)\,(z_3 - z)} \Pi_0(z), \quad (11)$$

$$\Pi_3(z) = \frac{\beta \Pi_0(z)}{\lambda\,(z_3 - z)}, \quad (12)$$

where

$$D_i = (-1)^{i-1} \frac{[\lambda\,(\lambda + \nu_3 - \lambda z_i) + \beta\,(\lambda + \nu_1 - \lambda z_i)]\,(\lambda + \nu_2 - \lambda z_i)}{\lambda^2\,(z_2 - z_1)\,(z_3 - z_i)}, i = 1,2,$$

$$(13)$$

$$D_3 = \frac{\beta\,(\nu_1 - \nu_3)\,(\nu_2 - \nu_3)}{\lambda^2\,(z_1 - z_3)\,(z_2 - z_3)}, \tag{14}$$

$$g(z) = \lambda\,(\lambda - \lambda z + \nu_3) + \beta\,(\lambda - \lambda z + \nu_1), \tag{15}$$

$$f_p(z) = az^2 + (b + p\lambda\alpha\nu_2)\,z + c - p\alpha\nu_2\,(\lambda + \nu_1), \tag{16}$$

with

$$a = \lambda^2\,(\alpha + \mu), \tag{17}$$

$$b = -\lambda\,(\alpha + \mu)\,(\lambda + \nu_2 + \nu_1), \tag{18}$$

$$c = \nu_1\nu_2\,(\alpha + \mu) + \lambda\,(\mu\nu_1 + \alpha\nu_2). \tag{19}$$

Further,

$$\tau_i = \frac{1}{z_i}, \; i = 1, 2, 3,$$

$$z_3 = \frac{\lambda + \nu_3}{\lambda}, \tag{20}$$

z_1, z_2 *are the roots of the equation* $f_p(z) = 0$ *which are different real numbers for all values of the system parameters. Finally,*

$$\pi_{0,0} = \Pi_0(1)\,(1 - \tau_1)^{\frac{D_1}{\alpha + \mu}}\,(1 - \tau_2)^{\frac{D_2}{\alpha + \mu}}\,(1 - \tau_3)^{\frac{D_3}{\alpha + \mu}}, \tag{21}$$

$$\Pi_0(1) = \frac{1 - \rho}{1 - \rho + \frac{\lambda}{\nu_1} + \frac{\beta}{\nu_3}}, \tag{22}$$

with

$$\rho = \frac{\lambda}{\alpha\overline{p} + \mu}\left(\frac{\mu}{\nu_1} + \frac{\alpha}{\nu_2}\right). \tag{23}$$

Proof. We multiply Eq. (7) by $q(z)$ and then sum Eqs. (6)–(8):

$$\lambda\,[\Pi_1(z) + q(z)\Pi_2(z) + \Pi_3(z)] = (\alpha\overline{p} + \mu)\,\Pi_0'(z). \tag{24}$$

Next, from (6)–(8) we express $\Pi_i(z)$ $(i = 1, 2, 3)$ in terms of $\Pi_0(z)$ and $\Pi_0'(z)$

$$\Pi_1(z) = \frac{\lambda\Pi_0(z) + \mu\Pi_0'(z)}{\lambda + \nu_1 - \lambda z}, \tag{25}$$

$$\Pi_2(z) = \frac{\alpha\Pi_0'(z)}{\lambda + \nu_2 - \lambda z}, \tag{26}$$

$$\Pi_3(z) = \frac{\beta\Pi_0(z)}{\lambda + \nu_3 - \lambda z}.$$

The last equation coincides with (12). Thus, we get the following differential equation for $\Pi_0(z)$

$$\Pi_0'(z) = \frac{\lambda g(z)\,(\lambda + \nu_2 - \lambda z)}{f_p(z)\,(\lambda - \lambda z + \nu_3)}\Pi_0(z), \tag{27}$$

where $g(z)$ and $f_p(z)$ are given by formulas (15)–(19). Substituting $\Pi_0'(z)$ from (27) into (25)–(26) we prove formulas (10)–(11).

Further, in order to find the roots of equation $f_p(z) = 0$ we consider its discriminant,

$$\begin{aligned} \Delta_p &= (b + p\lambda\alpha\nu_2)^2 - 4a\left[c - p\alpha\nu_2\left(\lambda + \nu_1\right)\right] \\ &= p^2\lambda^2\alpha^2\nu_2^2 + p\left[2b\lambda\alpha\nu_2 + 4a\alpha\nu_2\left(\lambda + \nu_1\right)\right] + \Delta, \end{aligned}$$

where

$$\Delta = b^2 - 4ac.$$

Applying formulas (17)–(19) it is not difficult to verify that

$$\Delta = \lambda^2\left(\alpha + \mu\right)\left[\alpha\left(\nu_2 - \nu_1 - \lambda\right)^2 + \mu\left(\nu_1 - \nu_2 - \lambda\right)^2\right]$$

and

$$\Delta_p = p^2\lambda^2\alpha^2\nu_2^2 + 2p\lambda^2\left(\alpha + \mu\right)\alpha\nu_2\left(\lambda + \nu_1 - \nu_2\right) + \Delta.$$

The last expression as a function of the parameter p has discriminant

$$\begin{aligned} \tilde{\Delta} &= \left(\alpha + \mu\right)^2\alpha^2\nu_2^2\left(\lambda + \nu_1 - \nu_2\right)^2 - \alpha^2\nu_2^2\left(\alpha + \mu\right)\left[\alpha\left(\nu_2 - \nu_1 - \lambda\right)^2 + \mu\left(\nu_1 - \nu_2 - \lambda\right)^2\right] \\ &= 4\lambda\alpha^2\nu_2^2\left(\alpha + \mu\right)\mu\left(\nu_1 - \nu_2\right). \end{aligned}$$

This means that when $\nu_1 < \nu_2$, Δ_p is positive for all values of p. When $\nu_1 \geqslant \nu_2$, then we can see that the function Δ_p is increasing for non negative values of p and since $\Delta_p|_{p=0} = \Delta > 0$ this implies that Δ_p is positive for positive values of p, in particular for $p \in [0, 1]$.

Thus, the function $f_p(z)$ has two different real roots z_1 and z_2 for all possible values of the system parameters. Then the differential Eq. (27) can be expressed in the form

$$\frac{\Pi_0'(z)}{\Pi_0(z)} = \frac{1}{(\alpha + \mu)}\left(\frac{D_1}{z_1 - z} + \frac{D_2}{z_2 - z} + \frac{D_3}{z_3 - z}\right),$$

where D_i $(i = 1, 2, 3)$ and z_3 are given by (13), (14) and (20). The solution of this equation is the function $\Pi_0(z)$ given by formulas (9) and (21). Finally, to prove Eqs. (22) and (23), we express $\Pi_i(1)$ $(i = 1, 2, 3)$ according to (10)–(12) and, as it is easy to verify that

$$f_p(1) = a + b + c - p\alpha\nu_2\nu_1 = \nu_2\nu_1\left(\alpha + \mu\right)\left(1 - \bar{p}\right),$$

with

$$\bar{p} = \lambda\left(\frac{\mu}{\nu_1\left(\alpha + \mu\right)} + \frac{\alpha}{\nu_2\left(\alpha + \mu\right)}\right) + \frac{\alpha p}{\alpha + \mu},$$

and that

$$g(1) = \lambda\nu_3 + \beta\nu_1,$$

we obtain

$$\Pi_1(1) = \lambda\frac{\left(1 - \bar{p}\right)\left(\alpha + \mu\right)\nu_1\nu_3 + \mu\left(\lambda\nu_3 + \beta\nu_1\right)}{\left(\alpha + \mu\right)\left(1 - \bar{p}\right)\nu_1^2\nu_3}\Pi_0(1), \qquad (28)$$

$$\Pi_2(1) = \frac{\alpha\lambda(\lambda\nu_3 + \beta\nu_1)}{(1 - \overline{\rho})(\alpha + \mu)\nu_1\nu_2\nu_3}\Pi_0(1), \tag{29}$$

$$\Pi_3(1) = \frac{\beta}{\nu_3}\Pi_0(1). \tag{30}$$

Substituting with these expressions in the normalizing condition

$$\Pi_0(1) + \Pi_1(1) + \Pi_2(1) + \Pi_3(1) = 1,$$

after some transformations we get

$$\Pi_0(1) = \frac{1 - \overline{\rho}}{1 - \overline{\rho} + \left(\frac{\lambda}{\nu_1} + \frac{\beta}{\nu_3}\right)\frac{\alpha\overline{\rho} + \mu}{\alpha + \mu}}.$$

The last expression, together with the equation

$$\frac{1 - \overline{\rho}}{\frac{\alpha\overline{\rho} + \mu}{\alpha + \mu}} = 1 - \rho,$$

where ρ is given by (23), proves formula (22). This finishes the proof of the theorem.

Corollary 1. *The necessary and sufficient condition for the stability of the system is*

$$\rho < 1.$$

Proof. Formula (21) shows that $\pi_{0,0}$ exists if and only if both roots of the equation $f_p(z) = 0$ are greater than 1. This in turn holds if and only if

$$\left| \begin{array}{l} -\frac{b + p\lambda\alpha\nu_2}{2a} = \frac{\lambda + \nu_1 + \nu_2}{2\lambda} - \frac{p\alpha\nu_2}{\lambda(\alpha + \mu)} > 1, \\ f_p(1) = \nu_2\nu_1(\alpha + \mu)(1 - \overline{\rho}) > 0. \end{array} \right.$$

The second inequality holds if and only if $\overline{\rho} < 1$ which, in turns holds if and only if $\rho < 1$. Now, it is not difficult to verify that when $\rho < 1$ the first inequality also holds and that $\pi_{0,0} > 0$.

3.2   Stationary Joint Distribution of the Server State and the Orbit Size

In this Section we derive explicit and recursive formulas for computing the stationary joint distribution $\pi_{i,j}$ ($i = 0, 1, 2, 3$, $j \epsilon \mathbb{Z}_+$) of the server state and the orbit size. Inverting formula (9), then applying (25), (26) and (12), it is not difficult to prove the next proposition.

Proposition 1. *The stationary distribution $\pi_{i,j}$ can be calculated by the following explicit expressions:*

$$\pi_{0,j} = \pi_{0,0}\left(\sum_{k=0}^{j}\sum_{l=0}^{j-k}\left(\frac{D_1}{\alpha + \mu}\right)_k\frac{\tau_1^k}{k!}\left(\frac{D_2}{\alpha + \mu}\right)_l\frac{\tau_2^l}{l!}\left(\frac{D_3}{\alpha + \mu}\right)_{j-k-l}\frac{\tau_3^{j-k-l}}{(j-k-l)!}\right),$$

$$\pi_{1,j} = \frac{1}{\lambda + \nu_1} \sum_{k=0}^{j} [\lambda \pi_{0,k} + (k+1)\mu \pi_{0,k+1}] \left(\frac{\lambda}{\lambda + \nu_1}\right)^{j-k},$$

$$\pi_{2,j} = \frac{1}{\lambda + \nu_2} \sum_{k=0}^{j} (k+1)\alpha \pi_{0,k+1} \left(\frac{\lambda}{\lambda + \nu_2}\right)^{j-k},$$

$$\pi_{3,j} = \frac{1}{\lambda + \nu_3} \sum_{k=0}^{j} \beta \pi_{0,k} \left(\frac{\lambda}{\lambda + \nu_3}\right)^{j-k},$$

where $\pi_{0,0}$ as well as the values of D_i, τ_i are given in Theorem 1, and $(x)_j$ denotes the Pochhammer symbol,

$$(x)_j = \begin{cases} 1, & if \quad j = 0, \\ x(x+1)...(x+j-1), & if \quad j \in \mathbb{N} = \{1, 2, ...\}. \end{cases}$$

In the next proposition we derive recursive formulas for calculation of $\pi_{i,j}$. They are more convenient than the explicit ones, presented in the previous proposition.

Proposition 2. *The stationary probabilities $\pi_{i,j}$ can be computed from the following recursive formulas:*

$$\pi_{0,j} = \frac{\lambda \left(\pi_{1,j-1} + \overline{p}\pi_{2,j-1} + p\pi_{2,j-2} + \pi_{3,j-1}\right)}{j \left(\alpha\overline{p} + \mu\right)}, j = 1, 2, ..., \qquad (31)$$

$$\left(\nu_1 + \frac{\alpha\lambda\overline{p}\nu_2}{\nu_2(\alpha\overline{p}+\mu) + \lambda\mu}\right)\pi_{1,j}$$

$$= \lambda \left(1 + \frac{\beta\mu(\lambda + \nu_2)}{[\nu_2(\alpha\overline{p}+\mu) + \lambda\mu](\lambda + \nu_3)}\right)\pi_{0,j} + \lambda\pi_{1,j-1} \qquad (32)$$

$$+ \frac{\lambda\mu(\nu_2 p + \lambda)}{\nu_2(\alpha\overline{p}+\mu) + \lambda\mu}\pi_{2,j-1} + \frac{\lambda^2\mu(\lambda + \nu_2)}{[\nu_2(\alpha\overline{p}+\mu) + \lambda\mu](\lambda + \nu_3)}\pi_{3,j-1},$$

$$\left(\nu_2 + \frac{\mu\lambda\nu_1}{\nu_1(\alpha\overline{p}+\mu) + \alpha\lambda\overline{p}}\right)\pi_{2,j}$$

$$= \frac{\alpha\lambda}{\nu_1(\alpha\overline{p}+\mu) + \alpha\lambda\overline{p}}\left(\lambda + \beta\frac{\lambda + \nu_1}{\lambda + \nu_3}\right)\pi_{0,j} + \frac{\alpha\lambda^2}{\nu_1(\alpha\overline{p}+\mu) + \alpha\lambda\overline{p}}\pi_{1,j-1} \qquad (33)$$

$$+ \frac{\lambda[\nu_1(\alpha+\mu) + \alpha\lambda]}{\nu_1(\alpha\overline{p}+\mu) + \alpha\lambda\overline{p}}\pi_{2,j-1} + \frac{\alpha\lambda^2(\nu_1 + \lambda)}{[\nu_1(\alpha\overline{p}+\mu) + \alpha\lambda\overline{p}](\lambda + \nu_3)}\pi_{3,j-1}.$$

$$\pi_{3,j} = \frac{\beta}{\lambda + \nu_3}\pi_{0,j} + \frac{\lambda}{\lambda + \nu_3}\pi_{3,j-1}, j = 0, 1, ...,$$

the last one coinciding with balance Eq. (4). Here $\pi_{2,-1} = 0$, and $\pi_{0,0}$ is given in Theorem 1.

Proof. The level crossing formula (31) follows from Eq. (24). Combining this formula with balance Eqs. (2)–(3), we get

$$\frac{\nu_1(\alpha\overline{p}+\mu) + \alpha\lambda\overline{p}}{\alpha\overline{p}+\mu}\pi_{1,j} = \lambda\pi_{0,j} + \frac{\lambda\mu(\pi_{2,j}\overline{p} + \pi_{3,j})}{\alpha\overline{p}+\mu}$$

$$+ \lambda\pi_{1,j-1} + \frac{\lambda\mu p\pi_{2,j-1}}{\alpha\overline{p}+\mu},$$

$$\frac{\nu_2(\alpha\overline{p}+\mu)+\lambda\mu}{\alpha\overline{p}+\mu}\pi_{2,j} = \frac{\alpha\lambda}{\alpha\overline{p}+\mu}\left(\pi_{1,j}+\pi_{3,j}\right) + \frac{\lambda(\alpha+\mu)}{\alpha\overline{p}+\mu}\pi_{2,j-1}.$$

Here we substitute $\pi_{3,j}$ according to balance Eq. (4),

$$\frac{\nu_1(\alpha\overline{p}+\mu)+\alpha\lambda\overline{p}}{\alpha\overline{p}+\mu}\pi_{1,j} = \left(\lambda + \frac{\beta\lambda\mu}{(\lambda+\nu_3)(\alpha\overline{p}+\mu)}\right)\pi_{0,j} + \frac{\lambda\mu\overline{p}\pi_{2,j}}{\alpha\overline{p}+\mu}$$
$$+ \lambda\pi_{1,j-1} + \frac{\lambda\mu p\pi_{2,j-1}}{\alpha\overline{p}+\mu} + \frac{\lambda^2\mu\pi_{3,j-1}}{(\alpha\overline{p}+\mu)(\lambda+\nu_3)},$$

$$\frac{\nu_2(\alpha\overline{p}+\mu)+\lambda\mu}{\alpha\overline{p}+\mu}\pi_{2,j} = \frac{\alpha\lambda\beta\pi_{0,j}}{(\lambda+\nu_3)(\alpha\overline{p}+\mu)} + \frac{\alpha\lambda\pi_{1,j}}{\alpha\overline{p}+\mu}$$
$$+ \frac{\lambda(\alpha+\mu)\pi_{2,j-1}}{\alpha\overline{p}+\mu} + \frac{\alpha\lambda^2\pi_{3,j-1}}{(\lambda+\nu_3)(\alpha\overline{p}+\mu)}.$$

Now we substitute $\pi_{2,j}$ from the second into the first and $\pi_{1,j}$ from the first - into the second of the last two equations and after some transformations we get

$$\left[(\alpha\overline{p}+\mu)\nu_1 + \alpha\lambda\overline{p} - \frac{\alpha\lambda^2\mu\overline{p}}{\nu_2(\alpha\overline{p}+\mu)+\lambda\mu}\right]\pi_{1,j}$$

$$= \lambda\left(\alpha\overline{p}+\mu+\frac{\beta\mu}{\lambda+\nu_3} + \frac{\alpha\beta\lambda\mu\overline{p}}{(\lambda+\nu_3)[\nu_2(\alpha\overline{p}+\mu)+\lambda\mu]}\right)\pi_{0,j} + \lambda\left(\alpha\overline{p}+\mu\right)\pi_{1,j-1}$$

$$+ \lambda\mu\left(p+\frac{\lambda\overline{p}(\alpha+\mu)}{\nu_2(\alpha\overline{p}+\mu)+\lambda\mu}\right)\pi_{2,j-1} + \frac{\lambda^2\mu}{(\lambda+\nu_3)}\left(1+\frac{\alpha\lambda\overline{p}}{\nu_2(\alpha\overline{p}+\mu)+\lambda\mu}\right)\pi_{3,j-1},$$

$$\left[\nu_2\left(\overline{p}\alpha+\mu\right)+\mu\lambda - \frac{\alpha\lambda^2\mu\overline{p}}{\nu_1(\alpha\overline{p}+\mu)+\alpha\lambda\overline{p}}\right]\pi_{2,j}$$

$$= \frac{\alpha\lambda}{\lambda+\nu_3}\left(\beta+\frac{\lambda[(\lambda+\nu_3)(\alpha\overline{p}+\mu)+\beta\mu]}{\nu_1(\alpha\overline{p}+\mu)+\alpha\lambda\overline{p}}\right)\pi_{0,j} + \frac{\alpha\lambda^2(\alpha\overline{p}+\mu)}{\nu_1(\alpha\overline{p}+\mu)+\alpha\lambda\overline{p}}\pi_{1,j-1}$$

$$\lambda\left(\alpha+\mu+\frac{\alpha\lambda\mu p}{\nu_1(\alpha\overline{p}+\mu)+\alpha\lambda\overline{p}}\right)\pi_{2,j-1} + \frac{\alpha\lambda^2}{\lambda+\nu_3}\left(1+\frac{\lambda\mu}{\nu_1(\alpha\overline{p}+\mu)+\alpha\lambda\overline{p}}\right)\pi_{3,j-1}.$$

It is easy to verify that each of the last two equations can be divided by $(\alpha\overline{p}+\mu)$, which leads to formulas (32) and (33). The proposition is proved.

4 Basic Performance Macro Characteristics

The basic macro characteristics of the steady state system performance are the server utilization,

$$P_u = 1 - \lim_{t\to\infty} P\left(S(t)=0\right),$$

and the moments of the orbit size, in particular the mean orbit size, $\lim_{t\to\infty} E\left[R(t)\right]$. The probabilities

$$\lim_{t\to\infty} P\left(S(t)=i\right) = \Pi_i(1), \quad i=0,1,2,3$$

can be calculated by formulas (22), (28)–(30). In this section we deal with the partial factorial moments $\{M_k^i; i=0,1,2,3, k\in\mathbb{Z}_+\}$, defined as

$$M_k^i = \sum_{j=k}^{\infty}(j-k+1)_k\pi_{i,j}.$$

Obviously,

$$M_0^i = \sum_{j=0}^{\infty} \pi_{i,j} = \Pi_i(1) = \lim_{t \to \infty} P(S(t) = i), \quad i = 0, 1, 2, 3,$$

presents the stationary server state distribution which we already know. Since

$$\Pi_i(1+z) = \sum_{k=0}^{\infty} \frac{M_k^i}{k!} z^k, \quad i = 0, 1, 2, 3,$$

we can obtain M_k^i from the coefficients of z^k in the series $\Pi_i(1+z)$. Using this property and Eqs. (9), (25), (26) and (12) we express $\Pi_i(1+z)$ as a convolution of 2 or 3 series. Converting these convolutions we prove the following proposition.

Proposition 3. *The partial factorial moments are given by the following explicit formulas*

$$M_k^0 = M_0^0 k! \sum_{j=0}^{k} \sum_{l=0}^{j-k} \frac{\left(\frac{D_1}{\alpha+\mu}\right)_j \left(\frac{D_2}{\alpha+\mu}\right)_l \left(\frac{D_3}{\alpha+\mu}\right)_{k-j-l}}{j! l! (k-j-l)! (z_1-1)^j (z_2-1)^l (z_3-1)^{k-j-l}},$$

$$M_k^1 = \frac{k!}{\nu_1} \sum_{j=0}^{k} \frac{\lambda M_j^0 + \mu M_{j+1}^0}{j!} \left(\frac{\lambda}{\nu_1}\right)^{k-j},$$

$$M_k^2 = \frac{\alpha k!}{\nu_2} \sum_{j=0}^{k} \frac{M_{j+1}^0}{j!} \left(\frac{\lambda}{\nu_2}\right)^{k-j},$$

$$M_k^3 = \frac{\beta k!}{\nu_3} \sum_{j=0}^{k} \frac{M_j^0}{j!} \left(\frac{\lambda}{\nu_3}\right)^{k-j}.$$

Now, similarly to the investigation of the stationary distribution we turn our attention to a recursive scheme for computing the factorial moments.

Proposition 4. *We have the following recursive formulas for the partial factorial moments:*

$$M_{k+1}^0 = \frac{\lambda}{\alpha \overline{p} + \mu} \left(M_k^1 + M_k^2 + M_k^3 + kp M_{k-1}^2\right), \tag{34}$$

$$(1-\rho) M_k^1 = \lambda \frac{\nu_2 \nu_3 (\alpha \overline{p} + \mu) + \mu \beta \nu_2 - \alpha \lambda \nu_3}{\nu_1 \nu_2 \nu_3 (\alpha \overline{p} + \mu)} M_k^0$$

$$+ k\lambda \frac{\nu_2(\alpha \overline{p} + \mu) - \alpha \lambda}{\nu_1 \nu_2 (\alpha \overline{p} + \mu)} M_{k-1}^1 + \frac{\lambda \mu k(\lambda + p \nu_2)}{\nu_1 \nu_2 (\alpha \overline{p} + \mu)} M_{k-1}^2 + \frac{\lambda^2 \mu k}{\nu_1 \nu_3 (\alpha \overline{p} + \mu)} M_{k-1}^3, \tag{35}$$

$$(1-\rho) M_k^2 = \frac{\alpha \lambda (\lambda \nu_3 + \beta \nu_1)}{\nu_1 \nu_2 \nu_3 (\alpha \overline{p} + \mu)} M_k^0$$

$$+ \frac{\alpha k \lambda^2}{\nu_1 \nu_2 (\alpha \overline{p} + \mu)} M_{k-1}^1 + \frac{\lambda k[\nu_1(\alpha + \mu) - \lambda \mu]}{\nu_1 \nu_2 (\alpha \overline{p} + \mu)} M_{k-1}^2 + \frac{\lambda^2 \alpha k}{\nu_2 \nu_3 (\alpha \overline{p} + \mu)} M_{k-1}^3, \tag{36}$$

$$\nu_3 M_k^3 - k\lambda M_{k-1}^3 = \frac{\beta M_k^0 + k\lambda M_{k-1}^1}{\nu_3}. \tag{37}$$

Here $M_{-1}^i = 0$ $(i = 0, 1, 2, 3)$, ρ is determened by formula (23).

Proof. We differentiate formulas (24)–(26) and (12) k times at $z = 1$ and obtain

$$(\alpha\overline{p} + \mu) M_{k+1}^0 = \lambda \left(M_k^1 + M_k^2 + M_k^3 + kp M_{k-1}^2 \right),$$

$$\nu_1 M_k^1 - k\lambda M_{k-1}^1 = \lambda M_k^0 + \mu M_{k+1}^0,$$

$$\nu_2 M_k^2 - k\lambda M_{k-1}^2 = \alpha M_{k+1}^0,$$

$$\nu_3 M_k^3 - k\lambda M_{k-1}^3 = \beta M_k^0.$$

The first of these equations gives formula (34), and the last - formula (37). We substitute according to the first equation into the next two,

$$[\nu_1 (\alpha\overline{p} + \mu) - \lambda\mu] M_k^1 = \lambda (\alpha\overline{p} + \mu) M_k^0 + \lambda\mu \left(M_k^2 + M_k^3 \right)$$
$$+ k\lambda (\alpha\overline{p} + \mu) M_{k-1}^1 + \lambda\mu kp M_{k-1}^2,$$

$$[\nu_2 (\alpha\overline{p} + \mu) - \alpha\lambda] M_k^2 = \alpha\lambda \left(M_k^1 + M_k^3 \right) + [\lambda k (\alpha\overline{p} + \mu) + \lambda\alpha kp] M_{k-1}^2,$$

and, replacing M_k^3 according to (37), obtain

$$[\nu_1 (\alpha\overline{p} + \mu) - \lambda\mu] M_k^1 = \lambda \left(\alpha\overline{p} + \mu + \frac{\mu\beta}{\nu_3} \right) M_k^0 + \lambda\mu M_k^2$$
$$+ k\lambda (\alpha\overline{p} + \mu) M_{k-1}^1 + \lambda\mu kp M_{k-1}^2 + \frac{\lambda^2 \mu k}{\nu_3} M_{k-1}^3,$$

$$[\nu_2 (\alpha\overline{p} + \mu) - \alpha\lambda] M_k^2 = \alpha\lambda \left(M_k^1 + \frac{\beta}{\nu_3} M_k^0 \right) + \lambda k (\alpha + \mu) M_{k-1}^2 + \frac{\lambda^2 \alpha k}{\nu_3} M_{k-1}^3.$$

Now we substitute M_k^2 from the second into the first, and M_k^1 from the first - into the second of these equations:

$$\left[\nu_1 (\alpha\overline{p} + \mu) - \lambda\mu - \frac{\alpha\lambda^2\mu}{\nu_2(\alpha\overline{p}+\mu)-\alpha\lambda} \right] M_k^1$$

$$= \lambda \left(\alpha\overline{p} + \mu + \frac{\mu\beta}{\nu_3} + \frac{\alpha\beta\lambda\mu}{\nu_3[\nu_2(\alpha\overline{p}+\mu)-\alpha\lambda]} \right) M_k^0 + k\lambda (\alpha\overline{p} + \mu) M_{k-1}^1$$

$$+ \lambda\mu k \left(p + \frac{\lambda(\alpha+\mu)}{\nu_2(\alpha\overline{p}+\mu)-\alpha\lambda} \right) M_{k-1}^2 + \frac{\lambda\mu k}{\nu_3} \left(1 + \frac{\alpha\lambda}{\nu_2(\alpha\overline{p}+\mu)-\alpha\lambda} \right) M_{k-1}^3,$$

$$\left[\nu_2 (\alpha\overline{p} + \mu) - \alpha\lambda - \frac{\alpha\lambda^2\mu}{\nu_1(\alpha\overline{p}+\mu)-\lambda\mu} \right] M_k^2$$

$$= \frac{\alpha\lambda}{\nu_3} \left(\beta + \frac{\lambda[\nu_3(\alpha\overline{p}+\mu)+\mu\beta]}{\nu_1(\alpha\overline{p}+\mu)-\lambda\mu} \right) M_k^0 + \frac{\alpha k\lambda^2(\alpha\overline{p}+\mu)}{\nu_1(\alpha\overline{p}+\mu)-\lambda\mu} M_{k-1}^1$$

$$+ \lambda k \left[(\alpha + \mu) + \frac{\alpha\lambda\mu p}{\nu_1(\alpha\overline{p}+\mu)-\lambda\mu} \right] M_{k-1}^2 + \frac{\lambda^2\alpha k}{\nu_3} \left(1 + \frac{\lambda\mu}{\nu_1(\alpha\overline{p}+\mu)-\lambda\mu} \right) M_{k-1}^3.$$

Here, like in the proof of Proposition 2 we can divide both equations by $(\alpha\overline{p} + \mu)$, which leads to the following formulas

$$\left(\nu_1 - \frac{\lambda\mu\nu_2}{\nu_2(\alpha\overline{p}+\mu)-\alpha\lambda} \right) M_k^1 = \lambda \left(1 + \frac{\mu\beta\nu_2}{\nu_3[\nu_2(\alpha\overline{p}+\mu)-\alpha\lambda]} \right) M_k^0$$

$$+ k\lambda M_{k-1}^1 + \frac{\lambda\mu k(\lambda+p\nu_2)}{\nu_2(\alpha\overline{p}+\mu)-\alpha\lambda} M_{k-1}^2 + \frac{\lambda^2\mu k\nu_2}{\nu_3[\nu_2(\alpha\overline{p}+\mu)-\alpha\lambda]} M_{k-1}^3,$$

$$\left(\nu_2 - \frac{\alpha\lambda\nu_1}{\nu_1(\alpha\overline{p}+\mu)-\lambda\mu}\right) M_k^2 = \frac{\alpha\lambda(\lambda\nu_3+\beta\nu_1)}{\nu_3[\nu_1(\alpha\overline{p}+\mu)-\lambda\mu]} M_k^0$$

$$+ \frac{\alpha k\lambda^2}{\nu_1(\alpha\overline{p}+\mu)-\lambda\mu} M_{k-1}^1 + \frac{\lambda k[\nu_1(\alpha+\mu)-\lambda\mu]}{\nu_1(\alpha\overline{p}+\mu)-\lambda\mu} M_{k-1}^2 + \frac{\lambda^2\alpha k\nu_1}{\nu_3[\nu_1(\alpha\overline{p}+\mu)-\lambda\mu]} M_{k-1}^3.$$

It is easy to verify that these equations can be presented in the form, given by (35) and (36), which finishes the proof of the proposition.

For $k = 1$ formulas (34)–(37) give a simple procedure to calculate the mean orbit size.

Remark 1. For $p = 0$ and $\nu_2 = \nu_3$ all results presented in this paper coinside with the results obtained by Dragieva and Phung-Duc [8] for the corresponding model without feedback, and with the same exponential distribution for outgoing calls of both types.

5 Conclusion and Future Work

In this paper we analyze $M/M/1/1$ retrial queue with two-way communication in which the server makes outgoing calls of two types - to the customers in orbit which are referred to as outgoing retrial calls, and to the customers outside the orbit. Durations of the outgoing calls of both types follow two distinct exponential distributions. In addition, after the service completion any outgoing retrial customer returns to the orbit with certain probability p. We derive explicit and recursive formulas for the stationary joint distribution of the server state and the orbit size and its factorial moments. This investigation could be extended by considering the corresponding model with feedback not only for the outgoing retrial calls. We are also planing to investigate the corresponding model in which the durations (service times) of the outgoing calls of both types follow two distinct arbitrary distributions. The corresponding model with finite population (a finite number of customers) could also be investigated.

References

1. Aguir, S., Karaesmen, E., Aksin, O., Chauvet, F.: The impact of retrials on call center performance. OR Spectrum **26**, 353–376 (2004)
2. Artalejo, J., Gómez-Corral, A.: Retrial Queueing Systems: A Computational Approach. Springer, Heidelberg (2008)
3. Artalejo, J., Phung-Duc, T.: Markovian retrial queues with two way communication. J. Ind. Manag. Optim. **8**, 781–806 (2012)
4. Artalejo, J., Phung-Duc, T.: Single server retrial queues with two way communication. Appl. Math. Model. **37**, 1811–1822 (2013)
5. Choi, B., Shin, Y.W., Ahn, W.C.: Retrial queues with collision arising from unslotted CSMA/CD protocol. Queueing Syst. **11**(4), 335–356 (1992)
6. Choi, B., Choi, K., Lee, Y.: M/G/1 retrial queueing systems with two types of calls and finite capacity. Queueing Syst. **19**, 215–229 (1995)

7. Deslauriers, A., L'Ecuyer, P., Pichitlamken, J., Ingolfsson, A., Avramidis, A.: Markov chain models of a telephone call center with call blending. Comput. Oper. Res. **34**, 1616–1645 (2007)

8. Dragieva, V., Phung-Duc, T.: Two-way communication M/M/1 retrial queue with server-orbit interaction. In: Proceedings of the 11th International Conference on Queueing Theory and Network Applications, (QTNA 2016) (ACM Digital Library), 7 pages (2016). doi:10.1145/3016032.3016049

9. Falin, G.: Model of coupled switching in presence of recurrent calls. Eng. Cybern. Rev. **17**, 53–59 (1979)

10. Falin, G., Artalejo, J., Martin, M.: On the single server retrial queue with priority customers. Queueing Syst. **14**, 439–455 (1993)

11. Falin, G., Templeton, J.: Retrial Queues. Chapman and Hall, London (1997)

12. Gómez-Corral, A., Phung-Duc, T.: Retrial queues and related models. Ann. Oper. Res. **247**(1), 1–2 (2016)

13. Kim, J., Kim, B.: A survey of retrial queueing systems. Ann. Oper. Res. **247**(1), 3–36 (2016)

14. Phung-Duc, T., Kawanishi, K.: Performance analysis of call centers with abandonment, retrial and after-call work. Perform. Eval. **80**, 43–62 (2014)

15. Sakurai, H., Phung-Duc, T.: Two-way communication retrial queues with multiple types of outgoing calls. TOP **23**(2), 466–492 (2015)

16. Sakurai, H., Phung-Duc, T.: Scaling limits for single server retrial queues with two-way communication. Ann. Oper. Res. **247**(1), 229–256 (2016)

17. Tran-Gia, P., Mandjes, M.: Modeling of customer retrial phenomenon in cellular mobile networks. IEEE J. Sel. Areas Commun. **15**, 1406–1414 (1997)

18. Van Do, T., Wochner, P., Berches, T., Sztrik, J.: A new finite-source queueing model for mobile cellular networks applying spectrum renting. Asia - Pac. J. Oper. Res. **31**, 14400004_1–14400004_19 (2014)

The Comparison of Structured Modeling and Simulation Modeling of Queueing Systems

Igor Yakimov[1], Alexander Kirpichnikov[2], Vladimir Mokshin[1,3(✉)],
Zuhra Yakhina[1], and Rustem Gainullin[2]

[1] Kazan National Research Technical University named after A. N. Tupolev KAI,
Karl Marx Street 10, 420111 Kazan, Republic of Tatarstan, Russian Federation
vladimir.mokshin@mail.ru
[2] Kazan National Research Technological University,
Karl Marx Street 68, 420015 Kazan, Republic of Tatarstan, Russian Federation
kirpichnikov@kstu.ru
[3] Automation and Control System Laboratory, Siemens Engineering Center,
Chetaev Street 18, 420111 Kazan, Republic of Tatarstan, Russian Federation
http://www.knrtu.ru
http://www.kai.ru

Abstract. The paper provides the description of 13 structured and simulation modeling systems: AnyLogic, Arena, Bizagi Modeler, Business Studio, Enterprise Dynamics, ExtendSim, Flexsim, GPSS W, Plant Simulation, Process Simulator, Rand Model Designer, Simio Simul8. The routes of dynamic objects movement in modeling systems in structured models built in these SSMS are visually represented. SSMS are compared according to structural models of $M/M/5$ queuing systems obtained in these SSMS and the difference of simulation modeling from analytical modeling results. The reliability was assessed by the values of mathematical expectation and standard deviations of quantity and time indexes. The paper aims to select SSMS for modeling probabilistic objects in conformity with the area the object refers to, consideration of simulation modeling results credibility, and users personal preferences as well.

Keywords: Simulated model · Analytical model · Queueing system · $M/M/5$

1 Introduction

Significant developments in simulation modeling (SM) have taken place recently; they are primarily related to the transference of SM specialized languages application to structural and simulation modeling systems (SSMS). SSMS allow users to abandon model programming and start model drawing. The programming process is the responsibility of software designers creating imitation subprograms of elements functioning in modeling objects; they are presented by specialized libraries.

© Springer International Publishing AG 2017
A. Dudin et al. (Eds.): ITMM 2017, CCIS 800, pp. 256–267, 2017.
DOI: 10.1007/978-3-319-68069-9_21

The monograph [1] gives a detailed review of SM systems as of late 1995; more than 20 years after its publication have passed and a great deal of new advanced SSMS have appeared. The article [2] provides more than 10 promising SSMS with their application ratio. This article compares 13 SSMS, accommodates their brief descriptions, supplies structural models of $M/M/5$ queueing system, identifies SM reliability results obtained in these systems and compares them with analytical modeling (AM) results of $M/M/5$ queueing system. Such approach further develops the study done by the authors of this article [3,4] and from their point of view, it will foster a qualified selection of modeling means within the subject area the modeling object related to, reliability of results and users personal preferences. 13 systems have been chosen from the set of SSMS given in [2]; there is no programming process for large scale users in them, their free versions are available on the Internet: AnyLogic, Arena, Bizagi Modeler, Business Studio, Enterprise Dynamics, ExtendSim, Flexsim, GPSS W, Plant Simulation, Process Simulator, Rand Model Designer, Simio Simul8.

1.1 The Brief Description of SSMS

AnyLogic. AnyLogic SSMS [5] was designed by the Russian company XJ Technologies. The first version of AnyLogic system 4.0 was created in 2003. AnyLogic 7.0 was made in 2014. AnyLogic SSMS includes graphical modeling language and allows the user to maximize created models with the help of Java. The relation of the concepts accepted in AnyLogic SSMS to the concepts accepted in the queueing system theory is the following: claims - entities, queues - queues, service machines - tasks. There are many references including [5].

Arena. Arena SSMS [6] was fashioned by Systems Modeling Corporation. Its first version appeared in 1993. In 2014 Arena SSMS 3.0 was developed. The foundation of Arena includes modeling metacompiler Siman and animation system Cinema Animation. The relation of the concepts accepted in the system to the concepts accepted in the queueing system theory is the following: claims - entities, queues-queues [7], service machines - tasks [8]. The key advantage of Arena SSMS is the possibility to transfer automatically from IDEF3 diagram, widely spread in BPwin [9], to a structured model in Arena SSMS.

Bizagi Modeler. Bizagi Modeler SSMS [10] was designed by a group of companies Object Management Group established in 1989. Bizagi Modeler SSMS was developed in 2007. In 2016 the 11th version of Bizagi BPM Suite was worked out. Business-process modeling notation is used to indicate modeled system elements (BPMN 2.0). The relation of the concepts accepted in Bizagi Modeler to the concepts accepted in the queueing system theory is the following: claims - messages, queues-queues, service machines - actions. Gateways are used to indicate the route selection of entities movements.

Business Studio. Business Studio SSMS [11] was elaborated by a group of companies Modern management technologies in 2004; in 2013 the version 4.0 was worked out. Business- process modeling notation is used to indicate modeled system elements (BPMN 2.0). The relation of the concepts accepted in Business Studio SSMS to the concepts accepted in the queueing system theory is the following: claims - messages, queues-queues, service machines - actions. Gateways are used to indicate the route selection of entities movements.

Enterprise Dynamics. Enterprise Dynamics SSMS was generated by InControl Simulation, founded in 1998. Enterprise Dynamics system was developed in 2004 [12]. The concepts accepted in Enterprise Dynamics SSMS are the following: claims–products, queue–queues, service machines–servers.

ExtendSim. The first version of ExtendSim SSMS [13] was designed by Imagine That Inc. in the beginning of 1987. It was one of the first programming products on the SM market which could create a simulated model by SSMS structural scheme. Currently ExtendSim SSMs is being upgraded. The last version was launched on the market in 2015. The relation of the concepts accepted in ExtendSim SSMS to the concepts accepted in the queueing system theory is the following: claims - elements, queues-queues, service machines activities [14].

Flexim. Flexim SSMS [15] was produced by Flxim Software Products Inc (ESP), set up in 1993. The year of Flexim development is 2003. The relation of the concepts accepted in Flexim SSMS to the concepts accepted in the queueing system theory is the following: claims - claims, queues-queues, service machines processors, memory [15].

GPSS W with Exceeded Editor. In 2014 Elina-Computer created an exceeded editor of simulated models for GPSS W SSMS [16]. Exceeded editor for mass user allows to abandon programming and to start drawing models. Software designers can create software for modeling new elements and add them into structured and simulated models. In addition, exceeded editor allows designers to plan simulated tests and process modeling findings. The relation of the concepts accepted in GPSS SSMS to the concepts accepted in the queueing system theory is the following: claims - transactions, queues - queues, service machines processors, memory [16].

Plant Simulation. Plant Simulation SSMS [17] has been supplied by Siemens PLM Software since 2007. In 2016 the 13th version of Plant simulation SSMS appeared. The relation of the concepts accepted in Plant Simulation SSMS to the concepts accepted in the queueing system theory is the following: claims - details, queues storage facilities, service machines occasional operations, in-parallel operations [17]. There is an opportunity to model manufacturing processes with liquid products.

Process Simulator. Process Simulator SSMS was elaborated by ProModel Corporation and appeared in the market in 2001 [18]. The last version of Process Simulator SSMS is version 9.3.0.2701 developed in 2016. The relation of the concepts accepted in Process Simulator SSMS to the concepts accepted in the queueing system theory is the following: claims - claims, queues - queues, service machines actions.

Rand Model Designer. A well-known Model Vision Studium SSMS changed its name, and since 2011 it has been called Rand Model Designer [19]. It was developed by MVSTUDIUM Group founded in 1997. The last version of Rand Model Designer (RMD) SSMS was designed in 2016 based on Modelica modeling language [19]. The relation of the concepts accepted in Rand Model Designer SSMS to the concepts accepted in the queueing system theory is the following: claims - transactions, queues - queues, service machines services.

Simio. Simio SSMS was generated in 2007 [20] by Simio LLC, established in 2005. The concepts accepted in Simio SSMS are the following: claims - agents (initiators), queues - queues, service machines - servers.

Simul8. The full version of Simul8 SSMS [21] was created by the corporation of the same name in the beginning of 2003. Currently Simul8 SSMS is being developed. The last version of this product appeared in 2017. Routs of claims movement and processes of their service are specified in details in Simul8 SSMS [21]. The concepts accepted in Simul8 SSMS are the following: claims - entities, queues - queues, servicing machines - work centers. Simul8 SSMS test version can be downloaded from the Internet free of charge; the trial period is up to 14 days.

2   Analytical Modeling of $M/M/5$ Queueing System

Analytical modeling of $M/M/5$ has been done by queueing system formula given in the book [22] for $M/M/m$ and has been calculated according to indicators of queueing system functioning. For calculations the quantity of servicing machines $m = 5$, average time between claims inflows $\bar{t}_{infl} = 10$ time units, average time of claim servicing $\bar{t}_{serv} = 30$ time unit were accepted.

1. Given density of claims inflows into the system:

$$\rho = \frac{\bar{t}_{serv}}{\bar{t}_{infl}} = \frac{30}{10} = 3$$

2. The probability of application absence in the system:

$$P_0 = \frac{1}{1 + \rho + \frac{\rho^2}{2!} + \frac{\rho^3}{3!} + \frac{\rho^4}{4!} + \frac{\rho^5}{4!(m - \rho)}} = \frac{1}{1 + 3 + \frac{9}{2} + \frac{27}{6} + \frac{81}{24} + \frac{243}{24(5 - 3)}} = 0.0466$$

3. Waiting probability:

$$P_{wait} = \frac{\rho^m \cdot P_0}{(m-1)!(m-\rho)} = \frac{3^5 \cdot 0.0466}{4! \cdot 2} = 0.236$$

4. Average quantity of claims in servicing machine:

$$\overline{m} = \rho = 3$$

5. Average quantity of claims in the queue:

$$\overline{l} = \frac{\rho \cdot P_{wait}}{m-\rho} = \frac{3 \cdot 0.236}{5-3} = 0.354$$

6. Average quantity of claims in the system:

$$\overline{k} = \frac{\rho(m-\rho+P_{wait})}{m-\rho} = \frac{3 \cdot (5-3+0.236)}{5-3} = 3.354$$

7. Average waiting time of claims in the queue:

$$\overline{t}_{wait} = \frac{\overline{t}_{serv} \cdot P_{wait}}{m-\rho} = \frac{30 \cdot 0.236}{5-3} = 3.542$$

8. Average time of claims inflows into the system:

$$\overline{t}_{inflow} = \frac{\overline{t}_{serv}(m-\rho+P_{wait})}{m-\rho} = \frac{30 \cdot (5-3+0.236)}{5-3} = 33.542$$

9. The dispersion of claims quantity in the servicing machine:

$$\sigma_m^2 = \rho(1-P_{wait}) = 3 \cdot (1-0.236) = 2.292$$

10. The dispersion of claims quantity in the queue:

$$\sigma_l^2 = \frac{(m+\rho) \cdot \overline{l}}{(m-p)^2} - \overline{l}^2 = \frac{(5+3) \cdot 0.354}{(5-3)^2} - 0.354^2 = 0.583$$

11. Covariation:

$$K_{ml} = \rho \cdot Pwait = 3 \cdot 0.236 = 0.708$$

12. The dispersion of claims quantity in the system:

$$\sigma_k^2 = \sigma_m^2 + \sigma_l^2 + 2K_{ml} = 2.292 + 0.583 + 1.416 = 4.291$$

13. The dispersion of claims servicing time in the servicing machine:

$$\sigma^2_{serv} = \overline{t}^2_{serv} = 30^2 = 900$$

14. The dispersion of claims waiting time in the queue:

$$\sigma^2_{serv} = \frac{\overline{t}^2_{serv} \cdot (2-P_{wait})}{(m-\rho)^2} = \frac{30^2 \cdot 0.236 \cdot (2-0.236)}{(5-3)^2} = 93.668$$

15. The dispersion of claims being time in the system:

$$\sigma_{time}^2 = \frac{\bar{t}_{serv}^2 \cdot ((m - \rho)^2 + P_{wait} \cdot (2 - P_{wait}))}{(m - \rho)^2}$$

$$= \frac{30^2 \cdot ((5 - 3)^2 + 0.236 \cdot (2 - 0.236))}{(5 - 3)^2} = 993.668$$

Table 1 provides the findings of $M/M/5$ queueing system analytical modeling

3 Structured Models of $M/M/5$ Queueing System

Structured models of M/M/5 queueing system were developed in AnyLogic, Arena, Bizagi Modeler, Business Studio, Enterprise Dynamics, ExtendSim, Flexsim, GPSS W, Plant Simulation, Process Simulator, Rand Model Designer, Simio and Simul8. Structured models of $M/M/5$ queueing system in AnyLogic, Flexsim and Plant Simulation are presented in Figs. 1, 2 and 3 as examples.

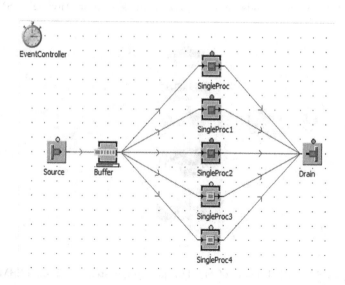

Fig. 1. Structured model of M/M/5 queuing system in Plant Simulation SSMS

Other structured models of $M/M/5$ queueing system are similar.
Based on the images of structured models of queueing system the following:

1. All 13 SSMS, the structured models of $M/M/5$ queueing system (QS) of which are given in drawings, illustrate the routes of claims movements with the help of arrows.
2. AnyLogic and Arena SSMS allow to specify the selection criteria of servicing machine from prescribed collection by the Select block.

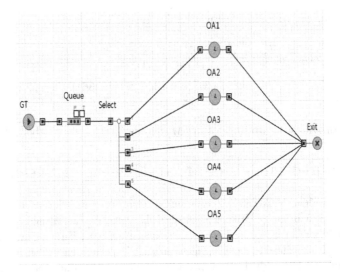

Fig. 2. Structured model of M/M/5 queueing system in AnyLogic SSMS

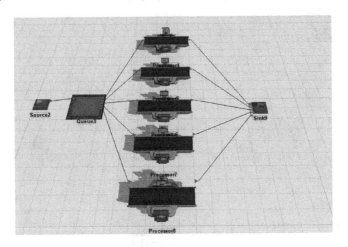

Fig. 3. Structured model of M/M/5 queueing system in Flexsim SSMS

3. The change criteria of claims movement routes can be indicated fully in Bizagi Modeler and Business Studio SSMS which employ the notation of BPMN business process modeling.
4. The modeled processes are shown conclusions can be made fully in AnyLogic system.

The basic graphical elements of SSMS allow us to draw the following conclusions:

1. Bizagi Modeler and Business Studio SSMS should be used for modeling probabilistic objects, element functioning in which can be described in a relatively

simple way. Structured models in these systems show the claims movement process in a modeled object in a clear view. Mastering Bizagi Modeler and Business Studio SSMS is not difficult.

2. It is preferable to apply AnyLogic SSMS for modeling probabilistic objects in which complex functioning processes of their elements must be shown. Structured models in this system show both claims movement in the modeled object and procedures of their processing. AnyLogic SSMS is much more difficult for mastering than Bizagi Modeler and Business Studio SSMS.

3. The list of graphical elements in Bizagi Modeler and Business Studio SSMS contains 21 elements. The list of graphical elements in AnyLogic includes 23 elements. 5–6 elements from the given SSMS are enough to start with.

3.1   Comparison of Simulation and Analytical Modeling Findings

The findings reliability assessment of simulation modeling (SM) of $M/M/5$ queueing system can be calculated by comparing their results with analytical modeling (AM) results according to the average value differences using formula:

$$\triangle_{ij} = \frac{(y_{ij}^* - y_i)}{y_i} \cdot 100, \tag{1}$$

where y_{ij}^2 is the estimation of i parameter calculated by simulation modeling findings in the j of SSMS; y_{ij}^* is the value of i parameter calculated by AM findings.

It is accepted for SM: the quantity of servicing machines $m = 5$, the average time between claims entering is $\bar{t}_{infl} = 10$ of time units, the average time of claims service is $\bar{t}_{serv} = 30$ of time units. The quantity of serviced claims is 25000.

Table 1 shows AM and SM in AnyLogic, Arena, Bizagi Modeler and Business Studio systems and evaluation of their differences in percentage terms by Eq. (1). SM in AnyLogic - SMAn, SM in Arena - SMA, SM in Business Studio - SMBS, SM in Bizagi Modeler SMBM.

Table 2 illustrates the findings of AM and SM in Enterprise Dynamics, ExtendSim, Flexsim GPSS W systems and their difference evaluation in percentage terms Eq. (1).

SM in Enterprise Dynamics has the abbreviation SMED, SM in ExtendSim - SMES, SM in Flexsim - SMF, SM in GPSS W - SMGPSS.

Table 3 identifies findings of AM and SM in Plant Simulation, Process Simulator, Rand Model Designer, Simio Simul8 systems and their difference evaluation in percentage terms Eq. (1).

In Table 3 7SM in Plant Simulation has the abbreviation SMPS, SM in Process Simulator - SMPrS, SM in Rand Model - SMRM, SM in Simio - SMS, SM in Simul8 SMSim8.

According to the findings shown in Tables 1, 2 and 3, the following conclusions can be made:

Table 1. The evaluation findings of main functioning indexes of $M/M/5$ queueing system in SSMS: AnyLogic, Arena, Bizagi odeler Business Studio by 6 parameters.

Name	AM	SMAn	Δ	SMA	Δ	SMBM	Δ	SMBS	Δ
$\bar{l}$	0.354	0.34	3.954	0.355	0.282	0.352	0.565	0.369	4.237
$\bar{m}$	3.00	3.09	3.00	3.152	5.067	3.075	2.5	3.125	4.167
$\bar{k}$	3.354	3.43	2.265	3.507	4.562	3.427	2.177	3.494	4.174
$\bar{t}_{wait}$	3.542	3.5	1.185	3.681	1.92	3.52	0.621	3.664	3.4
$\bar{t}_{serv}$	30.00	30.9	3.00	30.85	2.827	30.79	2.633	31.02	3.4
$\bar{t}_{inflow}$	33.542	34.4	2.557	34.531	2.325	34.31	2.289	34.684	3.405
Average difference in % by 6 tests			2.66		2.83		1.798		3.805

Table 2. The evaluation findings of the main functioning indexes of M/M/5 queueing system in SSMS: Enterprise Dynamics, ExtendSim, Flexsim and GPSS W by 6 parameters

Name	AM	SMED	Δ	SMES	Δ	SMF	Δ	SMGPSS	Δ
$\bar{l}$	0.354	0.342	3.389	0.346	2.26	0.366	3.389	0.346	2.312
$\bar{m}$	3.00	2.961	1.3	2.97	1.00	3.11	3.666	3.015	0.5
$\bar{k}$	3.354	3.303	1.521	3.316	1.133	3.476	3.637	3.362	0.239
$\bar{t}_{wait}$	3.542	3.474	2.033	3.441	2.851	3.679	3.733	3.463	2.23
$\bar{t}_{serv}$	30.00	30.095	0.317	29.58	1.401	31.12	3.867	30.17	0.567
$\bar{t}_{inflow}$	33.542	33.327	0.641	33.021	1.554	34.799	3.747	33.633	0.271
Average difference in % by 6 tests			1.534		1.7		3.673		1.02

Table 3. The evaluation findings of the main functioning indexes of M/M/5 queueing system in SSMS: Plant Simulation, Process Simulator, Rand Model Designer, Simio and Simul8 by 6 parameters

Name	AM	SMPS	Δ	SMPrS	Δ	SMRM	Δ	SMS	Δ	SMSim	Δ
$\bar{l}$	0.35	0.37	4.52	0.34	2.82	0.37	3.95	0.34	4.15	0.36	1.7
$\bar{m}$	3.00	3.05	1.53	2.98	0.73	2.95	1.77	2.98	0.64	3.0	0.03
$\bar{k}$	3.35	3.43	2.36	3.32	0.95	3.45	2.98	3.32	1.01	3.36	0.21
$\bar{t}_{wait}$	3.6	3.6	1.64	3.5	1.15	3.47	2.1	3.39	0.7	3.6	1.66
$\bar{t}_{serv}$	30.00	30.76	2.54	29.96	0.13	30.22	0.72	29.79	4.29	29.94	0.18
$\bar{t}_{inflow}$	33.54	34.0	1.37	33.46	0.24	34.13	1.75	33.18	1.08	33.55	0.01
Average difference in % by 6 tests			2.33		1.0		2.2		1.98		0.63

1. The compared findings are thought to be satisfactory if the SM and AM difference doesn't exceed 5%. According to this principle, one dissatisfactory evaluation has been obtained in Arena system. It can be thought that all findings are acceptable.
2. The average difference in percentage according to 6 tests for all 13 SSMS does not exceed 5%, so the findings are satisfactory.
3. SSMS according to SM test validity by the mean difference between AM and SM in percentage due to 6 tests can be selected by the graded list in which the average difference in percentage for SSMS is given: Simul8 (0.633), Process Simulator (1.004), GPSS W (1.020), Enterprise Dynamics (1.534), ExtendSim (1.700), Bizagi Modeler (1.798), Simio (1.978), Rand Model Designer (2.211), Plant Simulation (2.326), AnyLogic (2.660), Arena (2.830), Flexsim (3.673), Business Studio (3.805).

4 Conclusion

The paper provides the comparison of 13 SSMS which allow conventional users, not software designers, to create structured and simulated models without programming structured schemes of modeling objects. Software designers merely have to code element functioning of modeled objects and new ways of modeling finding processing. To compare SSMS structured, simulated and analytical models of queueing system (particularly $M/M/5$ queueing system) have been used. In accordance with obtained findings, the conclusion can be made. By virtue of these figures SSMS can be selected in accordance with the area the modeled object is related to and users preference.

1. All 13 SSMS, the structured models of $M/M/5$ queueing system (QS) of which are given in Figs. 1, 2 and 3, illustrate the routes of claims movements with the help of arrows. Based on these figures, SSMS can be selected in accordance with the area the modeled object is related to and users preference.
2. The criteria of claims movement route changes can be indicated fully in Bizagi Modeler and Business Studio SSMS, which employ notation of BPMN business process modeling. The list of graphical elements in BPMN can be used for creation structured models that are essential at the initial stage of learning these SSMS.
3. Modeled processes are indicated in detail in AnyLogic SSMS. The main graphical elements of Anylogic SSMS allow users to employ them as a hint in.
4. The average difference between SM and AM does not exceed 5%, so the findings are satisfactory.
5. By finding validation on average difference between AM and SM in percentage SSMS can be selected in compliance with the grading list, in which the average difference for SSMS is given in percentage: Simul8 (0.633), Process Simulator (1.004), GPSS W (1.020), Enterprise Dynamics (1.534), ExtendSim (1.700), Bizagi Modeler (1.798), Simio (1.978), Rand Model Designer (2.211), Plant Simulation (2.326), AnyLogic (2.660), Arena (2.830), Flexsim (3.673), Business Studio (3.805).

References

1. Kindler, E.: The modeling languages. In: Proceedings of the Workshop on Business Process Reference, BPRM (2005)
2. Borshev, A.V.: Simulation modeling: area state in 2015, tendency. In: Studies of the 7th Russian Scientific Practical Conference: Simulation Modeling. Theory and Practice, pp. 14–22 (2015). (in Russian)
3. Yakimov, I.M., Kirpichnokov, A.P., Zainullina, G.P., Yakhina, Z.T.: Validation evaluation of simulation modeling results by analytical modeling findings. Bull. Kazan Techn. Univ. **18**(6), 173–178 (2015). Kazan. (in Russian)
4. Yakimov, I.M., Kirpichnikov, A.P., Zainullina, G.P., Isaeva, Y.G., Yakhina, Z.T.: Informational system o simulation and analytical modeling of queueing system. Bull. Kazan Techn. Univ. **19**(5), 141–145 (2016). Kazan. (in Russian)
5. Mokshin, V.V., Yakimov, V.V.: System Modeling in AnyLogic: Tutorial Recomendation to Perform Laboratory Work. Shkola Publ., Kazan (2014). (in Russian)
6. Tayfur, A., Melamed, B.: Simulation Modeling and Analysis with ARENA. Elsevier, Inc., Amsterdam (2007)
7. Yue, W., Takahashi, Y., Takagi, H.: Advances in Queueing Theory and Network Applications. Springer, New York (2009). doi:10.1007/978-0-387-09703-9
8. Alfa, A.S.: Queueing Theory for Telecommunications. Springer, Boston (2010). doi:10.1007/978-1-4419-7314-6
9. Schriber, T.J.: Introduction to Simulation. Wiley, New York (1991)
10. Yakimov, I.M., Kirpichnikov, A.P., Mokshin, V.V., Alyautdinova, G.R., Paigina, L.R.: Simulation modeling of business processes in Bizagi Modeler. Bull. Kazan Technol. Univ. **18**(9), 236–239 (2015). Kazan. (in Russian)
11. Basic elements of queuering theory. Application to the Modelling of Computer Systems, France (2015)
12. Guerreiro, S., Vasconcelos, A., Tribolet, J.: Enterprise dynamic systems control enforcement of run-time business transactions. In: Albani, A., Aveiro, D., Barjis, J. (eds.) EEWC 2012. LNBIP, vol. 110, pp. 46–60. Springer, Heidelberg (2012). doi:10.1007/978-3-642-29903-2_4
13. Zabawa, J., Radosiński, E.: Comparison of discrete rate modeling and discrete event simulation. methodological and performance aspects. In: Świątek, J., Wilimowska, Z., Borzemski, L., Grzech, A. (eds.) Information Systems Architecture and Technology. AISC, vol. 523, pp. 153–164. Springer, Cham (2017). doi:10.1007/978-3-319-46589-0_12
14. Gross, D., Harris, C.M.: Fundamentals of Queueing Theory, 2nd edn. Wiley, New York (1985)
15. Garrido, J.M.: Introduction to flexsim. In: Garrido, J.M. (ed.) Object Oriented Simulation. Springer, Boston (2009). doi:10.1007/978-1-4419-0516-1_3
16. Devyatkov, V.V.: Exceeded Editor of GPSS World: Main Possibilities. Print-Service Publ., Moscow (2013). (in Russian)
17. Bangsow, S.: Manufacturing Simulation with Plant Simulation and Simtalk. Springer, Heidelberg (2010). doi:10.1007/978-3-642-05074-9
18. ProModel's Portfolio Simulation Solution. http://www.promodel.com
19. Tarasov, S.V.: Application experience of component modeling in Transas GroupS training system development for cargo-ballast and technological operations. Autom. Remote Control **77**(6), 1106–1114 (2016)

20. Masmoudi, M., Leclaire, P., Cheutet, V., Casalino, E.: Modelling and simulation of the doctors' availability in emergency department with SIMIO software. Case of study: Bichat-Claude Bernard hospital. In: Abbes, M.S., Choley, J.-Y., Chaari, F., Jarraya, A., Haddar, M. (eds.) Mechatronic Systems: Theory and Applications. LNME, pp. 119–129. Springer, Cham (2014). doi:10.1007/978-3-319-07170-1_12
21. Concannon, K., et al.: Simulation Modeling with SIMUL8, USA (2007)
22. Chakravarthy, S., Alfa, A.S.: A finite capacity queue with Markovian arrivals and two servers with group services. J. Appl. Math. Stoch. Anal. 7, 161–178 (1994)

A Sweep Method for Calculating Multichannel Queueing Systems

Yury I. Ryzhikov[⊠]

Institute for Informatics and Automation of the Russian Academy of Sciences,
39, 14-th Line VO, St. Petersburg 199178, Russian Federation
ryzhbox@yandex.ru

Abstract. The necessity is shown to design queueing systems with non-Markovian service time distribution and the big number of channels. A general characteristic is given on phase approximations. Techniques of iteration and matrix-geometric progression are discussed. A new (sweeping) method is suggested. Results of numerical calculations and the evaluation of theirs complexity are presented. Recommendations are proposed how methods discussed in this paper can be applied.

Keywords: Queueing systems · Big number of channels · Sweeping method

1 The Statement of the Problem

Because the production of the big integral circuits technology has achieved the fundamental physical limitations, a required computing performance can be attained only using multiprocessor and multicomputer systems. The basis of the methods of their design, data processing and transmission and performance evaluation is the queuing theory (QT), in which the apparent lack of attention is given to multi-channel systems. Note that these problems are also relevant for many other applications: industrial, transportation, healthcare, emergency services etc.

In this paper an overview of modern methods to calculate the multichannel systems is given, and a new (sweep) method is proposed. The analysis of the range of theirs applicability and computational efficiency is done. All discussions are applied to the most typical case of the *Poissonian* incoming flow (especially having in mind the calculation of service networks where the flows are exposed to multiple summation and random screening operations).

2 Phase Approximations for Multichannel Systems

The efficient numerical methods to calculate the multichannel queuing systems $M/M/n$, $GI/M/n$, and $M/D/n$ were known more than a half a century. Their further generalization (and thus more realistic analysis of the real systems)

© Springer International Publishing AG 2017
A. Dudin et al. (Eds.): ITMM 2017, CCIS 800, pp. 268–278, 2017.
DOI: 10.1007/978-3-319-68069-9_22

became possible only after the presentation of non-Markovian distributions in the form of successive (E_k), parallel (H_k) or combined (C_k, Ph) phase systems with exponentially distributed delay in the each phase. In [6] the possibility of a such approximation by the combination of Erlangian and Coxian distributions is considered, and there are 5 references in support of the adequacy of the equalizing three distribution moments only. In [3,7] Coxian approximation which uses three moments also was offered. Equalizing them, taking in account the low precision of the high order statistical moments of the parent distributions and the increasing complexity of approximations calculation, seems quite sufficient. However, in the above-mentioned works all parameters of approximations were assumed to be real, that restricts the range of realizable coefficients of variance: $v > 1/\sqrt{2}$. Acceptability of averaging service transition intensities proposed in [3] is questionable; calculations of its authors has not been confirmed by numeric results, and the experience of the author of this article refutes it.

In our opinion, a more valuable tool is the model $M/H_2/n$ which remove the mentioned restrictions. The operating of such system can be interpreted as a process of serving the flow of heterogeneous demands, there the demand type choice determines parameter of exponentially distributed service time. The microstate key indicates the number of each type of demands serving in the channels — see Figs. 1 and 2 for $M/H_2/3$ model.

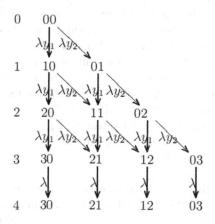

Fig. 1. Transitions on arrival

The demand arriving in the empty channel belongs with probability y_i to the i-th type. On the Fig. 2 at $j > n$ the parameter of i-th type servicing flow is $m_i\mu_i$, where m_i is the contents of the i-th "key" position. Completion of service leads to one of the overlying tier microstates with probabilities $\{y_i\}$ depending on the type of demand selected from the queue.

Similar diagrams can be drawn for Erlangian service E_k — see [8], as well as for the Coxian distribution. With regard to the choice of the approximation type one of the most frequently cited modern QT classics M. Neuts gives

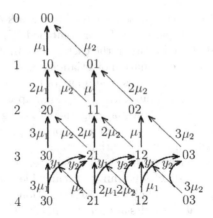

Fig. 2. Transitions on service

the "folklore" recommendation: when the variation coefficient $v > 1$, apply H_2-approximation, otherwise — the Erlangian. Because the order of Erlangian distribution is $k = 1/v^2$, this leads to extremely rapid growth in the number of tiers microstates by the number of channels (see corresponding table in [9]). Moreover, this distribution allows us to equalize only the first and approximately — the second moment, which induce an appreciable loss of precision. On the other hand, H_2-distribution, equalizing *three* moments, generates transition diagram having width $(n + 1)$ microstates only. But for $1/\sqrt{2} < v < 1$ it has *paradoxical* parameters (one of probabilities is more than unit, other is negative), and for $v < 1/\sqrt{2}$ the parameters become complex. The numerous calculations made by the author show that the "pathologies" mentioned above influence only the intermediate computation results, while the final ones have the traditional probabilistic sense and agree well for models $M/E_k/n$ and theirs approximating $M/H_2/n$. For these reasons, and taking into account the above arguments in favor of the Poissonian incoming flow model, we shall consider further $M/H_2/n$ model.

Below we present the common statement of the problem for all discussed methods. We denote by S_j the set of all possible system microstates then exactly j demands are in the system, and by σ_j — the number of elements in S_j. Further, in accordance with the transition diagram we construct the matrices of infinitesimal transitions intensities:

$A_j[\sigma_j \times \sigma_{j+1}]$ — in S_{j+1} (demand arrival),
$B_j[\sigma_j \times \sigma_{j-1}]$ — in S_{j-1} (completion of service),
$D_j[\sigma_j \times \sigma_j]$ — leaving the states of the tier j
(in square brackets the size of matrices is exposed).

We introduce the row vectors $\gamma_j = \{\gamma_{j,1}, \gamma_{j,2}, \ldots, \gamma_{j,\sigma_j}\}$ of the system (j, i) state probabilities, $j = 0, 1, \ldots$ Now we can write the vector-matrix equations for the balance of inter-state transitions

$$\gamma_0 D_0 = \gamma_1 B_1,$$
$$\gamma_j D_j = \gamma_{j-1} A_{j-1} + \gamma_{j+1} B_{j+1}, \qquad j = 1, 2, \ldots \tag{1}$$

System (1), supplemented by the normalizing condition, must be written component-wise. Even for models with bounded queue it is characterized by extremely high dimension, and the standard methods for solving systems of linear algebraic equations are ineffective for it.

3   Iterative Method

Takahashi and Takami [11] proposed an algorithm of iterative calculation for such systems, the central idea of which is the calculation of *conditional* (normalized to unit) probabilities of microstates $\{\bar{\gamma}_{j,i}^{(m)}\}$ for a fixed number of demands in the system (tier of the chart) and parallel computing of relations $x_j = p_{j+1}/p_j$, $j = 0, 1, \ldots$, for total probabilities. The calculation is performed for a limited number of tiers $j = \overline{0, N}$. In iteration number m the vector $\bar{\gamma}_j^{(m)}$ of conditional probabilities for each tier, when sweeping downward, is expressed through $\bar{\gamma}_{j-1}^{(m)}$ and $\bar{\gamma}_{j+1}^{(m-1)}$. When computing the last tier, the approximate equation closing the system is used:

$$\bar{\gamma}_{N+1}^{(m-1)} \approx \bar{\gamma}_{N-1}^{(m)}. \tag{2}$$

In [11], however, several key details of algorithm were not mentioned. The full scheme modified by author was described in [8,10]. In [9] its variants are discussed (choice of initial values for vectors $\{\bar{\gamma}_j\}$, the change of sweep direction). The conclusion was made from the numerical experiment results about the preference of binomial initial approximations to vectors of the microstates conditional probabilities and counting tiers of the chart from top to bottom.

4   Matrix-Geometric Progression

For computation of an *open* queueing systems the method of the *matrix-geometric progression* (MGP) proposed by Evans [5] and developed by M. Neuts and his followers looks very promising (see [2,4,13]). The idea of this method is to represent the vectors of microstates probabilities for a full-bused system by relationship as

$$\gamma_j = \gamma_n R^{j-n}, \qquad j = n, n+1, \ldots, \tag{3}$$

where R is the matrix progression denominator. We write down one of the equations of the system (1) for $j > n$, omitting the indices for the transition matrices being stabilized to this tier:

$$\gamma_j D = \gamma_{j-1} A + \gamma_{j+1} B. \tag{4}$$

Substituting expressions of the microstates probabilities vectors according to (3), we can rewrite (4) in the form

$$\gamma_{j-1} R D = \gamma_{j-1} A + \gamma_{j-1} R^2 B,$$

which implies that the progression denominator must satisfy the matrix quadratic equation

$$R^2 B - RD + A = 0. \tag{5}$$

Finding this denominator and having the microstates probabilities vector γ_n for the n-tier, we can calculate according to (3) the microstates probabilities for all $j > n$.

Implementations of MGP differ by methods of calculating the matrix denominator. We have investigate some variants of simple iteration and Newton's method and modes of calculating the probability vectors for the initial tiers $j = \overline{0, n-1}$. Detailed comparison of MGP variants see again in [9]. It is noted in particular that all these methods show the divergence of iterations in the denominator calculation for a number of channels about 30. Beside of that, the accumulation of errors during solving the system of linear algebraic equations for the unknown probabilities varying by many orders of magnitude leads to the appearance of negative initial probabilities.

To further comparing we have selected the MGP version using the simple iteration formula

$$R = A(D - RB)^{-1}. \tag{6}$$

5   Sweep Method

As it will be shown below by comparing the numerical results, the convergence of both described above methods is deteriorating rapidly by the number of channels. Accordingly, the real dimension of successfully solved tasks and the possibilities of apply these methods to practical queueing systems are constrained. These restrictions can be relaxed considerably by using the *sweep* method described below.

Sweep method has been used successfully during long time for solving the systems of linear algebraic equations with *scalar* coefficients. For QBD-processes (with transitions between microstates of only neighboring tiers) the global matrix contains the *matrices* of transition intensities — also on three diagonals, that gives reason to use appropriate analogies. Another principal singularity of our problem is the condition of (11) specifical for the queueing theory.

Consider the matrices for probability vectors $\{\gamma_j\}$ inverse recalculating according to

$$\gamma_j = \gamma_{j+1} F_j, \quad j = \overline{0, N-1}. \tag{7}$$

It follows from the first equation of (1)

$$F_0 = B_1 D_0^{-1} = B_1 / \lambda. \tag{8}$$

For the subsequent tiers of the same system we have

$$\gamma_j D_j = \gamma_{j-1} A_{j-1} + \gamma_{j+1} B_{j+1}.$$

Using (7), it can be rewritten as

$$\gamma_j (D_j - F_{j-1} A_{j-1}) = \gamma_{j+1} B_{j+1},$$

whence the recurrence rule to calculate matrices $\{R_j\}$ follows:

$$F_j = B_{j+1}(D_j - F_{j-1}A_{j-1})^{-1}, \quad j = \overline{1, n}. \tag{9}$$

Taking into account the rules of transition matrices formation for subsequent tiers, we have

$$F_j = B_{n+1}(D_n - \lambda F_{j-1})^{-1}, \quad j = \overline{n+1, N-1}. \tag{10}$$

Here N is limit index of calculated tiers, $N > n$.

We will denote the limits at $j \to \infty$ of the matrices, conditional vectors of microstates probabilities (normalized to unity within the tiers) and relations of the cumulative probabilities for adjacent tiers by the previous symbols, but without indices. If there is a $\gamma_\infty = \gamma$, then also limits of probabilities relations x and $z = 1/x$ exist, and it follows from (1)

$$\gamma = (x^{-1}\lambda\gamma + x\gamma B)D^{-1} = \gamma(x^{-1}\lambda I + xB)D^{-1} = \gamma Q.$$

We denote $(Q - I)_1$ the matrix obtained from $Q - I$ by replacing its first line to the unit one, and put $\delta_1 = \{1, 0, 0, \ldots 0\}^T$. Then $\det(Q - I) \neq 0$, and the desired vector can be obtained as the solution of the system of linear algebraic equations

$$\gamma(Q - I)_1 = \delta_1. \tag{11}$$

Performing the inverse sweeping for $j = N-1, N-2, \ldots, 0$ according to (7), it is possible to obtain scaled microstates probabilities vectors and for each j — the sums of components, i.e. scaled cumulative probabilities of tiers. Now the final renormalization of cumulated probabilities to unity is done.

The commonly used formula for limit ratio of adjacent probabilities for $A/B/n$ models is

$$x_\infty = \rho^{2/(v_A^2 + v_B^2)}. \tag{12}$$

In [12], it was shown that

$$\int_0^\infty e^{-n\omega t} \, dA(t) \cdot \int_0^\infty e^{\omega\tau} \, dB(\tau) = 1. \tag{13}$$

Founded from this condition $\omega = \omega^*$ gives the limit relation of adjacent probabilities

$$x_\infty = \int_0^\infty e^{-n\omega^* t} \, dA(t), \tag{14}$$

which is reduced for the exponential $A(t)$ to

$$x_\infty = \frac{\lambda}{\lambda + n\omega^*}. \tag{15}$$

Replacing service time distribution $B(\tau)$ by gamma-density with parameters $\{\beta, \mu\}$, which equalizes two moments, we can reduce Eq. (13) to iterative formula

$$w = \mu \left[1 - \left(\frac{\lambda}{nw + \lambda} \right)^{1/\beta} \right], \qquad (16)$$

which provides a convergent iterative process for the initial

$$w_0 = \frac{\lambda}{n} \left[\rho^{-2/(1+v_B^2)} - 1 \right]. \qquad (17)$$

It is easy to obtain particular versions of these formula, including for constant service time, which are required for universal subroutine calculating x_∞.

Let us compare (Table 1) the limiting probability ratio for 10-channel systems, obtained through approximation (A) according to [1] and by Takahashi (T) using (15) and (16).

Table 1. Estimates of limiting probabilities ratios

ρ	$\beta = 0.25$		$\beta = 1.0$		$\beta = 3.0$		$\beta = 10^9$	
	T	A	T	A	T	A	T	A
0.7	0.8623	0.8670	0.7000	0.7000	0.5945	0.5857	0.5089	0.4900
0.9	0.9582	0.9587	0.8999	0.9000	0.8549	0.8538	0.8128	0.8100

The value $\beta = 10^9$ of the gamma-distribution parameter practically corresponds to the degenerate distribution and was used to unify calculation scheme. It is possible to make the following conclusions from this table:

1. The results presented in the table correspond to qualitative expectations: with the increasing of β (decaying variance of corresponding distribution), the rate of probabilities decay increases.
2. The ratios of probabilities, which we are interesting in, do not depend on the number n of channels (the calculation was performed up to $n = 20$).
3. When $\alpha = \beta = 1$, i.e. for a Markovian system, they are equal to loading factor ρ. This confirms the correctness of computing.
4. The discrepancies in the results decrease on the system loading factor. Apparently, for systems with $\rho \geq 0.7$ much more simple calculation according to (12) can be used.

Table 2 presents the probabilities derived by the method Takahashi—Takami (T) and by approximation (12) — (A). The incoming flow is assumed to be Poissonian, number of channels $n = 10$. From this table, it follows that the recommendations of [1] on the calculation of stationary distribution of the number of demands in the system should regarded as a very rough, especially in the neighborhood of $j = n$.

Table 2. Relations $\{x_j\}$ for H_2-service

j	$\rho = 0.7$		$\rho = 0.9$		j	$\rho = 0.7$		$\rho = 0.9$	
	T	A	T	A		T	A	T	A
0	6.999	7.000	8.988	9.000	10	0.678	0.847	0.810	0.884
1	3.499	3.500	4.489	4.500	11	0.726	0.867	0.851	0.959
2	2.332	2.333	2.986	3.000	12	0.764	0.867	0.883	0.959
3	1.747	1.750	2.231	2.250	13	0.793	0.867	0.905	0.959
4	1.395	1.400	1.774	1.800	14	0.813	0.867	0.921	0.959
5	1.157	1.166	1.462	1.500	15	0.827	0.867	0.932	0.959
6	0.983	1.000	1.231	1.286	16	0.837	0.867	0.939	0.959
7	0.846	0.875	1.048	1.125	17	0.844	0.867	0.945	0.959
8	0.729	0.778	0.895	1.000	18	0.849	0.867	0.949	0.959
9	0.622	0.700	0.758	0.900	19	0.852	0.867	0.951	0.959

Table 3. Probabilities of states of the system $M/H_2/5$

j	$\beta = 3.0$			$\beta = 0.25$		
	Iter	MGP	SWP	Iter	MGP	SWP
0	1.2440e−2	1.2440e−2	1.2440e−2	1.3799e−2	1.3790e−2	1.3786e−2
1	5.0313e−2	5.0312e−2	5.0312e−2	5.4703e−2	5.4670e−2	5.4643e−2
2	1.0235e−1	1.0235e−1	1.0235e−1	1.0695e−1	1.0693e−1	1.0689e−1
3	1.4049e−1	1.4049e−1	1.4049e−1	1.3527e−1	1.3533e−1	1.3529e−1
4	1.4851e−1	1.4851e−1	1.4851e−1	1.2082e−1	1.2094e−1	1.2091e−1
5	1.3374e−1	1.3375e−1	1.3375e−1	7.7329e−2	7.7447e−2	7.7424e−2
6	1.0767e−1	1.0767e−1	1.0767e−1	5.6147e−2	5.6264e−2	5.6247e−2
7	8.2209e−2	8.2201e−2	8.2201e−2	4.4571e−2	4.4689e−2	4.4676e−2
8	6.1082e−2	6.1064e−2	6.1064e−2	3.7444e−2	3.7564e−2	3.7553e−2
9	4.4732e−2	4.4705e−2	4.4705e−2	3.2543e−2	3.2665e−2	3.2655e−2
10	3.2507e−2	3.2471e−2	3.2471e−2	2.8850e−2	2.8973e−2	2.8964e−2
11	2.3525e−2	2.3485e−2	2.3485e−2	2.5874e−2	2.5997e−2	2.5989e−2
12	1.6989e−2	1.6946e−2	1.6946e−2	2.3363e−2	2.3484e−2	2.3477e−2
13	1.2255e−2	1.2213e−2	1.2213e−2	2.1179e−2	2.1299e−2	2.1293e−2
14	8.8348e−3	8.7962e−3	8.7962e−3	1.9246e−2	1.9363e−2	1.9357e−2
15	6.3675e−3	6.3330e−3	6.3330e−3	1.7515e−2	1.7628e−2	1.7623e−2
16	4.5886e−3	4.5587e−3	4.5587e−3	1.5953e−2	1.6062e−2	1.6057e−2
17	3.3065e−3	3.2812e−3	3.2812e−3	1.4539e−2	1.4643e−2	1.4638e−2
18	2.3825e−3	2.3616e−3	2.3616e−3	1.3255e−2	1.3353e−2	1.3349e−2
19	1.7167e−3	1.6997e−3	1.6997e−3	1.2087e−2	1.2179e−2	1.2175e−2

Table 4. Labor consuming for the model $M/H_2/n$

n	$\beta = 3.0$			$\beta = 0.25$		
	Iter	MGP	SWP	Iter	MGP	SWP
5	48/0	11/0	0/0	120/0	42/0	0/0
10	94/0.016	20/0	0/0	202/0.031	55/0/016	0/0
20	170/0.078	34*/0.047	0/0.031	320/0.172	70/0.062	0/0.031
30	232/0.0312	-/-	0/0.094	500/0.672	80/0.219	0/0.078
50	-/-	-/-	0/0.469	778/4.859	-/-	0/0.485
70	-/-	-/-	0/1.578*	782/14.532	-/-	0/1.562
100	-/-	-/-	0/5.515*	762/78.500	-/-	0/5.500

6   Numerical Experiment

We present the results of calculation for $M/H_2/5$ system — Table 3. Tolerance $\varepsilon = 10^{-8}$ for iterative method "Iter" determines the maximum module of $\{x_j\}$ refinement, and for the method of matrix-geometric progression "MGP" — the norm of the progression denominator. The column "SWP" corresponds to sweep method. The agreement of the results should be considered as satisfactory, confirming the correctness of calculated dependencies and also their program implementation.

Now let us compare (Table 4) the complexity of tested methods by the number of iterations and the counting time (processor Intel Celeron, tact frequency 1.5 GHz). The number of tiers was assigned $n + 20$. In this table the zero time indicates labor consuming below the threshold of system clock (0.01 s). Dashes indicate cases of iterations divergence, and the asterisks — appearance of negative starting probabilities due to the accumulation of errors.

Method MGP in the real case ($\beta \geq 1$) still can be applied for $n = 30$, and in the complex one fails already for $n = 20$. The iterative method has much greater application scope (in the real case for at least 100 channels). A sweep method have labor consuming much smaller in comparison with its competitors discussed.

7   Conclusion

A new sweep method to calculate the multi-phase queueing systems was developed and its comparison was made with two previously known ones: iterative and matrix-geometric progression.

From the analysis of the computational schemes and comparison of calculation results (including those not presented in this article because of its limited volume) the following *conclusions* imply:

1. The sweep method is applicable only for QBD-processes (with transitions only between adjacent tiers of the diagram) and an unbounded queue length, but

in these conditions substantially exceeds its alternatives, particularly for large n. For H_2 service time distribution with real parameters, it is applicable for at least 100 channels, and in the complex case at the service time coefficient of variance $1/\sqrt{3} = 0.577$ — up to 50 channels.

2. The iterative method can be easily modified with respect to systems with the intensity of incoming flow depending on the number of demands in the system and to the systems with limited queue. It can be generalized to the system with batch arrivals. Due to work with the relative probabilities vectors and the presence of aggregation on each layer its accuracy almost does not depend on the number of accounted tiers. Since, with N increasing the convergence of calculations worsens, it makes sense to limit the number of chart tiers and get the state probabilities with senior sequence indices by multiplying the precedent ones on the limit value of x. For the real parameters of H_2-approximation, the method converges at 100 channels.

3. The method of matrix-geometric progression (MGP) is fundamentally applicable only for QBD-processes. It is useful when operating with infinite sums of the probabilities. Its convergence does not depend on the number of tiers and is restricted by the number of channels $n < 30$.

4. Some of these conclusions may be revised by increasing computing bit grid (they were carried out under twice bit grid). This increasing leads to significant growth of time consuming.

5. With the Poissonian input flow, we can compute for all considered methods the factorial moments $\{q_{[k]}\}$ of queueing length distribution and waiting time moments

$$w_k = q_{[k]}/\lambda^k, \qquad k = 1, 2, \ldots$$

By implementing its convolution with service time moments we receive moments of sojourn time and the possibility to compute by them the most important characteristic of the system operating performance — complementary distribution function.

6. All of these methods in principle can be generalized for recurrent input flow. The distribution of intervals between demands is again advisable be approximated by H_2-distribution.

Acknowledgments. The research described in the paper was partially supported by state research 0073-2014-009 and 0073-2015-0007.

References

1. Basharin, G.P., Bordukova, V.T.: Some exact and approximate results for the multiserver queuing systems with limited storage. Teletraffic Theory and Information Networks, 6–15 (1977). (in Russian)

2. Bocharov, P.P., Pechinkin, A.V.: Queueing Theory: Textbook. People Friendship University of Russia, Moscow (1995). (in Russian)

3. Brandwajn, A., Begin, T.: Preliminary results on simple approach to G/G/c-like queues. In: Proceeding of the 16th International Conference on Analytical and Stochastic Modeling Techniques and Applications, ASMTA, pp. 159–173 (2009)

4. Daigle, J.N.: Queuing Theory with Applications to Packet Telecommunication. Springer, Boston (2005)
5. Evans, R.D.: Geometric distribution in some two dimensional queuing systems. Operat. Res. **5**, 830–846 (1967)
6. Osogami, T., Harchol-Balter, M.: A Closed-form solution for mapping general distribution to minimal PH distributions. In: Proceeding of the 13th International Conference on Computer Performance Evaluation. Modeling Technique and Tools, Urbana, Il, USA, pp. 200–217 (2003)
7. Osogami, T., Harchol-Balter, M.: Necessary and sufficient conditions for representing general distributions by coxians. In: Proceeding of the 13th International Conference on Computer Performance Evaluation. Modeling Technique and Tools, Urbana, Il, USA, pp. 182–199 (2003)
8. Ryzhikov, Y.I.: Algorithm for calculating multichannel system with Erlangian service. Autom. Remote Control **5**, 30–37 (1980). (in Russian)
9. Ryzhikov, Y.I: Development and comparizon of methods for calculating multichannel queueing systems. In: Proceeding of the XII All-Russian Meeting on Control Systems, pp. 5208–5219. Institute of Control Sciences, Moscow (2014). (in Russian)
10. Ryzhikov, Y.I., Khomonenko, A.D.: Iterative method of calculating multichannel systems with arbitrary distribution of service time. Prob. Control Inf. Theor. **3**, 32–38 (1980). (in Russian)
11. Takahashi, Y., Takami, Y.: A Numerical method for the steady-state probabilities of a GI/G/c queuing system in a general class. J. of the Operat. Res. Soc. Japan **2**, 147–157 (1976)
12. Takahashi, Y.: Asymptotic exponentiality of the tail of the waiting time distribution in a Ph/Ph/c queue. Adv. Appl. Probab. **13**, 619–630 (1981)
13. Vishnevsky, V.M.: The theoretical basis to design computer networks. Technosphere, Moscow (2003). (in Russian)

Modeling End-to-End Business Processes of a Telecom Company with a BCMP Queueing Network

Natalia Yarkina[1], Natalia Popovskaya[1], Viktoriya Khalina[1(✉)],
Anna Gaidamaka[1], and Konstantin Samouylov[1,2]

[1] Department of Applied Probability and Informatics,
Peoples' Friendship University of Russia (RUDN University),
6 Miklukho-Maklaya Street, Moscow 117198, Russian Federation
{natyarkina,npopovskaya,aagajdamaka}@sci.pfu.edu.ru,
khalina_va@pfur.ru, samuylov_ke@rudn.university
[2] Institute of Informatics Problems,
Federal Research Center "Computer Science and Control" of the Russian Academy
of Sciences, 44-2 Vavilov Street, Moscow 119333, Russian Federation

Abstract. A thorough analysis of business processes allows a communication and digital service provider to reduce costs and to carry out digital transformation efficiently, which are important factors of the telecommunication business success. Massive numbers of randomly arriving customer requests, real-time service and standardized and highly automated procedures make telecommunication company business processes a good subject for analysis using queueing theory. Such an analysis is facilitated by the extensive body of standards, developed by the global telecommunication industry association TM Forum and addressing various service business management issues. We propose an approach to estimating certain important TM Forum business metrics and other business process measures using a BCMP network that combines stochastic models of several standard TM Forum end-to-end eTOM business flows. The steady-state probability distribution of the model is derived along with the expressions for a number of performance measures.

Keywords: TM Forum · Frameworx · eTOM · Business process framework · Metrics framework · Business process modelling · Workflow · Queueing network · BCMP network · Mean response time · Capacity planning

1 Introduction

Today's communication and digital service market is highly competitive. A thorough analysis of business processes allows the service provider to reduce costs and to carry out digital transformation efficiently, which are important factors of the telecommunication business success. Massive numbers of randomly arriving customer requests, real-time service and standardized and highly automated

© Springer International Publishing AG 2017
A. Dudin et al. (Eds.): ITMM 2017, CCIS 800, pp. 279–296, 2017.
DOI: 10.1007/978-3-319-68069-9_23

procedures make telecommunication company business processes a good subject for analysis using queueing theory. Moreover, such an analysis is facilitated by the extensive body of standards and recommendations, developed by the global telecommunication industry association TM Forum[1] and addressing various communication and digital service business management issues. TM Forum Business Process Framework, also largely known as eTOM, is one of the core models of the TM Forum Frameworx standards and best practices suite [1]. eTOM identifies and puts into a hierarchical framework virtually all activities a telecommunication company business process may include, which provides a comprehensive set of standard building blocks for business process flows. Furthermore, eTOM standards package includes document [2] offering a set of generic end-to-end business flows applicable to the majority of the companies in the industry. Although this work has not been finished yet, the most important business processes of a service provider – customer and network centric – have been described. Another TM Forum standard of interest to us, [3], contains the definitions of numerous business metrics permitting quantitative estimation of various aspects of telecommunication service provision.

Queueing theory methods and specifically queueing networks have been used for capacity planning and delay estimation in relation to various aspects of business operation, such as manufacturing [4], supply chain management [5], healthcare services [6], call centre operation [7], service automation and e-commerce [8]. The application of BCMP networks [9] to workflow analysis was addressed, for instance, in [10,11]. A stand-alone eTOM end-to-end business flow was modelled as an open BCMP network in [12]. Here, we go further and propose an approach to modelling several end-to-end business flows jointly. Such an approach allows to estimate not only the performance measures related to activity execution time, but also reflects resource sharing among business processes and delays due to such sharing. To illustrate this approach, we model several customer centric eTOM business flows as a single open BCMP network and show how the model can be used to evaluate a number of important TM Forum business metrics.

The remainder of the paper is structured as follows. In Sect. 2, a BCMP model of a single standard eTOM end-to-end business flow Complaint-to-Solution is presented for illustrative purposes. This section is largely based upon [12]. In Sect. 3, we introduce the mathematical notation of the joint model and present the model of five eTOM end-to-end business flows and its parameters. In Sect. 4, steady-state probabilities for the joint model are derived. In Sect. 5, we discuss system performance measures and link them to standard TM Forum business metrics. In Sect. 6, a simple numerical example is provided. Finally, Sect. 7 concludes the paper.

2   BCMP Model of a Stand-Alone Business Flow

Consider eTOM end-to-end business flow Complaint-to-Solution [2]. The process deals with customer complaints related to non-technical issues (such as

---

[1] https://www.tmforum.org/.

processing delays, quality of customer service, billing errors, etc.) and consists in identifying the source of the issue, initiating resolution (which may involve bill adjustment) and monitoring the progress. The reference diagram of the flow is depicted in Fig. 1 using standard eTOM process elements and BPMN (Business Process Model and Notation). In order to model this process with a BCMP network, we first assign the process activities to service stations, or queues, which will serve as network nodes. Service stations may correspond to the functional units of the company involved in the business flow (in this case, multiple-server FCFS (First Come First Served) queues are used) or to certain random delays (IS (Infinite Server) queues). For the process under analysis, the following service stations have been chosen:

1 – Customer service specialists (multiple-server FCFS);
2 – Automated customer interface (IS);
3 – Delay due to CRM information system access (IS);
5 – Billing specialists (multiple-server FCFS);
8 – Call-centre operators (multiple-server FCFS).

(Some numbers are omitted due to their use in the joint model introduced in the next section.) However, it should be noted that the choice of service stations or, for that matter, of functional units executing process activities is not specified in TM Forum standards and depends largely on the structure of the company under consideration and on the goals of the modelling effort. Figure 1 shows the partition of process activities among the service stations listed above. In BCMP networks, a job is served at a service station in a certain class and can change class with a given probability when going from one station to another. We shall use this property to route jobs through the network in a specific way, for instance

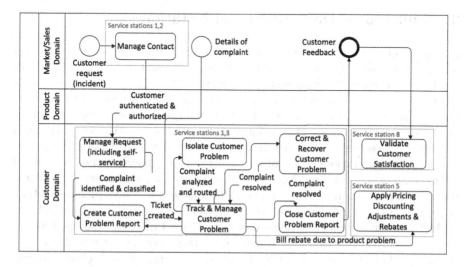

Fig. 1. Complaint-to-Solution business flow BPMN

we would like a job to loop back to station 3 with a given probability, but not more than once. Moreover, job classes may also be given business interpretation. We assume that jobs are served in class 3 at all service stations with the exception of service station 3, where they can be served in classes 3, 7 or 8. In terms of business semantics, class 3 corresponds to customer complaints (as it will be used in the joint model), class 7 corresponds to the requests that have been served twice at station 3, and, finally, class 8 corresponds to the complaints that have been already handled by the billing team. The resulting network model is shown in Fig. 2. Here, for each transition we indicate its source and destination service station and class in the form (service station, class); in case of branching, the corresponding probabilities are also indicated. More formally, the network routing is given in the form of routing matrix, as it is shown in Table 1.

Table 1. Routing matrix for Complaint-to-Solution BCMP network

(Service station, class)	$(1,3)$	$(2,3)$	$(3,3)$	$(3,7)$	$(3,8)$	$(5,3)$	$(8,3)$	Sink
Source	α_3	$1-\alpha_3$	0	0	0	0	0	
$(1,3)$	0	0	1	0	0	0	0	0
$(2,3)$	0	0	1	0	0	0	0	0
$(3,3)$	0	0	0	γ_2	0	$1-\gamma_2$	0	0
$(3,7)$	0	0	0	0	0	1	0	0
$(3,8)$	0	0	0	0	0	0	1	0
$(5,3)$	0	0	0	0	1	0	0	0
$(8,3)$	0	0	0	0	0	0	0	1

3   Joint Model Description and Notation

Let us introduce the formal notation for a joint BCMP model of a set of K business processes sharing certain resources/service stations. Let $\mathcal{M} = \{1, \ldots, M\}$ be the set of network's service stations, and let $\mathcal{R} = \{1, \ldots, R\}$ be the set of job classes. Service stations correspond either to the resources shared by the business processes under consideration (for example, customer service specialists (station 1), billing specialists (station 5), etc.), or to the delays common to some of these processes (such as a delay related to CRM information system access). All stations have infinite waiting room, however they may be of different types: we shall use $|M|c_i$-FCFS queues for shared resources (we denote the set of such stations $\mathcal{M}_{FCFS}$), and $|M|$IS queues for delays (denoted $\mathcal{M}_{IS}$), $\mathcal{M}_{FCFS} + \mathcal{M}_{IS} = \mathcal{M}$. Note that other types of service stations for which the BCMP theorem [9] holds can be used, for example a Processor Sharing (PS) station could correspond to a time-consuming data processing activity. Let $c_i \geq 1$ be the number of servers at FCFS stations, $i \in \mathcal{M}_{FCFS}$. Job classes mainly correspond to different customers requests (e.g., information request, complaint, etc.) and serve for routing.

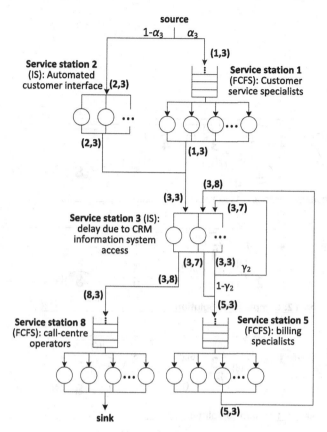

Fig. 2. BCMP model of the Complaint-to-Solution business flow

Now, let $\theta_{(i,r)(j,s)}$, $i,j \in \mathcal{M}$, $r,s \in \mathcal{R}$, denote the transition probability, i.e., the probability that a job that completes service in class r at station i will next require service in class s at station j. The routing matrix $\boldsymbol{\Theta} = \left[\theta_{(i,r)(j,s)}\right]$ defines a Markov chain (MC) with the state space $\mathcal{L} = \{(i,r),\ i \in \mathcal{M},\ r \in \mathcal{R}\}$. Transition probabilities are chosen so that the MC is decomposable into K ergodic subchains, each corresponding to one business flow. In other words, for the sake of simplicity of the modelling procedure, we assign job classes and transitions between them so that each business flow is modelled with its own ergodic subchain, although this requirement can be relaxed.

Let $\mathcal{L}_k$ denote the set of states in subchain $k = 1,\ldots,K$, $\mathcal{L} = \sum_{k=1}^{K} \mathcal{L}_k$. Each subchain has its Poisson arrival stream of rate $\lambda^{(k)}$, $k = 1,\ldots,K$. An arrival in stream k will enter service station i in class r with probability $q_{i,r}$, $\sum_{(i,r)\in\mathcal{L}_k} q_{i,r} = 1$, $k = 1,\ldots,K$. Finally, a job of class r that completes service at station i will depart the system with probability $1 - \sum_{(j,s)\in\mathcal{L}} \theta_{(i,r)(j,s)} =: \theta_{(i,r)(0)}$. Depending on the context, (0) denotes the network's source or sink node.

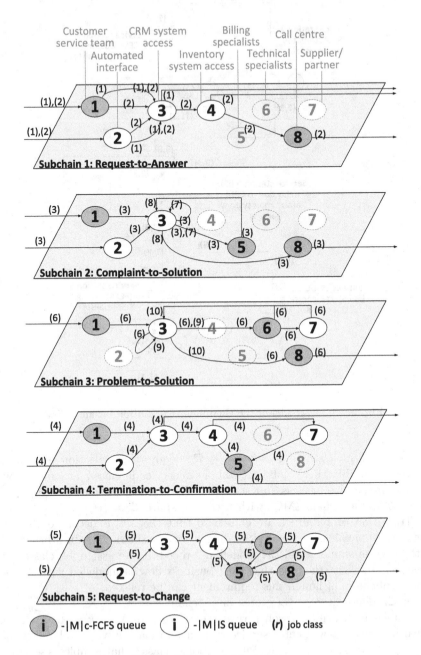

Fig. 3. Joint model transitions

Now, we apply the notation introduced above to describe a joint BCMP model of $K = 5$ end-to-end eTOM business flows: Request-to-Answer, Complaint-to-Solution, Problem-to-Solution, Termination-to-Confirmation and Request-to-Change. Please refer to [2] for the business flows description. The network contains $L = 8$ service stations, $\mathcal{M}_{FCFS} = \{1, 5, 6, 8\}$ and $\mathcal{M}_{IS} = \{2, 3, 4, 7\}$, and has $R = 10$ job classes. The subchains consist of the following MC states:

$$\mathcal{L}_1 = \{(1,1), (1,2), (2,1), (2,2), (3,1), (3,2), (4,2), (8,2)\},$$
$$\mathcal{L}_2 = \{(1,3), (2,3), (3,3), (3,7), (3,8), (5,3), (8,3)\},$$
$$\mathcal{L}_3 = \{(1,6), (3,6), (3,9), (3,10), (6,6), (7,6), (8,6)\},$$
$$\mathcal{L}_4 = \{(1,4), (2,4), (3,4), (4,4), (5,4), (7,4)\},$$
$$\mathcal{L}_5 = \{(1,5), (2,5), (3,5), (4,5), (5,5), (6,5), (7,5), (8,5)\}.$$

The general routing diagram for the joint model is depicted in Fig. 3. Tables 2 and 3 summarise the routing parameters of the network and provide details on their correspondence to the business processes under consideration.

Table 2. Job arrivals to the joint model

Subchain	Rate	Description	Destination MC state, probability
1	$\lambda^{(1)}$	Information and sales requests; the input stream is split into information requests coming to "human" CRM (probability $\alpha_{1,1}$), sales requests coming to "human" CRM (probability $\alpha_{1,2}$), information requests coming to automated CRM (probability $\alpha_{2,1}$), and sales requests coming to automated CRM	$(1,1)$, $q_{1,1} = \alpha_{1,1}$ $(1,2)$, $q_{1,2} = \alpha_{1,2}$ $(2,1)$, $q_{2,1} = \alpha_{2,1}$ $(2,2)$, $q_{2,2} = 1 - \alpha_{1,1} - \alpha_{1,2} - \alpha_{2,1}$
2	$\lambda^{(2)}$	Customer complaints about non-technical issues, e.g. billing errors; the input stream is split between "human" CRM (probability α_3) and automated CRM	$(1,3)$, $q_{1,3} = \alpha_3$ $(2,3)$, $q_{2,3} = 1 - \alpha_3$
3	$\lambda^{(3)}$	Customer problems (technical issues)	$(1,6)$, $q_{1,6} = 1$
4	$\lambda^{(4)}$	Termination requests; the input stream is split between "human" CRM (probability α_4) and automated CRM	$(1,4)$, $q_{1,4} = \alpha_4$ $(2,4)$, $q_{2,4} = 1 - \alpha_4$
5	$\lambda^{(5)}$	Change requests; the input stream is split between "human" CRM (probability α_5) and automated CRM	$(1,5)$, $q_{1,5} = \alpha_5$ $(2,5)$, $q_{2,5} = 1 - \alpha_5$

Table 3. MC states description and routing

MC state	Subchain	Description	Destination MC state, probability ($\theta_{(i,r)(j,s)} = 1$ if not indicated otherwise)
Service station 1: Customer service specialists (front desk, "human" CRM), M|M|c_1-FCFS queue			
$(1,1)$	1	Information request; with probability β_1, the request turns into a sales request, otherwise continues as information request	$(3,1)$, $\theta_{(1,1)(3,1)} = 1 - \beta_1$ $(3,2)$, $\theta_{(1,1)(3,2)} = \beta_1$
$(1,2)$	1	Sales request	$(3,2)$
$(1,3)$	2	Customer complaint (non-technical issues, e.g. billing errors)	$(3,3)$
$(1,4)$	4	Termination request	$(3,4)$
$(1,5)$	5	Change request	$(3,5)$
$(1,6)$	3	Customer problem (technical issues)	$(3,6)$
Service station 2: Automated customer interface (front desk, automated CRM), M|M|IS queue			
$(2,1)$	1	Information request; with probability β_2, the request turns into a sales request, otherwise continues as information request	$(3,1)$, $\theta_{(2,1)(3,1)} = 1 - \beta_2$ $(3,2)$, $\theta_{(2,1)(3,2)} = \beta_2$
$(2,2)$	1	Sales request	$(3,2)$
$(2,3)$	2	Customer complaint (non-technical issues, e.g. billing errors)	$(3,3)$
$(2,4)$	4	Termination request	$(3,4)$
$(2,5)$	5	Change request	$(3,5)$
Service station 3: delay related to CRM information system access, -M|M|IS queue			
$(3,1)$	1	Information request; is handled and leaves the system	(0)
$(3,2)$	1	Sales request; with probability γ_1, the sales request is carried on towards sales proposal development, otherwise (unable to fulfil) it leaves the system	$(4,2)$, $\theta_{(3,2)(4,2)} = \gamma_1$ (0), $\theta_{(3,2)(0)} = 1 - \gamma_1$
$(3,3)$	2	Customer complaint (first run); with probability γ_2, the complaint needs additional time at service station 3 (e.g., additional information from customer needed), otherwise it is carried on	$(3,7)$, $\theta_{(3,3)(3,7)} = \gamma_2$ $(5,3)$, $\theta_{(3,3)(5,3)} = 1 - \gamma_2$
$(3,4)$	4	Termination request; with probability γ_4, the request is carried on, otherwise, the customer is retained with win-back measures	$(4,4)$, $\theta_{(3,4)(4,4)} = \gamma_4$ (0), $\theta_{(3,4)(0)} = 1 - \gamma_4$

(Continued)

Table 3. *(Continued)*

MC state	Subchain	Description	Destination MC state, probability ($\theta_{(i,r)(j,s)} = 1$ if not indicated otherwise)
$(3,5)$	5	Change request	$(4,5)$
$(3,6)$	3	Customer problem (first run); with probability γ_3, the request needs additional time at service station 3 (e.g., additional information from customer needed), otherwise it is carried on	$(3,9)$, $\theta_{(3,6)(3,9)} = \gamma_3$ $(6,6)$, $\theta_{(3,6)(6,6)} = 1-\gamma_3$
$(3,7)$	2	Customer complaint (second run – delay due to complexity)	$(5,3)$
$(3,8)$	2	Customer complaint (handling after resolution)	$(8,3)$
$(3,9)$	3	Customer problem (second run – delay due to complexity)	$(6,6)$
$(3,10)$	3	Customer problem (handling after resolution)	$(8,6)$
Service station 4: delay due to customer subscription and service inventory access, -M|M|IS queue			
$(4,2)$	1	Sales request; with probability δ_1, the request is handed over to customer satisfaction evaluation, otherwise it leaves the system	$(8,2)$, $\theta_{(4,2)(8,2)} = \delta_1$ (0), $\theta_{(4,2)(0)} = 1 - \delta_1$
$(4,4)$	4	Termination request; with probability δ_4, the request requires supplier/partner involvement, otherwise it is carried on to the billing team	$(5,4)$, $\theta_{(4,4)(5,4)} = 1 - \delta_4$ $(7,4)$, $\theta_{(4,4)(7,4)} = \delta_4$
$(4,5)$	5	Change request; with probability δ_5, the request requires handling by the technical team, otherwise (handled by CRM) it is carried on directly to the billing team	$(5,5)$, $\theta_{(4,5)(5,5)} = 1 - \delta_5$ $(6,5)$, $\theta_{(4,5)(6,5)} = \delta_5$
Service station 5: billing specialists, - |M|c_5-FCFS queue			
$(5,3)$	2	Customer complaint (billing issues resolution)	$(3,8)$
$(5,4)$	4	Termination request	(0)
$(5,5)$	5	Change request	$(8,5)$
Service station 6: technical specialists, - |M|c_6-FCFS queue			
$(6,5)$	5	Change request; with probability ε_5, the request requires supplier/partner involvement, otherwise it is carried on to the billing team	$(5,5)$, $\theta_{(6,5)(5,5)} = 1-\varepsilon_5$ $(7,5)$, $\theta_{(6,5)(7,5)} = \varepsilon_5$

(Continued)

Table 3. *(Continued)*

MC state	Subchain	Description	Destination MC state, probability ($\theta_{(i,r)(j,s)} = 1$ if not indicated otherwise)
$(6,6)$	3	Customer problem (technical issue resolution); with probability ε_3, the problem is resolved by the technical team of the telco, otherwise supplier/partner is involved	$(3,10)$, $\theta_{(6,6)(3,10)} = \varepsilon_3$ $(7,6)$, $\theta_{(6,6)(7,6)} = 1-\varepsilon_3$
Service station 7: delay due to supplier/partner involvement, -|M|IS queue			
$(7,4)$	4	Termination request	$(5,4)$
$(7,5)$	5	Change request	$(5,5)$
$(7,6)$	3	Customer problem (technical issue resolution with supplier/partner)	$(3,10)$
Service station 8: call-centre operators, - |M|c_8-FCFS queue			
$(8,2)$	1	Sales request (customer satisfaction validation)	(0)
$(8,3)$	2	Customer complaint (customer satisfaction validation)	(0)
$(8,5)$	5	Change request (customer satisfaction validation)	(0)
$(8,6)$	3	Customer problem (customer satisfaction validation)	(0)

4   Steady-State Probabilities

The routing matrix of the network can be written in the block-diagonal form:

$$
\boldsymbol{\Theta} = \begin{bmatrix} \boldsymbol{\Theta}^{(1)} & & 0 \\ & \ddots & \\ 0 & & \boldsymbol{\Theta}^{(K)} \end{bmatrix},
$$

where $\boldsymbol{\Theta}^{(k)}$ is the routing matrix for subchain $k = 1,\ldots,K$. Let also denote $\boldsymbol{q}^{(k)} = (q_{i,r})_{(i,r)\in\mathcal{L}_k}$. Now, for each subchain, we can find the visit ratio $e_{i,r}$ of class r jobs to service station i by solving the set of equations

$$
e_{i,r} = \sum_{(j,s)\in L_k} e_{j,s}\theta_{(j,s)(i,r)} + q_{i,r}, (i,r) \in L_k, k = 1,...,K. \tag{1}
$$

Let n_{ir} be the number of jobs in class r at service station i. Let $\boldsymbol{n}_i = (n_{ir})_{r\in\mathcal{R}}$ denote the state of station i, $\mathcal{N} = \{\boldsymbol{n} = (\boldsymbol{n}_1, \ldots, \boldsymbol{n}_M)\}$. Also, let $n^{(k)} = \sum_{(i,r)\in\mathcal{L}_k} n_{ir}$ be the total number of jobs in subchain k, and let $n_i = \sum_{r\in\mathcal{R}} n_{ir}$ be the total number of jobs at station i. We assume that the service time at any station $i \in \mathcal{M}$ is exponentially distributed with parameter μ_i (independent on job classes). Then, in accordance with BCMP theorem [9], the steady-state probability of any state $\boldsymbol{n} \in \mathcal{N}$ equals

$$P(\boldsymbol{n}) = G^{-1} d(n) \prod_{i=1}^{M} g_i(n_i), \qquad (2)$$

where G is the normalising constant determined from the condition $\sum_{\boldsymbol{n} \in \mathcal{N}} P(\boldsymbol{n}) = 1;\ d\,(\boldsymbol{n}) = \prod_{k=1}^{K} \left(\lambda^{(k)}\right)^{n^{(k)}}$, and

$$g_i(\boldsymbol{n_i}) = \begin{cases} n_i! \displaystyle\prod_{r=1}^{R} \frac{1}{n_{ir}!}(e_{ir})^{n_{ir}} \prod_{j=1}^{n_i} \frac{1}{\mu_i min(c_i,j)}, & i \in \mathcal{M}_{FCFS} \\ \displaystyle\prod_{r=1}^{R} \frac{1}{n_{ir}!}\left(\frac{e_{ir}}{\mu_i}\right)^{n_{ir}}, & i \in \mathcal{M}_{IS}. \end{cases}$$

Furthermore, let $\rho_i = \frac{\lambda_i}{\mu_i}$, $i \in \mathcal{M}$, where

$$\lambda_i = \sum_{k=1}^{K} \lambda^{(k)} \sum_{r:(i,r)\in L_k} e_{i,r}, i \in M \qquad (3)$$

are the input stream rates at the service stations. Then, for the aggregate states $\tilde{\boldsymbol{n}} = (n_i)_{i\in\mathcal{M}}$, we have

$$P(\tilde{\boldsymbol{n}}) = \prod_{i=1}^{M} h_i(n_i), \qquad (4)$$

where

$$h_i(\boldsymbol{n_i}) = \begin{cases} \left(\displaystyle\sum_{m=0}^{c_i-1} \frac{\rho_i{}^m}{m!} + \frac{\rho_i{}^{c_i}}{c_i!}\frac{1}{1-\frac{\rho_i}{c_i}}\right)^{-1} \frac{\rho_i{}^{n_i}}{\prod_{m=1}^{n_i} min(m,c_i)}, & i \in \mathcal{M}_{FCFS}, \\ e^{-\rho_i}\frac{\rho_i{}^{n_i}}{n_i!}, & i \in \mathcal{M}_{IS}, \end{cases}$$

Note that for service stations $i \in \mathcal{M}_{FCFS}$ to be stable we require $\rho_i < c_i$.

5   Performance Measures

Formulae (2) and (4) can be used to obtain various measures of system performance. In particular, we are interested in the mean response time of the service stations, which allows to evaluate certain standard TM Forum business metrics, related to the processes under consideration (see Table 4).

The mean response time $\overline{T}_i$ of station $i \in \mathcal{M}$ can be obtained using Little's law $\overline{T}_i = \frac{\overline{n}_i}{\lambda_i}$, where $\overline{n}_i$ is the mean number of jobs at station i. Thus, we have

$$\overline{T}_i = \begin{cases} \dfrac{c_i\mu_i}{(c_i\mu_i-\lambda_i)^2} \dfrac{\rho_i{}^{c_i}}{c_i!}\left(\displaystyle\sum_{m=0}^{c_i-1} \frac{\rho_i{}^m}{m!} + \frac{\rho_i{}^{c_i}}{c_i!}\frac{1}{1-\frac{\rho_i}{c_i}}\right) + \frac{1}{\mu_i}, & i \in \mathcal{M}_{FCFS}, \\ \frac{1}{\mu_i}, & i \in \mathcal{M}_{IS} \end{cases} \qquad (5)$$

Table 4. Standard TM Forum business metrics that can be estimated with the five-process joint model

Metric ID	Metric name
28 (CM-CE-2a)	*Average Hold Time*
29 (CM-CE-2b)	*Average Handle Time*
36[a] (CM-OE-1f)	*$ Cost of Customer Management per Customer Request*
56 (A-CE-2a)	*# Minutes per Customer Incident Resolution, by Severity Type*
57 (A-CE-2b)	*# Minutes per Customer Incident Resolution, by Customer Type*
68[a] (A-OE-1f)	*$ Cost of Assurance per Service Problem Resolved*
69 (A-OE-2a)	*# Minutes per Service Problem Resolution*
70 (A-OE-2b)	*# Hours per Service Problem Resolution, by Process Type*
74 (A-OE-6a)	*# Problem Reports per NOC FTE*
96[a] (B-OE-3a)	*% Cost of Billing Errors, of Revenue Billed*
97 (B-OE-3b)	*# Days per Billing Error*

[a]The metric can be estimated via the number of servers and their utilisation.

Now, we can derive expressions for standard business metrics. For example, TM Forum business metric 28 is the average hold time when a customer contacts the company by telephone, which corresponds to the waiting time at service station 1 of the five-process joint model. Hence, the following formula can be used to estimate this measure:

$$I_{28} = \overline{T_1} - \frac{1}{\mu_1}. \tag{6}$$

Business metric 29 is the average time needed to handle any request, so it can be estimated as the mean response time of the network:

$$I_{29} = \frac{\sum\limits_{i \in M} \overline{n_i}}{\sum\limits_{k=1}^{K} \lambda^{(k)}}. \tag{7}$$

Business metric 69 is related to Problem-to-Solution business flow (subchain 3) and corresponds to the mean time between the creation of a trouble ticket and its closure upon the confirmation by the customer that the problem has been resolved. We assume that the trouble ticket is created at the moment of transition of a subchain 3 job from station 1 to station 3, and is closed when the job leaves station 8, although this may depend on the policy of the company. Thus, we can estimate the value of the metric as the sum of the stations' mean response times multiplied by the corresponding visit ratios:

$$I_{69} = \sum\limits_{i \in M \setminus \{1\}} \overline{T_i} \sum\limits_{r:(i,r) \in L_3} e_{i,r}. \tag{8}$$

Certain parameters of the model also can be based upon standard business metrics. Examples of these can be found in Table 5.

Table 5. Standard TM Forum business metrics related to five-process model parameters

Metric ID	Metric name	Related parameters
27 (CM-CE-1)	*% Customer Contacts Received, by Channel Type*	$\alpha_{1,1},\ \alpha_{1,2},\ \alpha_{2,1},\ \alpha_3,\ \alpha_4,\ \alpha_5$
31 (CM-CE-3a)	*# Customer Requests (x1000) per Customer*	$\lambda^{(k)},\ k = 1,\ldots,K$
32 (CM-CE-3b)	*% Customer Requests Received, by Request Type*	$\lambda^{(k)},\ k = 1,\ldots,K$
61 (A-CE-4c)	*% Problem Reports from Customers*	$\lambda^{(3)}$
84 (B-CE-4e)	*# Customer Contacts About Billing per Bill*	$\lambda^{(2)}$
95 (B-OE-2f)	*# Hours per Bill Processing Fault Resolution*	μ_5

6 Numerical Example

For the five-process model described in Tables 2 and 3, matrices $\Theta^{(k)}$ and vectors $q^{(k)}$ are explicitly listed in the Appendix. By using them in Eq. (1), we obtain the following explicit expressions for the visit ratios:

Subchain 1: $e_{1,1} = \alpha_{1,1},\ e_{1,2} = \alpha_{1,2},\ e_{2,1} = \alpha_{2,1},\ e_{2,2} = 1 - \alpha_{1,1} - \alpha_{1,2} - \alpha_{2,1},\ e_{3,1} = (1 - \beta_1)\alpha_{1,1} + (1 - \beta_2)\alpha_{2,1},\ e_{3,2} = \beta_1\alpha_{1,1} + \beta_2\alpha_{2,1},\ e_{4,2} = \gamma_1(\beta_1\alpha_{1,1} + \beta_2\alpha_{2,1}),\ e_{8,2} = \delta_1(\beta_1\alpha_{1,1} + \beta_2\alpha_{2,1}).$

Subchain 2: $e_{1,3} = \alpha_3,\ e_{2,3} = 1 - \alpha_3,\ e_{3,7} = \gamma_2,\ e_{3,3} = e_{3,8} = e_{5,3} = e_{8,3} = 1.$

Subchain 3: $e_{1,6} = e_{3,6} = e_{3,10} = e_{6,6} = e_{8,6} = 1,\ e_{3,9} = \gamma_3,\ e_{7,6} = 1 - \varepsilon_3.$

Subchain 4: $e_{1,4} = \alpha_4,\ e_{2,4} = 1 - \alpha_4,\ e_{3,4} = 1,\ e_{4,4} = \gamma_4,\ e_{5,4} = \gamma_4,\ e_{7,4} = \gamma_4\delta_4.$

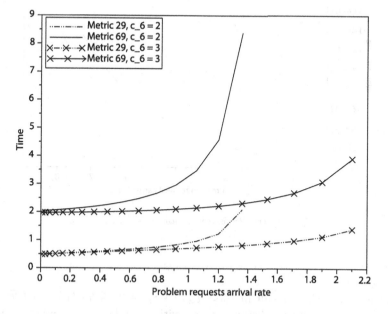

Fig. 4. Metrics 29 and 69 as functions of $\lambda^{(3)}$ for $c_6 = 2$ and $c_6 = 3$

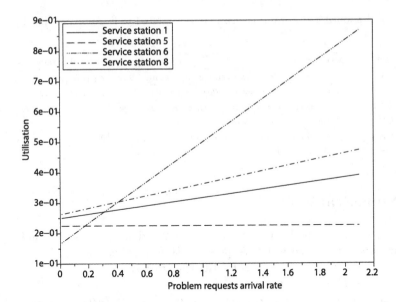

Fig. 5. Service station utilisations as functions of $\lambda^{(3)}$, $c_6 = 3$

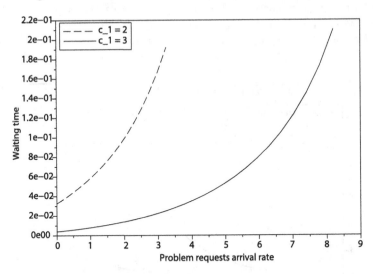

Fig. 6. Metric 28 as a function of $\lambda^{(3)}$, $c_1 = 2, 3$

Subchain 5: $e_{1,5} = \alpha_5$, $e_{2,5} = 1 - \alpha_5$, $e_{3,5} = e_{4,5} = e_{5,5} = e_{8,5} = 1$, $e_{6,5} = \delta_5$, $e_{7,5} = \varepsilon_5 \delta_5$.

In particular, these result in $I_{69} = \overline{T}_3 (2 + \gamma_3) + \overline{T}_6 + \overline{T}_7 (1 - \varepsilon_3) + \overline{T}_8$.

For simplicity, we consider the system with the following parameter values:

- transition probabilities $\alpha_{1,1} = \alpha_{1,2} = \alpha_{2,1} = 0.25$, $\alpha_3 = \alpha_4 = \alpha_5 = 0.5$, $\beta_1 = \beta_2 = 0.5$, $\gamma_1 = \gamma_2 = \gamma_3 = \gamma_4 = 0.5$, $\delta_1 = \delta_4 = \delta_5 = 0.5$, $\varepsilon_3 = \varepsilon_5 = 0.5$;
- service parameters $\boldsymbol{\mu}=(\mu_i)_{i \in \mathcal{M}} = (5, \ 20, \ 10, \ 20, \ 5, \ 1, \ 1, \ 5)$, $c_1 = 3$, $c_5 = 2$, $c_8 = 2$;
- input stream rates $\lambda^{(1)} = 5$, $\lambda^{(2)} = 1$, $\lambda^{(4)} = 0.5$, $\lambda^{(5)} = 1$.

Figure 4 shows metrics 29 and 69 as functions of $\lambda^{(3)}$, which is plotted on the X-axis. The metrics are computed for $c_6 = 2$ and $c_6 = 3$ and the curves have vertical asymptotes at $\lambda^{(3)} = 1.5$ and $\lambda^{(3)} = 2.5$ respectively. This is due to a bottleneck at service station 6, since the station utilisation $\frac{\lambda_6}{c_6 \mu_6} = \frac{\lambda^{(3)}+0.5}{c_6}$ equals 1 at these points. Service station utilisations for stations 1, 5, 6, and 8 are plotted in Fig. 5 (here $c_6 = 3$). Finally, Fig. 6 shows metric 28 (Average Hold Time) for $c_1 = 2$ and $c_1 = 3$.

7   Conclusions and Future Work

Combining several business processes in a single queueing network allows to take into consideration resource sharing, occurring at functional units of the company (such as the technical team, the billing team, call-centre operators, etc.). We have proposed an approach to estimating certain important TM Forum business metrics using a BCMP network that combines the models of several standard end-to-end eTOM business flows. The method can be applied to analyze the set of business processes of a particular service provider, in which case we summarise it as follows:

1. Standard eTOM process elements are used to make business flow diagrams of the processes under analysis. These can be largely based upon the reference eTOM business flows.
2. Common service stations corresponding to shared resources or delays are identified. Here, on the contrary, the structure and specific procedures of the company should be consulted.
3. Each business flow is modelled with a separate BCMP network by assigning its activities to the corresponding service stations.
4. The models are put together into a single BCMP networks as ergodic sub-chains.
5. Expressions for the business metrics of interest are derived. Here, TM Forum Metrics Framework can be used.
6. The values of the system parameters are determined based upon the corresponding business statistics and indicators and the metrics of interest are computed.

Future work will focus on modelling the whole set of eTOM end-to-end business flows, both customer and network centric, as well as on developing a model of a complex telecommunication service provider CRM taking into account automated and non-automated functions.

Acknowledgement. The publication was financially supported by the Ministry of Education and Science of the Russian Federation (the Agreement number 02.a03.21.0008), RFBR according to the research projects No. 15-07-03608 and No. 16-07-00766.

Appendix: Routing Matrices

See Tables 6, 7, 8, 9 and 10.

Table 6. Subchain 1: Request-to-Answer

$q^{(1)}$	$\alpha_{1,1}$	$\alpha_{1,2}$	$\alpha_{2,1}$	$1-\alpha_{1,1}-\alpha_{1,2}-\alpha_{2,1}$	0	0	0	0	
$\Theta^{(1)}$	$(1,1)$	$(1,2)$	$(2,1)$	$(2,2)$	$(3,1)$	$(3,2)$	$(4,2)$	$(8,2)$	(0)
$(1,1)$	0	0	0	0	$1-\beta_1$	β_1	0	0	0
$(1,2)$	0	0	0	0	0	1	0	0	0
$(2,1)$	0	0	0	0	$1-\beta_2$	β_2	0	0	0
$(2,2)$	0	0	0	0	0	1	0	0	0
$(3,1)$	0	0	0	0	0	0	0	0	1
$(3,2)$	0	0	0	0	0	0	γ_1	0	$1-\gamma_1$
$(4,2)$	0	0	0	0	0	0	0	δ_1	$1-\delta_1$
$(8,2)$	0	0	0	0	0	0	0	0	1

Table 7. Subchain 2: Complaint-to-Solution

$q^{(2)}$	α_3	$1-\alpha_3$	0	0	0	0	0	
$\Theta^{(2)}$	$(1,3)$	$(2,3)$	$(3,3)$	$(3,7)$	$(3,8)$	$(5,3)$	$(8,3)$	(0)
$(1,3)$	0	0	1	0	0	0	0	0
$(2,3)$	0	0	1	0	0	0	0	0
$(3,3)$	0	0	0	γ_2	0	$1-\gamma_2$	0	0
$(3,7)$	0	0	0	0	0	1	0	0
$(3,8)$	0	0	0	0	0	0	1	0
$(5,3)$	0	0	0	0	1	0	0	0
$(8,3)$	0	0	0	0	0	0	0	1

Table 8. Subchain 3: Problem-to-Solution

$q^{(3)}$	1	0	0	0	0	0	0	
$\Theta^{(3)}$	$(1,6)$	$(3,6)$	$(3,9)$	$(3,10)$	$(6,6)$	$(7,6)$	$(8,6)$	(0)
$(1,6)$	0	1	0	0	0	0	0	0
$(3,6)$	0	0	γ_3	0	$1-\gamma_3$	0	0	0
$(3,9)$	0	0	0	0	1	0	0	0
$(3,10)$	0	0	0	0	0	0	1	0
$(6,6)$	0	0	0	ε_3	0	$1-\varepsilon_3$	0	0
$(7,6)$	0	0	0	1	0	0	0	0
$(8,6)$	0	0	0	0	0	0	0	1

Table 9. Subchain 4: Termination-to-Confirmation

$q^{(4)}$	α_4	$1-\alpha_4$	0	0	0	0	
$\Theta^{(4)}$	$(1,4)$	$(2,4)$	$(3,4)$	$(4,4)$	$(5,4)$	$(7,4)$	(0)
$(1,4)$	0	0	1	0	0	0	0
$(2,4)$	0	0	1	0	0	0	0
$(3,4)$	0	0	0	γ_4	0	0	$1-\gamma_4$
$(4,4)$	0	0	0	0	$1-\delta_4$	δ_4	0
$(5,4)$	0	0	0	0	0	0	1
$(7,4)$	0	0	0	0	1	0	0

Table 10. Subchain 5: Request-to-Change

$q^{(5)}$	α_5	$1-\alpha_5$	0	0	0	0	0	0	
$\Theta^{(5)}$	$(1,5)$	$(2,5)$	$(3,5)$	$(4,5)$	$(5,5)$	$(6,5)$	$(7,5)$	$(8,5)$	(0)
$(1,5)$	0	0	1	0	0	0	0	0	0
$(2,5)$	0	0	1	0	0	0	0	0	0
$(3,5)$	0	0	0	1	0	0	0	0	0
$(4,5)$	0	0	0	0	$1-\delta_5$	δ_5	0	0	0
$(5,5)$	0	0	0	0	0	0	0	1	0
$(6,5)$	0	0	0	0	$1-\varepsilon_5$	0	ε_5	0	0
$(7,5)$	0	0	0	0	1	0	0	0	0
$(8,5)$	0	0	0	0	0	0	0	0	1

References

1. TM Forum Frameworx. https://www.tmforum.org/tm-forum-frameworx
2. TM Forum: Business Process Framework (eTOM) End-to-End Business Flows. Frameworx Standard GB921 Addendum E. R16.5.1, vol. 16.5.2 (2017)
3. TM Forum: TM Forum Metrics Definitions. Frameworx Best Practice GB988. R16.5.0, vol. 4.0.1 (2016)
4. Curry, G.L., Feldman, R.M.: Manufacturing Systems Modeling and Analysis, 2nd edn. Springer, Heidelberg (2011). doi:10.1007/978-3-642-16618-1
5. Bhaskar, V., Lallement, P.: Modeling a supply chain using a network of queues. Appl. Math. Model. **34**, 2074–2088 (2010). doi:10.1016/j.apm.2009.10.019
6. Gupta, D.: Queueing models for healthcare operations. In: Denton, B.T. (ed.) Handbook of Healthcare Operations Management Methods and Applications, vol. 184, pp. 19–44. Springer-Verlag, New York (2013). doi:10.1007/978-1-4614-5885-2-2
7. Gans, N., Koole, G., Mandelbaum, A.: Telephone call centers: tutorial, review, and research prospects. Manuf. Serv. Oper. Manag. **5**, 79–141 (2003). doi:10.1287/msom.5.2.79.16071
8. Menasce, D.A., Almeida, V.A.F.: Capacity planning an essential tool for managing Web services. IT Prof. **4**, 33–38 (2002). doi:10.1109/mitp.2002.1046642
9. Baskett, F., Chandy, K.M., Muntz, R.R., Palacios, F.G.: Open, closed, and mixed networks of queues with different classes of customers. J. ACM **22**, 248–260 (1975). doi:10.1145/321879.321887
10. Zhao, X.: Workflow simulation across multiple workflow domains. In: Hameurlain, A., Cicchetti, R., Traunmüller, R. (eds.) DEXA 2002. LNCS, vol. 2453, pp. 50–59. Springer, Heidelberg (2002). doi:10.1007/3-540-46146-9_6
11. Van der Aalst, W.M.P.: Re-engineering knock-out processes. Decis. Support Syst. **30**, 451–468 (2001). doi:10.1016/s0167-9236(00)00136-6
12. Samouylov, K., Gaidamaka, Y., Zaripova, E.: Analysis of business process execution time with queueing theory models. In: Dudin, A., Gortsev, A., Nazarov, A., Yakupov, R. (eds.) ITMM 2016. CCIS, vol. 638, pp. 315–326. Springer, Cham (2016). doi:10.1007/978-3-319-44615-8_28

The Multi-product Newsboy Problem with Price-Depended Demand and Fast Moving Items

Anna Kitaeva, Klimentii Livshits$^{(\boxtimes)}$, and Ekaterina Ulyanova

National Research Tomsk State University, Tomsk, Russia
kit1157@yandex.ru, kim47@mail.ru, katerina_tomsk@sibmail.com

Abstract. The present paper supposes that the newsboy sells multiple products, and in addition to determining optimal order quantities needs to determine the prices of each product. The demand for each product is a compound Poisson process that depends on the demands of other products. We suppose that the price-dependent intensity of the demand is large enough to use the normal approximation to the joint demand distribution. The goal of the retailer's decision is to maximize his profit. Equations for optimal retail price under optimal order quantity are obtained and approximate solutions are proposed. Also approximate distribution of the selling time of a large order is found.

Keywords: Newsboy problem · Multiple products · Fast moving items · Selling time distribution

1 Introduction

Models and methods of queueing theory; see Sharma [1] and Nazarov [2] are widely used in supply chain management to construct the models and analyze the decision making. The connection between queueing and inventory control theories models and methods can be clearly seen, for example, from the works by Schwarz and Daduna [3]. A review of queueing-inventory models can be found, for example, in Krishnamoorthy et al. [4], for more recent references see Manikandan and Nair [5].

The newsboy problem is a classic problem in the inventory management. It has been studied since the nineteenth century, see Edgeworth [6]. Nowadays it is widely used to analyze systems with perishable products in different fields; see reviews by Khouja [7], Qin et al. [8].

Originally, the problem was formulated for a single product with fixed selling price, but nowadays analogous models have been considered in multi-product setting; see Turken et al. [9]. Also the price is often treated as a decision variable influencing the demand. Whitin [10] was the first to analyze price-dependent demand and determine the order size and selling price simultaneously. The topic has been the theme of many papers; see reviews by Petruzzi and Dada [11], Yano and Gilbert [12], and Chen and Simchi-Levi [13].

© Springer International Publishing AG 2017
A. Dudin et al. (Eds.): ITMM 2017, CCIS 800, pp. 297–311, 2017.
DOI: 10.1007/978-3-319-68069-9_24

The aim of the paper is to maximize the retailer's expected profit in multi-product setting and to find the asymptotic distribution of the selling time, i.e. the amount of time it takes to sell the order. This distribution can be used for calculation of different characteristics of the selling process and estimation of the demand parameters in case of the censored samples; see, for example, Nahmias [14], and Kitaeva et al. [15].

2   Problem Statement

Let us consider a supply chain consisting of a supplier, a retailer, and customers. At the beginning of the selling period, the retailer purchases an order $q = [q_1, q_2...q_m]^T$, q_i is the i-th product order quantity, at a fixed price per units of each products. Denote $d = [d_1, d_2...d_m]^T$ the vector of corresponding procurement costs. The duration of the selling period is bounded by T. At the end of the period, we need to utilize the leftovers or lower the start prices to the next period. Let us denote $b = [b_1, b_2...b_m]^T$ the corresponding costs incurred.

Let the demand be a Poisson process with known price-and time-dependent intensity $\lambda(t, c)$, where $c = [c_1, c_2...c_m]^T$, $c_i > d_i$ is a retail price per unit of the i-th product, and the values of purchases $Z = [Z_{1j}, Z_{2j}...Z_{mj}]^T$ are i.i.d. continuous random vectors independent of the arrivals process with the mean equals to $a = [a_1, a_2...a_m]^T$ and covariance matrix $R = [R_{pq}]$.

The model under consideration can be interpreted as a periodic queuing system with two streams of arrivals: a periodic one with a period T with positive orders of value q entering the service (sale) and a stream of customers with intensity $\lambda(t, c)$ with negative orders (purchases) Z. Since at the end of each period the leftovers are recycled, we need to consider only one period.

3   Demand Distribution for Fast Moving Items

Let us denote $p(\cdot)$ PDF of Z, $X(t)$ a random customer demand at $[0, t]$, $p(\cdot)$ the probability density function of $X(T) = X$, and N a number of customers during T. PMF of random value N

$$P_N(n) = \frac{(\Lambda(T, c))^n}{n!} \exp(-\Lambda(T, c)), \qquad (1)$$

where $\Lambda(T, c) = \int_0^T \lambda(t, c)dt$. In this section we will consider time- and price-dependent demand rate.

Let us consider the case of fast moving items, that is we assume that $\Lambda(T, c) \to \infty$ as $T \to \infty$. Denote $P(\cdot|n)$ conditional PDF of

$$X = Z_1 + Z_2 + ... + Z_n, \qquad (2)$$

given $N = n$, then

$$P(x) = \sum_{n=0}^{\infty} P(x|n) \frac{(\Lambda(T, c))^n}{n!} \exp(-\Lambda(T, c)). \qquad (3)$$

Let $\phi_X(u) = E\{e^{ju^T X}\}, \phi_Z(u) = E\{e^{ju^T Z}\}$, be the characteristic functions of demand X and purchase Z. Then

$$\phi_X(u) = \sum_{n=0}^{\infty} \phi_Z(u)^n \frac{(\Lambda(T,c))^n}{n!} \exp(-\Lambda(T,c)) = \exp\left(-\Lambda(T,c) + \Lambda(T,c)\phi_Z(u)\right).$$

Consider

$$Y_i = \frac{X_i - \Lambda(T,c)a_i}{\sqrt{\Lambda(T,c)}}, i = \overline{1,m}. \tag{4}$$

The characteristic function of $Y = [Y_1, Y_2 ... Y_m]^T$

$$\phi_Y(u) = \exp\left(-j\sqrt{\Lambda(T,c)}u^T a - \Lambda(T,c) + \Lambda(T,c)\phi_Z\left(\frac{u}{\sqrt{\Lambda(T,c)}}\right)\right)$$

Let $\phi_Z(u)$ be a twice differential function. Then using Taylor expansion we obtain as $\Lambda(T,c) \to \infty$

$$\phi_Z\left(\frac{u}{\sqrt{\Lambda(T,c)}}\right) = 1 + \frac{1}{\sqrt{\Lambda(T,c)}}u^T a - \frac{1}{2\Lambda(T,c)}u^T Ru + o\left(\frac{1}{\Lambda(T,c)}\right).$$

and

$$\phi_Y(u) = \exp\left(-\frac{1}{2}u^T Ru + o\left(\frac{1}{\Lambda(T,c)}\right)\right). \tag{5}$$

Function $\phi_Y(u)$ corresponds to a normal vector with zero mean and covariance matrix R. It follows that $X = \Lambda(T,c)a + \sqrt{\Lambda(T,c)}Y$ has an approximate normal distribution with mean $\Lambda(T,c)a$ and covariance matrix $\Lambda(T,c)R$ as $\Lambda(T,c) >> 1$.

Consider, for example, the case of a single product, i.e. $m = 1$, when the purchases obey the Gamma distribution with parameters α and $n - 1$, where $n > 1$ is an integer; PDF of Z

$$p(z) = \frac{z^{n-1}}{\alpha^n(n-1)!}e^{-\frac{z}{\alpha}}.$$

Then from (3) we get

$$P(x) = \delta(x)e^{-\Lambda(T,c)} + \sum_{k=1}^{\infty} \frac{x^{kn-1}}{\alpha^{kn}(kn-1)!}\frac{(\Lambda(T,c))^k}{k!}\exp\left(-\frac{x}{\alpha}-\Lambda(T,c)\right), \tag{6}$$

where $\delta(\cdot)$ is the Dirac delta function. Figure 1 illustrates the normal approximation for Gamma batch size distribution for $n = 2, \alpha = 0.5, \Lambda(\cdot) = 10$ and $\Lambda(\cdot) = 50$. Here the solid lines correspond to the exact distribution, and the dashed lines correspond to the approximate normal distribution.

In Table 1 the Kolmogorov-Smirnov distances

$$\Delta = \max_x \left|P(x) - N\left(\alpha n \Lambda(T), \sqrt{\alpha^2 n(n+1)\Lambda(T)}\right)\right|$$

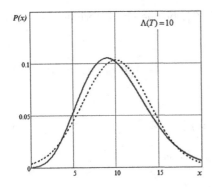 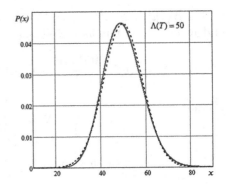

Fig. 1. The exact and approximate normal distributions of X, $\Lambda(T,c) = 10$ and $\Lambda(T,c) = 50$

Table 1. The Kolmogorov-Smirnov distance between exact distribution $P(\cdot)$ and its normal approximation

$\Lambda(T,c)$	5	10	50	100	200	1000
$\Delta, n = 2$	0.031	0.014	$2.604 \cdot 10^{-3}$	$1.277 \cdot 10^{-3}$	$6.303 \cdot 10^{-4}$	0
$\Delta, n = 3$	$0.019 \cdot 10^{-3}$	$8.756 \cdot 10^{-3}$	$1.615 \cdot 10^{-3}$	$7.941 \cdot 10^{-4}$	0	0

between the two distributions for $\alpha = 0.5, n = 2$ and $n = 3$, and different values of $\Lambda(T,c)$ are given.

For exponential batch size distribution the results of comparing the exact and approximate normal distributions of the demand are presented in Kitaeva et al. [16].

Let us consider the probability $P(q,T)$ that order $q = (q_1, q_2...q_m)^T$ have been sold to the end of the selling period

$$P(q,T) = \int_{q_1}^{\infty} \dots \int_{q_m}^{\infty} P(x_1, x_2 \dots x_m) dx_1 dx_2 \dots dx_m$$

$$= \frac{(\det R)^{-\frac{1}{2}}}{(2\pi \Lambda(T,c))^{\frac{m}{2}}} \int_{q_1 - \Lambda(T,c)a_1}^{\infty} \dots \int_{q_m - \Lambda(T,c)a_m}^{\infty} e^{\left(-\frac{x^T W x}{2\Lambda(T,c)}\right)} dx_1 dx_2 \dots dx_m$$

(7)

where $W = R^{-1}$.

Consider $P(q,T)$ assuming that $S_i = (\Lambda(T,c)a_i - q_i)/\sqrt{\Lambda(T,c)} \to \infty$ as $T \to \infty$, i.e. $\Lambda(T,c)$ increases faster than q_i.

Let A_i be the event that the demand for the i-th product is greater than q_i, i.e. there are lost sales. Then by the inclusion-exclusion formula

$$P(q,T) = P(A_1 A_2 \ldots A_m)$$

$$= 1 - \sum_{i=1}^{m} P(\overline{A}_i) + \sum_{i<j} P(\overline{A}_i \overline{A}_j) + \ldots + (-1)^m P(\overline{A}_1 \overline{A}_2 \ldots \overline{A}_m). \tag{8}$$

Using normal approximation

$$P(\overline{A}_i) = \frac{1}{\sqrt{2\pi \Lambda(T,c) R_{ii}}} \int_{-\infty}^{q_i} \exp\left(-\frac{(x - \Lambda(T,c)a_i)^2}{2\Lambda(T,c) R_{ii}}\right) dx$$

$$= \frac{\exp\left(-S_i^2/2\right)}{\sqrt{2\pi} S_i} \int_{-\infty}^{0} \exp\left(-\frac{x^2}{2S_i^2} + x dx\right)$$

we get

$$\lim_{T \to \infty} \sqrt{2\pi} S_i \exp\left(\frac{S_i^2}{2}\right) P(\overline{A}_i) = 1,$$

i.e. as $T \gg 1$

$$P(\overline{A}_i) \sim \frac{1}{\sqrt{2\pi} S_i} \exp\left(-\frac{S_i^2}{2}\right). \tag{9}$$

Taking all the terms in the expansion of (8), consider finally the last term

$$P(\overline{A}_1 \overline{A}_2 \ldots \overline{A}_m) = \frac{\det R^{-\frac{1}{2}}}{(2\pi \Lambda(T,c))^{\frac{m}{2}}} \int_{-\infty}^{q_1} \ldots$$

$$\int_{-\infty}^{q_m} \exp\left(-\frac{(x - \Lambda(T,c)a)^T W (x - \Lambda(T,c)a)}{2\Lambda(T,c)}\right) dx_1 \ldots dx_m$$

$$= \frac{\det R^{-\frac{1}{2}} \exp\left(-\frac{1}{2} S^T W S\right)}{(2\pi)^{\frac{m}{2}} \prod_{i=1}^{m} t_i} \int_{-\infty}^{0} \ldots$$

$$\int_{-\infty}^{0} \exp\left(-\frac{1}{2} \sum_{i,j}^{m} W_{ij} \frac{y_i y_j}{t_i t_j} + \sum_{i=1}^{m} y_i\right) dy_1 \ldots dy_m,$$

where $S = (S_1, S_2, ..., S_m)^T$, $t = (t_1, t_2, ..., t_m) = WS$. It follows

$$\lim_{\Lambda(T,c) \to \infty} P(\overline{A}_1 \overline{A}_2 \ldots \overline{A}_m)(2\pi)^{\frac{m}{2}} \det R^{\frac{1}{2}} \exp\left(\frac{1}{2} S^T W S\right) \prod_{i=1}^{m} t_i = 1,$$

or as $T \gg 1$

$$P(\overline{A}_1 \overline{A}_2 \ldots \overline{A}_m) \sim \frac{\exp\left(-\frac{1}{2} S^T W S\right)}{(2\pi)^{\frac{m}{2}} \det R^{\frac{1}{2}} \prod_{i=1}^{m} \sum_{j=1}^{m} W_{ij} S_j}. \tag{10}$$

It is easy to see that the main part of (8) is $1 - \sum_{i=1}^{m} P(\overline{A_i})$ as $T \gg 1$, i.e.

$$P(q,T) \approx 1 - \sum_{i=1}^{m} \frac{\sqrt{\Lambda(T,c)R_{ii}}}{\sqrt{2\pi}(\Lambda(T,c)a_i - q_i)} \exp\left(-\frac{(\Lambda(T,c)a_i - q_i)^2}{2\Lambda(T,c)R_{ii}}\right). \tag{11}$$

If q_i increases faster than $\Lambda(T,c)$ as $T \to \infty$, we receive analogously as $T \gg 1$

$$P(q,T) \approx \frac{\exp\left(-\frac{1}{2}\sum_{i,j=1}^{m} W_{ij}\frac{(q_i - \Lambda(T,c)a_i)(q_j - \Lambda(T,c)a_j)}{\Lambda(T,c)}\right)}{(2\pi)^{\frac{m}{2}} \det R^{\frac{1}{2}} \prod_{i=1}^{m} \sum_{j=1}^{m} W_{ij}\frac{q_j - \Lambda(T,c)a_j}{\sqrt{\Lambda(T,c)}}}. \tag{12}$$

Thus, as expected $P(q,T) \ll 1$ in this case for large T values, and $P(q,T) \approx 1$ if $\Lambda(T,c)$ increases faster than q_i.

4 Distribution of the Duration of the Large Order Selling

Here we will consider the intensity $\lambda(t,c) \equiv \lambda(c)$. The intensity dependence on the prices does not matter in the context of this section, so instead of $\lambda(c)$ we will use the notation λ. Denote $p_S(\cdot)$ PDF of S, a cost of one purchase. The mean of S $E\{S\}$ and the variance $Var\{S\}$ are

$$E\{S\} = m_1 = \sum_{i=1}^{m} c_i a_i, \quad Var\{S\} = \sigma^2 = \sum_{i=1}^{m} R_{ij} c_i c_j. \tag{13}$$

Let $T(s)$ be the amount of time it takes to sell goods for an amount equal to the money s, and

$$g(w,s) = E\{e^{-wT(s)}\} \tag{14}$$

be a moment generating function of $T(s)$.

Consider small time interval Δt. Denote ΔS the cost of the purchase during the interval. Then

$$T(s) = \Delta t + T(s - \Delta S),$$

and it follows

$$g(w,s) = e^{-w\Delta t} E_{\Delta S}\{g(w, s - \Delta S)\} = e^{(-w\Delta t)}$$

$$\times \left[(1 - \lambda\Delta t)g(w,s) + \lambda\Delta t \int_0^s g(w, s-x)p_S(x)dx + \lambda\Delta t \int_s^\infty p_S(x)dx\right] + o(\Delta t). \tag{15}$$

As $\Delta t \to 0$, we get

$$(\lambda + w)g(w,s) = \lambda \int_0^s g(w, s-x)p_S(x)dx + \lambda \int_s^\infty p_S(x)dx. \tag{16}$$

The initial moments $T_k(s)$ of $T(s)$ are defined by following relation

$$T_k(s) = (-1)^k g_w^{(k)}(0, s). \tag{17}$$

Taking derivative with respect to w and putting $w = 0$, and taking into account that $g(0, s) = 1$, we obtain an equation for $T_1(\cdot)$, the mean of $T(\cdot)$,

$$\lambda T_1(s) = 1 + \lambda \int_0^s T_1(s - x)p_S(x)dx. \tag{18}$$

The k-th, $k \neq 1$, initial moment of $T(\cdot)$, is defined by equation

$$\lambda T_k(s) = kT_{k-1}(s) + \lambda \int_0^s T_k(s - x)p_S(x)dx. \tag{19}$$

Consider the Laplace transforms of functions $p_S(\cdot)$ and $T_k(\cdot)$

$$X(u) = \int_0^\infty p_S(s)\exp(-us)ds, \quad F_k(u) = \int_0^\infty T_k(s)\exp(-us)ds. \tag{20}$$

From (18) we get

$$F_1(u) = \frac{1}{\lambda u(1 - X(u))},$$

and from (19)

$$\lambda F_k(u) = kF_{k-1}(u) + \lambda F_k(u)X(u).$$

It follows

$$F_k(s) = \frac{k!}{\lambda^k u(1 - X(u))^k}, \quad k = 1, 2, ...,$$

and

$$T_k(s) = \frac{k!}{2\pi j} \int_{c-j\infty}^{c+j\infty} \frac{\exp(us)}{\lambda^k u(1 - X(u))^k}du. \tag{21}$$

Let us consider the case $s \gg 1$. If $\mathrm{Re}u < 0$ then $|X(u) < 1|$ and $X(0) = 1$. Therefore the zeros of the function $1 - X(u)$ lie on the imaginary axis and in the left half-plane. By the residue theorem

$$\frac{1}{k!}T_k(s) = \mathrm{Res}\frac{\exp(us)}{\lambda^k u(1 - X(u))^k}\Big|_{u=0} + \sum_i \mathrm{Res}\frac{\exp(us)}{\lambda^k u(1 - X(u))^k}\Big|_{u=u_i \neq 0}.$$

All the residues, except for the residue at $u = 0$, contain exponential factor $e^{u_i s}$, where $\mathrm{Re}u_i > 0$, so the first term gives the main contribution to $T_k(s)$ as $s \gg 1$.
Denote

$$Y(u) = \frac{1 - X(u)}{u},$$

then

$$T_k(s) = \frac{1}{\lambda^k} \lim_{u \to 0} \frac{d^k}{du^k} \left[\frac{\exp(us)}{Y(u)^k} \right] \qquad (22)$$

as $s \gg 1$, and for $k = 1$

$$T_1(s) = \frac{s}{\lambda Y(0)} - \frac{\dot{Y}(0)}{\lambda Y(0)^2}.$$

Denote

$$m_k = \int_0^\infty s^k p_S(s)ds, k = 1, 2, ...,$$

then

$$T_1(s) = \frac{s}{\lambda m_1} + \frac{m_2}{2\lambda m_1^2}. \qquad (23)$$

For instance, consider the Gamma distribution of S with parameters α and $n-1$, where $n > 1$ is an integer.

Denote the Laplace transform of function $T_1(s)$

$$\Phi(\omega) = \int_0^\infty T_1(s)e^{-\omega s}ds.$$

It follows from (18) that

$$\Phi(\omega) = \frac{(1 + \alpha\omega)^n}{\lambda\omega((1 + \alpha\omega)^n - 1)},$$

and

$$T_1(s) = \frac{1}{2\pi j} e^{-\frac{s}{\alpha}} \int_{\sigma-j\infty}^{\sigma+j\infty} \frac{u^n}{\lambda(u-1)(u^n-1)} e^{\frac{s}{\alpha}u} du.$$

Let $u_k = \sqrt[n]{1} = \cos\frac{2(k-1)\pi}{n} + j\sin\frac{2(k-1)\pi}{n}$, $k = \overline{2, n}$, $u_k \neq 1$. Then

$$T_1(s) = \frac{1}{\lambda}\left(1 + \frac{s}{\alpha n} - \frac{1}{n}\sum_{k=2}^n \frac{1 + u_k e^{\frac{s(u_k-1)}{\alpha}}}{1 - u_k} \right). \qquad (24)$$

Figure 2 illustrates the accuracy of the approximation (23) for the Gamma distribution of S for $n = 5, \alpha = 0.5$ and $\alpha = 2$. Here the solid lines correspond to the exact formula (24), and the dashed lines correspond to the approximate formula (23), parameter $\lambda = 1$.

Analogously we receive the variance of $T(s)$

$$Var\{T(s)\} = D(s) = \frac{1}{\lambda^2}\left[\frac{sm_2}{m_1^3} - \frac{m_3}{3m_1^2} + \frac{5m_2^2}{4m_1^4} \right]. \qquad (25)$$

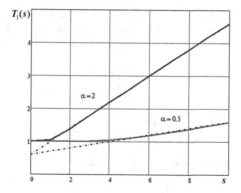

Fig. 2. The exact and approximate $T_1(s), \alpha = 0.5$ and $\alpha = 2, \lambda = 1$.

Let us denote

$$\Phi(w, u) = \int_0^\infty g(w, s) \exp(-us) ds.$$

From (16), we get

$$\Phi(w, u) = \lambda \frac{1 - X(u)}{u[w + \lambda - \lambda X(u)]},$$

and

$$g(w, s) = \frac{\lambda}{2\pi j} \int_{c-j\infty}^{c+j\infty} \frac{1 - X(u)}{u[w + \lambda - \lambda X(u)]} \exp(us) du. \qquad (26)$$

By the convolution theorem

$$g(w, s) = \int_0^s \phi(w, s - x) \int_x^\infty p_S(y) dy dx, \qquad (27)$$

where

$$\phi(w, s) = \frac{\lambda}{2\pi j} \int_{c-j\infty}^{c+j\infty} \frac{\exp(us)}{w + \lambda - \lambda X(u)} du. \qquad (28)$$

Consider the case when $s \gg 1$ and $w \ll 1$. Zeros of function $w + \lambda - \lambda X(u)$, $\mathrm{Re} w > 0$ that are defined by equation

$$\lambda(X(u) - 1) = w \qquad (29)$$

lie on the imaginary axis and in the left half-plane because for $\mathrm{Re} w > 0$ function $\frac{\exp(us)}{w + \lambda - \lambda X(u)}$ has to be analytical. As $s \gg 1$, the residue at zero gives the main contribution to the integral (28).

Note that if $w = 0$ then $u = 0$, and $X(0) - 1 = 0$, $\dot{X}(0) \neq 0$. It follows that the solution of (29) for $w \neq 0$ can be sought as the Burmann-Lagrange expansion

$$u_0 = \sum_{n=1}^{\infty} d_n w^n, \tag{30}$$

where

$$d_n = \frac{1}{n!} \lim_{z \to 0} \frac{d^{n-1}}{dz^{n-1}} \left[\frac{z^n}{(\lambda(X(z) - 1))^n} \right]. \tag{31}$$

According to the expansion we get as $w \ll 1$

$$u_0 = -\frac{1}{\lambda m_1} w + \frac{m_2}{2\lambda^2 m_1^3} w^2 + o(w^2). \tag{32}$$

By the residue theorem for $s \gg 1$ and $w \ll 1$

$$\phi(w, s) \sim -\frac{1}{\dot{X}(u_0)} \exp(u_0 s),$$

and

$$g(w, s) \sim -\frac{1}{\dot{X}(u_0)} \int_0^s \exp\left(u_0(s - x)\right) \int_x^{\infty} p_S(y) dy dx. \tag{33}$$

Consider random variable

$$V = \frac{T(s) - q^2}{q\delta},$$

where

$$q = \frac{s}{\lambda m_1}, \delta = \sqrt{\frac{m_2}{\lambda m_1^2}}. \tag{34}$$

It follows from (23) and (25) that $E\{V\} = 0$ and $Var\{V\} = 1$ as $s \gg 1$. The characteristic function of V

$$g_V(w, s) = \exp\left(\frac{wq}{\delta}\right) g\left(\frac{w}{q\delta}, s\right).$$

Taking into account (33) and (34) we get

$$g_V(w, s) \sim \frac{e^{\frac{w^2}{2}}}{\dot{X}\left(u_0\left(\frac{w}{\delta q}\right)\right)} \int_0^s \exp\left(-u_0\left(\frac{w}{\delta q}\right) x\right) \int_x^{\infty} p_S(y) dy dx.$$

Tending s to infinity we get

$$\lim_{s \to \infty} g_V(w, s) = \exp\left(\frac{w^2}{2}\right),$$

i.e. V is the standard normal random variable.

Thus, as $s \gg 1$ the selling time $T(s)$ has the normal distribution with parameters

$$T_1(s) = \frac{s}{\lambda m_1}, D(s) = \frac{sm_2}{\lambda^2 m_1^3} \qquad (35)$$

The cost of an order sold during the session with fixed probability

$$\alpha = P\{T(s) \leq T\} = \frac{1}{\sqrt{2\pi D}} \int_{-\infty}^{T} \exp\left(-\frac{(u - T_1)^2}{2D}\right) du$$

is defined by following equation

$$s = \frac{\lambda m_1}{4}\left[\sqrt{\frac{m_2}{\lambda m_1^2}\psi(\alpha)^2 + 4T} + \sqrt{\frac{m_2}{\lambda m_1^2}\psi(\alpha)}\right]^2,$$

where $\psi(\cdot)$ is the standard normal quantile function.

5   Price Optimization for Fast Moving Items

Let us denoted W the expected profit at the end of the session

$$W = -S + \sum_{j=1}^{m}\left[c_j q_j \int_{q_j}^{\infty} p_j(x)dx + \int_{0}^{q_j}(c_j x - b_j(q_j - x))p_j(x)dx\right], \qquad (36)$$

where $p_j(x)$ is the probability density function of a random customer demand for j-th product, and S is the cost of the order

$$S = \sum_{j=1}^{m} d_j q_j. \qquad (37)$$

The retailer is interested in determining an optimal value of $q = (q_1, q_2, ... q_m)$ and then corresponding value of $c = [c_1, c_2, ... c_m]^T$ by maximizing the expected profit. Differentiating (36) with respect to q_j we receive the following system of equations

$$\int_{0}^{q_j} p_j(x)dx = 1 - \frac{b_j + d_j}{b_j + c_j}. \qquad (38)$$

Taking into account that $d_j \leq c_j$, we see that the task has unique solution determined by the equation

$$q_j = \psi_j\left(1 - \frac{b_j + d_j}{b_j + c_j}\right), \qquad (39)$$

where $\psi_j(\cdot)$ is the inverse of the cumulative distribution function of demand for the product j.

The corresponding expected profit

$$W = \sum_{i=1}^{m} (c_i + b_i) \int_0^{q_i} x p_i(x) dx.$$ (40)

The dependence of the optimal batch volume for $m = 1$ and the density of demand distribution, determined by the relation (5) is shown in Fig. 3. The parameters of the density of the distribution of demand $\alpha = 0.5, n = 2$. Sales price $c = 8$, he cost of recycling the unit of production $s = 7$. As can be seen from the graphs given, the optimum value of q is determined mainly by the intensity of demand $\Lambda(T, c)$.

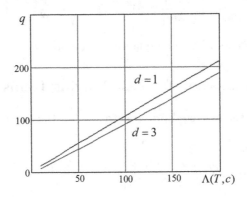

Fig. 3. Dependence of the optimal volume of the consignment of goods q on the value of $\Lambda(T, c)$

The majority of researches in the framework of multi-product newsvendor problem focus on determining the optimal stocking policy under some constraints.

Suppose that a budget limitation exists, that is, the overall order cost is fixed and is equal to S_0. Then we need to solve the task of profit maximization under the condition

$$S = \sum_{j=1}^{m} d_j q_j = S_0.$$ (41)

Solving the problem we get

$$\int_0^{q_j} p_j(x) dx = 1 - \frac{b_j + \mu d_j}{b_j + c_j},$$ (42)

where μ is the Lagrange multiplier defined by (41). Thus,

$$q_j = \psi_j \left(1 - \frac{b_j + \mu d_j}{b_j + c_j}\right),$$ (43)

and

$$f(\mu) = \sum_{j=1}^{m} d_j \psi_j \left(1 - \frac{b_j + \mu d_j}{b_j + c_j} \right) = S_0. \tag{44}$$

Function $\psi_j(\cdot)$ monotonically increases on $[0, 1]$ and $0 \leq \psi_j(\cdot) < \infty$, so the solution of (44) with respect to $\mu > 1$ exists.

The corresponding expected profit

$$W = (\mu - 1)S_0 + \sum_{j=1}^{m} (c_j + b_j) \int_0^{q_j} x p_j(x) dx > 0. \tag{45}$$

As $\Lambda(T, c) \gg 1$ we can consider PDF of $p_j(\cdot)$ as approximately normal with the mean $\Lambda(T, c)a_i$ and variance $\Lambda(T, c)R_{ii}$, and we can rewrite (39) and (43) in the following way correspondingly

$$q_j = \Lambda(T, c)a_j + \sqrt{\Lambda(T, c)R_{ii}} \psi \left(1 - \frac{b_j + d_j}{b_j + c_j} \right), \tag{46}$$

$$q_j = \Lambda(T, c)a_j + \sqrt{\Lambda(T, c)R_{ii}} \psi \left(1 - \frac{b_j + \mu d_j}{b_j + c_j} \right), \tag{47}$$

where $\psi(\cdot)$ is the standard normal quantile function.

The relative error of approximation for $\Delta = |q - q_{as}|/q$ for $m = 1$ as a function of the quantity $\Lambda(T, c)$, where q is calculated from formula (38), and q - by formula (46) for $b = 0.1$, $d = 2$, $c = 4$ is given in Table 2.

Table 2. The error of approximation

$\Lambda(T, c)$	5	10	20	50	100	200
Δ	0.074	0.035	0.017	$6.7 \cdot 10^{-3}$	$3.36 \cdot 10^{-4}$	$3.36 \cdot 10^{-4}$

As follows from the data presented, the approximation error practically decreases linearly with increasing $\Lambda(T, c)$.

Let the retail prices be proportional to the wholesale prices with the same coefficient, i.e., $c_i = \gamma d_j$, where $\gamma \geq 1$, and $\Lambda(T, c) = \lambda_0 T f(\gamma)$, where $f(\gamma)$ is a monotonically decreasing function, such that $f(0) = 1$ and $\gamma f(\gamma) \to 0$ as $\gamma \to \infty$.

We will consider the case $\Lambda(T, c) \gg 1$, or $\lambda_0 T \gg 1$, and we are interested in obtaining the optimal value of γ when the order quantities are defined by (46).

We can rewrite (40) as

$$W(\gamma) = \lambda_0 T \left[\sum_{i=1}^{m} a_i d_i f(\gamma)(\gamma - 1) \right.$$

$$\left. - \frac{\sqrt{f(\gamma)}}{\sqrt{2\pi\lambda_0 T}} \sum_{i=1}^{m} \sqrt{R_{ii}}(\gamma d_i + b_i) \exp \left(-\frac{1}{2}\psi^2 \left(1 - \frac{b_i + d_i}{b_i + \gamma_0 d_i} \right)^2 \right) \right]. \tag{48}$$

As the second term in (48) is of order $\frac{1}{\sqrt{\lambda_0}}$ for $\lambda_0 \gg 1$ the solution of the problem can be sought in the form

$$\gamma = \gamma_0 + \gamma_1 \frac{1}{\sqrt{\lambda_0 T}}$$

where γ_0 is determined from the condition $f(\gamma)(\gamma - 1) = \max$, and γ_1 when γ_0 is found from condition $\frac{dW(\gamma_1)}{d\gamma_1} = 0$.

6 Conclusion

Thus, we have considered a single period multi items inventory system when the demand is a compound Poisson process with price-and time-dependent intensity and a continuous joint batch sizes distribution with known means and covariance matrix independent from the arrival process.

We assume that the mean of the cumulative demand is large enough, that is the items are fast moving, to use the normal approximation to the demand distribution. Under this assumption we have obtained the main parts of the probabilities that an order have been sold to the end of the period assuming that the orders quantities increase faster than the corresponding means of the cumulative demand and vice versa.

As to the prices optimization we consider the case when the retail prices are proportional to the wholesale prices with the same coefficient and the intensity depends on the coefficient only and does not depend on time. Equation for the optimal coefficient maximizing the expected profit under optimal order quantities is obtained and approximate solution is proposed. The results are analogous to the single item case.

For time-independent intensity approximate distribution of the duration of the large order selling time is found without using the diffusion approximation of the demand process. As to compare to Kitaeva et al. [17], where the term "large order" is treated as large in size here it is treated as expensive.

References

1. Sharma, S.C.: Operation Research: Inventory Control and Queuing Theory. Series: DPH Mathematics. Discovery Publishing House Pvt. Ltd., New Delhi (2006)
2. Nazarov, A.A.: Asymptotic Analysis of Markovian Systems. Tom. State Univ. Publ. House, Tomsk (1991). (in Russian)
3. Schwarz, M., Daduna, H.: Queueing systems with inventory management with random lead times and with back ordering. Math. Meth. Oper. Res. 64, 383–414 (2006)
4. Krishnamoorthy, A., Lakshmy, B., Manikandan, R.: A survey on inventory models with positive service time. Opsearch 48(2), 153–169 (2011)
5. Manikandan, R., Sajeev, S.: Nair: M/M/1/1 queueing-inventory system with retail of unsatisfied customers. Commun. Appl. Anal. 21(2), 217–236 (2017)

6. Edgeworth, F.: The mathematical theory of banking. J. Roy. Statist. Soc. **51**, 113–127 (1888)
7. Khouja, B.: The single period (news-vendor) problem: literature review and suggestions for future research. Omega **27**, 537–553 (1999)
8. Qin, Y., Wang, R., Vakharia, A., Chen, Y., Hanna-Seref, M.: The newsvendor problem: review and directions for future research. Eur. J. Oper. Res. **213**, 361–374 (2011)
9. Turken, N., Tan, Y., Vakharia, A.J., Wang, L., et al.: The Multi-Product Newsvendor Problem: Review, Extension, and Directions for Future Research. In: Handbook of Newsvendor Problems: Models, Extensions and Applications, New York, vol. 176, pp. 3–39 (2012)
10. Whitin, T.M.: Inventory control and price theory. Manage. Sci. **2**(1), 61–68 (1955)
11. Petruzzi, N.C., Dada, M.: Pricing and the newsvendor problem: a review with extensions. Oper. Res. **47**(2), 183–194 (1999)
12. Yano, C.A., Gilbert, S.M.: Coordinated pricing and production/procurement decisions: a review. In: Eliashberg, J., Chakravarty, A. (eds.) Managing Business Interfaces: Marketing, Engineering, and Manufacturing Perspectives, pp. 63–103. Kluwer Academic Publishers, Dordrecht (2003)
13. Chen, X., Simchi-Levi, D.: Joint pricing and inventory management. In: Özer, O., Phillips, R. (eds.) The Oxford Handbook of Pricing Management, pp. 784–822. Oxford University Press, London (2012)
14. Nahmias, S.: Demand estimation in lost sales inventory systems. Naval Res. Logist. **41**, 739–757 (1994)
15. Kitaeva, A.V., Subbotina, V.I., Stepanova, N.V.: Estimating the compound poisson demand's parameters for single period problem for large lot size. In: Dolgui, A., Sasiadek, J., Zaremba, M. (eds.) The 15th IFAC Symposium on Information Control in Manufacturing, Ottawa, Canada, 11–13 May 2015, vol. 48(3), pp. 1357–1361. Elsevier, IFAC-PapersOnLine (2015)
16. Kitaeva, A.V., Subbotina, V.I., Zhukovskiy, O.I.: Compound poisson demand inventory models with exponential batch size's distribution. In: Dudin, A., Nazarov, A., Yakupov, R. (eds.) ITMM 2015. CCIS, vol. 564, pp. 240–249. Springer, Cham (2015). doi:10.1007/978-3-319-25861-4_21
17. Kitaeva, A.V., Zhukovskaya, A.O., Zmeev, O.A.: Compound Poisson demand with price-dependent intensity for fast moving items: price optimisation and parameters estimation. Int. J. Prod. Res. **55**(4), 4153–4163 (2016)

Two-Sided Truncations for a Class of Continuous-Time Markov Chains

Yacov Satin[1], Alexander Zeifman[1,2,3(✉)], Anna Korotysheva[1], and Ksenia Kiseleva[1,4]

[1] Vologda State University, Vologda, Russia
[2] Institute of Informatics Problems FRC CSC RAS, Moscow, Russia
[3] ISEDT RAS, Vologda, Russia
a_zeifman@mail.ru
[4] RUDN University, Moscow, Russia

Abstract. We consider nonstationary Markovian queueing models with batch arrivals and group services. We study the mathematical expectation of the respective queue-length process and obtain the bounds on the rate of convergence and error of truncation of the process.

Keywords: Nonstationary Markovian queueing model · Queue-length process · Batch arrivals · Group services · Truncation · Limiting characteristic · Bounds

1 Introduction

The problem of finding of the limiting characteristics for inhomogeneous Markov chains is both very important and very difficult. For the computation of the main limiting characteristics of the chain such their "quasi-stationary distribution" of state probabilities and the limiting mean, we need to solve some additional problems. Firstly, one should obtain explicit bounds on the rate of convergence, this is the subject of the series of papers, see for instance [2,4,13,17]. On the other hand, if the state space of the process is countable or finite but large, then we need to approximate the original chain by similar chains on the smaller state spaces. As a rule the authors deal with so-called north-west truncations, see [3,9,11,12,16,18]. More general approach for a class of homogeneous Markovian queues has been studied in [10]. There is a number of situations when we need consider another truncations with states $\{N_1, \ldots, N_2\}$, say. For instance, such situation is natural for the queue-length process in $M_t/M_t/S$ queue in the case of sufficiently large traffic intensity. As we know the first investigations of such truncations (for birth-death processes) have been published in [7,8]. In this paper we consider the corresponding problem for another class of inhomogeneous Markov chains. Namely, we deal with a class of chains which transition rates do not depend on the current state of the system and depend on the 'length of jump' and on the time moment only, see detailed discussion and bounds (the

© Springer International Publishing AG 2017
A. Dudin et al. (Eds.): ITMM 2017, CCIS 800, pp. 312–323, 2017.
DOI: 10.1007/978-3-319-68069-9_25

rate of convergence, perturbation bounds, north-west truncation bounds) in our previous papers [5, 6, 15].

Consider a nonstationary continuous-time Markovian queueing model on the state space $E = \{0, 1, \dots \}$ with possible batch arrivals and group services.

Let $X = X(t)$, $t \geq 0$ be a queue-length process for the queue.

Let $p_{ij}(s, t) = Pr\{X(t) = j\,|X(s) = i\}$, $i, j \geq 0$, $0 \leq s \leq t$ be transition probabilities for $X = X(t)$, and $p_i(t) = Pr\{X(t) = i\}$ be its state probabilities.

Throughout the paper we assume that

$$\Pr(X(t+h) = j/X(t) = i)$$

$$= \begin{cases} q_{ij}(t)\,h + \alpha_{ij}(t, h) & \text{if } j \neq i \\ 1 - \sum_{k \neq i} q_{ik}(t)\,h + \alpha_i(t, h) & \text{if } j = i, \end{cases}$$

where all $\alpha_i(t, h)$ are $o(h)$ uniformly in i, i.e. $\sup_i |\alpha_i(t, h)| = o(h)$.

We also assume $q_{i, i+k}(t) = \lambda_k(t)$, $q_{i, i-k}(t) = \mu_k(t)$ for any $k > 0$.

In other words, we suppose that the arrival rates $\lambda_k(t)$ and the service rates $\mu_k(t)$ do not depend on the length of queue. In addition, we assume that $\lambda_{k+1}(t) \leq \lambda_k(t)$ and $\mu_{k+1}(t) \leq \mu_k(t)$ for any k and almost all $t \geq 0$. Applying our standard approach (see details in [2, 13, 14]) we suppose in addition, that all intensity functions are locally integrable on $[0, \infty)$. Moreover we assume

$$L_\lambda(t) \leq L_\lambda < \infty, \quad L_\mu(t) \leq L_\mu < \infty,$$

for almost all $t \geq 0$, where

$$L_\lambda(t) = \sum_{i=1}^{\infty} \lambda_i(t), L_\mu(t) = \sum_{i=1}^{\infty} \mu_i(t).$$

Then the probabilistic dynamics of the process is represented by the forward Kolmogorov system:

$$\frac{d\mathbf{p}}{dt} = A(t)\mathbf{p}(t), \tag{1}$$

where

$$A(t) = \begin{pmatrix} a_{00}(t) & \mu_1(t) & \mu_2(t) & \mu_3(t) \cdots & \mu_r(t) & \cdots \\ \lambda_1(t) & a_{11}(t) & \mu_1(t) & \mu_2(t) \cdots \mu_{r-1}(t) \cdots \\ \lambda_2(t) & \lambda_1(t) & a_{22}(t) & \mu_1(t) \cdots \mu_{r-2}(t) \cdots \\ \cdots \\ \lambda_r(t) & \lambda_{r-1}(t) & \lambda_{r-2}(t) & \cdots & \cdots & a_{rr}(t) & \cdots \\ \cdots \end{pmatrix},$$

and $a_{ii}(t) = -\sum_{k=1}^{i} \mu_k(t) - \sum_{k=1}^{\infty} \lambda_k(t)$.

Throughout the paper by $\|\cdot\|$ we denote the l_1-norm, i.e. $\|\mathbf{x}\| = \sum |x_i|$, and $\|B\| = \sup_j \sum_i |b_{ij}|$ for $B = (b_{ij})_{i,j=0}^{\infty}$.

Let Ω be a set all stochastic vectors, i.e. l_1 vectors with nonnegative coordinates and unit norm. Hence we have $\|A(t)\| = 2 \sum_{k=1}^{\infty} (\lambda_k(t) + \mu_k(t)) \leq$

$2\left(L_{\lambda}+L_{\mu}\right)$ for almost all $t \geq 0$. Hence the operator function $A(t)$ from l_1 into itself is bounded for almost all $t \geq 0$ and locally integrable on $[0; \infty)$. Therefore we can consider (1) as a differential equation in the space l_1 with bounded operator.

It is well known (see [1]) that the Cauchy problem for differential equation (1) has a unique solutions for an arbitrary initial condition, and $\mathbf{p}(s) \in \Omega$ implies $\mathbf{p}(t) \in \Omega$ for $t \geq s \geq 0$.

Denote by $E(t, k) = E\{X(t) \,|\, X(0) = k\}$ the mean (the mathematical expectation) of the process at the moment t under the initial condition $X(0) = k$, and by $E_{\mathbf{p}}(t)$ the mathematical expectation (the mean) at the moment t under the initial probability distribution $\mathbf{p}(0) = \mathbf{p}$.

Recall that a Markov chain $X(t)$ is called *weakly ergodic*, if $\|\mathbf{p}^*(t) - \mathbf{p}^{**}(t)\| \to 0$ as $t \to \infty$ for any initial conditions $\mathbf{p}^*(0), \mathbf{p}^{**}(0)$, where $\mathbf{p}^*(t)$ and $\mathbf{p}^{**}(t)$ are the corresponding solutions of (1).

A Markov chain $X(t)$ has the *limiting mean* $\varphi(t)$ if $\lim_{t \to \infty}(\varphi(t) - E(t, k)) = 0$ for any k.

Our approach in general is based on the notion of logarithmic norm of a linear operator and a special similarity transformation of the matrix of intensities of the Markov chain considered.

Recall that the logarithmic norm of operator function $B(t)$ is defined as

$$\gamma(B(t)) = \lim_{h \to +0} h^{-1}\left(\|I + hB(t)\| - 1\right).$$

The important inequality

$$\|U(t, s)\| \leq \exp \int_s^t \gamma(B(u))\, du$$

holds, where $U(t, s) = U(t)U^{-1}(s)$ is the Cauchy operator of the corresponding differential equation

$$\frac{d\mathbf{z}}{dt} = B(t)\mathbf{z}(t).$$

2   Auxiliary Transformations

By introducing $p_i(t) = 1 - \sum_{j \neq i} p_j(t)$ from (1) we obtain the following equation

$$\frac{d\mathbf{z}}{dt} = B(t)\mathbf{z}(t) + \mathbf{f}(t), \tag{2}$$

where $\mathbf{f}(t) = (\mu_i, \mu_{i-1}, \ldots, \mu_1, \lambda_1, \lambda_2, \ldots)^T$,
$$\begin{array}{ccccc} & 0 & 1 & & i-1 & i+1 \end{array}$$
$$\mathbf{z}^T(t) = \begin{pmatrix} p_0, & p_1, & \ldots, & p_{i-1}, & p_{i+1}, & \ldots \end{pmatrix}.$$

Let D_1 be a matrix

$$
D_1 = \begin{array}{c}
\\ 0 \\ \\ i-2 \\ i-1 \\ i+1 \\ i+2 \\ i+3 \\ \\
\end{array}
\begin{array}{c}
\begin{array}{ccccccc}
0 & & i-2 & i-1 & i+1 & i+2 & i+3 \\
\end{array} \\
\left(\begin{array}{ccccccc}
-1 & \cdots & 0 & 0 & 0 & 0 & 0 & \cdots \\
\cdots & \cdots & \cdots & \cdots & \cdots & \cdots & \cdots & \cdots \\
-1 & \cdots & -1 & 0 & 0 & 0 & 0 & \cdots \\
-1 & \cdots & -1 & -1 & 0 & 0 & 0 & \cdots \\
0 & \cdots & 0 & 0 & 1 & 1 & 1 & \cdots \\
0 & \cdots & 0 & 0 & 0 & 1 & 1 & \cdots \\
0 & \cdots & 0 & 0 & 0 & 0 & 1 & \cdots \\
\cdots & \cdots & \cdots & \cdots & \cdots & \cdots & \cdots & \cdots \\
\end{array}\right)
\end{array}.
$$

Put $D_2 = diag\,(d_0, \ldots, d_{i-1}, d_{i+1}, d_{i+2}, \ldots)$, where $\{d_k\}$ be a sequence of positive numbers.

Denote by $D = D_2 D_1$.

Let l_{1D} be the space of sequences

$$l_{1D} = \{\mathbf{z} : \|\mathbf{z}\|_{1D} \equiv \|D\mathbf{z}\| < \infty\}.$$

For an operator function from l_1 to itself we have the simple formula of the logarithmic norm

$$\gamma(B(t)) = \sup_j \left(b_{jj}(t) + \sum_{i \neq j} |b_{ij}(t)| \right).$$

Hence we obtain the following bound for the logarithmic norm of operator function $B(t)$:

$$\gamma(B(t))_{1D} = \gamma(DB(t)D^{-1}) = \sup_{j \neq i}\{-\alpha_j(t)\} = -\alpha(t),$$

where $-\alpha_j(t)$ is the sum of all elements of j-th column of the matrix

$$
DBD^{-1} = \left(\begin{array}{cccccc}
a_{11} - \mu_1 & (\mu_1 - \mu_2)\frac{d_0}{d_1} & (\mu_2 - \mu_3)\frac{d_0}{d_2} & \cdots & (\mu_{i-2} - \mu_{i-1})\frac{d_0}{d_{i-2}} & \frac{d_0}{d_{i+1}}\mu_{i+1} & \cdots \\
\lambda_1 \frac{d_1}{d_0} & a_{22} - \mu_2 & (\mu_1 - \mu_3)\frac{d_1}{d_2} & \cdots & (\mu_{i-3} - \mu_{i-1})\frac{d_1}{d_{i-2}} & \frac{d_1}{d_{i+1}}\mu_i & \cdots \\
\lambda_2 \frac{d_2}{d_0} & \lambda_1 \frac{d_2}{d_1} & a_{33} - \mu_3 & \cdots & (\mu_{i-4} - \mu_{i-1})\frac{d_2}{d_{i-2}} & \frac{d_2}{d_{i+1}}\mu_{i-1} & \cdots \\
\cdots & & & & & & \\
\end{array}\right).
$$

Therefore

$$\|U(t,s)\|_{1D} \leq e^{-\int_s^t \alpha(u)\,du}.$$

We will suppose that there exist positive M and a such that

$$e^{-\int_s^t \alpha(u)\,du} \leq M e^{-a(t-s)}, \tag{3}$$

for any $0 \leq s \leq t$.

316   Y. Satin et al.

Put

$$g_j = \begin{cases} \sum_{j}^{i-1} d_j, \, j < i, \\ \sum_{i+1}^{j} d_j, \, j > i. \end{cases}$$

Consider the family of 'truncated' processes $X_\delta(t)$ on the state spaces $E_\delta = \{N_1, N_1 + 1, ..., N_2\}$, where $N_2 - N_1 = \delta$ and the respective birth rates are $\lambda_n(t), 0 < n \le \delta$ and death rates are $\mu_n(t), 0 < n \le \delta$. We denote by $\widetilde{A}(t)$ the corresponding transposed intensity matrix. Then the probabilistic dynamics of the truncated process is represented by the corresponding forward Kolmogorov system

$$\frac{d\widetilde{\mathbf{p}}}{dt} = \widetilde{A}(t)\mathbf{p}(t),$$

where

$$\widetilde{A}(t) = \begin{pmatrix} \widetilde{a}_{N_1,N_1}(t) & \mu_1(t) & \mu_2(t) & \mu_3(t) & \cdots & \mu_\delta(t) \\ \lambda_1(t) & \widetilde{a}_{11}(t) & \mu_1(t) & \mu_2(t) & \cdots & \mu_{\delta-1}(t) \\ \lambda_2(t) & \lambda_1(t) & \widetilde{a}_{22}(t) & \mu_1(t) & \cdots & \mu_{\delta-2}(t) \\ \cdots \\ \lambda_\delta(t) & \lambda_{\delta-1}(t) & \lambda_{\delta-2}(t) & \cdots & \cdots & \widetilde{a}_{N_2,N_2}(t) \end{pmatrix},$$

where $\widetilde{a}_{j,j}(t) = -\sum_{i=1}^{\delta-j} \lambda_i(t) - \sum_{i=1}^{j} \mu_i(t)$,

$$\widetilde{\mathbf{p}}(t) = (\widetilde{p}_{N_1}, \widetilde{p}_{N_1+1}, ..., \widetilde{p}_{N_2})^T.$$

By introducing $\widetilde{p}_i(t) = 1 - \sum_{j=N_1}^{i-1} \widetilde{p}_j(t) - \sum_{j=i+1}^{N_2} \widetilde{p}_j(t)$ we obtain the following equation

$$\frac{d\widetilde{\mathbf{z}}}{dt} = \widetilde{B}(t)\widetilde{\mathbf{z}}(t) + \widetilde{\mathbf{f}}(t),\qquad(4)$$

where $\widetilde{\mathbf{f}}(t) = (\mu_{i-N_1}, \mu_{i-N_1-1}, ..., \mu_1, \lambda_1, \lambda_2, ..., \lambda_{N_2-i})^T$,

$$\widetilde{\mathbf{z}}^T(t) = \begin{matrix} N_1 & N_1+1 & & i-1 & i+1 & & N_2 \\ (\widetilde{p}_{N_1}, & \widetilde{p}_{N_1+1}, & \cdots, & \widetilde{p}_{i-1}, & \widetilde{p}_{i+1}, & \cdots, & \widetilde{p}_{N_2}), \end{matrix}$$

$$\widetilde{B} = \begin{matrix} & N_1 & & i-1 & i+1 & & N_2 \\ N_1 \\ N_2 \end{matrix}\begin{pmatrix} \widetilde{a}_{N_1,N_1} - \mu_{i-N_1} & \cdots & \mu_{i-N_1-1} - \mu_{i-N_1} & \mu_{i-N_1+1} - \mu_{i-N_1} & \cdots & \mu_\delta - \mu_{i-N_1} \\ \cdots \\ \lambda_\delta - \lambda_{\delta-i} & \cdots & \lambda_{\delta-i+1} - \lambda_{\delta-i} & \lambda_{\delta-i-1} - \lambda_{\delta-i} & \cdots & \widetilde{a}_{N_2,N_2} - \lambda_{\delta-i} \end{pmatrix}.$$

Let $\widetilde{D}_1$ be a matrix

$$
\widetilde{D}_1 =
\begin{array}{c}
\\ N_1 \\ \\ i-2 \\ i-1 \\ i+1 \\ i+2 \\ i+3 \\ \\ N_2
\end{array}
\begin{array}{ccccccccc}
N_1 & & i-2 & i-1 & i+1 & i+2 & i+3 & & N_2 \\
\left(\begin{array}{ccccccccc}
-1 & \cdots & 0 & 0 & 0 & 0 & 0 & \cdots & 0 \\
\cdots & & \cdots & & \cdots & & \cdots & \cdots & \cdots \\
-1 & \cdots & -1 & 0 & 0 & 0 & 0 & \cdots & 0 \\
-1 & \cdots & -1 & -1 & 0 & 0 & 0 & \cdots & 0 \\
0 & \cdots & 0 & 0 & 1 & 1 & 1 & \cdots & 1 \\
0 & \cdots & 0 & 0 & 0 & 1 & 1 & \cdots & 1 \\
0 & \cdots & 0 & 0 & 0 & 0 & 1 & \cdots & 1 \\
\cdots & \cdots & \cdots & \cdots & \cdots & \cdots & \cdots & \cdots & \cdots \\
0 & \cdots & 0 & 0 & 0 & 0 & 0 & \cdots & 1
\end{array}\right)
\end{array}.
$$

Denote by $\widetilde{D}_2 = diag\left(\widetilde{d}_{N_1},\dots,\widetilde{d}_{i-1},\widetilde{d}_{i+1},\dots,\widetilde{d}_{N_2}\right)$, where $\widetilde{d}_k$ be positive numbers.

Put $\widetilde{D} = \widetilde{D}_2\widetilde{D}_1$.

Let $l_{1\widetilde{D}}$ be the space of sequences: $l_{1\widetilde{D}} = \{\widetilde{\mathbf{z}} : \|\widetilde{\mathbf{z}}\|_{1\widetilde{D}} \equiv \|\widetilde{D}\widetilde{\mathbf{z}}\|\}$.

Hence we have the following bound for the logarithmic norm of operator function $\widetilde{B}(t)$:

$$
\gamma(\widetilde{B}(t))_{1\widetilde{D}} = \gamma(\widetilde{D}\widetilde{B}(t)\widetilde{D}^{-1}) = \sup_{j\neq i}\{-\widetilde{\alpha}_j(t)\} = -\widetilde{\alpha}(t), \tag{5}
$$

where

$$
\widetilde{D}\widetilde{B}\widetilde{D}^{-1}
$$

$$
= \left(\begin{array}{ccccc}
\widetilde{a}_{N_1,N_1} - \mu_1 & (\mu_1-\mu_2)\frac{d_{N_1}}{d_{N_1+1}} & (\mu_2-\mu_3)\frac{d_{N_1}}{d_{N_1+2}} & \cdots & (\mu_{\delta-1}-\mu_\delta)\frac{d_{N_1}}{d_{N_2}} \\
(\lambda_1-\lambda_\delta)\frac{d_{N_1+1}}{d_{N_1}} & \widetilde{a}_{N_1+1,N_1+1} - \mu_2 & (\mu_1-\mu_3)\frac{d_{N_1+1}}{d_{N_1+2}} & \cdots & (\mu_{\delta-2}-\mu_\delta)\frac{d_{N_1+1}}{d_{N_2}} \\
(\lambda_2-\lambda_\delta)\frac{d_{N_1+2}}{d_{N_1}} & (\lambda_1-\lambda_{\delta-1})\frac{d_{N_1+2}}{d_{N_1+1}} & \widetilde{a}_{N_1+2,N_1+2} - \mu_3 & \cdots & (\mu_{\delta-3}-\mu_\delta)\frac{d_{N_1+2}}{d_{N_2}} \\
\cdots & & & & \\
(\lambda_{\delta-1}-\lambda_\delta)\frac{d_{N_2}}{d_{N_1}} & (\lambda_{\delta-2}-\lambda_{\delta-1})\frac{d_{N_2}}{d_{N_1+1}} & (\lambda_{\delta-3}-\lambda_{\delta-2})\frac{d_{N_2}}{d_{N_1+2}} & \cdots & \widetilde{a}_{N_2,N_2} - \lambda_1
\end{array}\right).
$$

3   Main Bounds

Let $V(t,s)$ is the Cauchy matrix of differential equation (4). Using the Eq. (5) we obtain the equation

$$
\|V(t,s)\|_{1\widetilde{D}} \leq e^{-\int_s^t \widetilde{\alpha}(u)\,du} \leq \widetilde{M}e^{-\widetilde{a}(t-s)},
$$

under the additional supposition of an existence positive $\widetilde{M}$ and $\widetilde{a}$ such that

$$
e^{-\int_s^t \widetilde{\alpha}(u)\,du} \leq \widetilde{M}e^{-\widetilde{a}(t-s)}, \tag{6}
$$

for any $0 \leq s \leq t$.

Moreover, there exists a constant $\widetilde{K}$ such that $\|\widetilde{\mathbf{f}}(t)\|_{1\widetilde{D}} \leq \widetilde{K}$ for any $t \geq 0$. Hence the following bound holds:

$$\|\widetilde{\mathbf{z}}(t)\|_{1\widetilde{D}} \leq \|V(t)\|_{1\widetilde{D}}\|\widetilde{\mathbf{z}}(0)\|_{1\widetilde{D}}$$
$$+ \int_0^t \|V(t,\tau)\|_{1\widetilde{D}}\|\widetilde{\mathbf{f}}(\tau)\|_{1\widetilde{D}}d\tau \leq \frac{\widetilde{K}\widetilde{M}}{\widetilde{a}} + \widetilde{M}e^{-\widetilde{a}t}\|\widetilde{\mathbf{z}}(0)\|_{1\widetilde{D}},$$

for any $t \geq 0$.

Let now $X(0) = i$. Then $\widetilde{\mathbf{z}}(0) = (0,0,...,0)^T$, $\|\widetilde{\mathbf{z}}(0)\|_{1\widetilde{D}} = 0$ and

$$\|\widetilde{\mathbf{z}}(t)\|_{1\widetilde{D}} \leq \frac{\widetilde{K}\widetilde{M}}{\widetilde{a}}.$$

Therefore $\widetilde{p}_j(t) \geq 0$ for any j, t. Then

$$\|\widetilde{D}\widetilde{\mathbf{z}}\| \geq (\widetilde{d}_{i+1} + \cdots + \widetilde{d}_n) \sum_{j=n}^{N_2} \widetilde{p}_i = \widetilde{g}_n \sum_{j=n}^{N_2} \widetilde{p}_i(t),$$

$$\|\widetilde{D}\widetilde{\mathbf{z}}\| \geq (\widetilde{d}_n + \cdots + \widetilde{d}_{i-1}) \sum_{j=N_1}^{n} \widetilde{p}_i = \widetilde{g}_n \sum_{j=N_1}^{n} \widetilde{p}_i(t),$$

hence

$$\sum_{j=n}^{N_2} \widetilde{p}_i(t) \leq \frac{\widetilde{K}\widetilde{M}}{\widetilde{a}\widetilde{g}_n},$$

and

$$\sum_{j=N_1}^{n} \widetilde{p}(t) \leq \frac{\widetilde{K}\widetilde{M}}{\widetilde{a}\widetilde{g}_n},$$

for any $t \geq 0$.

Put

$$\widetilde{v}_j = \begin{cases} 0 & j < N_1, \\ \min\left(1, \frac{\widetilde{M}\widetilde{K}}{\widetilde{a}\widetilde{g}_j}\right) & N_1 \leq j < i, \\ \min\left(1, \frac{\widetilde{M}\widetilde{K}}{\widetilde{a}\widetilde{g}_j}\right) & i > j \geq N_2 \\ 0 & j > N_2. \end{cases}$$

Let $\widehat{A} = (\widehat{a}_{m,n})_{m,n=0}^{\infty}$, where

$$\widehat{a}_{m,n} = \begin{cases} \widetilde{a}_{m,n} & \text{if } N_1 \leq m,n \leq N_2 \\ 0 & \text{otherwise} \end{cases}$$

Consider a system

$$\frac{d\widehat{\mathbf{p}}}{dt} = \widehat{A}(t)\widehat{\mathbf{p}}(t), \quad \widehat{\mathbf{p}}(t) = (\widehat{p}_1, \widehat{p}_2, \dots)^T.$$

Let $X(0) = i$, then we get $\widehat{p}_j = \widetilde{p}_j$ for $N_1 \leq j \leq N_2$ and $\widehat{p}_j = 0$ otherwise. Therefore, we have

$$\widehat{\mathbf{p}}(t) = (0, \ldots, 0, \widetilde{p}_{N_1}, \widetilde{p}_{N_1+1}, \ldots, \widetilde{p}_{N_2}, 0, 0, \ldots)^T.$$

By introducing $\widehat{p}_i(t) = 1 - \sum_{j=N_1}^{i-1} \widetilde{p}_j(t) - \sum_{j=i+1}^{N_2} \widetilde{p}_j(t)$ we obtain the following equation

$$\frac{d\widehat{\mathbf{z}}}{dt} = \widehat{B}(t)\widehat{\mathbf{z}}(t) + \widehat{\mathbf{f}}(t), \tag{7}$$

where $\widetilde{\mathbf{f}}(t) = \widehat{\mathbf{f}}(t)$,

$$\widehat{\mathbf{z}}^T(t) = \begin{pmatrix} 0 & \cdots & 0 & \widetilde{p}_{N_1}, & \ldots, & \widetilde{p}_{i-1}, & \widetilde{p}_{i+1}, & \ldots, & \widetilde{p}_{N_2} & 0 & \cdots \end{pmatrix},$$

with column labels $0 \quad N_1-1 \quad N_1 \quad i-1 \quad i+1 \quad N_2 \quad N_2+1$.

Rewrite Eq. (7) in the form

$$\frac{d\widehat{\mathbf{z}}}{dt} = \mathbf{B}(t)\widehat{\mathbf{z}}(t) + \widehat{\mathbf{f}}(t) + \left(\widehat{B}(t) - B(t)\right)\widehat{\mathbf{z}}(t).$$

Therefore one has from (2) and (7)

$$\frac{d(\widehat{\mathbf{z}} - \mathbf{z})}{dt} = \mathbf{B}(t)\left(\widehat{\mathbf{z}}(t) - \mathbf{z}(t)\right) + \left(\widehat{B}(t) - B(t)\right)\widehat{\mathbf{z}}(t)$$

and

$$\|\widehat{\mathbf{z}}(t) - \mathbf{z}(t)\|_{1D} \leq \|U(t,0)\|_{1D}\|\widehat{\mathbf{z}}(0) - \mathbf{z}(0)\|_{1D}$$
$$+ \int_0^t \|U(t,s)\left(\widehat{B}(s) - B(s)\right)\widehat{\mathbf{z}}(s)\|_{1D}\, ds.$$

Since $X(0) = i$, then $\widehat{\mathbf{z}}(0) = \mathbf{z}(0) = 0$, $\|\widehat{\mathbf{z}}(0) - \mathbf{z}(0)\|_{1D} = 0$ and

$$\|\widehat{\mathbf{z}}(t) - \mathbf{z}(t)\|_{1D} \leq \int_0^t \|U(t,s)\|_{1D} \left\|\left(\widehat{B}(s) - B(s)\right)\widehat{\mathbf{z}}(s)\right\|_{1D}\, ds.$$

Consider the expression $(B(s) - \widehat{B}(s))\widehat{\mathbf{z}}(s)$, we have:

$$(B(s) - \widehat{B}(s))\widehat{\mathbf{z}}(s) =$$
$$(\mu_{N_1}\widetilde{p}_{N_1} + \mu_{N_1+1}\widetilde{p}_{N_1+1} + \cdots + \mu_{N_2}\widetilde{p}_{N_2},$$
$$\mu_{N_1-1}\widetilde{p}_{N_1} + \mu_{N_1}\widetilde{p}_{N_1+1} + \cdots + \mu_{N_2-1}\widetilde{p}_{N_2},$$
$$\cdots,$$
$$\mu_1\widetilde{p}_{N_1} + \mu_2\widetilde{p}_{N_1+1} + \cdots + \mu_{N_2-N_1+1}\widetilde{p}_{N_2},$$
$$-\left(\sum_{j=1}^{N_1}\mu_j + \sum_{j=N_2-N_1+1}^{\infty}\lambda_j\right)\widetilde{p}_{N_1},$$
$$\cdots,$$

$$- \left(\sum_{j=N_2-N_1+1}^{N_2} \mu_j + \sum_{j=1}^{\infty} \lambda_j \right) \widetilde{p}_{N_2},$$
$$\lambda_{N_2-N_1+1} \widetilde{p}_{N_1} + \lambda_{N_2-N_1} \widetilde{p}_{N_1+1} + \cdots + \lambda_1 \widetilde{p}_{N_2},$$
$$\ldots)^T.$$

The following bounds hold in $1D$ norm:

$$\|(\widehat{B}(s) - B(s)\widehat{\mathbf{z}}(s)\|_{1D}$$
$$\leq \sum_{k=0, k\neq i-N_1}^{N_2-N_1} \left(g_{N_1+k} \left(\sum_{j=1+k}^{N_1+k} \mu_j + \sum_{j=N_2-N_1+1-k}^{\infty} \lambda_j \right) \right.$$
$$+ \sum_{j=0}^{N_1-1} g_j \mu_{N_1-j+k} + \left. \sum_{j=N_2+1}^{\infty} g_j \lambda_{j-N_1-k} \right) \widetilde{p}_{N_1+k}$$
$$\leq 2 \sum_{k=0, k\neq i-N_1}^{N_2-N_1} \left(\sum_{j=0}^{N_1-1} g_j \mu_{N_1-j+k} + \sum_{j=N_2+1}^{\infty} g_j \lambda_{j-N_1-k} \right) \widetilde{p}_{N_1+k}.$$

Let now there exist n_1 and n_2, such that $N_1 < n_1 < i < n_2 < N_2$, and

$$\sum_{j=0}^{N_1-1} g_j \mu_{N_1-j+n_1}(t) \leq \varepsilon, \qquad \sum_{j=N_2+1}^{\infty} g_j \lambda_{j-N_1-n_2}(t) \leq \varepsilon, \qquad (8)$$

for any $t \geq 0$.
Then we have

$$\|(\widehat{B} - B)\widehat{\mathbf{z}}\|_{1D} \leq 2(n_2 - n_1)\varepsilon$$
$$+ 2\left((n_1 - N_1)\widetilde{v}_{N_1+n_1-1} + (N_2 - n_2)\widetilde{v}_{N_1+n_2+1}\right)$$
$$\left(\sum_{j=0}^{N_1-1} g_j \mu_{N_1-j} + \sum_{j=N_2+1}^{\infty} g_j \lambda_{j-N_2} \right).$$

Let

$$\left(\sum_{j=0}^{N_1-1} g_j \mu_{N_1-j}(t) + \sum_{j=N_2+1}^{\infty} g_j \lambda_{j-N_2}(t) \right) \leq L,$$

for any $t \geq 0$.
Hence

$$\|\widehat{\mathbf{z}}(t) - \mathbf{z}(t)\|_{1D}$$
$$\leq \frac{2M}{\alpha} \left((n_2 - n_1)\varepsilon + L\left((n_1 - N_1)\widetilde{v}_{N_1+n_1-1} + (N_2 - n_2)\widetilde{v}_{N_1+n_2+1}\right) \right).$$

From inequality

$$\|\mathbf{p}^* - \mathbf{p}^{**}\| = |p_0^* - p_0^{**}| + \cdots + |p_{i-1}^* - p_{i-1}^{**}|$$

$$+ \left| \left(1 - \sum_{j \neq i} p_j^*\right) - \left(1 - \sum_{j \neq i} p_j^{**}\right) \right|$$

$$+ |p_{i+1}^* - p_{i+1}^{**}| + \cdots \leq \frac{2}{\min(d_{i-1}, d_{i+1})} \|\mathbf{z}^* - \mathbf{z}^{**}\|_{1D},$$

we obtain

$$\|\widehat{\mathbf{p}} - \mathbf{p}\| \leq \frac{4M}{\alpha \min(d_{i-1}, d_{i+1})} \left((n_2 - n_1)\varepsilon \right.$$
$$\left. + L\left((n_1 - N_1)\widetilde{v}_{N_1+n_1-1} + (N_2 - n_2)\widetilde{v}_{N_1+n_2+1}\right)\right).$$

On the other hand, from inequality

$$\|\mathbf{p}^* - \mathbf{p}^{**}\|_{1E} = |p_1^* - p_1^{**}| + \cdots + (i-1)|p_{i-1}^* - p_{i-1}^{**}|$$

$$+ i \left| \left(1 - \sum_{j \neq i} p_j^*\right) - \left(1 - \sum_{j \neq i} p_j^{**}\right) \right| + (i+1)|p_{i+1}^* - p_{i+1}^{**}| + \cdots$$

$$\leq \max_{j \neq i} \left(\frac{j+1}{g_j}\right) \|\mathbf{z}^* - \mathbf{z}^{**}\|_{1D},$$

we obtain

$$\|\widehat{\mathbf{p}} - \mathbf{p}\|_{1E} \leq \frac{2M}{\alpha} \max_{j \neq i} \left(\frac{j+1}{g_j}\right) \left((n_2 - n_1)\varepsilon\right.$$
$$\left. + L\left((n_1 - N_1)\widetilde{v}_{N_1+n_1-1} + (N_2 - n_2)\widetilde{v}_{N_1+n_2+1}\right)\right).$$

As a result, we obtain the following statement.

Theorem 1. *Let there exist two sequences $\{d_k\}$ and $\{\widetilde{d}_k\}$ of positive numbers which are decreasing if $k < i$ and increasing if $k > i$ such that inequalities (3), (6) and (8) hold. Then $X(t)$ is exponentially weakly ergodic, has the limiting mean, say, $E(t,0)$, and the following bounds of truncation error hold:*

$$\|\mathbf{p}(t) - \widehat{\mathbf{p}}(t)\| \leq \frac{4M}{\alpha \min(d_{i-1}, d_{i+1})} \left((n_2 - n_1)\varepsilon \right.$$
$$\left. + L\left((n_1 - N_1)\widetilde{v}_{N_1+n_1-1} + (N_2 - n_2)\widetilde{v}_{N_1+n_2+1}\right)\right)$$

and

$$|E(t,0) - \widehat{E}(t,0)| \leq \frac{2M}{\alpha} \max_{j \neq i} \left(\frac{j+1}{g_j}\right) \left((n_2 - n_1)\varepsilon \right.$$
$$\left. + L\left((n_1 - N_1)\widetilde{v}_{N_1+n_1-1} + (N_2 - n_2)\widetilde{v}_{N_1+n_2+1}\right)\right),$$

for any $t \geq 0$.

4   Conclusion

In this paper we have obtained truncation bounds of two-sided approximations for a class of Markovian queueing models with batch arrivals and group services. The development of methodology for other classes of inhomogeneous Markovian queues seems to be a promising direction of further research.

Acknowledgments. The work was supported by the Ministry of Education of the Russian Federation (the Agreement number 02.a03.21.0008 of 24 June 2016), by the Russian Foundation for Basic Research, project no. 15-01-01698.

References

1. Daleckij, J., Krein, M.G.: Stability of solutions of differential equations in Banach space. Am. Math. Soc. Transl. **43** (1974)
2. Granovsky, B.L., Zeifman, A.I.: Nonstationary queues: estimation of the rate of convergence. Queueing Syst. **46**, 363–388 (2004)
3. Masuyama, H.: Continuous-time block-monotone Markov chains and their block-augmented truncations. Linear Algebr. Appl. **514**, 105–150 (2017)
4. Meyn, S.P., Tweedie, R.L.: Stability of Markovian processes III: Foster-Lyapunov criteria for continuous-time processes. Adv. Appl. Probab. **28**, 518–548 (1993)
5. Satin, Y.A., Zeifman, A.I., Korotysheva, A.V., Shorgin, S.Y.: On a class of Markovian queues. Inform. Appl. **5**(4), 6–12 (2011). (in Russian)
6. Satin, Y.A., Zeifman, A.I., Korotysheva, A.V.: On the rate of convergence and truncations for a class of Markovian queueing systems. Theor. Probab. Appl. **57**(3), 529–539 (2013)
7. Satin, Y., Korotysheva, A., Kiseleva, K., Shilova, G., Fokicheva, E., Zeifman, A., Korolev, V.: Two-sided truncations of inhomogeneous birth-death processes. In: ECMS 2016 Proceedings (2016). doi:10.7148/2016-0663
8. Satin, Y., Korotysheva, A., Shilova, G., Sipin, A., Fokicheva, E., Kiseleva, K., Zeifman, A., Korolev, V., Shorgin, S.: Two-sided truncations for the $M_t|M_t|S$ queueing model. In: ECMS 2017 Proceedings (2017)
9. Seneta, E.: Non-negative Matrices and Markov Chains. Springer Science & Business Media, Heidelberg (2006)
10. Stepanov, S.N.: Markov models with retrials: the calculation of stationary performance measures based on the concept of truncation. Math. Comput. Model. **30**(3–4), 207–228 (1999)
11. Tweedie, R.L.: Truncation approximations of invariant measures for Markov chains. J. Appl. Probab. **35**, 517–536 (1998)
12. Zeifman, A.I.: Truncation error in a birth and death system. USSR Comput. Math. Math. Phys. **28**(6), 210–211 (1988)
13. Zeifman, A.I.: Upper and lower bounds on the rate of convergence for nonhomogeneous birth and death processes. Stoch. Process. Appl. **59**, 157–173 (1995)
14. Zeifman, A., Leorato, S., Orsingher, E., Satin, Y., Shilova, G.: Some universal limits for nonhomogeneous birth and death processes. Queueing Syst. **52**, 139–151 (2006)
15. Zeifman, A., Korolev, V., Satin, Y., Korotysheva, A., Bening, V.: Perturbation bounds and truncations for a class of Markovian queues. Queueing Syst. **76**, 205–221 (2014)

16. Zeifman, A., Satin, Y., Korolev, V., Shorgin, S.: On truncations for weakly ergodic inhomogeneous birth and death processes. Int. J. Appl. Math. Comput. Sci. **24**, 503–518 (2014)
17. Zeifman, A.I., Korolev, V.Y.: Two-sided bounds on the rate of convergence for continuous-time finite inhomogeneous Markov chains. Stat. Probab. Lett. **103**, 30–36 (2015)
18. Zeifman, A.I., Korotysheva, A.V., Korolev, V., Satin, Y.A.: Truncation bounds for approximations of inhomogeneous continuous-time Markov chains. Theor. Probab. Appl. **61**, 563–569 (2016)

Construction of the Stability Indicator of Wireless D2D Connection in a Case of Fractal Random Walk of Devices

Yuliya V. Gaidamaka[1,2](✉), Elisabeth P. Kirina-Lilinskaya[3], Yurii N. Orlov[1,3], Andrey K. Samouylov[1,4], and Dmitri A. Molchanov[1,4]

[1] Department of Applied Probability and Informatics,
Peoples' Friendship University of Russia (RUDN University),
Miklukho-Maklaya Street 6, Moscow 117198, Russian Federation
gaydamaka_yuv@rudn.university
[2] Institute of Informatics Problems, Federal Research Center,
"Computer Science and Control" of the RAS,
Vavilov Street 44-2, Moscow 119333, Russian Federation
[3] Deparment of Kinetic Equations, Keldysh Institute of Applied
Mathematics of RAS, Miusskaya Sq. 4, Moscow 125047, Russian Federation
[4] Department of Electronics and Communications Engineering,
Tampere University of Technology, Korkeakoulunkatu 10, 33720 Tampere, Finland

Abstract. Fractional Fokker-Plank kinetic equation is used for simulation of stochastic motion of transmitter and receiver devices in wireless networks with D2D-communications. The evolution equations for dispersion of signal-to-interference ratio value and for normalized SIR average value as an indicator of stability of D2D connection are derived with the use of this kinetic equation. Some numerical results are presented.

Keywords: D2D communication · Wireless network · Fractal stochastic motion · Kinetic evolution equation · Signal-to-interference ratio · SIR · SIR dispersion evolution · Stability indicator

1 Introduction

The telephone industry was one of important application of queuing theory from its emergence at the beginning of the 20th century [1] through its development connected, among other things, with the solution of problems in communication networks [2–4] to its new application in modern telecommunications networks, 5 Generation networks (5G) [5]. Traditionally in telecommunications networks the main task is to distribute the network resource (in 2G or 3G networks – fixed communication channels, in modern 4G or 5G networks – bandwidth, data rates, radio frequencies for wireless communication networks) between users (subscribers, devices and applications transmitting data). In terms of queuing theory, the network resource is modeled as devices, and users who request the

© Springer International Publishing AG 2017
A. Dudin et al. (Eds.): ITMM 2017, CCIS 800, pp. 324–335, 2017.
DOI: 10.1007/978-3-319-68069-9_26

resource as input flow of the queuing system. In most cases in so-called multiservice networks [6–9]) it is necessary to take into account the heterogeneous incoming request flows with different resource requirements of fixed volume for different types of requests. For modern networks, flows with resource requirements of random volume [10–12] should be considered, with either discrete or continuous and even changing in time resource requirements [11–14]. A separate task here is to model the input flow, which for the 5G networks is associated with a random event flow implemented as a non-equidistant time series. A random value here is the volume of resource requirement to service the request. In modern wireless networks [5,15,16], the amount of network resource (bandwidth of radio frequencies in GSM and UMTS networks, the number of physical resource blocks in LTE, LTE-A, WIMAX networks) allocated for data transmission by a wireless device depends significantly on the total interference, which is created by other devices transmitting data at the same radio frequency. In this case, the power of the interfering signal from each such device is a function of the distance between the interfering transmitter and the signal receiver, which change the location due to the mobility of users in the wireless networks. Thus, in order to estimate the resource requirement for data transmission between the transmitter and receiver, it is necessary to take into account the mobile user traffic patterns and be able to model the trajectories of their movement. Because of lack of experimental data, concerning to random moving in a public place, one should considered different models of variation of the distance between devices: from regular straight line motion to chaotic motion over fractal structure. According to [17,18] the commonly used mobility models in Mobile Ad hoc Network are two variants of the Random Waypoint model, namely the Random Walk model and the Random Direction model, Brownian Motion, more complicated Jump Brownian Motion [19] and Lévy Flight [20]. In this paper following by our previous works [21,22], we consider a fractal random walk of the transmitter and receiver in the model of D2D wireless communication. Fractal motion is characterized by spatial correlation between sequence of coordinates of moving points. Hence if we need to describe the effect of long term spatial correlation between the moving points, the non-local integrodifferential equation of evolution of point coordinates distribution function can be used as a mathematical model. So the equations with fractional derivatives are the natural instruments to analyze this effect, because they can be presented as integral conversions with non-local singular kernels.

2   A Fractal Random Walk Model

We shall use the following notations. Let α is an order of fractional derivative. We shall consider a case, when $0 < \alpha < 1$. The value of fractal derivative is a parameter of the concrete task. This value could be estimated by method, described in [23].

The fractional order derivative has a different interpretations (see e.g. [24]). It can be considered in the senses of Riemann-Liouville, Riesz-Feller, Caputo-Gerasimov, Grunwald-Letnikov and others. For its existence, the function f is required to belonged to L_p, $p > 1$ class of the variable x. For example, a symmetrical Riesz-Feller derivative on section $[a; b]$ (in the case of normalized random variable this section is $[0; 1]$) is represented by the formula

$$\frac{\partial^{2\alpha} f(x)}{\partial x^{2\alpha}} = \frac{1}{2 \cos \pi \alpha} \left(\left(D_{a+}^{2\alpha} f \right)(x) + \left(D_{b-}^{2\alpha} f \right)(x) \right), \tag{1}$$

where

$$\begin{aligned}
\left(D_{a+}^{2\alpha} f \right)(x) &= \frac{1}{\Gamma(m - 2\alpha)} \frac{d^m}{dx^m} \int_a^x f(y)(x - y)^{m - 2\alpha - 1} \, dy, \\
\left(D_{b-}^{2\alpha} f \right)(x) &= \frac{(-1)^m}{\Gamma(m - 2\alpha)} \frac{d^m}{dx^m} \int_x^b f(y)(y - z)^{m - 2\alpha - 1} \, dy, \\
m &= [Re(2\alpha)] + 1.
\end{aligned} \tag{2}$$

If $0 < \alpha < 1$, then m in (2) can be equal to 1 or 2. Let us consider a case $m = 1$. Then definitions (2) are rewritten as

$$\begin{aligned}
\left(D_{a+}^{2\alpha} f \right)(x) &= \frac{1}{\Gamma(1 - 2\alpha)} \frac{d}{dx} \int_a^x \frac{f(y)}{(x - y)^{2\alpha}} \, dy, \\
\left(D_{b-}^{2\alpha} f \right)(x) &= \frac{-1}{\Gamma(1 - 2\alpha)} \frac{d}{dx} \int_x^b \frac{f(y)}{(y - x)^{2\alpha}} \, dy
\end{aligned} \tag{3}$$

and for the case of normalized values $x \in [0; 1]$ we have

$$\frac{\partial^{2\alpha} f(x)}{\partial x^{2\alpha}} = \frac{1}{2\Gamma(1 - 2\alpha) \cos \pi \alpha} \frac{d}{dx} \left(\int_0^x \frac{f(\xi)}{(x - \xi)^{2\alpha}} d\xi - \int_x^1 \frac{f(\xi)}{(\xi - x)^{2\alpha}} d\xi \right). \tag{4}$$

If $m = 2$, instead of (4) we have

$$\frac{\partial^{2\alpha} f(x)}{\partial x^{2\alpha}} = \frac{1}{2\Gamma(2 - 2\alpha) \cos \pi \alpha} \frac{d^2}{dx^2} \left(\int_0^x \frac{f(\xi)}{(x - \xi)^{2\alpha - 1}} d\xi + \int_x^1 \frac{f(\xi)}{(\xi - x)^{2\alpha - 1}} d\xi \right). \tag{5}$$

The expressions (4) and (5) will be used below.

We suppose, that evolution equation for distribution function density (DFD) of coordinates differences obeys to fractional Fokker–Planck equation with spatial derivative of the 2α order. Let $X(t) = X(0) + \sum_{k=1}^{t} x_k$, $Y(t) = Y(0) + \sum_{k=1}^{t} y_k$ is a trajectory coordinates for a certain devise in a discrete moment of time t, and $\{x_k, y_k\}$ is a pare of independent coordinate differences for the time step k. We suppose, that values $\{x_k, y_k\}$ vary independently, but all of belong to a certain fractal set.

Plan of our investigation is following. First of all we construct ensemble of trajectories on various fractal sets. For each ensemble realization we calculate the value of non-linear functional signal-to-interference ratio (SIR) for an arbitrary pare of transmitter and receiver. We obtain also theoretical formulas for evolution of average over ensemble SIR value and evolution of SIR dispersion due to kinetic equation. These equations give us the instrument to analyze the evolution of indicator of connection stability, equals to the ratio of average SIR to its standard deviation. If this ratio less then unit, the connection is unstable even if the average value of SIR is more, then lower limit, needed for communication. We present some numerical illustration of this situation.

3   Kinetic Approach to Random Functional Analysis

The method of generation of a trajectory of a random process as a random walk on a fractal set was proposed in [23]. In this work the Fokker-Planck equation with fractional derivatives as model of an evolution of the distribution function which floats on such sets was used. Now we describe this method in brief.

At first, we suppose, that the motion of each spatial coordinate does not depend on each other. Although this assumption is somewhat artificial, it can be met with small movements. Then the evolution equation in one-dimensional coordinate space can be considered as a basic model. This equation has the form:

$$\frac{\partial f(x,t)}{\partial t} + \frac{\partial \left(u(x,t) f(x,t) \right)}{\partial x} = B(t) \frac{\partial^{2\alpha} f(x,t)}{\partial x^{2\alpha}}, \tag{6}$$

where $f(x,t)$ is a continuously differentiable function of coordinates and time, $u(x,t)$ is so-called drift velocity, which was defined in [25] as the average speed of mutual two-dimensional distribution

$$u(x,t)f(x,t) = \int F(x,v,t)v\,dv.$$

In the last formula the integration is performed over the entire domain of walk with taking into account the boundary conditions. Nonnegative diffusion coefficient $B(t)$ is determined by the formula (see [23,25])

$$B(t) = \frac{1}{2}\frac{d\lambda^2}{dt} - cov_{x,u}(t),$$
$$\lambda^2(t) = \int (x - \bar{x}(t))^2 f(x,t)dx, \tag{7}$$
$$\bar{x}(t) = \int x f(x,t)dx.$$

The values of drift and diffusion are treated to be known from the experimental data. The typical values were presented in [22,26].

Now we must note, that kinetic equation (6) in a finite region is not valid. The thing is that non-negative DFD $f(x,t)$ for all time moments t must be normalized to unit, but the Eq. (6) does not conserve this property. Let us consider this situation in detail. If we suppose, that $f(0,t) = f(1,t) = 0$ (zero boundary conditions), then after integrating by x we obtain from (6), that

$$\int\limits_0^1 \frac{\partial^{2\alpha} f(x,t)}{\partial x^{2\alpha}}dx = 0.$$

If α is digit, this equality is valid because of boundary conditions. In general case with the use of definition (4) or (5) we find, that this equality is equivalent to

$$\int_0^1 (f(x,t) + f(1-x,t)) \frac{dx}{x^{2\alpha}} = 0.$$

But the last equality can not be satisfied, excluding the case $f(x,t) \equiv 0$. Only if the integration region is $(-\infty, +\infty)$, the problem is absent. Thus the boundary problem can not be solved correctly in the frame of this approach. Nevertheless if we are interested in fractal motion during relatively short time interval, the boundary effects are negligible and we can consider the infinite spatial interval.

So if we now want to analyze any functional

$$V(t) = \int_{-\infty}^{\infty} v(x) f(x,t)\, dx, \tag{8}$$

where $v(x)$ is a function, which average value we are interested in, its derivative over the time is obtained from the Eq. (6) with the use of (4) or (5). For brevity we consider the first case. Then we have

$$\begin{aligned}
\frac{dV(t)}{dt} &= \int_{-\infty}^{\infty} v(x) \frac{\partial f(x,t)}{\partial t} dx \\
&= \int_{-\infty}^{\infty} v(x) \left(-\frac{\partial(u(x,t)f(x,t))}{\partial x} + B(t)\frac{\partial^{2\alpha} f(x,t)}{\partial x^{2\alpha}} \right) dx \\
&= \int_{-\infty}^{\infty} v'(x) \left(u(x,t)f(x,t) - \frac{B(t)}{Z(\alpha)} G(x,t) \right) dx, \\
\end{aligned} \tag{9}$$

$$Z(\alpha) = 2\Gamma(1-2\alpha)\cos\pi\alpha,$$

$$G(x,t) = \int_{-\infty}^{x>0} \frac{f(\xi)}{(x-\xi)^{2\alpha}} d\xi - \int_{x}^{+\infty} \frac{f(\xi)}{(\xi-x)^{2\alpha}} d\xi.$$

Thus the evolution of a functional (8) can be represented through the average flow of corresponding values. This method is applied further for average SIR evolution analysis.

4   The Average SIR and Dispersion

Now we consider the functional, which depends on the distance between moving points, where their provisions together forms an ensemble of trajectories of a random process. Admit that the coordinates $(X_i(t), Y_i(t))$ determines the position of the i-th point of trajectory on a discrete step in two-dimensional considered region. In the accidentally moving bodies system the number of points (i.e. paths) is equal to N. The distance between points on two trajectories in 2D-space defined by formula:

$$r_{ij}^2(t) = (X_i(t) - X_j(t))^2 + (Y_i(t) - Y_j(t))^2 . \tag{10}$$

Let us denote two-points communication function, which depends on the distance between that points as $\varphi_{ij} \equiv \varphi\,(r_{ij})$. In case of tasks of communication between the transceivers the power function is usually chosen. Assume that

$$\varphi_{ij} \equiv \varphi\,(r_{ij}) = 1/r_{ij}^2 . \tag{11}$$

Select a specific pair of points, for example, 1 and 2. The SIR functional between a given pair of points at each time moment is determined by:

$$S\,(\mathbf{r}_1, \mathbf{r}_2) = \frac{\varphi_{12}}{\sum\limits_{j=3}^{N} \varphi_{1j}} . \tag{12}$$

The sum in the denominator of formula (12) is the average value of the function coupling with the first point multiplied by N which is defined as

$$U\,(\mathbf{r}, t) = \int \varphi\,(|\mathbf{r} - \mathbf{r}'|)\,f\,(\mathbf{r}', t)\,d\mathbf{r}', \tag{13}$$

where $f\,(\mathbf{r}', t)$ is the density distribution of distances between points in the present field. Let us introduce the distance $\mathbf{r} = \mathbf{r}_{12}$ and then consider all other points in the second particle's reference system. The quality of communication between the two points is determined by the average value of the functional $S(\mathbf{r}, t)$ due to the arbitrary pair of points. With the accuracy of $o(1/N)$ formula (12) can be represented as

$$S(\mathbf{r}, t) = \frac{\varphi(\mathbf{r})}{N U(\mathbf{r}, t)} . \tag{14}$$

The average value of (14) over ensemble is defined as

$$q(t) = \int S(\mathbf{r}, t) f(\mathbf{r}, t) dr. \tag{15}$$

The SIR dispersion for ensemble of trajectories is defined by formula

$$\sigma^2(t) = \int (S(\mathbf{r}, t) - q(t))^2 f(\mathbf{r}, t) dr. \tag{16}$$

We present below the results of modeling the mean SIR $q(t)$, the standard deviation of SIR $\sigma(t)$, and the value

$$\mu(t) = \frac{q(t)}{\sigma(t)} . \tag{17}$$

The question of variability of the average value of SIR with non-stationary trajectory of subscribers emerges. To resolve the matter, the corresponding evolution equation should be obtained. It is the main goal of this paper.

5 The SIR Dispersion Evolution Equation

Now we derive the evolution equation of the average value of the functional (14) in the frame of the model equation (6). Let us consider in detail one-dimensional case. Because of non-linearity of functional (15) with respect to DFD $f(x,t)$ we construct sufficiently complex expression. From (14)–(15) we have

$$N\frac{dq}{dt} = \int_{-\infty}^{\infty} \left(\frac{\varphi(x)}{U(x,t)} \frac{\partial f(x,t)}{\partial t} - \frac{\varphi(x)}{U^2(x,t)} \frac{\partial U(x,t)}{\partial t} f(x,t) \right) dx. \qquad (18)$$

Substituting the expression of derivative $\frac{\partial f(x,t)}{\partial t}$ from (6), we obtain in a same way as (9)

$$\int_{-\infty}^{\infty} \frac{\varphi(x)}{U(x,t)} \frac{\partial f(x,t)}{\partial t} dx = \int_{-\infty}^{\infty} \left(u(x,t)f(x,t) - \frac{B(t)}{Z(\alpha)}G(x,t) \right) \cdot \frac{\partial}{\partial x} \left(\frac{\varphi(x)}{U(x,t)} \right) dx. \qquad (19)$$

Analogously

$$\frac{\partial U(x,t)}{\partial t} = \int_{-\infty}^{\infty} \varphi(|x-y|) \frac{\partial f(y,t)}{\partial t} dy$$

$$= \int_{-\infty}^{\infty} \frac{\partial \varphi(|x-y|)}{\partial y} \left(u(y,t)f(y,t) - \frac{B(t)}{Z(\alpha)}G(y,t) \right) dy \qquad (20)$$

$$= -\frac{\partial}{\partial x} \int_{-\infty}^{\infty} \varphi(|x-y|) \left(u(y,t)f(y,t) - \frac{B(t)}{Z(\alpha)}G(y,t) \right) dy.$$

Finally we transform the expression (18) to the following form:

$$N\frac{dq}{dt} = \int_{-\infty}^{\infty} \left(u(x,t)f(x,t) - \frac{B(t)}{Z(\alpha)}G(x,t) \right) \frac{\partial}{\partial x} \left(\frac{\varphi(x)}{U(x,t)} \right) dx + \int_{-\infty}^{\infty} \frac{\varphi(x)f(x,t)}{U^2(x,t)}$$

$$\cdot \frac{\partial}{\partial x} \left(\int_{-\infty}^{\infty} \varphi(|x-y|) \left(u(y,t)f(y,t) - \frac{B(t)}{Z(\alpha)}G(y,t) \right) dy \right) dx. \qquad (21)$$

This equation nonlinearly depends on the distribution function of the points, i.e. on the trajectories of points density of the ensemble. Consequently, in order to assess the contributions of each of the members in relation to a particular practical situation, the numerical distribution statistics simulation, where statistics are directly related to the $S(r,t)$ and $q(t)$ distribution, is an urgent task. Equation (21) is a general model of the evolution of average SIR values in approximation of Fokker–Planck equation. We see, that drift $u(x,t)$ and diffusion $B(t)$ act on average SIR separately.

From (16) it follows, that

$$\frac{d\sigma^2(t)}{dt} = -2q(t)\frac{dq(t)}{dt} + 2\int S(x,t)\frac{\partial S(x,t)}{\partial t}f(x,t)dx + \int S^2(x,t)\frac{\partial f(x,t)}{\partial t}dx. \tag{22}$$

The first term in (22) is taken to be known from (21).
From (14) and (20) it follows, that

$$\frac{\partial S(x,t)}{\partial t} = -\frac{\varphi(x)}{NU^2(x,t)}\frac{\partial U(x,t)}{\partial t}$$

$$= \frac{\varphi(x)}{NU^2(x,t)}\cdot\frac{\partial}{\partial x}\int_{-\infty}^{\infty}\varphi(|x-y|)\left(u(y,t)f(y,t) - \frac{B(t)}{Z(\alpha)}G(y,t)\right)dy,$$

so that the second term in (22) takes the form

$$2\int S(x,t)\frac{\partial S(x,t)}{\partial t}f(x,t)dx = \frac{2}{N^2}\int\frac{\varphi^2(x)}{U^3(x,t)}$$

$$\cdot\left(\frac{\partial}{\partial x}\int_{-\infty}^{\infty}\varphi(|x-y|)\left(u(y,t)f(y,t) - \frac{B(t)}{Z(\alpha)}G(y,t)\right)dy\right)f(x,t)dx. \tag{23}$$

The third term in (22) is transformed with the use of (9):

$$\int S^2(x,t)\frac{\partial f(x,t)}{\partial t}dx$$

$$= 2\int_{-\infty}^{\infty}S(x,t)\frac{\partial S(x,t)}{\partial x}\left(u(x,t)f(x,t) - \frac{B(t)}{Z(\alpha)}G(x,t)\right)dx$$

$$= \frac{2}{N^2}\int_{-\infty}^{\infty}\left(\frac{\varphi(x)}{U^2(x,t)}\frac{d\varphi(x)}{dx} - \frac{\varphi^2(x)}{U^3(x,t)}\frac{\partial U(x,t)}{\partial x}\right)$$

$$\cdot\left(u(x,t)f(x,t) - \frac{B(t)}{Z(\alpha)}G(x,t)\right)dx. \tag{24}$$

This formulas (21), (23), (24) determine the evolution of SIR dispersion in one-dimensional case. For random fractal walk in square one should replace $\frac{\partial}{\partial x} \to grad_x$, $u(x,t) \to \mathbf{u}(x,y,t)$.

6   Some Numerical Examples

We shall consider two situations: (1) a relatively small average SIR value and $\mu(t) > 1$ showing a sufficiently good quality of communication, in contrast to situation (2), when a large average SIR value with a small $\mu(t)$ correspond to a significantly lower connection reliability. So the value of $\mu(t)$ could be treated as an indicator of connection reliability for D2D wireless network communication. The time periods, when $\mu(t) > 1$, are the periods with stable connection, and it is interesting to obtain the distribution function of μ and also the distribution function of periods with stable or unstable D2D connection. For this purpose we need to analyze the evolution equation for $q(t)$ and $\sigma(t)$. But theoretical formulas are complicated. So we may use direct modeling of SIR and statistics, connecting with them.

The typical example of trajectories of points are shown in Fig. 1. The trajectory, according to the method, developed in [23, 26], was obtained for the fractal walk on a two-dimensional ternary Cantor set (Fig. 1).

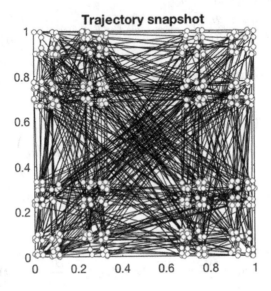

Fig. 1. Example of 10 trajectories of points on the ternary Cantor set in R^2

The corresponding SIR trajectory for an arbitrary pair of subscribers with their total number $N = 10$ is shown in Fig. 2. The values of SIR are presented on the vertical axis of this figure, horizontal axis corresponds to a time moment with a unit discrete step. The critical value of connection is equal to $s^* = 0.01$. In this example the average SIR over the generated sample is equal to 0.05, but the value of $\mu = q/\sigma = 0.7$. It appears, that in this case the SIR trajectory is below the critical level during 30% of time. For this case $2\alpha = 1.262$ (Cantor set).

Another example is presented at Fig. 3 for Menger set, $2\alpha = 1.893$. Here the average value of SIR is less then in previous case, but indicator μ is relatively large: $q = 0.03$, $\mu = 3$. The drift and diffusion parameters in (6) are the same in both cases. We see, that in the last case the connection for this sample of trajectory is without any break down.

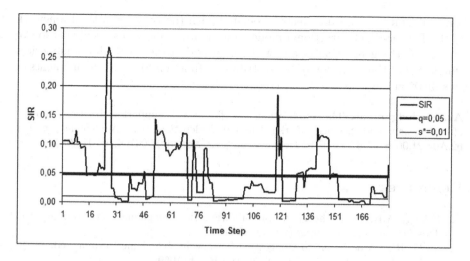

Fig. 2. Example of SIR trajectory on ternary Cantor set in R^2 in various time discrete points; the case of unstable connection

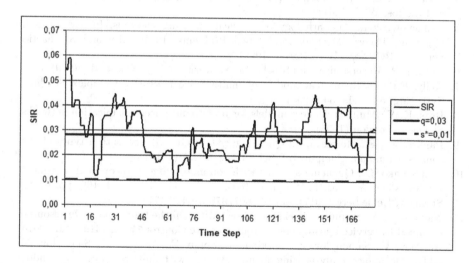

Fig. 3. Example of SIR trajectory on ternary Menger set in R^2 in various time discrete points; the case of stable connection

7   Conclusion

Thus, in this paper the equations of evolution of new indicators of reliability of D2D connections in the conditions of fractal random walk of subscribers are obtained. Since the investigated SIR functional is non-linear, the equations of evolution of the moments of its distributions turn out to be very cumbersome and have meaning only as constructions that make it possible to influence in a meaningful way certain effects associated with the transfer in the framework of the Fokker-Planck fractional model equation. A more convenient method of analysis is the generation of an ensemble of trajectories that allows numerical simulation of the evolution of the distribution function SIR and the moments of this distribution.

Acknowledgement. The publication was financially supported by the Ministry of Education and Science of the Russian Federation (the Agreement number 02.A03.21.0008) and by RFBR (research projects No. 16-07-00766, 17-07-00845).

References

1. Erlang, A.K.: The theory of probabilities and telephone conversations. Nyt Tidsskrift for Matematik **20**(B), 33–39 (1909)
2. Syski, R.: Introduction to Congestion Theory in Telephone Systems, ed. by R. Syski. Edinburgh and London: Oliver and Boyd (1960)
3. Benes, V.E.: Mathematical Theory of Connecting Networks and Telephone Traffic. Academic Press, New York (1965)
4. Sesia, S., Baker, M., Toufik, I.: LTE-the UMTS Long Term Evolution: From Theory to Practice. Wiley (2011)
5. Dshalalow, J.H.: An anthology of classical queueing methods. In: Advances in Queueing Theory, Methods, and Open Problems (Probability and Stochastics Series). CRC Press, Boca Raton (1995)
6. Kelly, F.P.: Reversibility and Stochastic Networks. Wiley, New York (1979)
7. Kelly, F.P.: Mathematical models of multiservice networks. Complex Stochast. Syst. Eng. **54**, 221–234 (1995)
8. Ross, K.W.: Multiservice loss models for broadband telecommunication networks. Springer, London (1995)
9. Pankratova, E., Moiseeva, S.: Queueing system GI—GI—∞ with n types of customers. Commun. Comput. Inf. Sci. **564**, 216–225 (2015)
10. Tikhonenko, O.: Queueing systems with common buffer: a theoretical treatment. In: Kwiecien, A., Gaj, P., Stera, P. (eds.) CN 2011. CCIS, vol. 160, pp. 61–69. Springer, Heidelberg (2011). doi:10.1007/978-3-642-21771-5_8
11. Naumov, V.A., Samuilov, K.E., Samuilov, A.K.: On the total amount of resources occupied by serviced customers. Autom. Remote Control **77**(8), 1419–1427 (2016)
12. Naumov, V., Samouylov, K., Yarkina, N., Sopin, E., Andreev, S., Samuylov, A.: LTE performance analysis using queuing systems with finite resources and random requirements. In: Proceedings of the 7th International Congress on Ultra Modern Telecommunications and Control Systems (ICUMT-2015), Brno, Czech Republic, pp. 100–103. IEEE, New Jersey (2015)
13. Sopin, E., Samouylov, K., Vikhrova, O., Kovalchukov, R., Moltchanov, D., Samuylov, A.: Evaluating a case of downlink uplink decoupling using queuing system with random requirements. In: Galinina, O., Balandin, S., Koucheryavy, Y. (eds.) NEW2AN/ruSMART -2016. LNCS, vol. 9870, pp. 440–450. Springer, Cham (2016). doi:10.1007/978-3-319-46301-8_37
14. Samoylov, K., Sopin, E., Shorgin, S.: Queuing systems with resources and signals and their application for performance evaluation of wireless networks. Inf. Appl. Sci. J. **11**(3), 50–58 (2017)

15. Mouly, M., Pautet, M.B., Foreword By-Haug, T.: The GSM System for Mobile Communications. Telecom Publishing (1992)
16. Holma, H., Toskala, A. (eds.): WCDMA for UMTS: Radio Access for Third Generation Mobile Communications. Wiley, New York (2005)
17. Orlov, Y., Kirina-Lilinskaya, E., Samuylov, A., Ometov, A., Moltchanov, D., Gaimamaka, Y., Andreev, S., Samouylov, K.: Time-dependent SIR analysis in shopping malls using fractal-based mobility models. In: Koucheryavy, Y., Mamatas, L., Matta, I., Ometov, A., Papadimitriou, P. (eds.) WWIC 2017. LNCS, vol. 10372, pp. 16–25. Springer, Cham (2017). doi:10.1007/978-3-319-61382-6_2
18. Orlov, Y., Zenyuk, D., Samuylov, A., Molchanov, D., Andreev, S., Romashkova, O., Gaidamaka, Y., Samouylov, K.: Time-dependent SIR modeling for D2D communication in indoor deployments. In: Proceedings of 29th European Conference on Modelling and Simulation (ECMS), pp. 726–731 (2017)
19. Zenyuk, D.A., Orlov, Y.N.: Fractional Diffusion Equation and Non-stationary Time-series Modeling. MIPT, Moscow (2016). (in Russian)
20. Samko, S.G., Kilbas, A.A., Marichev, O.I.: Integrals and derivatives of fractional order and some applications. Minsk, Science and Technology (1987). (in Russian)
21. Orlov, Y.: Kinetic methods for investigating time-dependent temporal series Moscow, MIPT (2014). (in Russian)
22. Zenyuk, D.A., Mitin, N.A., Orlov, Y.: Modeling of random walk on Cantor set. Preprints KIAM of RAS 31 (2013). (in Russian)
23. Hida, T.: Brownian motion. In: Brownian Motion, pp. 44–113. Springer, New York (1980)
24. Rhee, I., Shin, M., Hong, S., Lee, K., Kim, S.J., Chong, S.: On the levy-walk nature of human mobility. IEEE/ACM Trans. Netw. (TON) 19(3), 630–643 (2011)
25. Musolesi, M., Mascolo, C.: Mobility models for systems evaluation. Middlew. Netw. Eccentric Mob. Appl. 3, 43–62 (2009)
26. Camp, T., Boleng, J., Davies, V.: A survey of mobility models for ad hoc network research. Wireless Commun. Mob. Comput. 2(5), 483–502 (2002)

Algorithm of Balance Equations Decomposition and Investigation of Poisson Flows in Jackson Networks

Gurami Tsitsiashvili[1,2]([✉])

[1] Institute of Applied Mathematics of Far Eastern Branch of RAS,
Vladivostok, Russia
[2] Far Eastern Federal University, Vladivostok, Russia
guram@iam.dvo.ru

Abstract. In this paper a decomposition of a solution of balance equations for intensities of flows departing from nodes of the Jackson network is constructed. Procedures of the decomposition are based on a definition of classes of cyclically equivalent nodes and on a construction of acyclic directed graph consistent with the Jackson network and these classes. Then classes of cyclically equivalent nodes are arranged in the acyclic graph accordingly with maximal ways lengths from the source node to all others nodes classes. An algorithm of maximal ways lengths calculation is represented as an analogy of the Floyd-Steinberg algorithm of minimal ways lengths calculation.

Sets of independent stationary Poisson flows departing from nodes of Jackson network are enumerated. These sets are defined by non-return sets in the acyclic directed graph which is composed of cyclic equivalence classes of the Jackson network nodes. Special algorithm of an enumeration of non-return nodes sets is constructed. This algorithm is constructed accordingly with partial order of the equivalence classes in the acyclic directed graph of the equivalence classes of nodes.

Keywords: The Jackson network · A system of balance equations · A class of cyclic equivalent nodes · A directed graph · A non-return set of nodes

1 Introduction

Balance equations play a significant role in mathematical economics, in queuing theory, in calculation of complex chemical processes etc. The balance equations appear in a linear economic model (Leontief model) and can be used for analysis and planning activities of enterprises, industries and the economy as a whole. For an inter branch a balance of production and distribution of products is characteristic detailed analysis of the production relationships between industries in supply of raw materials, fuel, auxiliary materials, electricity, services of cargo transport and logistics. Inter-industry linkages are usually in the form of a chess table of

© Springer International Publishing AG 2017
A. Dudin et al. (Eds.): ITMM 2017, CCIS 800, pp. 336–346, 2017.
DOI: 10.1007/978-3-319-68069-9_27

costs and distribution of output by sectors of the national economy, and costs are shown in columns, and the distribution of products in the rows of the table.

In economics mathematical methods are the most important and sometimes the only possible tool for the study of complex economic objects and the construction of hypotheses. Mathematical methods and models used in practice, economic decisions are mainly used: to aggregate the available information and presenting it in convenient form for analysis; to identify key trends, their characteristics and forecasting; to conduct model experiments; to optimize the process of selecting the best solutions; for choosing the best option solutions. According to the complexity of mathematical models of economic objects and phenomena can be conditionally divided into several main types.

1. Functional models, which express, as a rule, direct dependencies between the known (exogenous) and the unknown (endogenous) variables. Necessary to build the model parameters are determined on the basis of normative data or statistical methods. Models of this type, based on economic models, statistical data are used for in-depth analysis of the processes of socio-economic development. With their help it is the forecasting of the main indicators of production, consumption, demographic projections and forecasts of development of science, technology and the environment.
2. The model, expressed through systems of equations for the endogenous variables, express usually balance relations between various economic indicators and are used for finding balanced planning decisions (e.g., model of inter branch balance). Their application is quite extensive, for example in planning.
3. Models of the optimization type. The major part of such models is also a system of equations or inequalities endogenous variables. It is necessary to find a solution to this system, which would give the optimum (i.e. maximum or minimum depending on setting specific tasks), the value of some economic indicator. The basic models of this type are of a linear programming problem.
4. Simulation models occupy a special place in the list of model types and are typically used in the composition of the man-machine games or simulation systems. This is one of the most powerful tools of mathematical modelling used in the analysis of the functioning and synthesis of structures of difficult systems, the management of which is associated with decision-making under uncertainty. They give rational and complete description of the real relations between economic operators. The simulation method is applied primarily to dynamic processes, the study of which in other ways is extremely difficult. The application area of simulation methods is very wide: from studies of the processes of management of ecological communities, ecological systems, to solve the problems of complex automatic control devices, design technology design, problem solving, resource management, research, queuing systems and networks, historical processes, etc. Especially the role of simulation in the experimental validation of proposals related to structural changes, modernization of economic mechanisms and other improvements that are not amenable to formal quantitative description. Simulation models are primarily very

accurate reflection of the economic process or phenomenon, therefore the resulting mathematical problems are quite complex, they contain non-linear and stochastic (probabilistic) dependencies and variables. The main way of solving such problems lies in using in a special way ordered alternative calculations on the PC.

5. Systems and complexes of interrelated models relating to the above types. The development of systems models to accurately reflect the various aspects of planning and operation of economic objects. The problem of finding agreed solutions in the models is a serious mathematical challenge.

The balance are closely connected with the transportation problem of linear programming. Transportation problem is a mathematical problem of finding the optimal distribution of supply homogeneous goods (cargo, medium) between points of departure and destination under the given numerically expressed cost (costs, expenses) for transportation. The general solution of the initially described methods of linear algebra like linear programming problem of a special kind. The transportation problem can be represented in writing in the form of a rectangular table.

This problem was formalized by the French mathematician Gaspar-Monge in 1781. According to Alexander Schrijver, the first who studied the transportation problem mathematically, was A. N. Tolstoy of the USSR. In 1930 he published his work on finding the minimal total mileage in a train carriage, where it was used redistributive cycles. According to Gass, the task of this type in western literature was first staged Hitchcock in 1941 and discussed in detail by Koopmans, who worked as a member of the joint Committee of transport during the Second world war, when a shortage of cargo vessels was a critical bottleneck.

As a problem of linear programming (specification of the simplex method), it first was considered by John Danzig. Other the process of calculation ("method of simultaneous solution of direct and dual problems") was proposed by Ford and Fulkerson in 1956. The way of solving the transportation problem (method of potentials) in the USSR was published by Kantorovich and Gavurin in 1949 and earlier in their book "Linear programming, its applications and generalizations". George Danzig refers to the publications of Kantorovich in 1939 and 1942 and the subsequent article of 1949 containing, as he believed, in a complete theory of the problem of transportation, together with incomplete computing algorithm, written in practical language. Unfortunately, in his view, these works were little known in the USSR and beyond. In contrast, Kantorovich himself in his memoirs, 1987 argued that the university immediately published his article and it was sent to fifty people's commissariats. According to Danzig, the computer program of the simplex method for solving transportation problem was first developed in 1950 for the machine, SEAC, and the program for the General simplex method in 1951 under the leadership of A. Order from the U.S. air force and A. J. Hoffman of the Bureau of standards.

The basis of the balance equations solution is the algorithm of material and heat balance of the network for example a column. Thus vapor-liquid equilibrium, the kinetics of mass transfer and hydrodynamics of streams create a separate complicated task. The use of different methods of the description of phase equilibrium, kinetics and hydrodynamics leads to a change in certain factors or dependencies in balance sheet ratios. But does not change the general algorithm for the solution of the balance equations. The convergence conditions are subject to change, if not to be violated. Diverse solution methods for equations of balance attest to the difficulties of developing a universal algorithms that guarantee the convergence of the different methods of descriptions of individual phenomena. It is possible to spread this approach to analysis and calculation of flow intensities in protein networks which consist of a large number of proteins.

In this paper a decomposition procedure of balance equations solution is considered. Such consideration is closely connected with balance equations in the queuing theory [1,2]. The graph theory balance equations representation is based on the articles [3–6] in which Poisson flows in open (Jackson) networks and their canonical decomposition are analysed. Sets of independent stationary Poisson flows departing Jackson network nodes are connected with non-return sets and are based on the Burke theorem [7,8]. Analysis of some queuing networks with feedbacks are considered in [9–14].

2   Formulation of Problem

In this paper opened Jackson queuing network S with Poisson input flow which has intensity λ_0 and consists of nodes $k = 0, 1, \ldots, m$, is considered. Customers motions in this network are defined by route matrix $\Theta = ||\theta_{i,j}||_{i,j=0}^m$, where $\theta_{i,j}$ is a probability of customer transition after service in node i to node j, $\theta_{0,0} = 0$, node 0 is a source of customers arriving the network and a run-off for customers departing the network. In the node i there are $l_i \leq \infty$ servers, their service times are independent and have exponential distribution with the parameter μ_i, $i = 1, \ldots, m$. In each node i a queue is unrestricted and customers are served accordingly with discipline FIFO.

Assume that the route matrix $\Theta = ||\theta_{i,j}||_{i,j=0}^m$ is indecomposable, that is for

$$\forall\, i,\, j \in \{0, \ldots, m\}\ \exists\, i_1, \ldots, i_r \in \{0, \ldots, m\}:\ \theta_{i,i_1} > 0, \theta_{i_1,i_2} > 0, \ldots, \theta_{i_r,j} > 0.$$

Then for fixed $\lambda_0 > 0$ the system of linear algebraic equation for intensities of flows departing the network nodes

$$\lambda_k = \lambda_0 \theta_{0,k} + \sum_{t=1}^m \lambda_t \theta_{t,k},\ k = 1, \ldots, m \tag{1}$$

has single solution $(\lambda_1, \ldots, \lambda_m)$ with $\lambda_1 > 0, \ldots, \lambda_m > 0$, [2, p. 13].

The system (1) is called the system of balance equations and plays large role in a formulation and a proof of the Jackson product theorem [1], widely used in the queuing theory. If $\lambda_i < l_i \mu_i$, $i = 1, \ldots, m$, then discrete Markov process $(n_1(t), \ldots, n_m(t))$, $t \geq 0$, describing numbers of customers in the network nodes has limit distribution (independent on initial conditions) represented in the form

$\prod_{i=1}^{m} p_i(n_i)$, where $p_i(n_i)$ is limit distribution of customers number in isolated l_i-server queuing system with Poisson input flow with an intensity defined by $\lambda_0, \ldots, \lambda_m$.

In [3] the Jackson network is confronted with directed graph G which has edges set corresponding positive elements of the route matrix. A concept of non-return set of graph nodes G is defined. In this paper graph G is factorized by a relation of cyclic equivalence and sets of independent stationary Poisson flows departing network nodes are enumerated. A decomposition of a solution of balance equations (1) into sub systems of balance equations corresponding classes of the cyclic equivalence is made and an existence and an uniqueness of their solutions is proved.

A problem of a factorization of the graph G by a relation of its nodes cyclic equivalence is considered. An accelerated algorithm of this problem solution based on sequential introduction of new nodes into directed graph and a recalculation of its equivalence classes and partial order between them is represented.

3   Decomposition of System of Balance Equations

3.1   Construction of Directed Graph Corresponding Jackson Network and Classification of Its Nodes

Construct directed graph G by the route matrix as follows. Define the nodes set U of the graph G by the equality $U = \{0_*, 1, \ldots, m, 0_{**}\}$ and introduce zero-one matrix $A = ||a_{i,j}||_{i,j \in U}$ by the equalities

$$a_{0_*,0_*} = a_{0_{**},0_{**}} = 1, \ a_{0_*,0_{**}} = a_{0_{**},0_*} = 0, \ a_{i,0_*} = a_{0_{**},i} = 0, \ i = 1, \ldots, m,$$

$$a_{i,j} = 1 \Leftrightarrow \theta_{i,j} > 0, \ a_{0_*,i} = 1 \Leftrightarrow \theta_{0,i} > 0, \ a_{i,0_{**}} = 1 \Leftrightarrow \theta_{i,0} > 0, \ i,j = 1, \ldots, m.$$

Then the set V of the graph G edges satisfies the equalities $V = \{(i,j), \ i,j \in U : a_{i,j} = 1\}$.

On the set U define the partial order $i \succeq j$ if in the graph G there is a way from the node i to the node j. If the route matrix Θ is indecomposable then we have the following statement.

Remark 1. It is obvious that the condition of the matrix Θ in decomposability is equivalent the following relation for the graph G nodes:

$$0_* \succeq i \succeq 0_{**}, \ i = 1, \ldots, m. \tag{2}$$

On the nodes set U define the relation of the cyclical equivalence $i \sim j \Leftrightarrow i \succeq j, \ j \succeq i$. Further put $\mathbf{i}$ the class of nodes cyclically equivalent in the graph G to the node i. Denote $\mathbf{U}$ the set of equivalence classes where $\mathbf{0}_* = \{0_*\}$, $\mathbf{0}_{**} = \{0_{**}\}$. On the set $\mathbf{U}$ it is possible in natural way to spread partial order $\succeq$:

$$\mathbf{0}_* \succeq \mathbf{i} \succeq \mathbf{0}_{**}, \ \mathbf{i} \in \mathbf{U}. \tag{3}$$

Connect classes $\mathbf{i}$, $\mathbf{j} \in \mathbf{U}$ by an edge if there are nodes $p \in \mathbf{i}$, $q \in \mathbf{j}$, so that the edge $(i, j) \in G$. It is simple to check that obtained factor-graph $\mathbf{G}$ is acyclic.

Remark 2. The condition (3) means an absence of sub networks of the network S which customers do not arrive and an absence of sub networks from which customers depart (and so it is possible their permanent accumulation).

3.2 Calculation of Maximal Ways Lengths in Acyclic Directed Graph

Consider now a decomposition of the balance equations system (1) based on equivalence classes in the set U. This decomposition procedure is based on a concept of maximal ways lengths in between the class $\mathbf{0}_*$ and other classes in the graph $\mathbf{G}$. In [15] an algorithm of a calculation of maximal way length from single maximal (by partial order relation $\succeq$) node 0 to any other nodes of acyclic directed graph is constructed. This algorithm is an analogy of the Floyd-Worshall algorithm of a calculation of shortest ways lengths between graph nodes [16–18].

Consider acyclic directed graph T with nodes set U_T and edges set $V_T = \{0, \ldots, m_T\}$. Assume that in the graph T for any node $i \in U_T$ there is a way from the node 0 to the node i. For any node $i \in U_T$ define maximal way length l_i from the node 0 to the node i, $l_0 = 0$.

To construct a calculation of l_i, $1, \ldots, m_T$, introduce the matrix $D^1 = ||d^1_{i,j}||^{m_T}_{i,j=0}$ with $d^1_{i,i} = 0$, $i = 0, \ldots, n$, $d^1_{i,j} = \infty$ if $(i, j) \notin V_T$, $d^1_{i,j} = 1$ if $(i, j) \in V_T$. So if a pair of nodes is not connected by an edge then $d^1_{i,j} = \infty$.

Construct an analogy of the Floyd-Steinberg algorithm to calculate a matrix of maximal ways lengths between nodes of acyclic directed graph T. Denote $D^k = ||d^k_{i,j}||^{m_T}_{i,j=0}$, $k = 0, \ldots, n$ in which $d^k_{i,j}$ is maximal length of ways between the nodes i, j passing through intermediary nodes $0, 1, \ldots, k$ if such ways exist. But if such ways do not exist then $d^k_{i,j} = \infty$. It is simple to prove the following statement.

Theorem 1. *Matrices* $D^k = ||d^k_{i,j}||^{m_T}_{i,j=0}$, $k = 2, \ldots, m_T$ *satisfy recurrent relations*

$$d^k_{i,j} = \max(d^{k-1}_{i,j}, d^{k-1}_{i,k} + d^{k-1}_{k,j}) \text{ if } \max(d^{k-1}_{i,j}, d^{k-1}_{i,k} + d^{k-1}_{k,j}) < \infty, \quad (4)$$

$$\text{else } d^k_{i,j} = \min(d^{k-1}_{i,j}, d^{k-1}_{i,k} + d^{k-1}_{k,j}). \quad (5)$$

The matrix $D^m = ||d^m_{i,j}||^{m_T}_{i,j=0}$ *define maximal lengths of ways between the graph* T *nodes if such ways exist. In a case of an absence of such ways corrisponding matrix elements equal the infinity.*

3.3 Conditional Solution of Balance Equations Sub Systems

Using this algorithm it is possible to calculate maximal way length $\mathbf{L(i)}$ from the node $\mathbf{0}_*$ to arbitrary node $\mathbf{i}$ in the graph $\mathbf{G}$. Denote $R = \max(\mathbf{L(i)} : \mathbf{i} \in \mathbf{U})$, then from [15, Theorem 1] we have that for any s, $1 \leq s \leq R$, there is the equivalence class $\mathbf{i} \in \mathbf{U}$ satisfying the equality $\mathbf{L(i)} = s$. And if $(\mathbf{i}, \mathbf{j})$ is an edge in the graph $\mathbf{G}$ then $\mathbf{L(i)} < \mathbf{L(j)}$, $\mathbf{L(0_*)} = 0$.

For any class $\mathbf{i} \in \mathbf{U}$ extract from the system (1) the sub system

$$\lambda_k = \sum_{\mathbf{j}: \, \mathbf{L(j)}<\mathbf{L(i)}} \sum_{t \in \mathbf{j}} \lambda_t \theta_{t,k} + \sum_{t \in \mathbf{i}} \lambda_t \theta_{t,k}, \quad k \in \mathbf{i}. \tag{6}$$

Introduce an auxiliary node $0^{\mathbf{i}}$ and denote

$$\lambda_{0^{\mathbf{i}}} = \sum_{\mathbf{j}: \, \mathbf{L(j)}<\mathbf{L(i)}} \sum_{t \in \mathbf{j}} \lambda_t \sum_{k \in \mathbf{i}} \theta_{t,k}, \tag{7}$$

$$\theta_{0^{\mathbf{i}},k} = \frac{\sum_{\mathbf{j}: \, \mathbf{L(j)}<\mathbf{L(i)}} \sum_{t \in \mathbf{j}} \lambda_t \theta_{t,k}}{\lambda_{0^{\mathbf{i}}}}, \quad \theta_{0^{\mathbf{i}},0^{\mathbf{i}}} = 0,$$

$$\theta_{k,0^{\mathbf{i}}} = \sum_{\mathbf{j}: \, \mathbf{L(j)}>\mathbf{L(i)}} \sum_{t \in \mathbf{j}} \theta_{k,t} = 1 - \sum_{t \in \mathbf{i}} \theta_{k,t}, \quad k \in \mathbf{i}. \tag{8}$$

Then the system (6) may be rewritten as follows

$$\lambda_k = \lambda_{0^{\mathbf{i}}} \theta_{0^{\mathbf{i}},k} + \sum_{t \in \mathbf{i}} \lambda_t \theta_{t,k}, \quad k \in \mathbf{i}. \tag{9}$$

Theorem 2. *For fixed $\lambda_{0^{\mathbf{i}}} > 0$ the system (6) has single solution $(\lambda_k, \ k \in \mathbf{i})$ with all positive components.*

Proof. The system (9) is defined by the matrix $\|\theta_{k,t}\|_{k,t \in \mathbf{i} \cup 0^{\mathbf{i}}}$ which is stochastic in an accordance with Formulas (7) and (8) (all its elements are non negative and sums of each line elements equal one). Consequently from Formula (3) this matrix is indecomposable. Consequently from [2, p. 13] we obtain Theorem 2 statement.

3.4   Decomposition of System (1) Solution

To calculate the vector $(\lambda_k : \ k \in \mathbf{i})$ from Formula (7) it is necessary to assume that all $\lambda_t : \ t \in \mathbf{j}$, $\mathbf{L(j)} < \mathbf{L(i)}$ are known. As single equivalence class in $\mathbf{U}$ satisfying the equality $\mathbf{L(i)} = 0$, is the class $\mathbf{0_*}$ then the value $\lambda_{\mathbf{0_*}} = \lambda_0$ is known. Consequently from Theorem 2 it is possible to solve systems (9) for all $\mathbf{i} : \mathbf{L(i)} = 1$.

Knowing $\lambda_t : \ t \in \mathbf{j}$, $\mathbf{L(j)} < 2$, it is possible to calculate now all $\lambda_k : \ k \in \mathbf{i}$, $\mathbf{L(i)} = 2$. Analogously knowing all $\lambda_t : \ t \in \mathbf{j}$, $\mathbf{L(j)} < p < R$, we may calculate all $\lambda_k : \ k \in \mathbf{i}$, $\mathbf{L(i)} = p$. So it is possible to divide sequentially by p the system (1) into sub systems (9) and solve them recurrently.

4 Independent Stationary Poisson Flows in Jackson Network

4.1 Representation of Independent Stationary Poisson Flows by Non-return Sets of Nodes

Assume that in the Jackson network S the ergodicity condition $\lambda_i < l_i\mu_i$, $k = 1, \ldots, m$, is true and initial distribution of customers numbers in network nodes coincides with stationary one [1]. Call a set of nodes $W \subseteq U$, $0_* \in W$, non-return if from any node which does not belong W there is not any edge to node belonging W. Then all flows departing nodes of non-return set W and arriving nodes which do not belong the set W are independent and Poisson [3]. So an enumeration of all non-return sets W is of large interest.

4.2 Description of Non-return Sets by Classes of Cyclically Equivalent Nodes

Lemma 1. *Any non-return set W including node $k \in \mathbf{i}$ includes all nodes of equivalence class $\mathbf{i}$ also.*

Proof. Indeed assume that nodes $k, t \in \mathbf{i}$ belong non-return set W then in the graph G there is a way from node t to node k, completely passing through the class $\mathbf{i}$. This way contains an edge (t', k') so that $t' \notin W$, $k' \in W$. But then the set W is not non-return. A contradiction proves Lemma 1 statement.

Lemma 2. *For any $\mathbf{i} \in \mathbf{U}$ an aggregation $U_{\mathbf{i}}$ of all equivalence classes $\mathbf{j} : 0_* \succeq \mathbf{j} \succeq \mathbf{i}$ is non-return set. Any non-return set W which contains equivalence class $\mathbf{i}$ contains the set $U_{\mathbf{i}}$ also.*

Proof. Assume that $U_{\mathbf{i}}$ is not non-return set and so there is a class $\mathbf{k} \cap U_{\mathbf{i}} = \emptyset$ and an edge from $\mathbf{k}$ to $U_{\mathbf{i}}$. Then $\mathbf{k} \subseteq U_{\mathbf{i}}$ and this contradiction proves that the set $U_{\mathbf{i}}$ is non-return.

Assume now that non-return set W contains a class $\mathbf{i}$ and there is a class $\mathbf{k} \subseteq U_{\mathbf{i}}$ which does not belong the set W. Then from any node of the class $\mathbf{k}$ (which does not belong the set W) there is a way γ to arbitrary node of the class $\mathbf{i}$ (belonging the set W). Consequently the way γ has an edge $(t', r') : t' \notin W$, $r' \in W$. This contradiction proves that $U_{\mathbf{i}} \subseteq W$.

Denote $\mathbf{W}$ the set of equivalence classes belonging non-return set W and put $\mathbf{I}$ the set of minimal by partial order $\succeq$ classes in $\mathbf{W}$.

Theorem 3. *Any non-return set W may be represented as follows*

$$W = \bigcup_{\mathbf{i} \in \mathbf{W}} U_{\mathbf{i}} = \bigcup_{\mathbf{i} \in \mathbf{I}} U_{\mathbf{i}}. \tag{10}$$

Proof. From the definition of non-return set (in our case finite set) we have that an aggregation of finite number of non-return sets is non-return also. So any non-return set W may be represented as follows $W = \bigcup_{i \in W} U_i$, that gives the first equality in Formula (10). The second equality is obtained from the definition of non-return set U_i.

5 Accelerated Sequential Algorithm of Calculation of Equivalence Classes and Their Partial Order

The author with his colleagues constructed algorithm of calculation of the set $\mathbf{U}$ and the partial order $\succeq$ on this set [19,20]. This algorithm is based on sequential including into graph G new node and edges connected with it and recalculation of the set $\mathbf{U}$ of classes and zero-one matrix representing partial order between them. In numerical experiment with matrix which have dimensionality about few thousands it is shown that this algorithm is significantly faster (in few orders) than traditional algorithm based on max-min product of the graph G adjacency matrix.

Assume that $\|d_{i,j}\|_{i,j=1}^{U}$ is adjacency matrix of the graph G. Construct a sequential algorithm of the graph G factorization by the equivalence relation $\sim$ and zero-one matrix of partial order between equivalence classes.

On the step 0 there is the single node 0_* creating the single cluster $\mathbf{0}_*$ and the clusters set $\mathbf{K} = \{\mathbf{0}_*\}$. Introduce the matrix $\mathbf{a} = \|\mathbf{a}(\mathbf{p}, \mathbf{q})\|_{\mathbf{p}, \mathbf{q} \in \mathbf{K}}$, characterizing the relation of partial order $\succeq$: $\mathbf{a}(\mathbf{i}, \mathbf{j}) = 1 \Leftrightarrow \mathbf{i} \succeq \mathbf{j}$. On the step 0 this matrix is defined by the equality $\mathbf{a}(\mathbf{0}_*, \mathbf{0}_*) = 1$.

5.1 Recalculation of Equivalence Classes

Assume that on the step $t-1$ we have the set of equivalence classes $\mathbf{K}$ and the matrix $\mathbf{a}$ of their partial order. Each class $\mathbf{k} \in \mathbf{K}$ is indexed by maximal number of its node $k \in U$. Then on the step t for new node t define the sets

$$P = \{k \in \{1, ..., t-1\} : d_{t,k} = 1\}, \quad Q = \{k \in \{1, ..., t-1\} : d_{k,t} = 1\}.$$

For each node i on the step $t-1$ we have its equivalence class $\mathbf{i}$. Denote by $\mathbf{P}$, $\mathbf{Q}$ the sets of equivalence classes in the sets P, Q relatively and define
$\mathbf{K_P} = \bigcup_{\mathbf{p} \in \mathbf{P}} \{\mathbf{k} \in \mathbf{K} : a(\mathbf{p}, \mathbf{k}) = 1\}$, $\mathbf{K_Q} = \bigcup_{\mathbf{q} \in \mathbf{Q}} \{\mathbf{k} \in \mathbf{K} : a(\mathbf{k}, \mathbf{q}) = 1\}$,
$\mathbf{A} = \mathbf{K_P} \cap \mathbf{K_Q}$, $\mathbf{A_1} = \mathbf{K_P} \setminus \mathbf{A}$, $\mathbf{A_2} = \mathbf{K_Q} \setminus \mathbf{A}$, $\mathbf{A_3} = \mathbf{K} \setminus (\mathbf{A_1} \cup \mathbf{A_2} \cup \mathbf{A})$, $t = t \cup \{k \in \mathbf{k} : \mathbf{k} \in \mathbf{A}\}$

Then new node t and nodes of the set $\{k \in \mathbf{k} : \mathbf{k} \in \mathbf{A}\}$ create new equivalence class $\mathbf{t}$, and the set of classes $\mathbf{K}$ is transformed as follows $\mathbf{K} := (\mathbf{K} \setminus \mathbf{A}) \cup \mathbf{t}\}$.

5.2 Recalculation of Partial Order Matrix

Recalculation of the matrix $\mathbf{a}$ on renewed set of equivalence classes $\mathbf{K}$ may be made by the rule:

$$\mathbf{a}(\mathbf{t}, \mathbf{i}) := 1, \ \mathbf{i} \in \mathbf{A_1} \cup \mathbf{t}, \ \mathbf{a}(\mathbf{i}, \mathbf{j}) := 1, \ \mathbf{i} \in \mathbf{A_2}, \ \mathbf{j} \in \mathbf{A_1} \cup \mathbf{t},$$

$$a(\mathbf{i,j}) := 0, \ \mathbf{i} \in \mathbf{A_1}, \ \mathbf{j} \in \mathbf{A_2} \cup \mathbf{t} \cup \mathbf{A_3}, \ \cdot$$

$$a(\mathbf{i,j}) := 0, \ \mathbf{j} \in \mathbf{A_2}, \ \mathbf{i} \in \mathbf{A_3} \cup \mathbf{t}, \ \mathbf{a(t,i)} = \mathbf{a(i,t)} := 0, \ \mathbf{i} \in \mathbf{A_3}.$$

All other components of renewed matrix $\mathbf{a}$ coincide with previous ones defined on the step $t-1$. The matrix $\mathbf{a}$ consequently has the following cell structure (see Table 1). In the table value 1 in some cells means that all components of these cells (rectangular submatrices) equal 1, value 0 in some cells means that all components of these cells equal 0, in all other cells meanings of cells components on the step $t-1$ repeat on the step t. Table 1 show why sequential algorithm gives so strong acceleration of calculations.

Table 1. Transformation on step t of matrix $\mathbf{a}$ defining partial order $\succeq$.

matrix a	classes of $\mathbf{A_1}$	class t	classes of $\mathbf{A_2}$	classes of $\mathbf{A_3}$
classes of $\mathbf{A_1}$	meanings on step $t-1$	0		
class t			0	
classes of $\mathbf{A_2}$	1		meanings on step $t-1$	
classes of $\mathbf{A_3}$	meanings on step $t-1$	0		meanings on step $t-1$

6   Conclusion

In this paper some algorithms of the graph theory allow to construct as decomposition of balance equations solution so to describe a class of non-return sets in the graph characterizing transitions in open queueing network. These algorithms are: the accelerated sequential algorithm of calculation of equivalence classes and their partial order, the factorization on open network graph by cyclic equivalence, the conditional solution of balance equations sub systems, the calculation of maximal ways lengths in acyclic directed graph, the decomposition of balance equations solution, the description of non-return sets by classes of cyclically equivalent nodes. These algorithms may be used in separate graph theory problems but they may give necessary results for queuing networks only in their complex.

Now it is possible to put a problem to analyse as queueing networks with different types of customers so protein networks with different types of matters moving along them.

Acknowledgments. This paper is supported by Russian Fund for Basic Researches, project 17-07-00177.

References

1. Jackson, J.R.: Networks of waiting lines. Oper. Res. **5**(4), 518–521 (1957)
2. Basharin, G.P., Tolmachev, A.L.: Queuing network theory and its applications to the analysis of information-computing networks. In: Itogi Nauki Tech. Ser. Teor. Veroiatn. Mat. Stat. Teor. Kibern. vol. 21, pp. 3–119. VINITI, Moscow (1983) (In Russian)
3. Beutler, F.J., Melamed, B.: Decomposition and customer streams of feedback networks of queues in equilibrium. Oper. Res. **26**(6), 1059–1072 (1978)
4. Melamed, B.: On Poisson traffic processes in discrete-state Markovian systems with applications to queueing theory. Adv. Appl. Probab. **11**(1), 218–239 (1979)
5. Melamed, B.: Characterization of Poisson traffic streams in Jackson queueing networks. Adv. Appl. Probab. **11**(2), 422–438 (1979)
6. Melamed, B.: On the reversibility of queueing networks. Stoch. Process. Appl. **13**(2), 227–234 (1982)
7. Burke, P.J.: The output of a queueing system. Oper. Res. **4**(6), 699–704 (1956)
8. Burke, P.J.: The dependence of sojourn times in tandem M/M/s queues. Oper. Res. **17**(4), 754–755 (1969)
9. Walrand, J.: Filtering formulas and the ·/M/1 queue in a quasireversible network. Stochastics **6**(1), 1–22 (1981)
10. Walrand, J.: On the equivalence of flows in networks of queues. J. Appl. Probab. **19**(1), 195–203 (1982)
11. Walrand, J.: Poisson flows in single class open networks of quasireversible queues. Stoch. Process. Appl. **13**(3), 293–303 (1982)
12. Walrand, J., Varaiya, P.: Flows in queueing networks: a martingale approach. Math. Oper. Res. **6**(3), 387–404 (1981)
13. Walrand, J., Varaiya, P.: Sojourn times and the overtaking condition in Jacksonian networks. Adv. Appl. Probab. **12**(4), 1000–1018 (1980)
14. Walrand, J., Varaiya, P.: Interconnections of Markov chains and quasi-reversible queueing networks. Stoch. Process. Appl. **10**(2), 209–219 (1980)
15. Tsitsiashvili, G.S., Osipova, M.A.: Stationary flows in acyclic queuing networks. Appl. Math. Sci. **11**(1), 23–30 (2017)
16. Cormen, T.H., Leiserson, C.E., Rivest, R.L.: Introduction to Algorithms, 1st edn. MIT Press and McGraw-Hill, Cambridge (1990)
17. Levitin, A.V.: Overcoming of the limitations: the method of bisection. In monograph: Algorithms. In: Introduction to Design and Analysis, Chap. 11, pp. 349–353. Williams, Moscow (2006) (In Russian)
18. Belousov, A.I., Tkachev, S.B.: Discrete Mathematics. MGTU, Moscow (2006). (In Russian)
19. Tsitsiashvili, G.: Sequential algorithms of graph nodes factorization. Reliab. Theor. Appl. **8**(4), 30–33 (2013)
20. Tsitsiashvili, G., Osipova, M.A., Losev, A.S.: Algorithms of graph clusterization. Vestnik of Voronej State University. Series of mathematics 1, 145–149 (2016). (In Russian)

Retrial Queue M/G/1 with Impatient Calls Under Heavy Load Condition

Ekaterina Fedorova[1](✉) and Konstantin Voytikov[2]

[1] Tomsk State University, Tomsk, Russia
moiskate@mail.ru
[2] Moscow Institute of Physics and Technology (State University), Moscow, Russia
voytikovk@gmail.com

Abstract. In the paper, the retrial queueing system of $M/GI/1$ type with impatient calls is considered. The delay of calls in the orbit has exponential distribution and the impatience time of calls in the system is dynamical exponential. Asymptotic analysis method is proposed for the system studying under a heavy load condition. The theorem about the gamma form of the asymptotic probability distribution of the number of calls in the orbit is formulated and proved. During the study, the expression for the system throughput is obtained. Numerical examples compare asymptotic, exact and simulation based distributions.

Keywords: Retrial queueing system · Impatient calls · Asymptotic analysis · Heavy load

1 Introduction

Retrial queueing systems (or queueing systems with repeated calls) are mathematical model of real systems, where unserved calls perform repeated attempts to get a service after a random time. There are such examples in telecommunication networks, mobile systems, call centres, etc. [1–3].

The first papers about retrial queues were devoted to practical problems and influence of repeated attempts to telephone traffics [4–7].

The main results and comprehensive description of retrial queues are contained in the books [8,9].

Nowdays, there are many papers devoted to investigations of retrial queueing systems with different structure and to solutions of different practical and theoretical problems. But the majority of studies are performed numerically, via computer simulation [9–11] or using matrix methods [12–14] and etc. Analytical results were obtained for only simplest models, e.g. a system with Poisson arrivals or the exponential distribution of service law [8].

Asymptotic and approximate methods were applied in the papers [8,15–18] and etc. Characteristics of performance of retrial queueing systems with Poisson input process under heavy, light loads and long delay were also studied [8,18–21].

A. Dudin et al. (Eds.): ITMM 2017, CCIS 800, pp. 347–357, 2017.
DOI: 10.1007/978-3-319-68069-9_28

Retrial models with impatience was considered by Cohen [5], Falin [8], Yang [22], Krishnamoorthy [23], etc. [24–28]. In these papers, the impatience is understood as follows: an arriving call joints the orbit with some probability p and leaves the system with the probability $1-p$. We research a different model which was not been considered early.

In this paper, we use the asymptotic analysis method, that gives analytical result for different types of queueing systems and networks, in particular with non Poisson arrivals [29–31]. We expand the results obtained previously for retrial queues without loss (patience time equals infinity), where the gamma form of the probability distribution of number of calls in the orbit under heavy load limit condition was proved [32, 33].

2 Mathematical Model

Let us consider a retrial queueing system of $M/GI/1$ type with impatient calls. The structure of the model is presented in Fig. 1.

The input process is Poisson with rate λ. There is one server with the service time distribution function $B(x)$. If a call arrives when the server is free, the call occupies it for the service. Otherwise, the call goes to the orbit, where it stays during a random time distributed exponentially with rate σ. After the delay, the call makes an attempt to reach the server again. If it is free, the call occupies it, otherwise the call instantly returns back to the orbit. From the orbit calls can leave the system after a random time distributed exponentially with dynamical rate α/i, where i is a number of calls in the orbit at this moment.

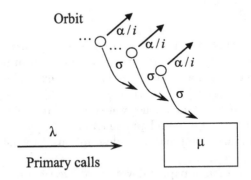

Fig. 1. Retrial queue $M/GI/1$ with impatient calls

Denote the random process that describes the number of calls in the orbit as $i(t)$, the random process of remaining service time as $z(t)$, and the random process that defines server states as

$$k(t) = \begin{cases} 0, & \text{if the server is free,} \\ 1, & \text{if the server is busy.} \end{cases}$$

The problem is to find the probability distribution of the number of calls in the orbit. The process $i(t)$ is not Markovian, therefore we consider the multidimensional process $\{k(t), i(t), z(t)\}$. Let us denote the probability distribution of the process as $P_0(i,t) = P\{k(t) = 0, i(t) = i\}$ and $P_1(i, z, t) = P\{k(t) = 1, i(t) = i, z(t) < z\}$. We write the following system of equations:

$$\begin{cases}
P_0(0, t + \Delta t) = P_0(0, t)(1 - \lambda \Delta t) + P_0(1, t)\alpha \Delta t + P_1(0, \Delta t, t) + o(\Delta t), \\
P_1(0, z + \Delta t, t + \Delta t) = P_1(0, z, t)(1 - \lambda \Delta t) - P_1(0, \Delta t, t) \\
\quad + P_0(1, t)\sigma \Delta t B(z) + P_0(0, t)\lambda \Delta t B(z) + P_1(1, z, t)\alpha \Delta t + o(\Delta t), \\
P_0(i, t + \Delta t) = P_0(i, t)(1 - \lambda \Delta t)(1 - i\sigma \Delta t)(1 - \alpha \Delta t) \\
\quad + P_0(i + 1, t)\alpha \Delta t + P_1(i, \Delta t, t) + o(\Delta t), \\
P_1(i, z + \Delta t, t + \Delta t) = P_1(i, z, t)(1 - \lambda \Delta t)(1 - \alpha \Delta t) - P_1(i, \Delta t, t) \\
\quad + P_1(i - 1, z, t)\lambda \Delta t + P_0(i, t)\lambda \Delta t B(z) \\
\quad + P_0(i + 1, t)(i + 1)\sigma \Delta t B(z) + P_1(i + 1, z, t)\alpha \Delta t + o(\Delta t),
\end{cases} \tag{1}$$

where $i = 1, 2, \ldots$

Let $\Delta t \to 0$ in Eq. (1), we obtain

$$\begin{cases}
\dfrac{\partial P_0(0, t)}{\partial t} = \dfrac{\partial P_1(0, 0, t)}{\partial z} - \lambda P_0(0, t) + \alpha P_0(1, t), \\[2mm]
\dfrac{\partial P_1(0, z, t)}{\partial t} = \dfrac{\partial P_1(0, z, t)}{\partial z} - \dfrac{\partial P_1(0, 0, t)}{\partial z} - \lambda P_1(0, z, t) \\
\quad + \sigma P_0(1, t)B(z) + \lambda P_0(0, t)B(z) + \alpha P_1(1, z, t), \\[2mm]
\dfrac{\partial P_0(i, t)}{\partial t} = \dfrac{\partial P_1(i, 0, t)}{\partial z} - (\lambda + i\sigma + \alpha)P_0(i, t) + \alpha P_0(i + 1, t), \\[2mm]
\dfrac{\partial P_1(i, z, t)}{\partial t} = \dfrac{\partial P_1(i, z, t)}{\partial z} - \dfrac{\partial P_1(i, 0, t)}{\partial z} - (\lambda + \alpha)P_1(i, z, t) \\
\quad + \lambda P_1(i - 1, z, t) + \lambda P_0(i, t)B(z) \\
\quad + (i + 1)\sigma P_0(i + 1, t)B(z) + \alpha P_1(i + 1, z, t).
\end{cases}$$

Considering the system in a steady-state regime, we derive the following system of Kolmogorov equations for stationary probabilities $P_0(i)$ and $P_1(i, z)$:

$$\begin{cases}
\dfrac{\partial P_1(0, 0)}{\partial z} - \lambda P_0(0) + \alpha P_0(1) = 0, \\[2mm]
\dfrac{\partial P_1(0, z)}{\partial z} - \dfrac{\partial P_1(0, 0)}{\partial z} - \lambda P_1(0, z) + \sigma P_0(1)B(z) \\
\quad + \lambda P_0(0)B(z) + \alpha P_1(1, z) = 0, \\[2mm]
\dfrac{\partial P_1(i, 0)}{\partial z} - (\lambda + i\sigma + \alpha)P_0(i) + \alpha P_0(i + 1) = 0, \\[2mm]
\dfrac{\partial P_1(i, z)}{\partial z} - \dfrac{\partial P_1(i, 0)}{\partial z} - (\lambda + \alpha)P_1(i, z) + \lambda P_1(i - 1, z) \\
\quad + \lambda P_0(i)B(z) + (i + 1)\sigma P_0(i + 1)B(z) + \alpha P_1(i + 1, z) = 0,
\end{cases} \tag{2}$$

$i = 1, 2, \ldots$

3   Asymptotic Analysis Under Heavy Load Condition

We introduce partial characteristic functions:

$$H_0(u) = \sum_i e^{jui} P_0(i), \quad H_1(u,z) = \sum_i e^{jui} P_1(i,z), \qquad (3)$$

where j is an imaginary unit.

Substituting functions (3) into Eq. (2), the following equation system is obtained:

$$\begin{cases} b\dfrac{\partial H_1(u,0)}{\partial z} - (\rho + \alpha b) H_0(u) + j\sigma b \dfrac{\partial H_0(u)}{\partial u} \\ \quad + \alpha b e^{-ju} H_0(u) = \alpha b P_0(0)(e^{-ju} - 1), \\[2mm] b\dfrac{\partial H_1(u,z)}{\partial z} - b\dfrac{\partial H_1(u,0)}{\partial z} - e^{-ju} j\sigma b \dfrac{\partial H_0(u)}{\partial u} B(z) + \rho H_0(u) B(z) \\ \quad + \rho(e^{ju} - 1) H_1(u,z) + \alpha b(e^{-ju} - 1) H_1(u,z) = \alpha b P_1(0,z)(e^{-ju} - 1), \end{cases} \qquad (4)$$

where $\rho = \lambda b$ is the system load parameter, b is a mean of the service time.

System (4) is solved by the asymptotic analysis method under limit condition of a heavy load $\rho \to S$, where S is the system throughput (the supremum of the load value when the stationary regime exists for the retrial queue).

Theorem 1. *Let $i(t)$ be a number of calls in the orbit in the retrial queue $M/GI/1$ with impatient calls in the stationary regime, then*

$$\lim_{\rho \to S} M e^{jw(S-\rho)i(t)} = \left(1 - \frac{jw}{\beta(S-\rho)} \right)^{-\gamma}$$

where $\beta = \dfrac{2b^2}{b_2 + 2b^2(S-1)}$, $\gamma = 1 + \dfrac{\beta}{\sigma b}$, b is the mean of the service time, b_2 is the second moment of the service time and $S = 1 + \alpha b$ is the system throughput.

To prove the theorem, we introduce the following notation:

$$\begin{aligned} \rho &= S - \varepsilon, \quad u = \varepsilon w, \\ H_0(u) &= \varepsilon F_0(w,\varepsilon), \quad H_1(u,z) = F_1(w,\varepsilon,z), \\ P_0 &= \varepsilon \pi_0, \quad P_1 = \varepsilon \pi_1(z), \end{aligned}$$

where $\varepsilon \to 0$. Substituting these notation into system (4), we obtain

$$\begin{cases} b\dfrac{\partial F_1(w,0,\varepsilon)}{\partial z} + j\sigma b \dfrac{\partial F_0(w,\varepsilon)}{\partial w} - (S - \varepsilon + \alpha b)\varepsilon F_0(w,\varepsilon) \\ \quad + \alpha b e^{-j\varepsilon w} \varepsilon F_0(w,\varepsilon) = \alpha b \varepsilon \pi_0 (e^{-j\varepsilon w} - 1), \\[2mm] b\dfrac{\partial F_1(w,z,\varepsilon)}{\partial z} - b\dfrac{\partial F_1(w,0,\varepsilon)}{\partial z} - e^{-j\varepsilon w} j\sigma b \dfrac{\partial F_0(w,\varepsilon)}{\partial w} B(z) \\ \quad + (S - \varepsilon)\varepsilon F_0(w,\varepsilon) B(z) + (S - \varepsilon)(e^{jew} - 1) F_1(w,z,\varepsilon) \\ \quad + \alpha b(e^{-j\varepsilon w} - 1) F_1(w,z,\varepsilon) = \alpha b \varepsilon \pi_1(z)(e^{-j\varepsilon w} - 1) \end{cases} \qquad (5)$$

Denoting limits $F_k(w, z) = \lim\limits_{\varepsilon \to 0} F_k(w, z, \varepsilon)$, $F_1(w) = \lim\limits_{z \to \infty} F_1(w, z)$ and using expansions

$$F_k(w, z, \varepsilon) = F_k(w, z) + \varepsilon \cdot f_k(w, z) + O(\varepsilon^2), \tag{6}$$

the following system of the asymptotic equations can be derived from system (5):

$$
\begin{cases}
b\dfrac{\partial F_1(w, 0)}{\partial z} = j\sigma b F_0'(w), \\[2mm]
b\dfrac{\partial F_1(w, z)}{\partial z} - b\dfrac{\partial F_1(w, 0)}{\partial z}(1 - B(z)) = 0, \\[2mm]
b\dfrac{\partial f_1(w, 0)}{\partial z} - SF_0(w) + j\sigma b f_0'(w) = 0, \\[2mm]
b\dfrac{\partial f_1(w, z)}{\partial z} - b\dfrac{\partial f_1(w, 0)}{\partial z} + jwj\sigma b F_0'(w)B(z) - \\[1mm]
\quad j\sigma b f_0'(w)B(z) + SF_0(w)B(z) + (S - \alpha b)jw F_1(w, z) = 0.
\end{cases} \tag{7}
$$

Then summing up Eq. (5), we obtain the following equation for $z \to \infty$:

$$-j\sigma b\frac{\partial F_0(w, \varepsilon)}{\partial w} + \alpha b\varepsilon F_0(w, \varepsilon) - (S - \varepsilon)e^{j\varepsilon w} F_1(w, \varepsilon) + \alpha b F_1(w, \varepsilon) = \alpha b\varepsilon(\pi_1 + \pi_0).$$

Substitute expansions (6) and write equalities for the members with equal powers of ε:

$$
\begin{cases}
-j\sigma b F_0'(w) - SF_1(w) + \alpha b F_1(w) = 0, \\
-j\sigma b f_0'(w) + \alpha b F_0(w) + (1 - Sjw)F1(w) \\
-Sf_1(w) + \alpha b f_1(w) = \alpha b(\pi_1 + \pi_0).
\end{cases} \tag{8}
$$

The asymptotic characteristic function of the probability distribution of the number of calls in the orbit $h(u)$ under the heavy load condition can be presented as

$$h(u) = Me^{jw(S-\rho)i(t)} = F_1(w) + O(\varepsilon) \approx F_1(w). \tag{9}$$

So, it is necessary to obtain the function $F_1(w)$ from Eqs. (7)–(8). The derivation is performed in four stages.

Stage 1. Let the function $F_1(w, z)$ have the form:

$$F_1(w, z) = A(z) \cdot \Phi(w). \tag{10}$$

Thus $F_1(w) = \Phi(w)$.

From the second equation of system (7), it is easy to show that

$$A(z) = A'(0)\int_0^z (1 - B(x))dx,$$

where $A'(0) = 1/b$.

Stage 2. From the first equation of System (7), we have the following expression

$$j\sigma F_0'(w) = -\frac{1}{b}\Phi(w).\tag{11}$$

Substituting formulas (10)–(11) into the first equation of system (8), the value of the system throughput is obtained:

$$S = 1 + ab.$$

So, the stationary regime for this retrial queue exists when $\rho < \alpha b$ or $\lambda < 1/b$.

Stage 3. From the third equation of system (7), we have

$$j\alpha b f_0'(w) = -b\frac{\partial f_1(w, 0)}{\partial z} + S F_0(w).$$

Perform some transformations in the forth equation of system (7).

$$f_1(w, z) = \frac{\partial f_1(w, 0)}{\partial z} \int_0^z (1 - B(x))dx - \frac{jw}{b}\Phi(w) \int_0^z (A(x) - B(x))dx.$$

Let us find the solution $f_1(w, z)$ in the form:

$$f_1(w, z) = \frac{jw}{b}\Phi(w)\nu(z).$$

Then we obtain

$$f_1(w) = jw\Phi(w)\nu'(0) - \frac{jw}{b}\Phi(w) \int_0^\infty (A(x) - B(x))dx,$$

where $\int_0^\infty (A(x) - B(x))dx = b - \frac{1}{2b}b_2.$

On the one hand, we have

$$f_1(w) = jw\Phi(w)\nu'(0) - \frac{jw}{b}\Phi(w)(b - \frac{1}{2b}b_2).$$

On the other hand, the following expression holds

$$f_1(w) = f_1(w, \infty) = \frac{jw}{b}\Phi(w)\nu(\infty).$$

Comparing these expressions, it can obtained that

$$b\nu'(0) - \nu(\infty) = b - \frac{1}{2b}b_2.\tag{12}$$

Stage 4. Substitute formulas (11)–(12) into the last equation of system (8):

$$(ab - S)F_0(w) + (1 - Sjw)\Phi(w) + \frac{jw}{b}\Phi(w)(b - \frac{1}{2b}b_2) = ab(\pi_1 + \pi_0).$$

Then we differentiate this equation. After some transformations, it is easy to obtain the following equation

$$\Phi'(w)(\beta - jw) - \Phi(w)\frac{\gamma}{\beta} = 0, \tag{13}$$

where

$$\beta = \frac{2b^2}{b_2 + 2b^2(S-1)}, \quad \gamma = 1 + \frac{\beta}{\sigma b}.$$

The solution of Eq. (13) has the form

$$\Phi(w) = C\left(1 - \frac{jw}{\beta}\right)^{-\gamma}.$$

Turning back to expressions (10) and (9), we finally obtain the following function:

$$h(u) = C\left(1 - \frac{ju}{\beta(S - \rho)}\right)^{-\gamma}, \tag{14}$$

where $C = 1$ due to the normalization requirement, q.e.d.

4 Numerical Analysis

Let us present some numerical examples to demonstrate the applicability area of the asymptotic results. There are no explicit formulas for the retrial queue $M/GI/1$ with impatient calls. Therefore, we analyze the accuracy of the obtained gamma-approximation (14) by comparing it with exact distribution for particular case of exponential service (system $M/M/1$) and simulation results for system with non-exponential service.

For the system $M/M/1$ we can obtain an exact probability distribution of number of calls in the orbit by numerical solving of Eq. (2).

Let the system parameters be

$$\mu = 1/b = 1, \sigma = 1, \lambda = \mu\rho.$$

Introduce a notation $\delta = \rho/S$, hence $0 < \delta < 1$. So, we compare asymptotic and exact distributions for different values of parameters δ and α, using the Kolmogorov distance between respective cumulative distribution functions:

$$\Delta = \max_{0 \le i < \infty}\left|\sum_{\nu=0}^{i} D_\nu - \sum_{\nu=0}^{i} P_\nu\right|,$$

where D_ν and P_ν are an exact and an asymptotic probability distributions respectively.

In Fig. 2, there are examples of comparison of the asymptotic and the exact distribution densities. Values of the Kolmogorov distance for these examples are presented in the Table 1.

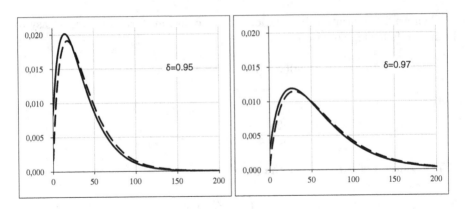

Fig. 2. Comparisons of the asymptotic (dashed line) and the exact (solid line) probability densities for $\alpha = 0.1$

Table 1. Kolmogorov distances between asymptotic and exact distributions

	$\alpha = 0$	$\alpha = 0.01$	$\alpha = 0.05$	$\alpha = 0.10$
$\delta = 0.90$	0.124	0.122	0.116	0.112
$\delta = 0.95$	0.062	0.060	0.054	0.052
$\delta = 0.97$	0.037	0.035	0.030	0.030

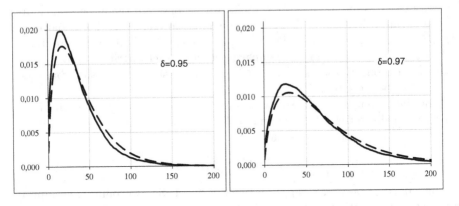

Fig. 3. Comparisons of the asymptotic (dashed line) and the empiric (solid line) probability densities

Consider an example of the system $M/GI/1$ with gamma-distribution of service law with form and inverse scale parameters both equal to 0.75, $\sigma = 1$, $\alpha = 0.1$, $\lambda = \rho$. The results are presented in Fig. 3. The Kolmogorov distances are equal to 0.723 and 0.497 for cases of $\delta = 0.95$ and $\delta = 0.97$ respectively.

If we suppose the Kolmogorov distance equal to 0.05 and less as acceptable accuracy of a result, then we can draw a conclusion that the approximation (14) can be applied for cases $\delta \geq 0.097$.

5   Conclusions

In this regard, the mathematical model of the retrial queue $M/GI/1$ with impatient calls (impatience time is distributed exponentially with dynamical rate α/i) is considered in the paper. For its studying, we propose the asymptotic method under a heavy load condition. It is proved that the asymptotic probability distribution of the number of calls in the orbit has the gamma form with obtained parameters. Also, expression for the system throughput is derived. Numerical analysis allows to draw a conclusion about an applicability area of the asymptotic result.

The future studies can be devoted to analysis of retrial queues with impatient calls, general law of service distribution and non-Poisson arrival processes.

Acknowledgments. The publication was financially supported by RFBR according to the research project No. 16-31-00292 mol-a.

References

1. Kuznetsov, D.Y., Nazarov, A.A.: Analysis of non-Markovian models of communication networks with adaptive protocols of multiple random access. Automation Remote Control **62**(5), 124–146 (2001)
2. Nazarov, A.A., Tsoj, S.A.: Common approach to studies of Markov models for data transmission networks controlled by the static random multiple access protocols. Automatic Control Comput. Sci. **4**, 73–85 (2004)
3. Aguir, M.S., Karaesmen, F., Aksin, O.Z., Chauvet, F.: The impact of retrials on call center performance. OR Spectrum **26**, 353–376 (2004)
4. Wilkinson, R.I.: Theories for toll traffic engineering in the USA. Bell Syst. Technical J. **35**(2), 421–507 (1956)
5. Cohen, J.W.: Basic problems of telephone trafic and the influence of repeated calls. Philips Telecommun. Rev. **18**(2), 49–100 (1957)
6. Elldin, A., Lind, G.: Elementary telephone trafic theory. Ericsson Public Telecommunications (1971)
7. Gosztony, G.: Repeated call attempts and their effect on traffic engineering. Budavox Telecommun. Rev. **2**, 16–26 (1976)
8. Falin, G.I., Templeton, J.G.C.: Retrial Queues. Chapman & Hall, London (1997)
9. Artalejo, J.R., Gomez-Corral, A.: Retrial Queueing Systems: A Computational Approach. Springer, Heidelberg (2008)
10. Ridder, A.: Fast simulation of retrial queues. In: Third Workshop on Rare Event Simulation and Related Combinatorial Optimization Problems, Pisa, pp. 1–5 (2000)
11. Neuts, M.F., Rao, B.M.: Numerical investigation of a multiserver retrial model. Queueing Syst. **7**(2), 169–189 (2002)

12. Dudin, A.N., Klimenok, V.I.: Queueing system $BMAP/G/1$ with repeated calls. Math. Comput. Modell. **30**(3–4), 115–128 (1999)
13. Kim, C.S., Mushko, V.V., Dudin, A.N.: Computation of the steady state distribution for multi-server retrial queues with phase type service process. Ann. Oper. Res. **201**(1), 307–323 (2012)
14. Gomez-Corral, A.: A bibliographical guide to the analysis of retrial queues through matrix analytic techniques. Ann. Oper. Res. **141**, 163–191 (2006)
15. Pourbabai, B.: Asymptotic analysis of $G/G/K$ queueing-loss system with retrials and heterogeneous servers. Int. J. Syst. Sci. **19**, 1047–1052 (1988)
16. Yang, T., Posner, M.J.M., Templeton, J.G.C., Li, H.: An approximation method for the $M/G/1$ retrial queue with general retrial times. Eur. J. Oper. Res. **76**, 552–562 (1994)
17. Diamond, J.E., Alfa, A.S.: Approximation method for $M/PH/1$ retrial queues with phase type inter-retrial times. Eur. J. Oper. Res. **113**, 620–631 (1999)
18. Anisimov, V.V.: Asymptotic analysis of reliability for switching systems in light and heavy trafic conditions. Recent Advances in Reliability Theory, pp. 119–133 (2000)
19. Aissani, A.: Heavy loading approximation of the unreliable queue with repeated orders. In: Actes du Colloque Methodes et Outils d'Aide 'a la Decision. Bejaa, pp. 97–102 (1992)
20. Stepanov, S.N.: Asymptotic analysis of models with repeated calls in case of extreme load. Problems Inf. Transmission **29**(3), 248–267 (1993)
21. Sakurai, H., Phung-Duc, T.: Scaling limits for single server retrial queues with two-way communication. Ann. Oper. Res. **247**(1), 229–256 (2015)
22. Yang, T., Posner, M., Templeton, J.: The $M/G/1$ retrial queue with non-persistent customers. Queueing Syst. **7**(2), 209–218 (1990)
23. Krishnamoorthy, A., Deepak, T., Joshua, V.: An $M/G/1$ retrial queue with non-persistent customers and orbital search. Stoch. Anal. Appl. **23**, 975–997 (2005)
24. Kim, J.: Retrial queueing system with collision and impatience. Commun. Korean Math. Soc. **4**, 647–653 (2010)
25. Fayolle, G., Brun, M.: On a system with impatience and repeated calls. In: Queueing theory and its applications: liber amicorum for J.W. Cohen, North Holland, Amsterdam, pp. 283–305 (1988)
26. Martin, M., Artalejo, J.: Analysis of an $M/G/1$ queue with two types of impatient units. Adv. Appl. Probability **27**, 840–861 (1995)
27. Aissani, A., Taleb, S., Hamadouche, D.: An unreliable retrial queue with impatience and preventive maintenance. In: Proceedings of 15th Applied Stochastic Models and Data Analysis, ASMDA 2013. Matar (Barcelona), Spain, pp. 1–9 (2013)
28. Kumar, M., Arumuganathan, R.: Performance analysis of single server retrial queue with general retrial time, impatient subscribers, two phases of service and Bernoulli schedule. Tamkang J. Sci. Eng. **13**(2), 135–143 (2010)
29. Nazarov, A.A., Lyubina, T.V.: The non-Markov dynamic RQ system with the incoming MMP flow of requests. Automation Remote Control **74**(7), 1132–1143 (2013)
30. Nazarov, A.A., Moiseev, A.N.: Analysis of an open non-Markovian $GI - (GI|\infty)^K$ queueing network with high-rate renewal arrival process. Problems Inf. Transmission **49**(2), 167–178 (2013)
31. Pankratova, E., Moiseeva, S.: Queueing system $map/m/\infty$ with n types of customers. In: Dudin, A., Nazarov, A., Yakupov, R., Gortsev, A. (eds.) ITMM 2014. CCIS, vol. 487, pp. 356–366. Springer, Cham (2014). doi:10.1007/978-3-319-13671-4_41

32. Moiseeva, E., Nazarov, A.: Asymptotic analysis of RQ-systems $M/M/1$ on heavy load condition. In: Proceedings of the IV International Conference Problems of Cybernetics and Informatics, PCI 2012, Baku, pp. 164–166. IEEE (2012)
33. Fedorova, E.: The second order asymptotic analysis under heavy load condition for retrial queueing system MMPP/M/1. In: Dudin, A., Nazarov, A., Yakupov, R. (eds.) ITMM 2015. CCIS, vol. 564, pp. 344–357. Springer, Cham (2015). doi:10.1007/978-3-319-25861-4_29

Analysis of Queueing System
with Resources and Signals

Konstantin Samouylov[1,2], Eduard Sopin[1,2(✉)], and Olga Vikhrova[1]

[1] Applied Probability and Informatics Department,
Peoples' Friendship University of Russia (RUDN University),
6 Miklukho-Maklaya Street, Moscow 117198, Russian Federation
{samuylov_ke,sopin_es,vikhrova_og}@rudn.university
[2] FRC CSC, RAS, 44-2 Vavilov Street, Moscow 119333, Russian Federation

Abstract. We analyze a multi-server queueing system with limited resources and signals. A customer requires a random value (RV) of the shared resources. Each customer generates a flow of signals triggering the resource reallocation process that make a customer release the occupied amount of resources and request a new RV of resources instead.

Considering users are constantly moving within the signal coverage area we describe a model of a wireless network where resources have to be reallocated due to changes in the requirements. We assume that user session cannot be interrupted because of lack of the resources in the instant of the resource reallocation.

Keywords: Limited resources · Random requirements · Flow of signals · Queuing system · Loss system · User motion

1 Introduction

We continue to analyze a class of queueing systems with limited resources and random requirements [1]. In general case the number of servers is finite $N < \infty$ and customers share limited amount of resources $\mathbf{R} = \{R_1, ..., R_M\}$, $R_m < \infty$. Each arriving customer requires a vector $\mathbf{r} = \{r_1, ..., r_M\}$, where r_m is a RV of the m-type resource. RVs of the resource requirements are given by the cumulative distribution function (CDF) $F(\mathbf{x})$ that can be either discrete or continuous. The advantage of the system with random requirements is the versatility of the CDF $F(\mathbf{x})$ and the insensitivity of the probability characteristics to the service time distribution [2]. As an example, the CDF of resource requirements can be based on the distance distribution from a customer to the base station within the network coverage area and implement a signal propagation model to evaluate the pathloss [3]. The CDF $F(\mathbf{x})$ can follow from a scheduler policy as well [4].

The state space of the system with random requirements is described by the number of arrived customers and amount of the occupied resources by each customer in order to know the amount of resources that a customer has to release at the end of the service time. In [5] it has been proposed to simplify the state

© Springer International Publishing AG 2017
A. Dudin et al. (Eds.): ITMM 2017, CCIS 800, pp. 358–369, 2017.
DOI: 10.1007/978-3-319-68069-9_29

space and track the total amount of occupied resources instead of storing all vectors $\mathbf{r}_k$, $k \leq N$. The amount of released resources has been defined by a CDF $F_k(\mathbf{x}|\mathbf{y})$ given that k customers already occupy $\mathbf{y}$ resources.

The system with L classes of the requirements has been investigated in [6]. Considering the Poisson arriving process and exponential distribution of the service times it has been proved via analytical and simulation evaluation that probability characteristics of the simplified and initial systems are equal [7]. In order to apply analytical results to the performance evaluation of the modern heterogeneous wireless network we investigated the system with a discrete CDF $F(\mathbf{x})$ in [8].

In [9] we proved the system with L classes of the requirements can be simplify to the system with aggregated flow of customers with the mean-weighted requirement. However, the further numerical analysis was still a challenge because of the potential memory overflow during the normalization constant calculation using the k-fold convolutions of the probabilities $p_{\mathbf{r}}$ that a customer occupies $\mathbf{r}$ resources, $\mathbf{r} \leq \mathbf{R}$. To avoid that we have derived a recurrent algorithms for the normalization constant evaluation and proposed recurrent formulas of the probability characteristics [10].

We apply a queueuing system with limited resources and signals to the wireless network performance evaluation considering that users are constantly moving within the signal coverage area. While a user is approaching the edge of the cell or getting closer to the base station the amount of the required frequency resources may change.

Consider a wireless network model approximated by a circle with a base station located in the center of the circle. Users' equipment randomly distributed within the circle. A scheduler allocate the frequency resources among the users according to their current channel quality and service requirements. The channel quality depends on the base station transmitting power, proximity of other users affecting the signal-to-interference and noise ratio (SINR), etc., Fig. 1.

Fig. 1. A model of a wireless network

Thus, we consider each user requires a RV of the frequency resources. Users may constantly move during the ongoing session and request an additional amount of resources or release a part of them instead.

The number of the frequency resource blocks (RBs) that a scheduler is to allocate to the users in the next time-slot depends of the number of RBs that has already been allocated. However, we assume that all RVs of the resource requirements are mutually independent and follow the cumulative distribution function $F(\mathbf{x})$ and leave the case with dependent RVs for the future consideration.

As a first step, we back to the multi-server system with a single flow of customers and one-dimensional vector $\mathbf{r} = (r)$ of the resource requirements as it has been investigated in [11]. We extend the system by adding the flow of signals independent of the customer arriving process. When a signal arrives, one of the customer has to release occupied resources and request a new RV of resources. If there are not enough free resources to meet the requirement, the customer will be lost. However, a service provider would rather prefer to keep ongoing user sessions expecting the degradation in a service quality until it can be switched to another base station [12].

2   A System with Resources and Signals

Consider a system with $N < \infty$ servers, shared amount of the resources $R < \infty$ and Poisson arriving flows of customers and signals. A customer requires a RV $r \geq 0$ of the shared resources with a CDF $F(x)$. The service times are exponentially distributed independent RVs with the parameter μ.

The system behavior can be described by a Stochastic process (SP) $X(t) = \big(\xi(t), \big(\eta_0(t), ..., \eta_{\xi(t)}(t)\big)\big)$ where $\xi(t)$ is a number of customers, $\eta_i(t)$ is an amount of occupied resources by i-th customer.

Let's consider k is a number of customers in the system at a moment $t_i > 0$ and $r_1, ..., r_k$ for the amount of the occupied resources. During the interval (t_i, t_{i+1}):

1. a new customer may arrive with the rate λ and require j resources, if $k < N$ and $j \leq R - r_{\bullet}$, where $r_{\bullet} = r_1 + ... + r_k$, the customer will be accepted and occupy j resources with the probability $p_j = P(r_{k+1} = j)$, otherwise it will be lost.
2. an i-th customer may leave the system and release r_i resources.
3. a signal may arrive with rate γ, then an i-th customer has to release r_i resources and try to occupy the new amount of resources r_i^* with the probability $p_{r_i^*}$, if there is not enough resources to meet the requirement, the customer will be lost, Fig. 2.

The state probabilities of the SP $X(t)$

$$q_0 = \lim_{t \to \infty} P\{\xi(t) = 0, \eta_0(t) = 0\},$$

$$q_k(r_1, ..., r_k) = \lim_{t \to \infty} P\{\xi(t) = k, \eta_0(t) = 0, \eta_1(t) = r_1, ..., \eta_{\xi(t)}(t) = r_k\},$$

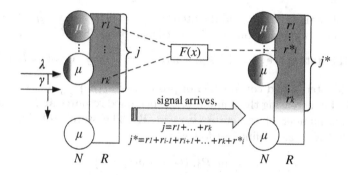

Fig. 2. Resource reallocation at a moment of a signal arriving

are the unambiguous solution of the system

$$\lambda q_0 \sum_{j=0}^{R} p_j = \mu \sum_{j=0}^{R} q_1(j) + \gamma \sum_{j=0}^{R} q_1(j) \left(1 - \sum_{j=0}^{R} p_j \right),$$

$$\left(\lambda \sum_{j=0}^{R-r_\bullet} p_j + k\mu + k\gamma \right) q_k(r_1, ..., r_k) =$$

$$= \lambda p_{r_k} q_{k-1}(r_1, ..., r_{k-1}) +$$

$$+ \mu \sum_{j=0}^{R-r_\bullet} [q_{k+1}(j, r_2, ..., r_{k+1}) + ... + q_{k+1}(r_1, ..., r_k, j)] +$$

$$+ \gamma \left[\sum_{j=0}^{R-r_\bullet} p_{r_1} q_k(j, r_2, ..., r_k) + ... + \sum_{j=0}^{R-r_\bullet} p_{r_k} q_k(r_1, ..., r_{k-1}, j) \right] +$$

$$+ \gamma \left(1 - \sum_{j=0}^{R-r_\bullet} p_j \right) \sum_{j=0}^{R-r_\bullet} [q_{k+1}(j, r_1, ..., r_{k+1}) + ... + q_{k+1}(r_1, ..., r_k, j)],$$

$$k(\mu + \gamma) q_N(r_1, ..., r_N) = \lambda p_{r_N} q_{N-1}(r_1, ..., r_{N-1}) +$$

$$+ \gamma \left[\sum_{j=0}^{R-r_\bullet} p_{r_1} q_N(j, r_2, ..., r_N) + ... + \sum_{j=0}^{R-r_\bullet} p_{r_N} q_N(r_1, ..., r_{N-1}, j) \right].$$

We simplify the SP $X(t)$ and denote the $\delta(t) = \eta_0(t) + \ldots + \eta_{\xi(t)}(t)$ as a total amount of the occupied resources. Thus, the simplified SP $X(t) = (\xi(t), \delta(t))$ is built over the state space $\mathcal{X} = \bigcup_{k=0}^{N} \mathcal{X}_k$, where $\mathcal{X}_k = \left\{ (k, r) : 0 \leq r \leq R, p_r^{(k)} > 0 \right\}$. Here $p_r^{(k)}$ denoted k-fold convolution of probabilities $\{p_r\}_{r \geq 0}$. Let's arrange the states in $\mathcal{X}_k$ by increasing the amount of the occupied resource and denote $I(k, r)$ the sequence number of the state (k, r) in $\mathcal{X}_k$. The state probabilities of the simplified SP $X(t)$ are defined as follows:

$$q_0 = \lim_{t \to \infty} P\{\xi(t) = 0, \delta(t) = 0\}, \tag{1}$$

$$q_k(r) = \lim_{t \to \infty} P\{\xi(t) = k, \delta(t) = r\}, (k, r) \in \mathcal{X}_k. \tag{2}$$

As we don't know the amount of the resources that a customer has to release at the departure, denote the RV of the released resource by i-th customer as ν_i, having the CDF $F_k(x|j) = P\{\nu_i \leq x | \xi(t) = k \, ; \delta(t) = j\}$. The probability that a customer releases i resources is $\frac{p_i p_{j-i}^{(k-1)}}{p_j^{(k)}}$.

The system of the equilibrium equations is given by

$$\lambda q_0 \sum_{j=0}^{R} p_j = \mu \sum_{j:(1,j) \in \mathcal{X}_1} q_1(j) + \gamma \sum_{j:(1,j) \in \mathcal{X}_1} q_1(j) \left(1 - \sum_{j=0}^{R} p_j \right), \tag{3}$$

$$\left(\lambda \sum_{j=0}^{R-r} p_j + k\mu + k\gamma \right) q_k(r) = \lambda \sum_{j \geq 0 : (k-1, r-j) \in \mathcal{X}_{k-1}} q_{k-1}(r-j) p_j +$$

$$+ (k+1)\mu \sum_{j \geq 0 : (k+1, r+j) \in \mathcal{X}_{k+1}} q_{k+1}(r+j) \frac{p_j p_r^{(k)}}{p_{j+r}^{(k+1)}} +$$

$$+ (k+1)\gamma \left(1 - \sum_{j=0}^{R-r} p_j \right) \sum_{j \geq 0 : (k+1, r+j) \in \mathcal{X}_{k+1}} q_{k+1}(r+j) \frac{p_j p_r^{(k)}}{p_{j+r}^{(k+1)}} +$$

$$+ k\gamma \sum_{j \geq 0 : (k,j) \in \mathcal{X}_k} q_k(j) \sum_{i=0}^{\min(j,r)} \frac{p_{j-i} p_i^{(k-1)}}{p_j^{(k)}} p_{r-i}, \tag{4}$$

$$1 \leq k \leq N - 1, (k, r) \in \mathcal{X}_k;$$

$$N(\mu + \gamma) q_N(r) = \lambda \sum_{j \geq 0 : (N-1, j) \in \mathcal{X}_{N-1}} q_{N-1}(r-j) p_j +$$

$$+ N\gamma \sum_{j:(N,j)\in \mathcal{X}_N}^{R} q_N(j) \sum_{i=0}^{\min(j,r)} \frac{p_{j-i}p_i^{(N-1)}}{p_j^{(N)}} p_{r-i}, \tag{5}$$

$$(N,r) \in \mathcal{X}_N.$$

The stationary probabilities (1)–(2) can be calculated by a numerical method as a solution of the matrix equations $\mathbf{q}^T \mathbf{A} = \mathbf{0}^T$, $\mathbf{q}^T \mathbf{1} = 1$. The infinitesimal matrix $\mathbf{A} = [a((i,j),(k,r))]$ of the SP $X(t)$ has a block tridiagonal structure with main diagonal blocks $\mathbf{\Psi}_0, \mathbf{\Psi}_1, ..., \mathbf{\Psi}_N$, upper diagonal blocks $\mathbf{\Lambda}_1, ..., \mathbf{\Lambda}_N$ and lower diagonal blocks $\mathbf{M}_0, ..., \mathbf{M}_{N-1}$:

$$A = \begin{pmatrix} \mathbf{\Psi}_0 & \mathbf{\Lambda}_1 & 0 & ... & 0 & 0 \\ \mathbf{M}_0 & \mathbf{\Psi}_1 & \mathbf{\Lambda}_2 & 0 & ... & 0 \\ 0 & \mathbf{M}_1 & \mathbf{\Psi}_2 & \mathbf{\Lambda}_3 & 0 & ... \\ ... & 0 & ... & ... & ... & 0 \\ 0 & ... & 0 & \mathbf{M}_{N-2} & \mathbf{\Psi}_{N-1} & \mathbf{\Lambda}_N \\ 0 & 0 & ... & 0 & \mathbf{M}_{N-1} & \mathbf{\Psi}_N \end{pmatrix}.$$

First blocks are $\mathbf{\Psi}_0 = -\lambda \sum_{j=0}^{R} p_j$, $\mathbf{\Lambda}_1 = (\lambda p_0, \dots, \lambda p_R)$ and

$$\mathbf{M}_0 = \left(\mu + \gamma \left(1 - \sum_{j=0}^{R} p_j \right), \dots, \mu + \gamma \left(1 - \sum_{j=0}^{R} p_j \right) \right)^T.$$

The remaining blocks have the following form:

$$\psi_n(I(n,i), I(n,j)) = \begin{cases} -\left[\lambda \sum_{k=0}^{R-i} p_k + n\mu + n\gamma \left(1 - \frac{p_{i-k}p_k^{(n-1)}}{p_i^{(n)}} p_{j-k} \right) \right], \\ \qquad i = j; \\ n\gamma \sum_{k=0}^{i} \frac{p_{i-k}p_k^{(n-1)}}{p_i^{(n)}} p_{j-k}, \quad i < j; \\ n\gamma \sum_{k=0}^{j} \frac{p_{i-k}p_k^{(n-1)}}{p_i^{(n)}} p_{j-k}, \quad i > j; \end{cases} \tag{6}$$

$$(n,i) \in \mathcal{X}_n, (n,j) \in \mathcal{X}_n, n = \overline{1, N-1};$$

$$\lambda_n(I(n,i), I(n,j)) = \begin{cases} \lambda p_{j-i}, & i \leq j \leq R; \\ 0, & j < i; \end{cases} \tag{7}$$

$$(n-1,i) \in \mathcal{X}_{n-1}, (n,j) \in \mathcal{X}_n, n = \overline{1, N-1};$$

$$\mu_n(I(n,i), I(n,j)) = \begin{cases} (n+1)\left[\mu \frac{p_{i-j}p_j^{(n)}}{p_i^{(n+1)}} + \gamma \left(1 - \sum_{k=0}^{R-j} p_k \right) \frac{p_{i-j}p_j^{(n)}}{p_i^{(n+1)}} \right], \\ \qquad j \leq i; \\ 0, j > i; \end{cases} \tag{8}$$

$$(n+1, i) \in \mathcal{X}_{n+1}, (n, j) \in \mathcal{X}_n, n = \overline{1, N-1};$$

$$\psi_N(I(N,i), I(N,j)) = \begin{cases} -\left[N\mu + N\gamma\left(1 - \frac{p_{i-k}p_k^{(N-1)}}{p_i^{(N)}}p_{i-k}\right)\right], & i = j; \\ N\gamma \sum_{k=0}^{i} \frac{p_{i-k}p_k^{(N-1)}}{p_i^{(N)}}p_{j-k}, & i < j; \\ N\gamma \sum_{k=0}^{j} \frac{p_{i-k}p_k^{(N-1)}}{p_i^{(N)}}p_{j-k}, & i > j; \end{cases} \tag{9}$$

$$(N, i) \in \mathcal{X}_N, (N, j) \in \mathcal{X}_N.$$

3   Application to Performance Analysis of Wireless Networks

We propose to analyze the model of a wireless heterogeneous network in terms of a multi-server queueing system with random resource requirements and signals. As a user session requirements may change while the user is moving within a cell. If a user is approaching the BS it requires less resources as well as a user will request additional amount of the resources distancing from the BS.

The system behavior follows the same SP $X(t) = (\xi(t), \delta(t))$ with the state space $\mathcal{X} = \bigcup\limits_{k=0}^{N} \mathcal{X}_k$. A customer arrives to the system with the rate λ and occupies an amount of required resources if there are enough free servers and available resources. At the end of the service time a customer leaves the system and releases a RV of resources following from the CDF $F_k(x|j)$.

The resource reallocation process is described by a Poison incoming flow of signals with rate γ. Unlike the general system investigated in the previous section we assume that a customer cannot be blocked due to the lack of the resources upon arrival of a signal.

Let's consider the transition from the state (k, j) to the state (k, r) at a moment of a signal arrived. In case $j > r$ a customer first release i resources with the probability $\frac{p_i p_{j-i}^{(k-1)}}{p_j^{(k)}}$, $j - r \leq i \leq j$, then it have to occupy $r - (j - i)$ resources with the normalized probability $\frac{p_{r-j+i}}{\sum\limits_{s=0}^{R-j+i} p_s}$ since a customer cannot occupy more than amount of the available resources. If $j < r$, a customer release i resources with the probability $\frac{p_i p_{j-i}^{(k-1)}}{p_j^{(k)}}$, $0 \leq i \leq j$ and then occupies $r - (j - i)$ resources with the probability $\frac{p_{r-j+i}}{\sum\limits_{s=0}^{R-j+i} p_s}$ for the $0 \leq i \leq j$.

The system of the steady-state equations differs from the equations of the general system with resources and signals:

$$\lambda q_0 \sum_{j=0}^{R} p_j = \mu \sum_{j:(1,j)\in\mathcal{X}} q_1(j), \tag{10}$$

$$\left(\lambda \sum_{j=0}^{R-r} p_j + k\mu + k\gamma\right) q_k(r) = \lambda \sum_{j\geq 0:(k-1,r-j)\in\mathcal{X}_{k-1}} q_{k-1}(r-j)p_j +$$

$$+ (k+1)\mu \sum_{j\geq 0:(k+1,r+j)\in\mathcal{X}_{k+1}} q_{k+1}(r+j)\frac{p_j p_r^{(k)}}{p_{j+r}^{(k+1)}} +$$

$$+ k\gamma \sum_{j\geq 0:(k,j)\in\mathcal{X}_k} q_k(j) \sum_{i=\max(0,j-r)}^{j} \frac{p_i p_{j-i}^{(k-1)}}{p_j^{(k)}} \frac{p_{r-j+i}}{\sum\limits_{s=0}^{R-j+i} p_s}, \tag{11}$$

$$1 \leq k \leq N-1, (k,r) \in \mathcal{X}_k;$$

$$(N\mu + k\gamma) q_N(r) = \lambda \sum_{j\geq 0:(N-1,r-j)\in\mathcal{X}_{N-1}} q_{N-1}(r-j)p_j +$$

$$+ N\gamma \sum_{j\geq 0:(N,j)\in\mathcal{X}_N} q_N(j) \sum_{i=\max(0,j-r)}^{j} \frac{p_i p_{j-i}^{(N-1)}}{p_j^{(N)}} \frac{p_{r-j+i}}{\sum\limits_{s=0}^{R-j+i} p_s}. \tag{12}$$

$$(N,r) \in \mathcal{X}_N.$$

Denote all non-zero blocks of the infinitesimal matrix $\mathbf{A}$. Blocks $\boldsymbol{\Psi}_0 = -\lambda \sum_{j=0}^{R} p_j$, $\boldsymbol{\Lambda}_1 = (\lambda p_0, \ldots, \lambda p_R)$ and $\mathbf{M}_0 = (\mu, \ldots, \mu)^T$ are represented by vectors while elements of the sub-matrices $\{\boldsymbol{\Psi}_n\}_{1\leq n\leq N}$, $\{\boldsymbol{\Lambda}_n\}_{2\leq n\leq N}$, $\{\mathbf{M}_n\}_{1\leq n\leq N-1}$ are given by:

$$\psi_n(I(n,i), I(n,j)) = \begin{cases} -\left[\lambda \sum\limits_{k=0}^{R-i} p_k + n\mu + n\gamma\right], & i = j; \\[2ex] n\gamma \sum\limits_{s=0}^{j} \frac{p_s p_{j-s}^{(k-1)}}{p_j^{(k)}} \frac{p_{i-j+s}}{\sum\limits_{k=0}^{R-j+s} p_k}, & i > j; \\[2ex] n\gamma \sum\limits_{s=j-i}^{j} \frac{p_s p_{j-s}^{(k-1)}}{p_j^{(k)}} \frac{p_{i-j+s}}{\sum\limits_{k=0}^{R-j+s} p_k}, & i < j; \end{cases}$$

$$(n, i) \in \mathcal{X}_n, (n, j) \in \mathcal{X}_n, n = \overline{1, N-1};$$

$$\lambda_n(I(n, i), I(n, j)) = \begin{cases} \lambda p_{j-i}, & i \le j \le R; \\ 0, j < i; \end{cases}$$

$$(n-1, i) \in \mathcal{X}_{n-1}, (n, j) \in \mathcal{X}_n, n = \overline{1, N-1};$$

$$\mu_n(I(n, i), I(n, j)) = \begin{cases} (n+1)\mu \dfrac{p_{i-j} p_j^{(n)}}{p_i^{(n+1)}}, j \le i \le R; \\ 0, j > i; \end{cases}$$

$$(n+1, i) \in \mathcal{X}_{n+1}, (n, j) \in \mathcal{X}_n, n = \overline{1, N-1};$$

$$\psi_N(I(n, i), I(n, j)) = \begin{cases} -[N\mu + N\gamma], & i = j; \\ N\gamma \displaystyle\sum_{s=0}^{j} \dfrac{p_s p_{j-s}^{(N-1)}}{p_j^{(N)}} \dfrac{p_{i-j+s}}{\sum_{k=0}^{R-j+s} p_k}, & i > j; \\ N\gamma \displaystyle\sum_{s=j-i}^{j} \dfrac{p_s p_{j-s}^{(N-1)}}{p_j^{(N)}} \dfrac{p_{i-j+s}}{\sum_{k=0}^{R-j+s} p_k}, & i < j; \end{cases}$$

$$(N, i) \in \mathcal{X}_N, (N, j) \in \mathcal{X}_N.$$

4   Numerical Analysis

As an example of the numerical analysis, we evaluate the blocking probability B

$$B = 1 - \sum_{k=0}^{N-1} \sum_{r:(k,r)\in\mathcal{X}_k} q_k(r) \sum_{j=0}^{R-r} p_j,$$

and average amount of the occupied resources b

$$b = \sum_{k=0}^{N} \sum_{r:(k,r)\in\mathcal{X}_k} r q_k(r).$$

The state probabilities (1)–(2) with a given matrix $\mathbf{A}$ are unambiguously determine the solution of the matrix equations $\mathbf{q}^T \mathbf{A} = \mathbf{0}^T$ and $\mathbf{q}^T \cdot \mathbf{1} = \mathbf{1}$. Denote vector $\mathbf{q}_0 = \{q_{0,0}\}$ and $\mathbf{q}_i = \{q_{i,0}, ..., q_{i,R}\}$, then the systems (3)–(5) and (10)–(12) are as follows:

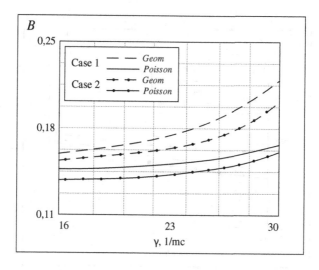

Fig. 3. Blocking probability

$$\mathbf{q}_0\boldsymbol{\Psi}_0 - \mathbf{q}_1^T\mathbf{M}_0 = \mathbf{0},$$

$$\mathbf{q}_i^T\boldsymbol{\Psi}_i - \mathbf{q}_{i+1}^T\mathbf{M}_i - \mathbf{q}_{i-1}^T\boldsymbol{\Lambda}_i = \mathbf{0}^T, i = 1,\ldots,N-1$$

$$\mathbf{q}_N^T\boldsymbol{\Psi}_N - \mathbf{q}_{N-1}^T\boldsymbol{\Lambda}_N = \mathbf{0}^T.$$

The matrix system (6)–(9) is solved in numbers using the LU decomposition algorithm [13].

The RVs of the resource requirements are considered to have a discrete distribution such as geometrical distribution with the parameter $p = \frac{1}{m+1}$ or Poisson distribution with the parameter m. Assuming that RVs are independent, we can easily get the direct formula of the k-fold convolution $p_j^{(k)}$ according to the properties of the distributions:

1. if $x \sim G(p)$ then $p_j^{(k)} = \binom{k+j-1}{k} p^j(1-p)^k$;

2. if $x \sim P(m)$ then $p_j^{(k)} = \frac{e^{-jm}}{j!}(jm)^j$.

Based on our previous experiment [11] we consider that base station can serve up to $N = 100$ active user session at the same time and all users share the $R = 100\%$ of radio resources. User's request to establish a session arrives with rate $\lambda = 16$ and average duration of a user session is $\mu^{-1} = 1$ mc. Figures 3 and 4 show the relation between blocking probability, average amount of occupied resources and a signal rate γ.

All parameters of the resource requirements distributions have to meet the average amount of the occupied resources $m = 5, 4$.

We evaluate the probability characteristics of the system assuming a user loss due to resource limitation (case 1) and lossless resource reallocation (case 2).

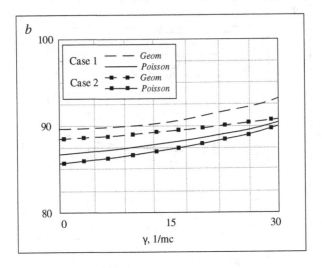

Fig. 4. Average amount of occupied resources

5   Conclusion

We have described and analyzed the general multi-server queueing system with random resource requirements and signals. A signal triggers the system transition between the states that can adequately describe the resource reallocation process for the ongoing user session in LTE network.

A simple numerical analysis has been proposed. For the future research we are going to investigate the dependence between RVs of resource requirements and consider a system with the vector of the occupied resources.

Acknowledgments. The publication was financially supported by the Ministry of Education and Science of the Russian Federation (the Agreement number 02.a03.21.0008), RFBR according to the research projects No. 16-37-60103, No. 16-07-00766 and No. 17-07-00845.

References

1. Naumov, V., Samouylov, K.: On the modeling of queueing systems with multiple resources. RUDN J. Math. Inform. Sci. Phys. **3**, 60–64 (2014). (in Russian)
2. Naumov, V., Samuylov, K., Sopin, E.: On the insensitivity of stationary characteristics to the service time distribution in queuing system with limited resources. In: IX International Workshop on Applied Problems in Theory of Probabilities and Mathematical Statistics Related to Modeling of Information Systems, Tampere, Finland, pp. 36–40 (2015)
3. Singh, S., Zhang, X., Andrews, J.: Joint rate and SINR coverage analysis for decoupled uplink downlink biased cell associations in HetNets. IEEE Trans. Wireless Commun. **14**(10), 5360–5373 (2015)

4. Sopin, E., Markova, E., Vikhrova, O., Ageev, K., Gudkova, I., Samouylov, K.: Performance analysis of M2M traffic in LTE network using queuing systems with random resource requirements. Automatic Control and Computer Sciences (in print)
5. Naumov, V., Samuoylov, K., Sopin, E., Andreev, S.: Two approaches to analysis of queuing systems with limited resources. In: Ultra Modern Telecommunications and Control Systems and Workshops (ICUMT-2014), pp. 485–488. IEEE (2014)
6. Naumov, V., Samuylov, A.: Queuing system with resource allocation of the random volume. RUDN J. Math. Inform. Sci. Phys. **2**, 38–45 (2015). (in Russian)
7. Naumov, V., Samouylov, K., Yarkina, N., Sopin, E., Andreev, S., Samuylov, A.: LTE performance analysis using queuing systems with finite resources and random requirements. In: 7th Congress on Ultra Modern Telecommunications and Control Systems ICUMT-2015, pp. 100–103. IEEE (2015)
8. Sopin, E., Samouylov, K., Vikhrova, O., Kovalchukov, R., Moltchanov, D., Samuylov, A.: Evaluating a case of downlink uplink decoupling using queuing system with random requirements. In: Galinina, O., Balandin, S., Koucheryavy, Y. (eds.) NEW2AN/ruSMART-2016. LNCS, vol. 9870, pp. 440–450. Springer, Cham (2016). doi:10.1007/978-3-319-46301-8_37
9. Samouylov, K., Sopin, E., Vikhrova, O.: On design of effective computing algorithm for blocking probability evaluation in systems with random requirements. In: 15th International Conference named after A. F. Terpugov, "Information Technologies and Mathematical Modelling" ITMM-2016, part 1, pp. 192–196. TGU, Tomsk (2016) (in Russian)
10. Samouylov, K., Sopin, E., Vikhrova, O., Sorgin, S.: Convolution algorithm for normalization constant evaluation in queuing system with random requirements. In: 12th International Conference of Numerical Analysis and Applied Mathematics ICNAAM-2016. AIP Publishing, USA (2017)
11. Samouylov, K., Sopin, E., Vikhrova, O.: Analyzing blocking probability in LTE wireless network via queuing system with finite amount of resources. In: Dudin, A., Nazarov, A., Yakupov, R. (eds.) ITMM 2015. CCIS, vol. 564, pp. 393–403. Springer, Cham (2015). doi:10.1007/978-3-319-25861-4_33
12. Ajay, D., Gurjar, D., Purohit, N.: An optimized network selection scheme for heterogeneous wireless networks. In: Sixth International Conference on Contemporary Computing (IC3), Noida, pp. 196–201 (2013)
13. Naumov, V., Samouylov, K., Gaydamaka, Y.: Multiplicative Solutions of the Finit Markov Chains. RUDN University, Moscow (2015). (in Russian)

Inventory Management System with On/Off Control of Input Product Flow

Anatoly Nazarov[1,2] and Valentina Broner[1,2(✉)]

[1] Tomsk State University, 36 Lenina Avenue, Tomsk 634050, Russian Federation
nazarov.tsu@gmail.com, valsubbotina@mail.ru
[2] Peoples Friendship University of Russia,
6 Miklukho-Maklaya Street, Moscow 117198, Russian Federation

Abstract. The subject of this paper is investigation of the inventory management systems mathematical model with following assumptions:
- Compound Poisson demand process with hyperexponential, Phase-type and arbitrary batch sizes distributions,
- On/Off control of piecewise-constant rate of input product flow.

Stationary distribution explicit expressions of inventory level accumulated in the system are obtained in this paper with hyperexponential and Phase-type distributions of demands purchases values. Fourier transform of the stationary probability density function is also determinated here in case of arbitrary distribution of demands purchases values. The results obtained in this paper are shown with illustrative numerical examples.

Keywords: Mathematical model · Inventory management · On/Off control · Hyperexponential distribution · PH-distribution · Arbitrary distribution

1 Introduction

There are different models of inventory management. The classical single-period problem of inventory management known as Newsvendor (Newsboy) Problem is one of the most widespread models, see Arrow et al. [1]; Silver et al. [11]; Khouja [7]; Qin et al. [12]; Gallego and Moon [4]; a handbook editing by Tsan-Ming Choi [3]; Kitaeva et al. [5,6].

Multi-period model is a generalization of single-period model. Multi-period inventory management models are considered in Zhang et al. [13], Mousavi et al. [8]. In [2] the multi-product multi-period inventory lot sizing with supplier selection problem are investigated.

Inventory management multi-period models with On/Off control are discussed in Nazarov and Broner [9,10]. In [9] multi-period problem is considered under following conditions: intensity of output product flow is piecewise-constant, demand purchases values have Erlang distribution. Inventory level probability density function approximation is provided in [10] for similarly mathematical model with arbitrary distribution of purchase values of demand.

© Springer International Publishing AG 2017
A. Dudin et al. (Eds.): ITMM 2017, CCIS 800, pp. 370–381, 2017.
DOI: 10.1007/978-3-319-68069-9_30

In this paper we consider single-product multi-period models of inventory management system with On/Off control and piecewise-constant rate of input product flow.

2   Mathematical Model

In this article we consider a mathematical model of inventory management (Fig. 1).

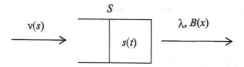

$$S$$

$$v(s) \qquad s(t) \qquad \lambda,\, B(x)$$

Fig. 1. Inventory management model

Let $s(t)$ is inventory level accumulated in the system at time t. Input product flow have piecewise-constant rate $\nu(s)$

$$\nu(s) = \begin{cases} \nu_1, s < S, \\ \nu_2, s \ge S, \end{cases}$$

where s is a value of the process $s(t)$, and S is some fixed threshold value.

We propose that demand occurs according to a Poisson process with constant intensity λ. The purchases values are independent, identically distributed random variables, having the distribution function $B(x)$.

Furthermore, assume that $s(t)$ can take negative values stand for the situation in which the customer expects when the necessary values of inventory is accumulated.

The condition of steady-state regime existence for the system can be determined from

$$\nu_1 > \lambda b > \nu_2, \tag{1}$$

where b is the first moment of function $B(x)$.

2.1   Problem Statement

Denotes

$$P(s) = \frac{\partial P\{s(t) < s\}}{\partial s}.$$

By construction the process $s(t)$ is Markovian with continuous time t and a continuous set of values $-\infty < s < \infty$.

Therefore we can write

$$P(s + \nu(s)\Delta t) = P(s)(1 - \lambda\Delta t) + \lambda\Delta t \int_0^\infty P(s + x)dB(x).$$

Using series expansion, we write the Kolmogorov equation for the stationary probability density function $P(s)$

$$\nu(s)P'(s) + \lambda P(s) = \lambda \int\limits_0^\infty P(s+x)dB(x), s \neq S, -\infty < s < \infty. \qquad (2)$$

The solution $P(s)$ of Eq. (2) satisfied following boundary conditions

$$P(-\infty) = P(\infty) = 0. \qquad (3)$$

The plan of this paper is to identify some property of the solution $P(s)$ and to determine the function $P(s)$.

2.2   Probabilities R_1 and R_2

First let us introduce some notations

$$\int\limits_{-\infty}^{S} P(s)ds = R_1, \int\limits_{S}^{\infty} P(s)ds = R_2.$$

Below we formulate a proposition about probabilities R_1 and R_2.

Proposition 1. *The probabilities R_1 and R_2 are determined by following expression*

$$R_1 = \frac{\nu_1 - \lambda b}{\nu_1 - \nu_2}, R_2 = \frac{\lambda b - \nu_2}{\nu_1 - \nu_2}. \qquad (4)$$

Proof. We multiply Eq. (2) by s, and integrate the obtained equality

$$\int\limits_{-\infty}^{\infty} s\nu(s)P'(s)ds + \lambda \int\limits_{-\infty}^{\infty} sP(s)ds = \lambda \int\limits_{-\infty}^{\infty} s \int\limits_0^\infty P(s+x)dB(x)ds. \qquad (5)$$

Let us consider the right side of expression (5)

$$\lambda \int\limits_{-\infty}^{\infty} s \int\limits_0^\infty P(s+x)dB(x)ds = \lambda \int\limits_0^\infty \int\limits_{-\infty}^{\infty} sP(s+x)dsdB(x)$$

$$= [s+x=y] = \lambda \int\limits_0^\infty \int\limits_{-\infty}^{\infty} (y-x)P(y)dydB(x)$$

$$= \lambda \int\limits_{-\infty}^{\infty} \int\limits_0^\infty (y-x)dB(x)P(y)dy$$

$$= \lambda \int\limits_{-\infty}^{\infty} (y-b)P(y)dy = \lambda \int\limits_{-\infty}^{\infty} yP(y)dy - \lambda b \int\limits_{-\infty}^{\infty} P(y)dy.$$

Substituting this expression into (5), we obtain the equality

$$\int_{-\infty}^{\infty} \nu(s)sP'(s)ds + \lambda \int_{-\infty}^{\infty} sP(s)ds = \lambda \int_{-\infty}^{\infty} yP(y)dy - \lambda b \int_{-\infty}^{\infty} P(y)dy.$$

Using condition

$$\int_{-\infty}^{\infty} P(y)dy = 1,$$

we obtain

$$\int_{-\infty}^{\infty} \nu(s)sP'(s)ds = -\lambda b.$$

We write the integral in the left side of last equation as two integral

$$\int_{-\infty}^{S} \nu_1 sP'(s)ds + \int_{S}^{\infty} \nu_2 sP'(s)ds = -\lambda b.$$

Using integration by parts we get

$$\nu_1 SP_1(S) - \nu_2 SP_2(S) - \nu_1 \int_{-\infty}^{S} P(s)ds - \nu_2 \int_{S}^{\infty} sP_2(s)ds = -\lambda b.$$

It is easily proved that

$$\nu_1 SP_1(S) - \nu_2 SP_2(S) = 0,$$

hence

$$\nu_1 \int_{-\infty}^{S} P(s)ds + \nu_2 \int_{S}^{\infty} sP_2(s)ds = \lambda b.$$

Thus we get

$$\nu_1 R_1 + \nu_2 R_2 = \lambda b.$$

Tacking into account expression

$$R_1 + R_2 = 1,$$

we find the probabilities

$$R_1 = \frac{\nu_1 - \lambda b}{\nu_1 - \nu_2}, R_2 = \frac{\lambda b - \nu_2}{\nu_1 - \nu_2}.$$

The Proposition 1 is proved.

2.3   The Solution $P(s)$ of (2) for $s > S$

For $s > S$ Eq. (2) has the form

$$\nu_2 P'(s) + \lambda P(s) = \lambda \int_0^\infty P(s+x) dB(x), \, s > S. \tag{6}$$

We will find the solution of this equation in the form of exponential function

$$P(s) = Ce^{-\gamma(s-S)}, \, s > S. \tag{7}$$

Substituting (7) into (6), we obtain

$$\lambda - \nu_2 \gamma = \lambda \int_0^\infty e^{-\gamma x} dB(x), \tag{8}$$

which is a nonlinear equation for γ.

By virtue of condition (1), Eq. (8) has a unique positive solution. Obviously other solutions of Eq. (8) are extraneous. Substituting (7) into the expression for the probability R_2

$$R_2 = \int_S^\infty P(s) ds,$$

we obtain

$$R_2 = \int_S^\infty Ce^{-\gamma(s-S)} ds = \int_S^\infty e^{-\gamma x} dx = \frac{C}{\gamma},$$

then we can write expression that defines the value of the parameter C of the function $P(s)$

$$C = \gamma R_2 = \gamma \frac{\lambda b - \nu_2}{\nu_1 - \nu_2}, \tag{9}$$

Thus, we can write the solution $P(s)$ of Eq. (2) in the following form

$$P(s) = \begin{cases} P_1(s), s < S, \\ Ce^{-\gamma(s-S)}, s > S, \end{cases} \tag{10}$$

where the function $P_1(x)$ for $x < 0$ will be defined below.

3   The Solution $P_1(s)$ of (2) for $s < S$

The main objective of this section is to try to find the solution $P(s)$ of Eq. (2). Equation (2) for $s < S$ has the form

$$\nu_1 P_1'(s) + \lambda P_1(s) = \lambda \int_0^{S-s} P_1(s+x) dB(x) + \lambda \int_{S-s}^\infty P_1(s+x) dB(x). \tag{11}$$

3.1 Case Hyperexponential Distribution $B(x)$

Lets the values of purchases be independent and identically distributed random variables having m-th order hyperexponential distribution

$$B(x) = \sum_{k=1}^{m} q_k \left(1 - e^{-\mu_k x}\right), \tag{12}$$

with positive parameters $\mu_k > 0$ and $q_k > 0$

$$\sum_{k=1}^{m} q_k = 1. \tag{13}$$

Taking into account (7) and (12), rewrite (11)

$$\nu_1 P_1'(s) + \lambda P_1(s)$$
$$= \sum_{k=1}^{m} q_k \mu_k \left\{ \lambda \int_0^{S-s} P_1(s+x) e^{-\mu_k x} dx + C e^{\mu_k s} \frac{\lambda}{\mu_k + \gamma} e^{-\mu_k S} \right\}. \tag{14}$$

It is necessary to find the solution $P_1(s)$ of (14) which will be defined in the Theorem 1.

Before formulating theorem about the function $P_1(s)$ we consider the equation

$$\nu_1 z + \lambda = \lambda \sum_{k=1}^{m} q_k \frac{\mu_k}{\mu_k - z}. \tag{15}$$

Equation (15) can be transformed to the algebraic equation of degree $m + 1$, consequently Eq. (15) has $m+1$ roots. Obviously $z = 0$ is a root of this equation.

For the other roots $z = z_k, k = \overline{1, m}$ of the Eq. (15) we formulate following lemma.

Lemma 1. *If the condition (1) is satisfied*

$$\nu_1 > \lambda b$$

all the roots $z = z_k, k = \overline{1, m}$ of Eq. (15) are real and positive.

The proof is omitted.

We now prove theorem about form of the function $P(s)$.

Theorem 1. *If $B(x)$ is hyperexponential distribution (12), then the solution $P_1(s)$ of Eq. (11) has form*

$$P_1(s) = C \sum_{n=1}^{m} x_n e^{z_n(s-S)}, s < S, \tag{16}$$

where z_n are positive roots of Eq. (15), x_n are components of the vector $\mathbf{X}$. This vector is a solution to a system of linear algebraic equations

$$\mathbf{AX} = \mathbf{h}, \tag{17}$$

where A_{kn} are elements of the matrix $\mathbf{A}$, h_k are elements of the vector $\mathbf{h}$. The elements A_{kn} and h_k have following form

$$A_{kn} = \frac{1}{\mu_k - z_n}, h_k = \frac{1}{\mu_k + \gamma}, k = \overline{1,m}, n = \overline{1,m}, \tag{18}$$

normalizing constant C is determined by (9).

The proof is left to the reader.

Hence the problem of investigating the mathematical model of the inventory management system with On/Off control and hyperexponential distribution of demands purchases values is completely solved.

3.2   Case PH-distribution $B(x)$

Consider PH-distribution of demand purchases values

$$B(x) = 1 - \beta e^{\mathbf{G}x}\mathbf{E}, \tag{19}$$

where $\beta_k > 0$ and

$$\beta \mathbf{E} = 1, \tag{20}$$

$\mathbf{G}$ is subgenerator matrix Markov chain that determines the Phase-type distribution.

Theorem 2. *If $B(x)$ is Ph-distribution (19) and equation*

$$\nu_1 z + \lambda = \lambda \beta (\mathbf{G} + z\mathbf{I})^{-1}\mathbf{GE} \tag{21}$$

has n simple roots with positive real parts, then solution $P_1(s)$ of Eq. (11) has form

$$P_1(s) = C \sum_{n=1}^{m} x_n e^{z_n(s-S)}, s < S, \tag{22}$$

where $z = z_n$, $n = \overline{1,m}$ are roots of Eq. (21), $x_n, n = \overline{1,m}$ are solutions to a system of equations

$$\left(\sum_{n=1}^{m} x_n(\mathbf{G} + z_n\mathbf{I})^{-1} - (\mathbf{G} - \gamma\mathbf{I})^{-1}\right)\mathbf{GE} = 0, \tag{23}$$

normalizing constant C is determined by the expression (9).

Proof. Solution $P_1(s)$ of the Eq. (11) will be find in the form (24).
Substituting (21) and (24) into (11) we obtain the equation

$$\sum_{n=1}^{m} x_n e^{z_n(s-S)} \left\{\nu_1 z_n + \lambda - \lambda\beta(\mathbf{G} + z\mathbf{I})^{-1}\mathbf{GE}\right\}$$

$$= \beta e^{\mathbf{G}(S-s)}\left(\lambda \sum_{n=1}^{m} x_n(\mathbf{G} + z_n\mathbf{I})^{-1} + \lambda(\mathbf{G} - \gamma\mathbf{I})^{-1}\right)\mathbf{GE}.$$

By equating to zero the coefficients in the linear combination of exponents $e^{z_n(s-S)}$ in this expression, we get

$$\nu_1 z_n + \lambda - \lambda\beta(\mathbf{G} + z\mathbf{I})^{-1}\mathbf{GE}, n = \overline{1, m}.$$

Obviously that this expression and (23) have the same form. Consequently z_n are the roots of the Eq. (23).

Continuing in the same way, we obtain

$$\left(\lambda\sum_{n=1}^{m} x_n(\mathbf{G} + z_n\mathbf{I})^{-1} + \lambda(\mathbf{G} - \gamma\mathbf{I})^{-1}\right)\mathbf{GE} = 0.$$

Theorem is proved.

4   The Solution $P_1(s)$ of Eq. (11) for $s < S$ with Arbitrary Functions $B(x)$

To solve Eq. (11) under the condition of arbitrary distribution, we use the Fourier transform method.

We multiply Eq. (2) by $e^{ju(s-S)}, j = \sqrt{-1}$ and integrate the obtained equality, then we get

$$\int_{-\infty}^{\infty} \nu(s)e^{ju(s-S)}P'(s)ds + \lambda \int_{-\infty}^{\infty} e^{ju(s-S)}P(s)ds = \lambda \int_{-\infty}^{\infty} e^{ju(s-S)} \int_{0}^{\infty} P(s+x)dB(x)ds. \tag{24}$$

Transform the right-hand side of this expression

$$\lambda \int_{-\infty}^{\infty} e^{ju(s-S)} \int_{0}^{\infty} P(s+x)dB(x)ds$$

$$= \lambda \int_{0}^{\infty} \int_{-\infty}^{\infty} e^{ju(s-S)}P(s+x)dsdB(x) = [s+x=y]$$

$$= \lambda \int_{0}^{\infty} \int_{-\infty}^{\infty} e^{ju(y-x-S)}P(y)dydB(x) = \lambda \int_{0}^{\infty} e^{ju(x-S)}dB(x) \int_{-\infty}^{\infty} e^{juy}P(y)dy.$$

Let us consider following form of function $P(s)$

$$P_1(s) = \begin{cases} P(S), s < S, \\ 0, s > S, \end{cases} \quad P_2(s) = \begin{cases} 0, s < S, \\ P(S), s > S, \end{cases}$$

and denotes Fourier transform for functions $P_1(s)$ and $P_2(s)$

$$P_1^*(u) = \int_{S}^{\infty} e^{juy}P_1(y)dy, P_2^*(u) = \int_{S}^{\infty} e^{juy}P_2(y)dy,$$

we find integral in right side of Eq. (24)

$$\lambda \int\limits_{-\infty}^{\infty} e^{ju(s-S)} \int\limits_{0}^{\infty} P(s+x)dB(x)ds$$

$$= \lambda e^{-juS} \int\limits_{0}^{\infty} e^{jux}dB(x) \left(P_1^*(u) + P_2^*(u)\right).$$

Similarly we can write expression for integral in left side of Eq. (24)

$$\int\limits_{-\infty}^{\infty} e^{ju(s-S)}\lambda(s)P(s)ds = \lambda e^{-juS} \left(P_1^*(u) + P_2^*(u)\right).$$

Using the method of integration by parts we can write

$$\int\limits_{-\infty}^{\infty} \nu(s)e^{ju(s-S)}P'(s)ds$$

$$= \nu_1 P_1(u) - \nu_2 P_2(u) - jue^{-juS} \left(\nu_1 P_1^*(u) + \nu_2 P_2^*(u)\right),$$

It possible to prove that $\nu_1 P_1(u) - \nu_2 P_2(u) = 0$, then we have equation for Fourier transform of function $P_1(s)$

$$P_1^*(u) = -\frac{\left\{\lambda - ju\nu_2 - \lambda \int\limits_0^{\infty} e^{jux}dB(x)\right\}}{\left\{\lambda - ju\nu_1 - \lambda \int\limits_0^{\infty} e^{jux}dB(x)\right\}} P_2^*(u).$$

Taking into account (7, 8, 9) we conclude that Fourier transform $P_2^*(u)$ is determined by

$$P_2^*(u) = \int\limits_{S}^{\infty} e^{juy}P_2(y)dy = Ce^{\gamma S} \int\limits_{S}^{\infty} e^{(ju-\gamma)y}dy = -\frac{C}{ju-\gamma}e^{juS},$$

it follows that

$$P_1^*(u) = \frac{\left\{\lambda - ju\nu_2 - \lambda \int\limits_0^{\infty} e^{jux}dB(x)\right\}}{\left\{\lambda - ju\nu_1 - \lambda \int\limits_0^{\infty} e^{jux}dB(x)\right\}} \frac{C}{ju-\gamma}e^{juS}.$$

Therefore, we define the Fourier transform of the function $P_1(s)$. Using the inverse Fourier transform, we obtain the expression that is determining the function $P_1(s)$. Hence, the function $P_1(s)$ is known for arbitrary distribution $B(x)$.

Nevertheless, the inverse Fourier transform can not be computed numerically for all kinds of functions $B(x)$, so the exact solutions obtained for hyperexponential and PH-distributions are essential.

Using $\nu_1 S P_1(S) - \nu_2 S P_2(S) = 0$, we have following expression

$$P_1(S) = \frac{\nu_2}{\nu_1} P_2(S),$$

thus the probability density function $P(s)$ of process $s(t)$ is continuous for all values of $s \neq S$, but at the point $s = S$ function $P(s)$ is discontinuous $P_1(S) \neq P_2(S)$.

5 Numerical Experiments

In this section, numerical results are obtained for case of Gamma distribution $B(x)$.

We assume that demands purchases values have Gamma distribution. For this case, it is necessary to apply the results obtained in Sect. 4.

Let the gamma distribution have parameters of the form α and the scale β. Further assume that $\alpha = \beta = 3$, then the average value will be equal to one.

Solution $P(s), s > S$ of Eq. (2) is defined by explicit expression (7) with following values of parameters

$$\gamma = 0.361, C = 0.181.$$

Solution $P(s), s < S$ of Eq. (2) is defined by inverse Fourier transform

$$P_1(s) = \frac{1}{2\pi} \int_{-\infty}^{\infty} e^{-jus} P_1^*(u) du.$$

The stationary distribution of the inventory level accumulated in the system is shown in Fig. 2.

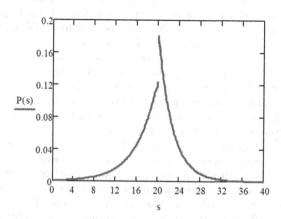

Fig. 2. Probability density function $P(s)$ for Gamma distribution $B(x)$

6 Conclusion

In this paper we presented the mathematical model of multi-period inventory management system with On/Off control under following conditions:

- input product flow rate is piecewise-constant,
- intensity of the output product flow is constant.

Our main result is the following. the explicit expressions for the stationary distributions of inventory level is obtained in the cases of the hyperexponential and the phase-type distributions of demands purchases values. The stationary distributions of inventory level is determined in terms of the inverse Fourier transform for arbitrary distribution $B(x)$. Thus the problem of investigating the mathematical model of the inventory management system with On/Off control is completely solved.

Acknowledgments. The publication was financially supported by the Ministry of Education and Science of the Russian Federation (the Agreement number 02.a03.21.0008).

References

1. Arrow, K.J., Harris, T.E., Marschak, J.: Optimal inventory policy. Econometrica **19**(3), 205–272 (1951)
2. Cardenas-Barron, L.E., Gonzalez-Velarde, J.L., Trevino-Garza, J.: A new approach to solve the multi-product multi-period inventory lot sizing with supplier selection problem. Comput. Oper. Res. **64**, 225–232 (2015)
3. Choi, T.-M. (ed.): Handbook of Newsvendor Problems: Models, Extensions and Applications. Springer, New York (2012)
4. Gallego, G., Moon, I.: The distribution free newsboy problem: review and extensions. J. Oper. Res. Soc. **44**, 825–834 (1993)
5. Kitaeva, A., Subbotina, V., Zmeev, O.: The newsvendor problem with fast moving items and a compound poisson price dependent demand. In: 15th IFAC Symposium on Information Control Problems in Manufacturing INCOM 2015, (IFAC-PapersOnLine), vol. 48, pp. 1375–1379. Elsevier (2015)
6. Kitaeva, A., Subbotina, V., Stepanova, N.: Estimating the compound poisson demand's parameters for single period problem for large lot size. In: 15th IFAC Symposium on Information Control Problems in Manufacturing INCOM 2015, (IFAC-PapersOnLine), vol. 48, pp. 1357–1361. Elsevier (2015)
7. Khouja, M.: The single-period (news-vendor) problem: literature review and suggestionsfor future research. OMEGA-INT J. **27**(5), 537–553 (1999)
8. Mousavia, S.M., Hajipoura, V., Niakib, S.T.A., Alikar, N.: Optimizing multi-item multi-period inventory control system with discounted cash flow and inflation: two calibrated meta-heuristic algorithms. Appl. Math. Model. **37**(4), 2241–2256 (2013)
9. Nazarov, A., Broner, V.: Inventory management system with erlang distribution of batch sizes. In: Dudin, A., Gortsev, A., Nazarov, A., Yakupov, R. (eds.) ITMM 2016. CCIS, vol. 638, pp. 273–280. Springer, Cham (2016). doi:10.1007/978-3-319-44615-8_24

10. Nazarov, A.A., Broner, V.I.: Resource control for physical experiments in the Cramer-Lundberg model. Russian Phys. J. **59**(7), 1024–1036 (2016)
11. Silver, E.A., Pyke, D.F., Peterson, R.: Inventory Management and Production Planning and Scheduling. Wiley, New York (1998)
12. Qin, Y., Wang, R., Vakharia, A., Chen, Y., Hanna-Seref, M.: The newsvendor problem: review and directions for future research. Eur. J. Oper. Res. **213**, 361–374 (2011)
13. Zhang, D., Xu, H., Wu, Y.: Single and multi-period optimal inventory control models with risk-averse constraints. Eur. J. Oper. Res. **199**, 420–434 (2009)

Semi-Analytical Methods for Complex Optimization of Non-Markov Queueing Networks

Vladimir N. Zadorozhnyi$^{(\boxtimes)}$ and Maksim A. Kornach

Omsk State Technical University, Omsk, Russia
zwn2015@yandex.ru, mkornach@gmail.com

Abstract. Effective analytic-imitational methods for complex optimization of routing matrices and node efficiencies in non-Markov queueing networks are proposed, and the optimization is to be carried out by the minimum of the mean time required for a claim to pass through the network. The characteristic feature of the method is a prompt and precise calculation of the target function, based on the varied transition probabilities. Consideration is given to the possibility to represent the roads in traffic networks as multiserver systems, in which the servicing intensities depend on the load coefficients. The method for complex optimization of networks with such multiserver nodes is developed. The application results are given.

Keywords: Non-Markov queueing networks · Variable service intensity · Gradient optimization methods · Monte Carlo simulation · Traffic networks

1 Introduction

In the general case, simulation is used to optimize non-Markov queueing networks. Given that, if the number of varied parameters is more than ten, optimization is almost impossible without gradient methods. Moreover, the gradient calculation in simulation is hindered by the stochastic errors in the target function estimates, evaluated by Monte Carlo method [1]. The problem of the gradient calculation at the queueing networks simulation, however, can be solved by applying a semi-analytical approach [2–5].

When optimizing the queueing networks, the varied parameters can be represented by the nodes performances [6], the amount of channels in the nodes [7], the spooling capacities [8], the transition probabilities, or a combination of these several parameter types [1,9]. In real networks total resource used to vary the given parameters tends to be limited, therefore, the optimization problems are usually put as search for optimal resource distribution of the given kind [10] or as an optimal combination of diverse resource distributions. Target functions tend to include such quality indicators as

© Springer International Publishing AG 2017
A. Dudin et al. (Eds.): ITMM 2017, CCIS 800, pp. 382–397, 2017.
DOI: 10.1007/978-3-319-68069-9_31

- mean time of a claim passing through the network (mean response time),
- probability of exceeding acceptable response time,
- probability of losing a claim in network nodes,
- spooling capacities,
- total number of the channels in the nodes, etc.

Developing new efficient methods for non-Markov queueing networks optimization is rather relevant in the field of computer [10–12] and traffic [13–15] networks design.

Significant obstacles occur at queueing networks optimization when varied parameters include transition probabilities [1], i.e. at the routing matrix optimization. The present work proposes effective methods for complex optimization of routing matrices and node efficiencies in non-Markov queueing networks.

2   Combined Optimization Problem for Transition Probabilities and Node Efficiencies

Using the problem with varied parameters including routing matrix as an example, let us consider open non-Markov queueing networks optimization through simultaneous redistribution of the network nodes' efficiency and its transition probabilities [1].

The network receives a recurrent flow of claims with intensity Λ. The lengths of the intervals between claims entering the network are described by distribution function (d.f.) $A(t)$. From the input flow, a claim with probability p_{0j} proceeds to the i-th node, $i = \overline{1, n}$. Independent claim servicing time in any of K_i channels in i-th node has d.f. $B_i(t)$. After servicing in i-th node, the claim selects one of the nodes $j = \overline{1, n}$ at random, according to the specified transition probabilities p_{ij}, to continue its route or (with probability p_{0j}) leaves the network (Fig. 1). Probabilities $p_{ij}(i, j = \overline{0, n})$ are set by a routing matrix $P = \|p_{ij}\|$.

Mean time E of passing through an open network (mean response time) can be expressed by formula:

$$E = \sum_{i=1}^{n} \alpha_i \left(w_i + \frac{1}{\mu_i} \right),\tag{1}$$

where α_i is the mean number (frequency) of the visits at i-th node a claim does when passing through the network, w_i is the mean waiting time at the queue at i-th node, $\mu_i = b_i^{-1}$ is the servicing intensity in the channel at i-th node (consequently, b_i is the mean servicing time at i-th node).

Frequencies α_i are uniquely determined by probabilities p_{ij}:

$$\alpha_i = \sum_{j=1}^{n} \alpha_j p_{ji}, \quad i = \overline{0, n}, \quad \alpha_0 \equiv 1.\tag{2}$$

Then, through α_i the intensities $\lambda_i = \Lambda \alpha_i$ of the flows entering the nodes and their load coefficients $p_i = \lambda_i / (\mu_i K_i)$, $i = \overline{1, n}$ are determined subsequently.

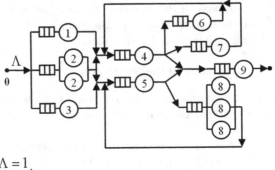

$\Lambda = 1$,

$p_{0,1} = 0.2$, $p_{0,2} = 0.3$, $p_{0,3} = 0.5$, $p_{2,4} = 0.7$, $p_{2,5} = 0.3$,
$p_{4,6} = 0.3$, $p_{4,7} = 0.4$, $p_{4,9} = 0.3$, $p_{5,8} = 0.9$, $p_{5,9} = 0.1$

Fig. 1. Example of an open queueing network

In the general case, the values of w_i for (1) are estimated by simulation and are effectively used for optimization of the efficiency distribution among the nodes by directing hyperboles method [6].

A more general problem of combined optimal distribution of efficiency and transition probabilities is formulated in [1] as follows. Resource M of the network efficiency as a function $\overrightarrow{\mu} = (\mu_1, \ldots, \mu_n)$ of the vector for the servicing intensities in nodes $i = \overline{1, n}$ is set as $M(\overrightarrow{\mu}) = \sum_{i=1}^{n} c_i \mu_i^{\beta_i}$, where $c_i > 0$ are cost coefficients, and $\beta_i > 0$ are nonlinearity coefficients. The varied transition probabilities given in a fixed order are the coordinates for vector $\overrightarrow{p_v}$ with the dimension m.

We need to find vectors $\overrightarrow{\mu} = \overrightarrow{\mu}_{opt}$ and $\overrightarrow{p_v} = \overrightarrow{p_v}_{opt}$ that realize the minimum of the mean response time:

$$E(\overrightarrow{\mu}, \overrightarrow{p_v}) = \sum_{i=1}^{n} \alpha_i(\overrightarrow{p_v}) \left(w_i(\overrightarrow{\mu}, \overrightarrow{p_v}) + \frac{1}{\mu_i} \right) \longrightarrow \min_{\overrightarrow{\mu}, \overrightarrow{p_v}} \tag{3}$$

and lie in the following admissible solutions region

$$M(\overrightarrow{\mu}) = \sum_{i=1}^{n} c_i \mu_i^{\beta_i} = M^* = const; \quad 0 \le \rho_i \le 1; \quad \sum_{j=0}^{n} p_{ij} = 1, \quad (i = \overline{0, n});$$

$$p_{vi}^{min} \le p_{vi} \le p_{vi}^{max}, \quad (i = \overline{1, m}). \tag{4}$$

The method for solving problem (3) and (4), developed and tested in [1], comprises a cycle of the iterations, drawing the varied parameters to solution $(\overrightarrow{\mu}_{opt}, \overrightarrow{p_v}_{opt})$. Each iteration consists of two steps. At the first step at fixed $\overrightarrow{p_v}$, the distribution $\overrightarrow{\mu} = (\mu_1, \ldots, \mu_n)$ of resource M^* by the network nodes is optimized due to directing hyperboles method [6]. At the second step at found $\overrightarrow{\mu}$, vector $\overrightarrow{p_v}$ for transition probabilities is optimized. Then, the iterations of vectors $\overrightarrow{\mu}$ and $\overrightarrow{p_v}$ step-by-step optimization are repeated till the specified breakpoint condition appears. At vector optimization step $\overrightarrow{p_v}$ partial derivatives (p.d.)

$\partial E/\partial p_{jk}$ of time E are calculated by all vector components $\overrightarrow{p_v}$, these derivatives $\partial E/\partial p_{jk}$ calculation amounts to calculating derivatives $\partial \alpha(\overrightarrow{p_v})/\partial p_{jk}$. In fact, differentiating equation (1) by any transition probability p_{jk}, we get equation

$$
\begin{aligned}
\frac{\partial E}{\partial p_{jk}} &= \sum_{i=1}^{n} \frac{\partial(\alpha_i/\mu_i)}{\partial p_{jk}} + \sum_{i=1}^{n} \frac{\partial(\alpha_i w_i)}{\partial p_{jk}} \\
&= \sum_{i=1}^{n} \frac{1}{\mu_i} \frac{\partial \alpha_i}{\partial p_{jk}} + \sum_{i=1}^{n} \frac{\partial \alpha_i}{\partial p_{jk}} w_i + \sum_{i=1}^{n} \alpha_i \frac{\partial w_i}{\partial p_{jk}},
\end{aligned}
\tag{5}
$$

from which in [16] approximation is derived

$$
\frac{\partial E}{\partial p_{jk}} = \sum_{i=1}^{n} \left(\frac{1}{\mu_i} - \mu_i \frac{\partial w_i}{\partial \mu_i} \right) \frac{\partial \alpha_i}{\partial p_{jk}},
\tag{6}
$$

in which the values of all variables and derivatives, excluding $\partial \alpha_i/\partial p_{jk}$, are known after vector $\overrightarrow{\mu}$ optimization step (by directing hyperboles method). For accurate calculation of p.d. $\partial E/\partial p_{jk}$, an advanced reduction method for semi-Markov process graph is used in [1]. All p.d. $\partial \alpha_i/\partial p_{jk}$ are calculated due to this method during successive graph reductions, accompanied by recursive recomputation of p.d. from superpositions of numerous variables functions. Let us demonstrate that calculating p.d.$\partial \alpha_i/\partial p_{jk}$ can be facilitated significantly, if it is reduced to solving auxiliary systems of linear algebraic equations (SLAE).

3   Calculation of Frequency p.d. α_i by Transition Probabilities

Since the problem of calculating p.d. $\partial \alpha_i/\partial p_{jk}$ is the one of calculating p.d. for solutions α_i of SLAE (2) by this SLAE coefficients p_{ij}, it is enough to demonstrate how to calculate p.d. for solutions $\overrightarrow{x}$ of random SLAE $\mathbf{A}\overrightarrow{x} = \overrightarrow{b}$ by its coefficients $\mathbf{A}, \overrightarrow{b}$.

NB 1. In this part, denomination b_i possesses a local meaning of free equation terms, not coinciding with the meaning they have in the rest of the article.

Let there be given SLAE with a single solution:

$$
\begin{cases}
a_{11}x_1 + a_{12}x_2 + \ldots + a_{1n}x_n = b_1 \\
a_{21}x_1 + a_{22}x_2 + \ldots + a_{2n}x_n = b_2 \\
\quad \vdots \\
aa_{n1}x_1 + a_{n2}x_2 + \ldots + a_{nn}x_n = b_n
\end{cases}
\tag{7}
$$

and it is necessary to calculate p.d. $\partial x_i/\partial a_{jk}$ for all solutions x_i of this system by its chosen coefficient a_{jk}. To calculate the desired p.d. accurately, this SLAE equation should be differentiated by the selected coefficient a_{jk}. As a result we get:

$$
\begin{cases}
a_{11}\frac{\partial x_1}{\partial a_{jk}} + \cdots + a_{1k}\frac{\partial x_k}{\partial a_{jk}} + \cdots + a_{1n}\frac{\partial x_n}{\partial a_{jk}} = 0 \\
\cdots \\
a_{j1}\frac{\partial x_1}{\partial a_{jk}} + \cdots + a_{jk}\frac{\partial x_k}{\partial a_{jk}} + \cdots + a_{jn}\frac{\partial x_n}{\partial a_{jk}} = -x_k \\
\cdots \\
a_{n1}\frac{\partial x_1}{\partial a_{jk}} + \cdots + a_{nk}\frac{\partial x_k}{\partial a_{jk}} + \cdots + a_{nn}\frac{\partial x_n}{\partial a_{jk}} = 0.
\end{cases}
\tag{8}
$$

All n p.d. $\partial x_i / \partial a_{jk}$ are now unknown in SLAE (8) and, therefore, can be determined as its solutions. Matrix $\mathbf{A}$ of SLAE coefficients (8) coincides with matrix $\mathbf{A}$ of the initial SLAE (7). The column of the right parts contains zeros everywhere except row j, where value $(-x_k)$ is put. Therefore, we get a simple method to calculate p.d. for all SLAE solutions x_i by its selected coefficient a_{jk}: (s1) solve initial SLAE (7); (s2) solve auxiliary SLAE (8), derived from the original SLAE (7) by substituting all elements of the column $\overrightarrow{b}$ with zeros, except element b_j, that is substituted by the value of $(-x_k)$, known after the step (s1).

Here, j is the number of the row with a free term not equal to zero; it is defined by the first index of the selected coefficient a_{jk}. Number k of element x_k, determining a non-zero free term, is set by the second index of coefficient a_{jk}.

Solution $(\dot{x}_1, \ldots, \dot{x}_n)$ of the auxiliary SLAE from step (s2), is the desired vector $\left(\frac{\partial x_1}{\partial a_{jk}}, \ldots, \frac{\partial x_n}{\partial a_{jk}} \right)$.

NB 2. When calculating p.d. for the solutions of the initial SLAE (7) by the selected coefficient b_j, the auxiliary SLAE differs from SLAE (8) only by the fact that $b_j = 1$.

As the numerical experiments have shown, the method of auxiliary SLAE allows one to calculate relevant SLAE with high accuracy and speed. For example, in EXCEL p.d. is calculated by this method with 15 sharp significant decimal digits. The solution is calculated almost momentarily, compared to the simulation steps. Therefore, to solve SLAE consisting of 100 equations in Excel (by matrix $\mathbf{A}$ inversion method) takes just a share of a second, together with the results presentation.

4   Example of Optimal Vector $\overrightarrow{p_v}$ Refinement

The problem of optimal distribution of efficiency and transition probabilities for the network in Fig. 1 is solved in [1]. The probabilities are optimized by the graph reduction method. The exact statement of the optimization problem is given in [1]. As a result of the optimization, the mean response time lowered in [1] from $E = 25.5$ to $E = \mathbf{3.1}$.

When preparing the current article, the experiment was made, in which the reduction method was substituted by that of auxiliary SLAE. The resulting solution was more accurate: time E decreased to **2.697**. Only transition probabilities distribution was refined. Instead of optimal values $p_{0,1} = 0.2286, p_{0,2} = 0.3159, p_{2,4} = 0.7075, p_{4,6} = 0.2, p_{4,7} = 0.3, p_{5,8} = 0.5$, given in [1], optimal values $p_{0,1} = 0.1, p_{0,2} = 0.2, p_{2,4} = 0.6, p_{4,6} = 0.2, p_{4,7} = 0.3, p_{5,8} = 0.5$ were obtained. Other transition probabilities (Fig. 1) were uniquely determined by the mentioned ones.

5 SE-Networks Optimization Problem and Its Applied Significance

Let us examine a segment of urban road network in Fig. 2. In this network roads 9-5, 9-6 and 9-7 are unidirectional, while the rest are bidirectional. The roads in all directions have two lanes.

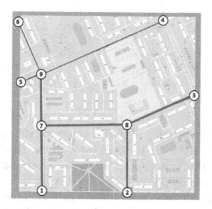

Fig. 2. Example of the simulated sector of the road network

Using this road network as an example, let us demonstrate the possibility of complex optimization (by efficiency and transition probabilities) of transport networks by queueing theory methods. For this purpose, the network roads will be represented by multiserver queueing systems. Let number K of channels in every multiserver system be equal to the corresponding road capacity, i.e. the maximal amount of the medium-sized cars that can drive on the road at the same time, i.e. they can be "serviced" by it parallely in time. Routing matrix **P** of the resulting network with multiserver nodes-roads is given in Table 1. In practice, transition probabilities of such routing matrices can be calculated by examining the road networks under investigation.

It is known that the speed of the cars decreases at growing traffic density. This dependence is taken into consideration thanks to the roads fundamental diagrams (FDs). There are several generally accepted representations of FDs. Let us define FD as a function $\delta(\rho)$ of the road utilization coefficient ρ:

$$\delta(\rho) = \frac{v(\rho)}{v_0}, \qquad (9)$$

Table 1. Transition probabilities of the road network sector

Road i	Road j													
	0	7-1	2-8	9-7	7-8	3-8	1-7	7-9	8-7	8-2	8-3	9-5	9-6	9-4
0			0.3			0.3	0.4							
7-1							1							
2-8									0.5	0.2	0.3			
9-7		0.3			0.2			0.5						
7-8									0.5	0.2	0.3			
3-8									0.5	0.2	0.3			
1-7		0.3			0.2			0.5						
7-9				0.2								0.3	0.2	0.3
8-7		0.3			0.2			0.5						
8-2			1											
8-3						1								
9-5	1													
9-6	1													
9-4	1													

where $\rho = \lambda b/K$, λ is the intensity of the cars entering the road, $b = b(\rho) = l/v(\rho)$ is the mean time it takes the car to drive through the road, l is the road length $(0 \leq \rho \leq 1)$; $v(\rho)$ is the mean speed at the road at utilization coefficient ρ; $v_0 = v(0)$ is the basic mean speed of the cars driving on an empty road.

NB 3. The capacities of the nodes-roads in the network are large enough, therefore, in case of a significant road utilization (which is of the largest interest), the random number of cars in the corresponding multiserver node has a distribution close to that of Gauss [12]. It permits an approximate calculation of stationary modes and their characteristics optimization based on the mean number of cars on the roads and their mean speed. Consideration of such approximation justifies the use of ratio $b(\rho) = l/v(\rho)$ (examining dense traffic flows, one can see that the speed of the cars is almost the same in a dense flow). In the general case, ratio $b(\rho) = l/v(\rho)$ is inadequate, since means b and v are treated as ensemble means for the cars having different values of speed. In the general case, however, this problem can be eliminated by determining FDs momentarily by the ratio of basic mean time $b_0 = b(0)$ of the road passage at $\rho = 0$ and mean time $b = b(\rho)$, depending on the load.

As a rule, FDs $\delta(\rho)$ (9) are monotone non-increasing functions, therefor $0 \leq \delta(\rho) \leq 1$. According to (9) at known FD, actual mean speed $v(\rho)$ on the road is determined by formula $v(\rho) = v_0\delta(\rho)$. As for real roads, their FDs are calculated just like routing matrices, i.e. by full-scale measurements on the roads.

For the present work objectives, it is possible to use simplified FD, the same one for all multiserver network nodes. As an example, function

$$\delta(\rho) = \frac{1}{1 + \rho^2},$$

shown by the upper line in Fig. 3, will be selected as FD.

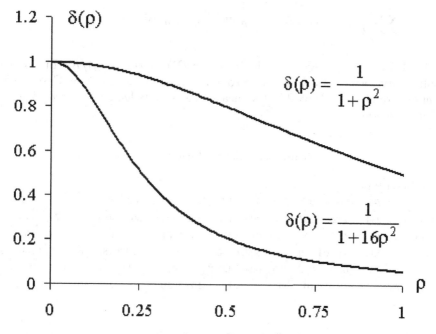

Fig. 3. Examples of roads' FDs

Main parameters of the roads are given in Table 2. The last row has the values of basic mean speed v_{0i}, their low levels are explained by bad quality of the roads. Financial resource $\Phi = \$63820$ is provided for repairing the roads. The network optimization problem involves finding such distribution of the provided resource among the roads that will allow mean time E of the car passing through the network to be minimal. Moreover, it is necessary to take into consideration that after the road repair the private transport drivers may change the previous routes of the road passage to the new, more beneficial ones. Transport services may affect the routes as well, including those of public transport. Such route optimization overlaps the distribution optimization for the financial resource Φ^* after its realization. Therefore, it is logical to optimize the resource distribution Φ^* at the same time with the routing matrix $\mathbf{P}$. The next optimization problem arises, that is roughly similar to problem (3) and (4). It is necessary to find the vector $\overrightarrow{\gamma} = \overrightarrow{\gamma}_{opt}$ of speed increased thanks to the repair and routing matrix $\mathbf{P} = \mathbf{P}_{opt}$, that provide the minimal mean response time at given Λ

$$E(\overrightarrow{\gamma}, \mathbf{P}) = \sum_{i=1}^{n} \alpha_i(\mathbf{P}) \frac{l_i}{(v_{0i} + \gamma_i)\delta(\rho_i)} \longrightarrow \min_{\overrightarrow{\gamma}, \mathbf{P}} \qquad (10)$$

and lying in the following admissible solutions region

$$\Phi(\overrightarrow{\gamma}) = \sum_{i=1}^{n} \phi_i(\gamma_i) = \sum_{i=1}^{n} c_i l_i \gamma_i^{\beta_i} = \Phi^*; \quad 0 \le \rho_i \le 1;$$

$$\sum_{j=0}^{n} p_{ij} = 1, \quad (i = \overline{0, n}); \quad p_{ij}^{min} \le p_{ij} \le p_{ij}^{max}, \quad (i, j = \overline{1, n}). \quad (11)$$

In (10) $\rho_i = \lambda_i b_i / K_i = \alpha_i \Lambda b_i / K_i = \alpha_i(\mathbf{P}) \Lambda l_i / (v_{0i} + \gamma_i) / K_i$. In (11) all nonlinearity coefficients β_i are assumed approximately equal to one. Cost coefficients c_i are given in Table 2 in the measurement units $[dollar/(km/h)]/km = dollar \times h/km^2$.

Table 2. Initial road parameters

Parameter	Road i												
	7-1	2-8	9-7	7-8	3-8	1-7	7-9	8-7	8-2	8-3	9-5	9-6	9-4
Length, m	234	242	257	234	305	176	242	257	305	484	207	72	176
Capacity	47	48	51	47	61	35	48	51	61	97	41	14	35
c_i	399.42	414.04	440.6	1398	523.3	305.7	1412	1437	524.3	1824	1352	1123	130
α_i	0.75	0.74	0.25	0.5	0.96	42005	45658	42736	0.44	0.66	0.375	0.25	0.375
v_{0i}, km/h	10	10	10	10	10	10	10	10	10	10	10	10	10

In (11) in limitations $p_{ij}^{min} \le p_{ij} \le p_{ij}^{max}$ for changes in the routing matrix we set that all p_{ij}^{min} are equal to $p_{ij}^0 - 0.1$ and all p_{ij}^{max} are equal to $p_{ij}^0 + 0.1$, where p_{ij}^0 are initial values of transition probabilities given in Table 1. However, the probabilities set by zeros and ones do not vary. In applied problems, it is more logical to set the limitations for changes in the transition probabilities as the limitations for frequencies α_i [9]. As numerical experiments demonstrate, this way of setting the limitations has no effect on the efficiency of the proposed approach.

The described herein networks with multiserver nodes, simulating the transport network roads, will be further called the networks with suppressed efficiency (SE-networks).

6   SE-Networks Optimization in a Low Load Mode

In practice, resource Φ^* distribution by multiserver nodes in SE-networks (by the network roads) can be optimized for a low load mode at $\rho \to 0$, i.e. for the mode when transport drives on empty roads, without taking into consideration the possibility of the routing matrix modification, if the reasons for it are considered insignificant. In this case the optimization problem is notably simplified and solved by analytical methods. The main question considering the legitimacy of such approach is how target function E of the optimized network will change when the road load will be considerable as, for example, during rush hours before the weekend.

This simplified optimization problem will be solved and the quality of the simplified solution will be evaluated on the example of SE-network constructed in the previous section of the article. As a result of the simplification, problem (10) and (11) is set as follows. It is necessary to find vector $\overrightarrow{\gamma} = \overrightarrow{\gamma}_{opt}$ for speed increased due to the road repair that provides the minimal mean response time

$$E(\overrightarrow{\gamma}) = \sum_{i=1}^{n} \alpha_i \frac{l_i}{(v_{0i} + \gamma_i)} \longrightarrow \min_{\overrightarrow{\gamma}} \qquad (12)$$

and lies in the following admissible solutions region

$$\Phi(\overrightarrow{\gamma}) = \sum_{i=1}^{n} \phi_i(\gamma_i) = \sum_{i=1}^{n} c_i l_i \gamma_i^{\beta_i} = \Phi^*. \qquad (13)$$

Problem (12) and (13) is easily solved by Lagrange multiplier method and at all $\beta_i = 1$ the optimal solution is determined by expressions

$$\gamma_i = \sqrt{\frac{\alpha_i}{\lambda c_i}} - v_{0i}, \quad i = 1, \ldots, n; \quad \lambda = \left(\frac{\sum_{i=1}^{n} l_i \sqrt{c_i \alpha_i}}{\sum_{i=1}^{n} l_i c_i v_{0i} + \Phi^*} \right). \qquad (14)$$

Here, λ is Lagrange multiplier; denomination λ has such a meaning only in formulae (14).

Using the network parameters given in Table 2, from formula (14) we get optimal added values of speed (in km/h) 42.59, 41.31, 18.91, 12.95, 41.98, 64.45, 26.11, 23.57, 25.16, 13.09, 10.21, 8.11, and 10.62 in roads 7-1, 2-8, ..., 9-4. With due consideration of the base speed, as a result of the road repair, the mean values of the speed on the roads are 52.59, 51.31, 28.91, 22.95, 51.98, 74.45, 36.11, 33.57, 35.16, 23.09, 20.21, 18.11, and 20.62 (km/h).

Now it is time to consider how this solution differs from a simpler one, according to which resource Φ^* is distributed among the roads uniformly, i.e. proportionally to the road lengths. Given that, each road is provided with $\phi_i(\gamma_i) = c_i l_i \gamma_i = \frac{l_i}{\sum l_i} \Phi^* = \frac{l_i}{3.191} \Phi^*$ dollars and therefore we get added speed $\gamma_i = \Phi^*/(3.191 c_i)$. Together with the base speed, the mean values of speed equal to 60.07, 58.30, 55.39, 24.30, 48.22, 75.43, 24.16, 23.91, 48.15, 20.96, 24.79, 27.82, and 25.39 (km/h) are achieved due to the repair of roads 7-1, 2-8, ..., 9-4.

To compare the quality of the two examined solutions in a wide range of loads, the corresponding simulation experiments with the traffic network are carried out. The traffic network model in terms of GPSS "simulation models publication language" [17] is shown in Fig. 4.

The fragments in Fig. 4 are sufficient to reconstruct a whole simulation model with the help of Tables 1 and 2. The model reproduces the following characteristic features of traffic flow motions, overlooked by SE-network:

- the speed of the car (transact) entering the road is determined by the actual (i.e. random) road's density of load at the given time moment;

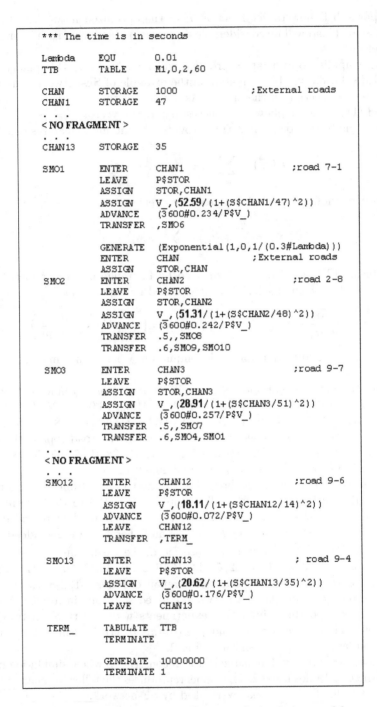

```
*** The time is in seconds

Lambda     EQU        0.01
TTB        TABLE      M1,0,2,60

CHAN       STORAGE    1000              ;External roads
CHAN1      STORAGE    47
. . .
< NO FRAGMENT >
. . .
CHAN13     STORAGE    35

SMO1       ENTER      CHAN1                          ;road 7-1
           LEAVE      P$STOR
           ASSIGN     STOR,CHAN1
           ASSIGN     V_,(52.59/(1+(S$CHAN1/47)^2))
           ADVANCE    (3600#0.234/P$V_)
           TRANSFER   ,SMO6

           GENERATE   (Exponential(1,0,1/(0.3#Lambda)))
           ENTER      CHAN              ;External roads
           ASSIGN     STOR,CHAN
SMO2       ENTER      CHAN2                          ;road 2-8
           LEAVE      P$STOR
           ASSIGN     STOR,CHAN2
           ASSIGN     V_,(51.31/(1+(S$CHAN2/48)^2))
           ADVANCE    (3600#0.242/P$V_)
           TRANSFER   .5,,SMO8
           TRANSFER   .6,SMO9,SMO10

SMO3       ENTER      CHAN3                          ;road 9-7
           LEAVE      P$STOR
           ASSIGN     STOR,CHAN3
           ASSIGN     V_,(28.91/(1+(S$CHAN3/51)^2))
           ADVANCE    (3600#0.257/P$V_)
           TRANSFER   .5,,SMO7
           TRANSFER   .6,SMO4,SMO1
. . .
< NO FRAGMENT >
. . .
SMO12      ENTER      CHAN12                         ;road 9-6
           LEAVE      P$STOR
           ASSIGN     V_,(18.11/(1+(S$CHAN12/14)^2))
           ADVANCE    (3600#0.072/P$V_)
           LEAVE      CHAN12
           TRANSFER   ,TERM_

SMO13      ENTER      CHAN13                         ; road 9-4
           LEAVE      P$STOR
           ASSIGN     V_,(20.62/(1+(S$CHAN13/35)^2))
           ADVANCE    (3600#0.176/P$V_)
           LEAVE      CHAN13

TERM_      TABULATE   TTB
           TERMINATE

           GENERATE   10000000
           TERMINATE  1
```

Fig. 4. Fragments of the traffic network simulation model

– blocks may arise at attempt of moving to the next road (if the car cannot move to the chosen road due to its congestion, it does not leave the current road).

The model provides a fair amount of more precise correspondences with real traffic flows, including the possibility to consider the durations of traffic lights phases. The present article does not aim to enumerate and justify these correspondences.

Bold-faced type is used to highlight mean speed $v_{0i} + \gamma_i$, calculated with formula (14) at resource Φ^* distribution optimization in a low load mode. At the load growth, the characteristics of the traffic network resulting from such optimizations are determined by simulation runs with increased values for parameter Lambda.

In Fig. 5 the comparison is given for $E(\Lambda)$-characteristics of SE-network, optimized in a low load mode (fine line E_1'), and the corresponding traffic network (bold curve E_1). At changeable Λ characteristic E_1' is calculated by formula (10) at added speed (14) and matrix $\mathbf{P}$ specified in Table 1. Characteristic E_1 results from the simulation runs of the model in Fig. 4. Curves E_2 and E_2' are similarly obtained characteristics of the traffic network and SE-network in which resource Φ^* is distributed among the roads in proportion with their length, i.e. equally per each kilometer of the road. Intensity Λ is given as the number of the cars per second, time E is in seconds. The comparison of the obtained characteristics shows that networks optimized due to zero load possess much better throughputs than those with unequally distributed resource. In a stationary node, the mean travel time in optimized networks is less, and they go into a nonstationary mode at larger intensity Λ of the incoming flow. Consequently, if during rush hours the traffic jams occur, they will happen later, will be shorter, and will end faster.

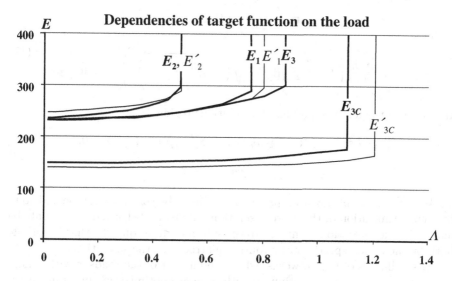

Fig. 5. Changes in the quality of the optimized networks with the load growth

Characteristics E_3, E_{3C} and E'_{3C}, shown in Fig. 5, are reached thanks to the complex optimization taking the load into consideration and using the possibility of changing transition probabilities.

7   SE-Networks Optimization in a High Load Mode

At zero intensity λ_i of the cars entering road i, positive load coefficient $\rho_i > 0$ results. This leads to a lower value of FD $\delta_i(\rho_i) = \delta(\rho_i)$ (Fig. 3) and decreased actual mean speed $v_i(\rho_i) = v_{0i}\delta(\rho_i)$ on the road. Taking FD into consideration, let us form an equation for load coefficient ρ_i of a random road:

$$\rho_i = \frac{\lambda_i b_i}{K_i} = \frac{\lambda_i}{K_i}\frac{l_i}{v_{0i}\delta(\rho_i)} = \frac{\rho_{0i}}{\delta(\rho_i)} = \rho_{0i}(1 + \rho_i^2), \qquad (15)$$

where l_i is the road length, $\rho_{0i} = \lambda_i l_i/(v_{0i}K_i)$ is a virtual load coefficient, that could result, if the car speed stayed the same with the load growth. Equation (15) is approximately carried out in a stationary mode.

The solution of quadratic equation (15) has a form of:

$$\rho_i = \frac{1}{2\rho_{0i}} - \sqrt{\frac{1}{(2\rho_{0i})^2} - 1} = \frac{v_{0i}K_i}{2\lambda_i l_i} - \sqrt{\left(\frac{v_{0i}K_i}{2\lambda_i l_i}\right)^2 - 1}, \qquad (16)$$

where $\lambda_i = \alpha_i \Lambda$. This solution is correct (it corresponds to a stationary mode) at $\rho_{0i} \le 1/2$ (in this case $\rho_i \le 1$).

Using solution (16), target function (10) is expressed through SE-network parameters, and the complex optimization problem is formulated as follows.

It is necessary to find speed vector $\overrightarrow{\gamma} = \overrightarrow{\gamma}_{opt}$ and routing matrix $\mathbf{P} = \mathbf{P}_{opt}$, providing minimal mean response time $E(\overrightarrow{\gamma}\mathbf{P})$ at given $\Lambda > 0$:

$$E(\overrightarrow{\gamma}, \mathbf{P}) = \sum_{i=1}^{n} \frac{\alpha_i(\mathbf{P})l_i}{v_{0i}+\gamma_i}(1 + \rho_i^2)$$

$$= \sum_{i=1}^{n} \frac{\alpha_i(\mathbf{P})l_i}{v_{0i}+\gamma_i}\left(1 + \left(\frac{v_{0i}K_i}{2\alpha_i(\mathbf{P})l_i} - \sqrt{\left(\frac{v_{0i}K_i}{2\alpha_i(\mathbf{P})l_i}\right)^2 - 1}\right)^2\right) \to \min_{\overrightarrow{\gamma},\mathbf{P}} \qquad (17)$$

and lying in the following admissible solutions region

$$\sum_{i=1}^{n} c_i l_i \gamma_i^{\beta_i} = \Phi^*; \quad 0 \le \rho_i \le 1; \quad \sum_{j=0}^{n} p_{ij} = 1, \quad (i = \overline{0,n});$$

$$p_{ij}^{min} \le p_{ij} \le p_{ij}^{max}, \quad (i,j = \overline{1,n}). \qquad (18)$$

This complex optimization problem can easily be solved by a repeated step-by-step optimization of the current solution only by added speed γ_i and only by matrix $\mathbf{P}$; such iterations converge to the solution rather quickly, that is optimal both by the added speed and by routing matrix $\mathbf{P}$ at the same time.

At small sizes of the networks, when the number of the roads is within hundred, the networks can be optimized with standard optimization programs using numerical differentiation to calculate the gradients. Thus the time consumptions for PC of medium power equal from a share of a second to several seconds. When queueing SE-networks include hundreds of nodes, the target function gradients should be accurately calculated by the transition probabilities. Such calculation is quickly performed by the method of auxiliary SLAE proposed above.

Solving problem (17) and (18) at some $\Lambda > 0$, we find such resource Φ^* distribution among repaired roads, that provides better results at the entering traffic flow intensity equal to Λ.

In Fig. 5 traffic network characteristic E_3 is calculated by simulation modeling at parameters γ_i, obtained when solving solution (17) and (18) at $\Lambda = 0.75$ cars per second and a fixed matrix $\mathbf{P}$, specified in Table 1.

Characteristic E_{3C} is calculated by traffic network simulation modeling at parameters γ_i and $\mathbf{P}$, obtained when solving problem (17) and (18) at $\Lambda = 0.75$ cars per second, at varied values of speed γ_i and at varied probabilities p_{ij} in the range of ± 0.1 (i.e. at the parameters obtained by a complex network optimization).

Curve E'_{3C} is characterized by the corresponding SE-network and is calculated without simulation modeling, directly through the expression of the target function (17).

Here is solution $\overrightarrow{\gamma} = \overrightarrow{\gamma}_{opt}$, $\mathbf{P} = \mathbf{P_{opt}}$, of problem (17) and (18), obtained for $\Lambda = 0.75$ and providing characteristic E'_{3C} for SE-network and characteristic E_{3C} for the corresponding traffic network (Fig. 5): $\overrightarrow{\gamma}_{opt}=$ 46.63, 44.15, 13.60, 7.54, 33.43, 80.24, 39.16, 26.09, 21.28, 7.37, 9.63, 16.63, 19.25; $\mathbf{P_{opt}}$ is given in Table 3.

Table 3. Transition probabilities of matrix $\mathbf{P_{opt}}$

Road i	Road j													
	0	7-1	2-8	9-7	7-8	3-8	1-7	7-9	8-7	8-2	8-3	9-5	9-6	9-4
0			0.3			0.2	0.5							
7-1							1							
2-8									0.6	0.2	0.2			
9-7		0.3			0.1		0.6							
7-8									0.6	0.2	0.2			
3-8									0.6	0.2	0.2			
1-7		0.3			0.1		0.6							
7-9				0.1								0.2	0.3	0.4
8-7		0.3			0.1		0.6							
8-2			1											
8-3					1									
9-5	1													
9-6	1													
9-4	1													

Comparing characteristics E_{3C} and E'_{3C} with the others, given in Fig. 5, it is easy to notice that complex optimization, oriented on a high load and taking into consideration the possibility of partial routes changing (adaptation), allows one to decrease the mean time of passing through the network significantly and to increase its throughput.

Economic impact, achieved by the traffic network optimization with quick numerical optimization of the corresponding SE-networks, is quite apparent.

Equally effective solutions are obtained in case of more "stiff" FDs as well, including the case when FD is described by the function, shown with a lower line in Fig. 3.

8   Conclusion

Effective analytic-imitational and semi-numerical methods for complex optimization of routing matrices and node efficiencies in non-Markov queueing networks are developed. Mean network passage time is used as target function. The characteristic feature of the proposed methods is a prompt and precise calculation of the target function partial derivatives, based on the varied transition probabilities.

Traffic network optimization method is developed for stationary modes in a wide range of traffic load changes. The method is based on traffic networks approximation by SE-networks with multiserver nodes; their efficiency drops with the load growth and subsequent optimization of these SE-networks.

Effective methods for complex optimization of queueing SE-networks are proposed. At small sizes of SE-networks, the optimization can be carried out with the existing standard optimization programs using numerical differentiation for gradient calculation. When queueing SE-networks include hundreds of nodes, the target function gradients should be accurately calculated by the transition probabilities. Such calculation is quickly performed by the method of auxiliary SLAE proposed above.

References

1. Zadorozhnyi, V.N.: Optimization of uniform non-Markov queueing networks using resources and transition probabilities redistribution. In: Dudin, A., Gortsev, A., Nazarov, A., Yakupov, R. (eds.) ITMM 2016. CCIS, vol. 638, pp. 366–381. Springer, Cham (2016). doi:10.1007/978-3-319-44615-8_32
2. Kleijnen, J.P.C.: Statistical Techniques in Simulation, Part 1. Marcel Dekker, New York (1974)
3. Johnson, M.E., Jackson, J.: Infinitesimal perturbation analysis: a tool for simulation. J. Oper. Res. Soc. 40(3), 134–160 (1989)
4. Rubinstein, R.Y.: Sensitivity analysis of computer simulation models via the efficient score. Oper. Res. 37, 72–81 (1989)
5. Suri, R., Zazanis, M.: Perturbation analysis gives strongly consistent sensitivity estimates for the M/G/1 queue. Manag. Sci. 34, 39–64 (1988)

6. Zadorozhnyi, V.N.: Optimizing uniform non-Markov queueing networks. Autom. Remote Control **71**(6), 1158–1169 (2010). doi:10.1134/S0005117910060172. ISSN 0005-1179

7. Zadorozhnyi, V.N.: Channels distribution in non-Markovian queueing networks. Omsk Sci. Bull. **1**(87), 5–10 (2010)

8. Zadorozhnyi, V.N.: Simulation modeling of fractal queues. In: Dynamics of Systems, Mechanisms and Machines (Dynamics 2014), pp. 1–4 (2015). doi:10.1109/Dynamics.2014.7005703

9. Tsitsiashvili, G.: Parametric and structural optimization of the queuing network throughput. Autom. Remote Control **68**(7), 1177–1185 (2007). doi:10.1134/S0005117907070065

10. Kleinrock, L.: Queueing Systems: Computer Applications, vol. 2. Wiley Interscience, New York (1976). 576 pages

11. Vishnevskiy, V.M.: Theoretical Bases of Designing Computer Networks. Technosphere, Moscow (2003). 512 pages

12. Moiseev, A.A., Nazarov, A.N.: Infinitely Linear Systems and Queuing Networks. NTL Publishing House, Tomsk (2015). 240 pages. (in Russian)

13. Haight, F.A.: Mathematical Theories of Traffic Flow. Academic Press, New York (1963). 225 pages

14. Daganzo, C.F.: Urban gridlock: macroscopic modeling and mitigation approaches. Transp. Part B **41**, 49–62 (2007)

15. Zadorozhnyi, V.N., Kornach, M.A.: Optimization of transport queueing networks on the basis of the method of directing hyperbole. In: 2016 International Siberian Conference on Control and Communications, SIBCON. Control of the Large-Scale Systems, Moscow, Russia, 12–14 May 2016. doi:10.1109/SIBCON.2016.7491716

16. Zadorozhnyi, V.N.: Analytical-Simulation Research of Queuing Systems and Networks. Omsk State Technical University Publishing House, Omsk (2010). 280 pages. (in Russian)

17. GPSS World Reference Manual. Minuteman Software, 5th edn. Holly Springs, NC, USA (2009)

Author Index

Printed in the United States
By Bookmasters